A Photographic Guide to the BIRDS OF INDIA

And the Indian Subcontinent, including Pakistan, Nepal, Bhutan, Bangladesh, Sri Lanka & the Maldives

Bikram Grewal, Bill Harvey & Otto Pfister

Princeton University Press
Princeton and Oxford

PROVINCES OF PAKISTAN
a Baluchistan
b Sindh
c Punjab
d North-West Frontier Province
AFGHANISTAN
IRAN
PAKISTAN
Arabian Sea
MALDIVES
SRI LANK
R. Indus
R. Jhelum
R. Chenab
R. Ravi
R. Beas
R. Sutlej
R. Yamuna
R. Luni
R. Chambal
R. Saraswati
R. Narmada
R. Tapti
R. Godavari
R. Bhim
R. Krishna
R. Payaswani
R. Kaveri
a
b
c
d
1
2
3
4
6
7
8
9
10
11
12
13
14
15
16
17

STATES OF INDIA

1 Jammu & Kashmir
2 Punjab
3 Himachal Pradesh
4 Uttaranchal
5 Uttar Pradesh
6 Haryana
7 Delhi
8 Rajasthan
9 Gujarat
10 Madhya Pradesh
11 Maharashtra
12 Goa
13 Lakshadweep
14 Karnataka
15 Kerala
16 Tamil Nadu
17 Andhra Pradesh
18 Chattisgarh
19 Orissa
20 Jharkhand
21 Bihar
22 West Bengal
23 Sikkim
24 Assam
25 Arunachal Pradesh
26 Nagaland
27 Manipur
28 Mizoram
29 Tripura
30 Meghalaya
31 Andaman & Nicobar Islands

Map not to scale.

The international borders as shown in this map are neither authentic nor purported to be correct.

Published in the United States and Canada
by Princeton University Press,
41 William Street, Princeton, New Jersey 08540

First published by Periplus Editions (HK) Ltd.

ISBN 0-691-11496-X

Library of Congress Control Number 2002111545

This book has been composed in
Sabon on QuarkXpress

www.pupress.princeton.edu

Printed in Singapore

1 3 5 7 9 10 8 6 4 2

CONTENTS

Acknowledgements

The authors would like to thank the following for their support and help in the publishing of this book:

Eric Oey and the staff at Periplus Editions, Nikhil Devasar, Sumit Sen, Lt. Gen. R. K. Gaur, Sharad Gaur, Harish Aggarwal, Sudhir Horo, Manju Singhal, Neeraj Aggarwal, Ayesha Sarkar, Clement M. Francis, Sujan Chatterjee, Sunita Chaudhry, Kamal Sahai, Arfin Zukoff, Ron Saldino, Vijay Chowdhary, Sudhir Oswal, Vivek Menon, Rao Inderjit Singh, Morten Strange, Nikhat Grewal, Raghuvir Khare, Arshiya Sethi, Deepshikha Singh, Pauline Hayes, Bittu Sahgal, Tim Loseby, Simon Harrap, Goren Eckstrom, Peter Morris, A. V. Manoj, Milan Bhatta-charya, Mohit Aggarwal, S. Kartikeyan, Aasheesh Pittie, K. B. Singh, Carol Inskipp, Dhananjai Mohan, Sheila Chhabra, Vijay Cavale, Toby Sinclair, Gertrud Denzau, M. B. Khriahna, Clive Harris, Martin Kelsey, Rajesh Bist, R. Vijaykumar Thondaman, Sudhir Vyas, Vivek Tiwari, Sanjeeva Pandey, Samiha Grewal, Keya Khare, Dhanraj Malik, Robert Kirk, Pradeep Sachdeva, Emil Wendel, Naseer Kitchloo, Kaoosi Sethna, Tsering Tashi, Thakur Dalip Singh, Suresh Elamon, Pia Sethi, Maggie Harvey, Shefali Upadhyay, Petrick William, Sekhar Raha, S. A. Hussain, Gillian Wright, Tim Inskipp, Krys Kazmierczek, Rattan Singh, Liza Cherian, S. Balachandran, Joanna Van Gruisen, Asad Rahmani, Jan Willem den Besten, Nigel Redman, Samir Bakshi and especially Alpana Khare.

ENJOYING BIRDWATCHING

If you have opened this book you are probably interested in birds. Your interest may be limited to trying to put a name to the birds you see casually from your window. Or you may be, or are about to be, one of the growing number of people who watch birds regularly. They range from professional ornithologists and serious world travellers through to the garden birdwatcher who takes no notes and makes no lists. Most birdwatchers (or birders) go out regularly, try to identify everything they see and hear and keep notes, or at least lists. Noting down what you see not only becomes a personal record, but might also be a useful contribution to wider scientific knowledge. It is essential if you cannot identify the bird immediately. It is risky to keep what you see in your head and then try and remember it as you consult other books or expert friends. It is all too easy to distort your memory. It is much better to note down, and even sketch, your observations at the time you make them.

Birdwatchers in Bharatpur

There is still something of the collector in all of us, but fortunately the collecting of dead birds and eggs has been replaced by listing. Almost every birder keeps lists—life lists, country lists, trip lists, site lists and even garden lists. Personally adding a species to one of your lists is one of the little pleasures of birding. There are a few birders (known colloquially as twitchers) who are solely interested in adding to their lists and spend their time (and money) travelling their country (and the world) looking at species they have never seen before. But many more people prefer to enjoy their birds in comfort, whether they have seen them before or not. Such birders often have favourite sites they visit regularly.

Birdwatching is best done singly or in small groups. Birds are wild animals and although some can be very confiding, most do not tolerate the inevitable extra noise that large groups of people make. It is often more productive to go out with one or two friends. You share your knowledge and multiply the number of eyes searching. If you are a beginner, you may well

Little Rann of Kutch

want to join a small group in order to learn. But try not to rely solely on a more expert companion. One of the joys of birding is identifying the birds yourself. On pp. 459–460 there is a list of national and local bird societies, including several very active e-mail based groups, which you can join.

Birdwatching often involves walking around a chosen site and the gentle exercise is an added attraction to many birders. But it is often much more productive to sit quietly and still and let the birds come to you. You might choose a good vantage point giving you a wide view of a wetland. Or you might sit by a fruiting or flowering forest tree or watch a small puddle where shy forest birds come to drink and bathe. With this book and your notepad and pen you can identify what you see at leisure.

In many parts of our region the birds are unusually confiding, since most of the religions in the region have a deep respect for wildlife. Several species even nest in houses and many come to habitation for food. In north India it is still possible to see the endangered and decreasing Sarus Crane nesting in small swamps on the edge of villages. The Indian Peafowl is a familiar sight on village roofs and Cattle Egrets often seek out the safety of village trees to nest in. For the beginner, this feature of our birdlife makes the task of identifying birds a little easier.

Birds are everywhere and there are no rules about where you should watch them. Town parks and less intensive agricultural land can yield many species, but if you want to see as much as possible on your outing, seek out an area of wetland or natural forest. There are wildlife sanctuaries and forest reserves throughout the region that can be very productive, particularly if you are allowed to walk in them. The more adventurous will want to make special trips to the most famous ones. Every year the region welcomes groups of birdwatchers from across the globe to share our enjoyment in our rich avifauna. A very useful book, for India at least, is Krys Kazmeirczak and Raj Singh's *A Birdwatchers Guide to India* (2001) which gives details of what you can see and how to see it in many of India's best sites. There is a similar guide for Nepal—Carol Inskipp's *A Birdwatchers' Guide to Nepal*

(1988)—and *Where to Watch Birds in Asia* by N. Wheatley (1996) gives some bird-watching sites in other countries.

It is not strictly necessary to have binoculars or a telescope to watch birds, but it does make it a lot easier to see them. Take expert advice before you choose your optical aids. Binoculars are described, for example, in terms of 7 x 50 or 10 x 40. The first figure is the magnification and the second the exit (front) lens diameter in millimetres. For birdwatching you should choose a magnification of between 7 and 10. Anything larger reduces light gathering and magnifies hand shake. Choose the largest possible diameter to increase light gathering, which is important at dawn and dusk and in forests. A 7 x 50 pair of binoculars, for instance, has nearly twice as much light gathering as a 10 x 40 and the extra magnification of the latter does not compensate. If you birdwatch at large wetlands or on the coast you may want a telescope. It is always more effective to use it on a tripod. Try to get the best telescope you can afford, and it should be prismatic. The same principles apply as in binoculars, but light gathering is even more important. It is best not to go above a magnification of 30 x although some birders use zoom lenses.

Birdwatching is one of the fastest growing leisure interests in the world today. The Royal Society for the Protection of Birds in Britain has well over a million members. Although the number of active birders resident in our region cannot approach that, the number is growing, particularly among young people. For instance, the North India Bird Network based in Delhi was only started in 2001 and a year later had over 300 active members. With this book and your binoculars you too can discover the joy and relaxation that bird watching can bring you. Enjoy!

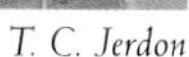

T. C. Jerdon

Brian Hodgson

Edward Blyth

A. O. Hume

ORNITHOLOGY IN THE REGION

Over 1300 of the world's 9700 or so species of birds are found in the Indian subcontinent. If sub-species are included then the number exceeds 2000. Of these 142 are endemic, which means they are restricted to the region. New species are continually added to the list because they are from neighbouring countries and are discovered within our region, or they are vagrants or stragglers from further afield, or are known sub-species reclassified as species.

There are a variety of reasons for this abundance. Primarily, the region covers a wide altitudinal range, extending from a long coastline to the Himalayas, which are the world's highest mountains. The region also has a very varied climate which brings with it diverse vegetation. Northwestern India and its contiguous desert area in Pakistan go almost rainless, while the evergreen forests of northeastern India, Bhutan and Nepal are some of the wettest areas in the world. The Himalayan range, which starts as the Karakorams in Afghanistan, extends through Pakistan, the northern part of India and then into Nepal and Bhutan, produces almost arctic conditions where alpine flowers blossom. In addition, the unique geographical position of this area is that it forms the overlap between the Oriental or Indo-Malayan, the Palearctic and the Afro-tropical regions. As a result, birds from all of these three regions occur.

The peoples of the region have lived for thousands of years in close contact with its rich wildlife. The 4000-year-old Hindu scripture, the Rig-Veda, refers to some 20 species of birds, but its anonymous compilers must have been familiar with many more. In the 3rd century BC, King Devanampiyatissa of Sri Lanka established one of the world's first wildlife sanctuaries. The great Mughals maintained royal menageries and revelled in hunting on a grand scale. But they were also meticulous in their observations of wildlife. The Mughal Empire began to disintegrate at the beginning of the 18th century and the establishment of British power gave enormous scope to officers in the police, civil, forest and armed services to observe the region's plentiful bird life. In the 18th century, several European countries sent expeditions to India and China. One such expedition had Pierre Sonnerat as naturalist and today the Grey Jungle Fowl lives under the name *Gallus sonnerratii*. Another traveller, Jean-Baptiste Leschenault de la Tour, had a forktail named after him, *Enicurus leschenaulti*.

As British and other colonial powers grew, there were increasing numbers of such officers, whose jobs required much less crippling paperwork

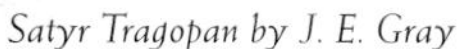

Satyr Tragopan by J. E. Gray

Spotted Forktail by J. Gould

than those of their successors today. The first major effort was by a Madras-based surgeon Edward Buckley who described and drew 22 birds seen around the East India Company's Fort St. George. Pioneering work was done by Brian Hodgson, T. C. Jerdon and Edward Blyth, often called the founders of Indian ornithology. Jerdon's *Birds of India*, published in 1862–64, was based on the work of all three men, assisted by a loyal group of field workers. Jerdon's *Fifty Illustrations from Indian Ornithology* was the first serious attempt to illustrate birds of the region. Similarly, John Gideon Loten (of Loten's Sunbird fame) documented the birds of Sri Lanka. Perhaps more important (but often overlooked) was the collection of bird illustrations, done by local artists, in the collection of Maj. General Hardwicke, which was later published as a folio of plates by J. E. Gray and called *Illustrations of Indian Zoology*. Bird illustrations improved tremendously when John Gould produced, in 1832, *A Century of Birds from the Himalayan Mountains* and, from 1850, the magnificent six-volume *Birds of Asia*.

The next major advance in ornithological knowledge came with the arrival of Allan Octavius Hume. He is widely known as a founder of the Indian National Congress, but is also called the Pope of Indian Ornithology. For over a decade, he and his team collected birds for study over most of the subcontinent. Hume also edited and published twelve volumes of bird observations between 1873 and 1888. These volumes, collectively known as *Stray Feathers*, are valuable reference works even today. After *Stray Feathers* ceased publication, the *Ibis*, which is published by the British Ornithological Union, and the *Journal of the Bombay Natural History Society* (*JBNHS*) became the main repositories of ornithological information of this region.

A year after the last *Stray Feathers* was published, Eugene W. Oates and W. T. Blandford produced the first volume of *Fauna of British India: Birds*. Three more volumes were published in the following nine years. These were the most significant reference works on Indian ornithology for at least twenty years. They included detailed observations from parts of the Indian region not covered by previous works. The next major work was by E. C. Stuart Baker, an Indian police officer for nearly twenty years. Some of Baker's classic early work appeared first in the journal of the BNHS, including *Game Birds of India, Burma and Ceylon*. But Baker's most notable works are the eight bird volumes of the second edition of the *Fauna of British India*, published between 1922 and 1931, and the *Nidification of Birds of the Indian Empire*, which were published between 1932 and 1935.

Bill Harvey with Salim Ali in Madras in 1982

The 20th century saw an increasing number of talented and dedicated ornithologists in India, but the most celebrated and long-lasting partnership was between Salim Ali and Dillon Ripley. On an early field trip, they conceived the idea of their *Handbook of the Birds of India and Pakistan*. The first step towards compiling a handbook was an up-to-date checklist, provided by Ripley's *A Synopsis of the Birds of India and Pakistan*, published by the BNHS in 1961, and a 2nd edition in 1982. The first volume of the Handbook was published in 1968, and the tenth and last in 1974. The Handbook, listing 2060 birds, remains the standard and most exhaustive work on Indian birds. In 1987 it became available in a compact edition. Salim Ali also published several other books, including *The Book of Indian Birds* (now in its 12th edition), as well as many invaluable regional guides.

Modern bird guides started in India rather late with the publication, in 1980, of Martin Woodcock's *Collins Handguide to the Birds of the Indian Sub-continent*. The first photographic guide by Bikram Grewal was published in 1993; a 3rd edition appeared in 2000. But arguably the most significant book to be published in recent times is Grimmett, Inskipp and Inskipp's *Birds of the Indian Subcontinent* (1998) and its subsequent offspring, *The Pocket Guide to the Birds of the Indian Subcontinent* (1999). These, with Krys Kazmierczak and B. van Perlo's *A Field Guide to the Birds of the Indian Subcontinent* (2000), truly ushered in the age of modern Indian bird books. The Bibliography on pp. 450–458 gives full details of these and other publications.

ORIENTAL REGION

P. L. Sclater, in the 19th century, studied the world's birds and divided the planet into six bio-geographic realms or regions. This was later modified to apply to all animals. The Oriental Region covers South and Southeast Asia, with the Himalayas separating it from the Palearctic to the north. Leafbirds are found exclusively in this region and broadbills nearly so. In general, the region's birds have closest affinities with those of tropical Africa.

The Indian sub-region has further been divided into seven different areas, in which characteristic types of birds are found.

The northernmost of these areas is the Himalayas, which forms an arc over 2500 km long and 150–400 km broad. The Himalayan mountains form roughly three parts, the foothills or Sivaliks to the south, the Himachal or lower mountains, and the Himadri or high Himalayas to the north. The Ladakh plateau (home to the Black-necked Crane), with an average elevation of 5300 m, occupies a large portion of the Indian state of Jammu and Kashmir and consists of steppe country with mountain lakes. Birds like the Bar-headed Goose, the Great Crested Grebe and the Brown-headed Gull breed there in summer. The state of Himachal Pradesh and the Kumaon and Garhwal regions of the hill state of Uttaranchal lie to the west of Nepal, which falls almost entirely within the central Himalayas. Further east, the increased rainfall give the eastern Himalayas of Bhutan and Sikkim a very different range of species from those in the west. A total of 21 species are endemic to the eastern Himalayas. Many of them are under threat from loss of suitable habitat. The fate of the Chestnut-breasted Hill Partridge is uncertain, as is that of the Rusty-throated Wren-babbler. The western Himalayas hold 11 endemic species, including the probably extinct Mountain Quail. The Western Tragopan is under threat in both Pakistan and India,

The Himalayas

Moss forest, the Himalayas

as is the Kashmir Red-breasted Flycatcher. The Central Himalayas have three endemic birds, including Nepal's only endemic, the Spiny Babbler. In spite of the constant threats, the Himalayas still contain extensive moss, rhododendron, conifer and broadleafed forests.

Thar Desert

Mustard fields, northern plains

The Northwest covers most of Pakistan, the flat plains of the Indian Punjab and Haryana and the semi-arid and arid Thar Desert of Rajasthan in the west. The five rivers water the Punjab (after which it takes its name), spanning both India and Pakistan. The Sindh province of Pakistan is irrigated by the mighty Indus River. Those areas without irrigation have to rely on the perennially deficient rainfall, but local grasses have adapted to this and, after a monsoon shower, even the desert blooms. Much of the area is, in fact, sparse thorn scrub rather than true desert. Among the numerous desert birds found in this area are many shared with the dry country further west. The Thar Desert ultimately joins the Rann of Kutch, a large salt wasteland adjoining the Arabian Sea.

North India comprises the rich alluvial Gangetic plain, which is watered by the River Ganga (Ganges) and her tributaries from the Himalayas. This region extends up to an altitude of 1000 m in the north, so that it includes the low foothills of the Himalayas and the terai of India and Nepal. This was once a marshy area covered with dense forest. Much of the terai has been cleared for farming, but some of the forests which still exist contain a fantastic variety of birdlife. The terai contains some of India and Nepal's most productive birdwatching sites, including the Corbet and Chitwan National Parks.

Peninsular India, which is bordered in the northwest by the Aravallis, in the north by the Vindhya mountains, on the west by the Arabian Sea and on the east by the Bay of Bengal, makes up the largest physiographic division of India. The central plateaux of this area, known as the Deccan, rise to over

Terai grasslands

Western Ghat forest

1000 m in the south, but hardly exceed 500 m in the north. The peninsula has some wonderful landscapes with hills and huge boulders littering the countryside and large areas of forest. It was in this sub-region that the Forest Owlet and the Jerdon's Courser were recently rediscovered. Great rivers like the Godavari, Krishna and Kaveri rise in the west of the peninsula and flow into the Bay of Bengal. In the northeast lies the under-studied Eastern Ghats which share some species with the Northeast region. The steep escarpments of the Western Ghats stand between the plateau and the low-lying coastal strip and thus take the full force of the Southwest monsoon.

The Southwest region, though within the peninsula, is a distinct avi-faunal region. Some of its birds, like the Spiderhunters, Fairy Bluebirds and Laughingthrushes, bear strong affinities with those found in the Northeast and in Myanmar. The highest hills are the Nilgiris or Blue Mountains, most of which are now under eucalyptus, tea and other plantation crops. Tea is also the main crop of the Annamalai or Elephant Mountains of Kerala, while cardamom and other spices are grown at lower elevations. The most interesting part of this area is the forests of the Wynad, where Kerala, Karnataka and Tamil Nadu meet. The Western Ghats support 16 endemic species, four of which are now considered threatened. Large-scale tree-felling is a growing cause for concern in this region.

The Northeast and Bangladesh region consists of the delta of the Ganga and Brahmaputra, with its tidal estuaries, sandbanks, mudflats, mangrove forests and islands. Further upstream are lands drained, and occasionally flooded, by these great rivers and their tributaries. Bangladesh has some of the highest human population densities in the world and most of the country is now farmed. The Northeast region extends northwards to include the states of Arunachal Pradesh, Mizoram, Meghalaya and Nagaland, as well as the Kingdom of Bhutan. As you progress eastwards, the birdlife begins to have increasingly strong affinities with the Indo-Chinese sub-region. There are three species endemic to the Assam low-lands and all are endangered, the Manipur Bush Quail being more at risk than the Marsh Babbler and the Black-breasted Parrotbill.

Tea plantation, Assam

The Andaman and Nicobar Islands, within the Indian state, have 17 endemic birds, six of which are found only in the Andamans and six only on the Nicobars. The four species under serious threat are the Nicobar Megapode, the Narcodam Hornbill, the Andaman Crake and the Nicobar Bulbul.

The Andamans

Sri Lanka is a remarkable country for birds. Although small, it has a wide range of climate and habitat and over 400 species and sub-species, including at least 26 endemic species. Many of the island's birds are identical to those found in India, although, for example, vultures have not been able to cross the Palk Strait. Sri Lanka is usually divided into two zones, the dry plains of the north (comparable to much of Tamil Nadu) and the mountainous, wet, central region including the coast around the capital Colombo, comparable to the Western Ghats and sharing many species with the southwestern India region. Recent political trouble and the loss of habitat have resulted in several species coming under threat, particularly the Green-billed Coucal and the Sri Lanka Whistling Thrush. Recent supplanting of natural forest with monoculture tree species has not helped although, remarkably, new species continue to be discovered; the most recent was a new owl, found in the Wet Zone.

The Sri Lankan coast

A jheel in summer

Flooded forest

HABITATS

While many common species extend over large areas of the Oriental Region, others are limited not just to a region but to a specific habitat, such as some birds of the conifer hill forest and the grasslands.

The habitat of our region can be broadly divided into wetlands, marine environments, scrub, forests, grasslands and agriculture. There are also specialised desert, marine island and montane habitats. There is, naturally, much overlap within the habitats and several species need to utilise different habitats at different times, for instance, for breeding and feeding, summering and wintering.

Wetlands: Shallow lagoons, inland jheels and bheels, marshes and rivers are particularly rich habitats for birds. The region supports abundant wetlands of different types, including mountain lakes, freshwater and brackish marshes, water storage reservoirs, village ponds and flooded forests. Many are only temporary, appearing after the monsoon rains, but drying up before the next rain. They all provide breeding and feeding grounds for resident species, while huge numbers of migratory waterbirds congregate during the winter months. Several species, like bitterns and crakes, conceal themselves in the reedbeds, which are an important habitat for a number of local species. Numerous birds of prey can be found near water, hunting these great concentrations of potential prey.

Wetland degradation, pollution, dam building and the draining of marshlands to meet human needs for farming and housing have reached alarming proportions and are affecting populations of birds like the Spot-billed Pelican and the Indian Skimmer. Amongst the most important remaining wetlands in the region are Harike in the Indian Punjab, Chilika in Orissa, the Chasma Barrage and Haleji Lake in Pakistan and the Keoladeo Ghana Sanctuary in Bharatpur in Rajasthan. The last is the only wintering site of the Siberian Crane, now reduced to a single pair at the last count. The wetlands of Assam support the rare White-winged Wood Duck and were probably the last home of the Pink-headed Duck, which is now almost certainly extinct.

West Indian coast

Mangroves, Sunderbans

Marine environments: The mangrove forests along some coastlines are also under grave threat. Those in the Sunderbans of West Bengal and Bangladesh are the most extensive in the world. Typical bird species include the Indian Reef Heron, seldom found far inland, which feeds mainly on mudskippers. Several typical coastal kingfishers, like the Collared and the Brown-winged, are now becoming rarer. Limited data is available, but the coastal mudflats and estuaries are important staging and wintering areas for migrant waders. Saline and mud flats and saltpans created by man are also important. Oceanic birds like Skuas, Boobies and Petrels are usually only passage migrants or accidental visitors to our shores. Breeding colonies of the seabirds are confined mostly to the Lakshwadeep Islands where they suffer persecution from egg-collectors. The Black-naped Tern and the Saunder's Tern nest on the Maldives, a coral-based archipelago.

Scrub: Poor soil condition is the most important reason for scrub to develop, often in semi-desert or high altitude desert. This habitat occurs along forest edges or where forest has been cleared. Areas around wetlands are often vegetated with scrub because the soil is saline or waterlogged. While only a handful of birds are restricted to scrub, many more are found in scrub where it is adjoining or in wetlands, or at the edge of forests. One of the more interesting scrub specialities is the Jerdon's Courser, only known from Andhra Pradesh. Another is the Stoliczka's Bushchat, which inhabits desert

Himalayan forest

Deccan scrub

scrub and whose nest has never been found. Its status and range are still unclear. Three were recorded recently in Haryana, well to the east of its normal sites in the Thar desert.

Forest: The region contains several different types of forest. Apart from the coastal mangroves, inland are wet evergreen, dry deciduous and desert thorn forests. Hill and mountain forests contain mixed broadleaf, moist oak and rhododendron and dry coniferous forests of pine, cedar and fir. Most forests are now declining at an alarming pace, even where they are protected. Over-exploitation of forest resources has reduced forest cover to less than 10 per cent of the region. However, Bhutan and certain parts of northeastern India still maintain good forest density. Typical of the forests of lowland north India is the Sal, while Teak is the characteristic dominant tree species further south. Even planted forest, particularly when well-grown or neglected, can provide a good habitat but the ubiquitous, exotic Eucalyptus and Mesquite are much less supportive of birds than are native tree species.

Forest birds are more difficult to see, especially when they are concealed in the tree canopy. They are best seen in clearings, flying from tree to tree or on the edge of forests. Frequently, assorted species form hunting groups or parties and move together through the forest searching for food. Finding and watching such groups is one of the best ways of seeing forest birds.

Grasslands: The most important grasslands in the region are those occurring along the foothills of the Himalayas, the floodplains of the Brahmaputra and Indus rivers, the arid grasslands of the Thar Desert and those in peninsular India. These grasslands are in grave danger. The need for more farmland has meant that huge areas are being converted, drained and overgrazed. Birds like the Lesser Florican and the Swamp Francolin are threatened. The Bristled Grassbird, the Rufous-vented Prinia and the Finn's Baya are grassland birds which may soon be extinct in our region.

Agriculture: Although less than 50 per cent of the region is cultivated, the biggest threat to natural habitats is from the ever-increasing need to bring more and more land under cultivation to meet the region's need for food. After all, a quarter of humanity shares the region with its birds. Increasing acreages are being cultivated with rice, wheat, millet, pulses and mustard as well as other crops. Every day, large areas of forest and grassland are cleared and marshes are drained. While some birds like doves and francolins (and even, very locally, the Indian Bustard) can thrive in cultivated land, every day natural habitat declines, putting more pressure on the already fragile habitat.

Rhododendron

Rice paddy

CLIMATE

Climate is the major determinant of vegetation and hence the habitats of birds. Directly and indirectly it determines their food supplies as it governs the fruiting and flowering of trees, the seeding of grasses and the abundance of invertebrates. It also governs the agricultural practices which provide food for many bird species. The size of the Indian subcontinent means there is much variation in the climate, but there are three overriding influences. The first is latitude, which determines temperatures; broadly speaking, it is hotter the further south towards the equator you go. The second is altitude, which also determines temperature; it gets cooler the higher you climb. Altitude also determines rainfall patterns as it provokes precipitation from the clouds. The Himalayan and Western Ghat ranges are particularly important in this respect and their wind-facing slopes are among the wettest places in the world. The third, and strongest, influence is the monsoon—

The monsoon breaking

seasonal rainfall from the oceans.

People have different descriptions for the various seasons in the region, and indeed they do vary. Generally speaking, there is a cooler winter season from November to February, a hot, mainly dry season from March to June, a hot, wet season from July to September (depending on the timing of the Southwest monsoon) and a warm post-monsoon period (autumn to some) in October–November. The Southwest monsoon reaches the west coast in June. It drops much rain on the seaward side of the Western Ghats and swings round, picking up more moisture from the Bay of Bengal, to travel northwestwards. The high Himalayas deflect it in this direction and this encourages major precipitation, particularly in the east, where Cheerapunji is reputed to have the highest average annual rainfall in the world at 1300 cm.

The monsoon travels rapidly up the northern plains, quelling the pre-monsoon dust, transforming the hot, dry landscape and flooding the bheels and jheels. It normally reaches Delhi at the end of June and Pakistan in July. In some years the monsoon peters out and parts of the Northwest regularly experience drought. Indeed, the Thar Desert, straddling India and Pakistan, is a desert because the monsoon carries little moisture by the time it reaches there. There is a second, Northeast, monsoon, of less influence, that strikes the southeastern coasts from October to December, but its impact is usually limited to this area. The monsoons vary in strength from year to year and, where they fail, this may prevent birds breeding successfully. There are also, less seasonally, other sources of rainfall such as cyclonic storms, particularly on the Bay of Bengal coasts, convectional rain, particularly in the north when temperatures get very high, and rain or even snow from the northern mountains in winter, which may penetrate south into the plains.

In much of the lowland peninsula and the Northeast the temperature range is fairly small. Kerala, for instance, has a very equable, if hot and humid, climate. In the north of the region the range is much greater. Winter temperatures in Delhi can fall as low as 5 °C (or even lower) while in the stifling heat of May and June, they regularly top 45 °C. The contrast is even greater in the Thar Desert and parts of Pakistan where winter temperatures drop to zero; in summer they may exceed 50 °C. There is also a wide diurnal range in cloudless conditions. There is a considerable annual range in the Himalayas, where the climate is decidedly temperate. It becomes arctic in the highest peaks where snow is perpetual.

BIRD MOVEMENTS

Very few bird species are completely sedentary. Birds wander accidentally and deliberately, and can turn up almost anywhere. Some move deliberately in search of favoured food, for instance when particular trees are fruiting or flowering or when their wetland habitat dries up. In extreme conditions, desert birds may wander out of their normal habitat. But even more species make regular seasonal movements in response to anticipated weather changes. Long-distance migrants visit the region for our winter from central and northern Asia, and even eastern Europe. Some arrive as early as July and leave as late as May, but the main arrivals are in September–October, with departures in March–April. Prominent among them are the ducks, geese and waders that throng the wetlands, accompanied by numerous migratory predators. Some of these long-distance migrants, possibly younger birds, do not move out in the spring and may over-summer in the region. Others return very early if their breeding attempt in the north fails. Altitudinal migrants move into adjoining foothills and plains before winter strikes and food supplies get low. A number of Kashmir and Ladakh species such as the Bar-headed Goose, the Ruddy Shelduck and the Citrine Wagtail move quite far south into India for the winter. Local migrants may change habitat after breeding or move north from more southerly wintering grounds to breed. The Golden Oriole, the Koel and the three commoner bitterns are characteristic of the latter.

Migrating cranes

There are also a number of species that may only be seen passing through a particular place. You would be hard-pressed to see a Blyth's Reed or Greenish Warbler in Delhi in mid-winter although they are both numerous in spring and autumn. Both species spend the winter further south

Bird-ringing near Delhi

in the peninsula. Several passage migrants are *en route* to and from Africa. The most remarkable is the Amur Falcon, which breeds mainly in northeast Asia and winters in southern Africa, while the Pied Cuckoo winters in East Africa and returns to India to breed during the monsoon.

Bird-banding or ringing, carried out mainly under the auspices of the Bombay Natural History Society (their address is on the rings), has taught us much about long-distance migration, but much remains to be learnt about local movements and whether they are changing.

BREEDING

Most species breed annually and many smaller birds raise more than one brood in a year. The timing of breeding is related to food supplies since the maximum amount of food needs to be available for the young to fledge successfully. This usually means that climate is the main determinant. Since climate varies through the region, so do breeding seasons, although a few species breed throughout the year. Most birds breed immediately prior to the main monsoon in their area, to take advantage of the increase in invertebrates that are the main food of their young. Even seed-eaters, such as weavers and sparrows, feed their young on insects. Some reedbed, sandbank and marshland ground-nesting species also breed before the onset of the main monsoon to avoid getting flooded out. Most larger waterbirds (including cormorants, egrets and the Spot-billed Pelican) breed during and at the end of the monsoon when more water means more prey. It is interesting that, near Delhi, reedbed-nesting Purple Herons start breeding in early May (the monsoon arrives in late June) while tree-nesting individuals nearby don't start until July. Many raptors breed in the winter months, presumably because of the abundance of migrants. Birds breeding at a high altitude normally do so only in the brief summer months.

Birds make all sorts of nests, of all sorts of complexity and at all heights in vegetation and on the ground. The skilfully woven grass nests of weavers are well known, while the tailorbirds literally stitch two leaves together to make a hanging pouch. Many ground-nesting birds don't build a nest, but scrape a slight hollow or lay their eggs straight onto the ground. The cuckoos simply deposit their eggs in the nests of host species and play no part in rearing their young. Woodpeckers, bee-eaters and sand martins excavate their own holes with considerable skill, while others utilise existing holes or even man-made nest-boxes.

Brown-headed Gull nesting colony

White-throated Thrush at nest

Most young birds are born blind and featherless and are completely dependent on their parents or foster parents. Many larger ground birds such as ducks, francolin and sandgrouse have precocial young that are well covered in down and sighted. They can run and feed (and in the case of wildfowl, swim) soon after they hatch. This is an adaptation to enable them to leave their very vulnerable ground nest as soon as they hatch. The degree to which both parents incubate the eggs and/or rear the young varies between species. In a number, the male plays a minimal role. In the case of the buttonquails and the Painted Snipe, all the female does is mate and lay the eggs. The more dowdy males incubate them and rear the young single-handedly, while the partners move on to mate with other males. Why this should have evolved is not fully understood.

The study of the breeding behaviour of birds is endlessly fascinating. But those pursuing it must take care not to disturb nesting birds, particularly during the hot season when exposed eggs and young will quickly fry if not protected by a parent bird. Among the biggest dangers to nesting birds are the predatory House and Large-billed Crows. These birds keep careful watch on possible nest sites and, if a human being shows any interest, monitor their movements. They devour the eggs or the young as soon as the coast is clear.

EXTREME RARITIES

Jerdon's Courser

One of the most significant and exciting achievements of Indian ornithology in recent years was the rediscovery of the Jerdon's Courser in 1986. The last authentic record was in 1900. Jerdon's, or Double-Banded, Courser *Rhinoptilus (Cursorius) bitorquatus* was first recorded by Dr T. C. Jerdon, an Indian Army Medical Officer, in 1848. Blandford recorded it in 1867 and 1871 and the last record was by Howard Campbell in 1900. Always a rare bird, these few sightings were restricted to a few river valleys in Andhra Pradesh. Despite surveys by outstanding ornithologists such as Whistler and Kinnear in 1929 and 1931, Salim Ali in 1932 and post-war surveys by the Bombay Natural History Society (BNHS), no birds were sighted. In 1985, Bharat Bhushan, a young BNHS scientist, surveyed the dry scrub hill country where Jerdon first recorded the bird. In January 1986 the species was rediscovered and found to be largely nocturnal, which partly explains why it was overlooked.

The Forest Spotted Owlet *Athene blewitti*, one of India's most elusive birds, has also recently been rediscovered. After its last confirmed sighting in 1884, there were several attempts to re-locate this bird. In 1997, a team of ornithologists came upon one in the Dhule region of Maharashtra. Col. Richard Meinertzhagen's claim to have sighted the bird in Gujarat in 1914 was later discredited and there had been no definite sightings of the bird for 113 years. It has now been found in other sites in Maharashtra and Madhya Pradesh. It is speculated that the owlet's sighting on the low foothills might be a response to the rapid destruction of lowland forest and that its continued existence might still be in jeopardy. However, much of north-central India has still not been systematically explored for the species.

Two other bird species that have not been recorded for the last 70 years are the Pink-headed Duck *Rhodonessa caryophyllacea* and the Himalayan Quail *Ophrysia superciliosa*. The Pink-headed Duck is thought to have been extinct in the wild since about 1926. Originally it was thinly distributed through the wetlands, swamps and wilderness areas that existed around the confluence of the Ganga and Brahmaputra rivers. Little is known about it and only a few skins exist in museum collections. It is probable that the last living specimen died around 1935, in a private collection in England.

Even less is known about the Himalayan Quail, which was first reported in 1846 and has not been seen since 1876. The few specimens that were collected during the 19th century were found between 1500 and 1850 m in the hills around Mussoorie, which are now in Uttaranchal. Judging by their thick, soft plumage it is possible they spent the summers at much higher altitudes and only wintered in the Himalayan foothills.

Pink-headed Duck

Only one specimen of the Large-billed Reed Warbler (*Acrocephalus orinus*) exists. It was collected by A. O. Hume in the Sutlej Valley in Himachal Pradesh in the mid-1800s. Until 2002 it was thought likely to be an aberrant form of another known species, but recent research shows this is not so. Its status remains unknown, but its short wings suggest it is (or was) a resident or short-distance migrant. It may now be extinct.

Himalayan Quail

Illegal wood poaching

Deforestation

THREATS TO BIRDS IN INDIA

Deforestation

The single most important reason for the precarious state of our fauna and flora is the destruction of their natural habitats. The statistics do not make happy reading: over half of India's natural forests are gone; one-third of the wetlands drained; 70 per cent of the water bodies polluted; 40 per cent of the mangroves destroyed and almost all of the grasslands converted to agriculture to feed India's population.

The latest satellite imagery of India's forests shows the cover is down to 17 per cent, only 12 per cent primary. This deforestation is not confined to a special type of forest. None is spared. Dry and moist deciduous, thorn scrub, alpine and coastal vegetation have all come under the axe, plough and bulldozer.

The felling of the tall trees of the moist deciduous and evergreen forests has seriously affected the hornbills in particular, including the Great, the Pied and the Malabar Pied. Other birds that have suffered from the felling of large trees are both the Adjutant Storks. They simply have no alternative place in which to nest.

The massive deforestation of the evergreen forests in the islands of Andaman and Nicobar has affected the status of the Nicobar Pigeon and the Megapode, the only one of this special group of birds found in our territorial limits. The Andamanese population of the Nicobar Pigeon is on its last legs. The tendency to replace coastal vegetation with coconut plantations is one of the reasons why the Megapode is under pressure.

The Himalayan range is the loftiest and youngest mountain system in the world. It runs 2500 km from the northwest to the southeast with extensions in the Northeast. In breadth it is about 250–400 km. This range is the catchment area of some of the largest rivers of the subcontinent. They flow through ten states, including the newly formed hill state of Uttaranchal and the Darjeeling area of West Bengal. They are among the greatest rivers in the world.

The destruction of the Himalayan forests is a major cause of concern. The policy to replace the natural forests with pine plantations does not bode well for several Himalayan species. The recent political trouble and the subsequent setting of fires in the forest have affected ground-nesting birds, including gravely endangered families like the pheasants. These forests hold the largest sustainable population of the Western Tragopan in India. One wonders what the future holds for them.

The greatest tragedy of hill development is the conversion of natural forests into timber "mines" to feed forest-based industries. This has happened as the sources of raw material for forest-based industries have been exhausted and most of the suitable land in the plains has become occupied by agriculture. The objective of the recent five-year plans in the forestry sector has been to bring more areas of hill forest into industrial plantations. For this purpose, valuable oak forests, as in Ringali and Tons in Uttar Pradesh and Thaltukhor and Mandi in Himachal Pradesh, have recently been converted into monoculture coniferous forests.

Landslides

Landslides are a natural phenomenon in the Himalayas, but the frequency and the intensity can be laid at the doors of ill-informed planners and civil engineers who ignore the local wisdom of terracing the hills. There are three natural factors that combine to cause landslides in the Himalayas: gravity, rain and tectonic movements. When a fourth factor, human

Illegal quarry

interference, is added to the first three, the combination is dramatic. As well as the rivers that flow down their courses to the plains, the Himalayan region is also drained by subsurface water flows. It is vital that this natural drainage is not interrupted, or the result will be waterlogging, followed by soil destabilisation which leads to landslides. Since the Himalayan mountains are young and still forming, tectonic plate movements render them exceedingly unstable. In such circumstances, dams, roads, mines and cement buildings in the higher mountains are all ill-advised. The problem is compounded during periods of heavy rain, when the sheer weight of water is added to that of loose rock and soil.

In Uttaranchal, experts such as Sundarlal Bahuguna suggest that recent landslides were triggered by deforestation coupled with the indiscriminate use of explosives which loosen the soil on which villagers have constructed terraces. He points to schemes such as the Tehri dam and the Vishnuprayag hydel project, where blasts are destroying village paths and even the Badrinath road. Damaging landslides are now occurring very frequently.

Intensive agriculture

Grasslands

While deforestation of tropical forests is often reported in the press, the death of grasslands goes largely unrecorded. These exceptionally diverse ecosystems have been ravaged by man who uses every method to convert them into agriculture, urban housing and dammed reservoirs. In short, they have been mutilated beyond recognition. It is no small wonder that fabulous birds of the grasslands like the Indian Bustard, the MacQueen's Bustard, the Bengal Florican and the Lesser Florican are amongst the most threatened birds of this region. Dr Asad Rahmani, who has done sterling work on these birds, revealed the alarming news that while the populations had stabilised in the 1980s they have recently shown a sharp decline. Some sanctuaries like Karera have none left. With a lot of attention on their larger cousins, smaller grassland birds like Finn's Weaver, Jerdon's Babbler and Hodgson's Bushchat are often overlooked. Their plight is as serious as that of the bustard family.

Wetlands

Like the grasslands, the wetlands of India face destruction. They are being drained and filled in at an alarming pace. Marshes are now human settlements. The Pink-headed Duck disappeared because of the destruction of the wetlands, combined with hunting. The White-winged Wood Duck is now in the same position and has the dubious distinction of being the most threatened waterbird in this part of the globe.

Ironically, the creation of wetlands by man can itself destroy natural wetland habitat. The classic example is in the Corbett National Park. Within a year of Project Tiger being launched in this park (amidst much fanfare), the controversial Kalagarh dam started to drown almost 10 per cent of the finest wet grasslands of this extraordinary park, including Patlidun which provided such good habitat for ungulates and birds. Some of the birds affected were the Red and Spotted Munias and the Black-breasted Weaver. These birds could once be seen in the thousands, as could the Common Myna. It is also being suggested that the submerged areas may once have been home to the Bengal Florican, which is no longer seen there.

Record duck shoot, Nepal

Hunting and trapping

The examples given above were not intended by man to destroy birds and their habitats. But some human activity, such as hunting and poaching, is more intentionally damaging. In theory, both are now illegal, but they continue, particularly in more remote areas. What began as essential food procurement for the hunter-gatherer became 'sport' for the maharajas and the British. The records of shoots at Bharatpur clearly enumerate the destruction that they wrought. Hunting is

7TH DEC 1936 — 2ND SHOOT
8TH FEB 1936 — 3RD SHOOT — 1683
9TH NOV 1936 — H.E. THE VICEROY LORD LINLITHGOW — 1415
5TH DEC 1937 — 2ND SHOOT — 1476
6TH FEB 1938 — 3RD SHOOT — 2568
12TH NOV — VISIT OF H.E. THE VICEROY LORD LINLITHGOW — MORNING 3044 — AFTERNOON 1229 — 4273
1938 4TH DEC — 2ND SHOOT CENTRAL INDIA HORSE — 2054

Hunting record, Bharatpur

Illegal bird trade

Live Great White Pelicans in travelling zoo

probably not a major threat today, but illegal netting certainly is. In many rural areas it is possible to see wild ducks, francolins and waders for sale for food. Travelling zoos add to the misery. Even falconry has caused recent depredations, especially in Pakistan. In India, the scandal in 1979, when permission was given to Arab sheikhs to hunt the endangered MacQueen's Bustard in the Thar Desert, shows that the threat still exists here.

Until 1991, when a total ban on the export of birds was introduced in India, the trade in birds was a major contributing factor to the precarious status of certain species. It is estimated that between 1970 and 1976, 13 million birds were exported, mostly munias and parakeets. Superficially, it looks as if the trade has much reduced. But recent research shows that a large number of birds are still caught and the trade exists, albeit on a reduced scale, both in India and abroad. In global terms it is estimated that between three and five million birds are traded annually. The mortality rate is two for every three birds caught. The calculation is that one dies during capture, one during transit and only the third ends up spending its life in a cage. The most popular birds caught are the munias, the Hill Myna and the parakeets. While the large Alexandrine parakeet is preferred, the Rose-ringed is the commonest species taken, usually as nestlings. More exotic species, like the Red-billed Leiothrix and the Silver-eared Mesia and the leafbirds, fetch large sums of money in the international market. The one loophole that exists in the current law is that while the export of Indian birds is banned, there is no restriction on the export of foreign (exotic) birds bred in India, like cockatoos and budgerigars. Unscrupulous dealers often manage to procure false documentation and use this loophole to export Indian species. They also paint native species (notably munias) to pass them off as exotics. Moreover, there is no legal control on the domestic trade in exotic birds.

Other birds, like hornbills and owls, are often used for local medicines. The neck hackles of the Grey Junglefowl are used for making "flies" for fishing. Partridges, watercocks and quails are used for the dubious sport of "cock fighting". But the most persecuted is the Common Peafowl, whose tail covert feathers are much admired by tourists and locals alike. On paper, only dropped feathers are picked up. In truth, thousands of peafowls die to feed the tourist trade.

Pollution

One area that needs immediate attention is the impact of pollutants on the region's environment. The residues of pesticides can be extremely long-lasting as they accumulate in increasing concentrations as they move up the food pyramids. Birds at the apex of the food chain, such as the raptors, are the most affected. DDT and other organochlorides cause metabolic changes in birds. These lead to the thinning of the eggshells, which in turn leads to the shells breaking before hatching. Aldrin, a popular pesticide used mostly on the wheat crop, can instantly kill even large birds like the Sarus Crane. Smaller birds, such as doves, that feed almost exclusively on grains are especially vulnerable.

Chemical dumping

Polluted stream

There is a need to find out what impact industries have on downstream areas. This includes tea, coffee and cardamom plantations and enterprises that release endocrine-disrupting chemicals, such as organochlorines. Without this information it is very difficult to encourage commercial plantations towards organic produce. There needs to be closer coordination with the Agriculture Ministries to promote non-chemical, non-toxic agriculture, particularly in and around our Protected Areas.

The major causes of water pollution are municipal sewage and industrial discharge, along with mineral run-off from agricultural land, which is mostly nitrogen and phosphates. It is estimated that only about 10 per cent of India's so-called potable water is treated. Even in the capital city, it is thought that only 25 per cent of the water is properly treated. Ordinary water filters cannot eliminate fluorides and arsenic. Offshore net fishing, aquaculture and prawn culture can also be major causes of coastal pollution, especially in Orissa where turtle breeding grounds are under great threat.

There was a time when no businessman in India was allowed to set up a large industry within 25 km of a national park or sanctuary. This has now been weakened. Statutory provisions now seek to prohibit the siting of industries within 7 km of a Protected Area. This leaves large mammals, such as elephants and tigers, virtually no space to expand into. Consider the scores of mines, dams, thermal plants, smelters, tourism projects and other similar destructive activities that are racing closer and closer to the heart of wild India, throwing a shadow on the future survival of species too numerous to list.

The holy Yamuna in Delhi

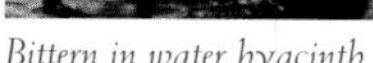

Bittern in water hyacinth

Water hyacinth flowers

Exotic Plants

Since such a large number of endangered animals live outside our Protected Areas, there needs to be a total ban on the conversion of all forest lands to monoculture of any sort. Tree monoculture does not encourage wildlife to flourish. Even forests in the control of the Revenue Department need to be kept free from such monoculture, which destroys diversity. Many World Bank-sponsored State Forestry projects that are currently converting natural ecosystems to commercial plantations include the application of chemical inputs and are contributing to the destruction of India's natural habitats.

How does the inadvertent introduction of exotic plant species like the water hyacinth, lantana and parthenium or Congress grass affect bird life? There has been some concern about the impact of lantana, particularly on ground-nesting birds like the nightjars. The Frogmouths, one of the rarest of birds, swoop to the ground to pick up insects. If lantana covers the ground, how are these birds to obtain their food? Conversely, certain birds like the bulbul seem to thrive on lantana but they greatly help disperse lantana seeds. These interactions need urgent study. The water hyacinth, introduced as an ornamental plant from South America, now clogs many northern lakes and rivers. Although some birds, such as the Purple Swamphen and the smaller bitterns, make good use of it, swimming birds are denied access to surface water and the de-oxidating effect of total vegetative cover kills off the aquatic life that many birds depend on.

If the extraordinary bird life of this region is to survive then it is clear that there are many human activities which need to be halted, regulated or reduced. Although the examples given above all come from India, most, if not all, of them apply to the other countries in the region.

HOW TO USE THIS BOOK

Area Covered

This book covers the countries of India, Pakistan, Nepal, Bhutan, Bangladesh, Sri Lanka and the Maldives. These are shown on the map on pp. 2–3 with the states of India and the provinces and administrative districts of Pakistan indicated as well as the larger rivers. Borders shown on this and all other maps in this book are neither correct nor authenticated and do not constitute an opinion on the part of the authors, publishers or distributors as to the position of any national or international boundaries.

Our region forms the western part of the Oriental Zoo-geographical Region or Realm, which is sometimes called the Indo-Malayan region. The rest of this faunal region covers Southeast Asia (including the Philippines and most of Indonesia) and southern China. West of the Indus basin in Pakistan and the Trans-Himalayan parts of India, Nepal and Bhutan are faunistically part of the Palearctic Faunal Region, which includes Europe, North Africa and the rest of Asia. Both resident and migratory species of this avifauna are included.

Nomenclature, Taxonomy and Sequence

The names and sequence of birds in formal lists change regularly with changing scientific opinion. It is unlikely that everyone will be happy with every version that is presented. However, one version has to be chosen.

We have followed the names and order in the *Annotated Checklist of the Birds of the Oriental Region* by Inskipp, Lindsey & Duckworth (1996) in the great majority of cases, although a few distinct sub-species are given different English names taken from other sources. There is still a debate about what constitutes the best order for Oriental birds and, particularly, which English name to use. We recognise that many birders are not happy with all the recent changes and prefer to use the English names they are most familiar with. But we felt we should be consistent with what is probably now the most widely accepted field guide in the region (*Birds of the Indian Subcontinent* by Grimmett, Inskipp & Inskipp, 1998) and what, in respect of most of the names, is followed in *A Field Guide to the Birds of the Indian Subcontinent* by Krys Kazmeirczak and B. van Perlo (2000). However, for those still using the old names, we have included most of them in brackets.

Ornithology is a vibrant, living science and new research leads to new species being described annually, mostly by redefining a sub-species as a species. The use of DNA analysis has raised all manner of questions, and a lively discussion exists about how different birds are actually related and what a species actually is. We are currently going through a phase of what is colloquially known as "splitting", where the trend is to give a sub-species specific rank if sufficient differences can be shown and they appear to have discrete breeding populations. The opposite trend, "lumping", tends to group as many closely related birds together as possible as one species. Debate continues over several groups, notably and most confusingly the large gulls, and this often means you cannot always put a name to a bird you have seen. Frequently, the problem is that for the birder in the field the differences between some sub-species can be considerable (see Kalij Pheasant and Wagtails). Using only a specific name may cause confusion when it is obvious the birds look so different in the field. We have included a number of different well-marked sub-species (both in photographs and text) for this reason, and in some cases used a distinct English name from the Handbook or elsewhere.

We have incorporated in the book some recently published or proposed changes (see Systematic List of Families and Species for scientific names). The Long-billed Vulture is now split into the Indian and the Slender-billed Vultures, the Eurasian Eagle Owl into the Eurasian and the Rock Eagle Owls, the Herring and Lesser Black-backed Gull into Yellow-legged, Armenian and Heuglin's Gulls in our list (with more changes pending!), the Sand Martin into Sand and Pale Martins, the Booted Warbler into Booted and Sykes's Warblers, the Pallas's Warbler into Pallas's and Lemon-rumped Warblers, the Yellow-browed Warbler into Yellow-browed and Hume's Warblers, the Lesser Whitethroat into Lesser, Hume's Lesser and Small Whitethroats, the Desert Warbler into Eastern and Western Desert Warblers, the Orphean Warbler into Eastern and Western Orphean Warblers and the Rufous-winged Bushlark into Rufous-winged and Bengal Bushlarks.

On pp. 461–500 there is a full Systematic List of all the species recorded in the region. It totals 1305 species (which is nearly 14 per cent of the total number of species in the world). Two of the species—the Himalayan Quail and the Pink-headed Duck—are probably extinct. The list includes recent acceptable additions such as the Long-billed Dowitcher, Pectoral Sandpiper, Grey-faced Buzzard, Rough-legged Buzzard, Eared Pitta, Swinhoe's Minivet, Chinese Leaf Warbler, Japanese Grosbeak and Pallas's Bunting as well as the changes noted above. The Olivaceous Warbler has been removed, as in our personal view the published description does not exclude Sykes's Warbler. We have added the Nicobar Scops Owl and the Large-billed Reed Warbler in the light of recent research. Similarly, we have accepted that the claimed Laced Woodpecker, in Bangladesh, is in fact a Streak-breasted Woodpecker. We have also included the Armenian Gull on the basis of our own observations in northwestern India in the last two years. Further research into the large gulls wintering in the region is likely to produce more species and sub-species. Given the position of the region and the steady improvement in field skills, it should be expected that other species will be added to the list on a regular basis.

There is little or no formal vetting procedure for claimed new sightings in most of the countries in our region. Rather, published records that go unchallenged are generally accepted, but not everybody publishes details of what they have seen. In the end it is a matter of personal choice what you accept as part of the avifaunal record and we may have erred on the side of generosity in our list.

Family and Genus

The letter F: is short for family. Families are groups of birds that are believed to be related through common origins. Each family has a scientific name (based on Latin), which ends in *-idae*. Some of the recent changes in how birds are grouped have lost us some familiar family names. But in practice, most birders do not use the formal scientific term and prefer to talk in terms of, for example, chats, thrushes, gulls and terns. Within families one or more genus and birds of the same genus are even more closely related. There is no consistent ending to the scientific name of a genus. Sometimes the genus are grouped into a sub-family, the scientific name of which ends in *–ini* or *-inae*.

The first of the words in the scientific name is the genus. The second word is the species name and the third is the sub-species name. Sub-species are often referred to as forms or races. If the third word is the same as the second, then that was the first sub-species in the species to be named.

If you see a final word or abbreviation (often in brackets) at the end of the name, this refers to the person who first published the scientific name being used. The most frequent is Linn or (Linn) which stands for Linneaus, the great 18th-century Swedish biologist who invented modern systematics. We do not include the "namer" in the scientific names given in this book.

Families are joined together in an even larger grouping called Orders (the scientific name always ends in -formes) and the Orders together constitute the kingdom of birds. Thus an Indian Robin in Delhi is called scientifically *Saxicoloides fulicata cambaiensis* (one in Chennai is *Saxicoloides fulicata fulicata* because it is a different sub-species and the first to be named. It is called the nominate sub-species). It belongs to the genus *Saxicoloides* and is popularly known as one of the chats. This genus is currently placed in the sub-family Saxicolini (which includes all the chats and wheatears), which in turn is part of the family Muscicapidae which includes thrushes and flycatchers. This family is in the order Passeriformes, which includes most of the smaller bird species popularly known as Song or Perching Birds (neither always accurate words!).

Size

The average length of the bird is given in centimetres following each bird name. Only where the size difference between male and female is considerable (e.g. the Indian Peafowl) are both sizes given. Generally, the larger a species, the more likely it is that there will be size variations. Note that it is not always the males that are larger. In raptors and some waders, the females are the larger of the two, sometimes considerably so, as in the hawks. The length given is meant as a guide only, to aid comparison with more familiar or similar species. Many species vary either way, and there may be differences between sub-species. Where variation is considerable, this is mentioned in the text.

Photographs

Each species is illustrated with at least one photograph. In many cases there are two photographs to illustrate different plumages or views (including flight shots). Most photographs are of adult males and the sex is not indicated. Often, female (or immature) plumages are illustrated where they are different from the male or the adult; the sex or age is indicated on the picture. In the case of some species with well-marked races, different races have their own section so that more than two photographs can be presented.

The majority of the photographs for this book were taken in the field in the region. Where they were taken elsewhere we have tried to ensure that the sub-species shown occurs in the region. A few of the very rare species were taken in captivity and a few are hand-held photographs of birds caught for migration studies (when the birds are ringed or banded and then released). We have tried to keep both types of location to a minimum, but felt it was better to include such pictures of often difficult species rather than omit them altogether.

Description: Bird Topography

The description includes the most important features of size, shape and plumage with particular emphasis on features not visible in the photographs and variations not illustrated. We have focused on the diagnostic features which are unique to the species. Often similar species are referred to, including over 100 species not covered elsewhere in this book. We use certain standard terms to describes parts of the bird and they are illustrated below. For simplicity we have often used the words upperparts (or above) to mean the upper surface of the entire body and underparts (or below) for the underparts. We have also referred to legs to include feet (or, more accurately, toes) and moustache for any stripes extending down from the bill (more accurately called variously malar, moustachial or sub-moustachial stripes!). We often refer to the bend of the wing as the shoulder and use flight feathers to mean both secondary and primary wing feathers. The Glossary on pp. 448–449 will help explain some of the terms used.

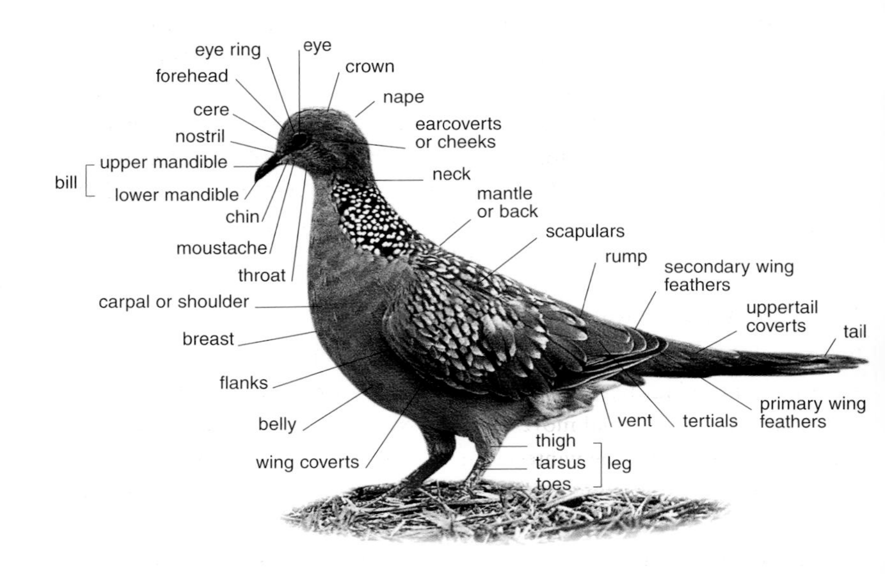

Voice

Songs and calls are often the best way of identifying a bird, but they are notoriously difficult to describe in words. Different linguistic cultures give rise to different verbalisations and, as individuals, we hear sounds differently. Just look at how the sound of a bird you are very familiar with is described in different books. We have based our descriptions on our own interpretations and careful reference to other texts. Where we disagree with others, we have used our own verbalisation. Capital letters represent loudness. We recognise that not everyone will agree with our choice. In many cases the same birds may have different songs or calls; we have tried to describe the most common. Several recordings of bird sounds are available (see the Bibliography for details), although quality varies considerably and care needs to be taken as there are regional variations in the calls of individual species.

Habits

Here we have described the preferred habitat (type of environment), food, feeding technique, behaviour, degree of sociability and where each bird nests. The essence of a bird is colloquially known as its "jizz", a term from the military to describe unidentified aircraft and ships and meaning "general impression and shape"; so strictly speaking it should by written "gis"! The aim has been to capture, with the description and voice, what the essence or "jizz" of each individual bird species is. Obviously there is much more to be said about each species and for those interested there is still no better reference than the *Handbook of the Birds of India and Pakistan* by Salim Ali & Dillon Ripley (1987). We have kept habitat descriptions simple and self-explanatory, but the Introduction and Glossary can be referred to if in doubt.

Distribution and Status

The maps give the range of the bird in the region. They are coded red for resident or breeding summer visitor and blue for non-breeding winter visitor. Relatively short altitudinal movements (moving to lower elevations in winter) are not shown. Where the wintering species only occurs on passage this is shown within the wintering range unless it covers a wide area. In such cases, it is not included in the range but the passage occurrence is mentioned in the text. In certain instances, some species are resident in part of the region and only breeding summer visitors elsewhere. Where these variations are not readily differentiated, the whole range is shown as red. If a species has both resident and wintering populations in the same area, the dominant status is shown and the other is mentioned in the text.

The ranges are based on our current knowledge. They may include places where the species has not actually been recorded, but where we think it is likely to occur. While we agree it is valuable to have maps based on actual records, much of the region is still not systematically covered and we simply do not know what could be there. Many feel some recently published maps give an impression of somewhat spurious accuracy as a result. We feel it is more helpful to indicate where a species is likely to be seen, always bearing in mind that it will only occur if there is suitable habitat.

We have not attempted to show isolated records outside the main range but have referred to where they have occurred (in general terms) in the text. In the case of vagrants (or stray and irregular visitors to the region), the broad area where the vagrancy occurred is shown in blue although it should be noted that re-occurrence would be exceptional. The abundance symbol and the text should be referred to in assessing the likelihood of seeing a particular species. Some species may only occur once every ten years at best.

The accompanying text summarises the status and the range. No attempt is made to specify altitudinal limits, as they tend to vary so much in the region according to aspect, latitude and longitude. If a species is endemic (confined to part of the region), this is mentioned. The world range of all non-endemic species is given in general terms to assist those who may have seen the species elsewhere. It is, however, important to note that the sub-species may not be the same.

Abundance Code

We have used the following abundance codes for our view of current status:

● Common. Encountered with at least 90 per cent certainty in preferred habitat.

◉ Fairly common. Encountered with 50–90 per cent certainty in preferred habitat.

⊙ Uncommon. Encountered with less than 50 per cent certainty in preferred habitat.

○ Rare. Encountered once a year or less in preferred habitat. This includes vagrant or stray visitors.

The text with the maps provides some modification of these broad categories, for instance by pointing out if a bird is local or probably over-looked. We also point out if there has been a known change in occurrence or discussions about the specific status. Our categorisation is based on our current knowledge and experience. We have used recent reports on the ornithological websites and trip reports by fellow birders to be completely up to date. Those interested in keeping abreast of the latest developments in the region should refer to the various websites or join the local email groups; details are given on pp. 459–460.

The authors and publisher would be happy to hear from readers and will endeavour to update the information in future editions. Please send us your comments at biks@vsnl.com.

Globally Threatened Status

We have included a code for globally threatened status based on the categories given in *Threatened Birds of the World* by BirdLife International (2000) where they apply to this region's birds:

● Critically endangered. Facing an extremely high risk of extinction in the wild in the immediate future.

◉ Endangered. Facing a very high risk of extinction in the wild in the near future.

⊙ Vulnerable. Facing a high risk of extinction in the wild in the medium-term future.

X Conservation-dependent. Dependent on a conservation programme in the absence of which it would be in one of the above categories within 5 years.

○ Near-threatened. Close to be being Vulnerable.

Some 140 species or over 10 per cent of those recorded in the region fall into one of these categories. For a further four, there are insufficient data to classify them. It is highly likely that the Mountain Quail and the Pink-headed Duck are, in fact, extinct though it should be noted that both Jerdon's Courser and the Forest Owlet have been rediscovered in recent years after long gaps. The Long-billed Reed Warbler will, in due course, be added.

The most striking thing about the list is the number of large birds included. For instance, every one of the six bustard species recorded in the region is globally threatened. Many globally threatened birds are very local restricted-range species, notably in the Western Ghats, Sri Lanka and the Andaman Islands. But there are a number of wintering endangered species, such as Greater Spotted and Imperial Eagles, whose welfare must be shared with countries outside the region. The most famous of these are the Siberian Cranes, reduced in the 21st century to an adult pair wintering in Bharatpur.

SNOW PARTRIDGE

Lerwa lerwa 38 cm F: Phasianidae

Joanna Van Gruisen

Description: A plump, high altitude, medium-sized, game bird with red bill and legs. Vermiculated black and white on head and upperparts. White below with heavy, chestnut streaking. Short tail. Sexes alike.

Voice: A repeated low whistle.

Habits: Inhabits steep rocky or grassy slopes near snowline. Feeds in pairs or small parties (coveys) on ground on seeds and vegetation. Very confiding where not hunted, often watching observer from rock. When disturbed flies downhill with clattering wings. Nests on ground.

Distribution: Local breeding resident in northern mountains from north Pakistan east to Arunachal Pradesh. Moves lower down in winter. Also occurs in Central Asia and China.

TIBETAN SNOWCOCK

Tetraogallus tibetanus 51 cm F: Phasianidae

Otto Pfister

Description: A large plump, high altitude, game bird with chestnut rump and tail. Grey-streaked white above with darker neck, crown and face stripes and white cheeks and throat. White below with bold black flank stripes. White secondaries obvious in flight. Sexes alike.

Voice: Noisy. An accelerating bubbling, a whistle and a *cur lee* call.

Habits: Inhabits rocky ridges, slopes and meadows above snowline. Feeds actively on ground on shoots and roots in pairs or coveys. Cocks tail. Runs uphill or flies a long distance down when disturbed. Nests on ground.

Distribution: Fairly common disjunct breeding resident in northern mountains from Ladakh to Arunachal Pradesh. Winters lower down. Also occurs in Tibet and China.

HIMALAYAN SNOWCOCK

Tetraogallus himalayensis 72 cm F: Phasianidae

Description: A very large, plump high altitude, game bird with grey rump and tail. Much larger than **Tibetan**. Grey with chestnut feather edgings and flank stripes. Head rather pale with chestnut neck and face lines and white neck patches and throat. White undertail coverts. White primaries, as opposed to secondaries in **Tibetan**, obvious in flight. Sexes alike.

Voice: Noisy. A whistled *curlee wi wi* and an accelerated chuckling.

Habits: Habitat and behaviour as **Tibetan** but usually at even higher altitudes.

RS Chundawat

Distribution: Locally common breeding resident in northern mountains from north Pakistan to Nepal. Moves lower down in winter.
Also occurs in Central Asia and Tibet.

CHUKAR (Chukar Partridge)

Alectoris chukar 38 cm F: Phasianidae

Description: A plump, medium-sized, dry country game bird with zebra-striped flanks. Pale sandy grey above, greyer on breast, tail and lower back. Chestnut outer tail. Creamy throat surrounded by black line. Prominent flank stripes and warm buff belly. Red bill and legs. Sexes alike.

Voice: A loud, rapidly repeated *chukchukchuk*.

Habits: Inhabits dry, stony and rocky hillsides and terraced cultivation. Feeds in pairs or coveys on ground on seeds, roots, shoots and invertebrates. Runs rapidly. Flies fast and low on whirring wings, usually downhill. Not shy where unmolested, often perching prominently on rocks. Males aggressive to each other when breeding. Nests on ground.

Otto Pfister

Distribution: Locally common breeding resident in northern mountains from western Pakistan east to Nepal. Very rare in accessible parts of Pakistan due to hunting. Some move lower down in winter.
Also occurs in E Europe, W, N and Central Asia and China.

Otto Pfister

BLACK FRANCOLIN (Black Partridge)

Francolinus francolinus 34 cm F: Phasianidae

Otto Pfister

Description: Dark, medium-sized game bird. Male buff-edged dark brown above including crown, barred paler on rump and tail. Chestnut under tail. Black below with heavy white spotting, white cheeks and a broad chestnut collar. Female duller. Streaked brown above, white-edged black below with chestnut hind neck.

Voice: Noisy. Rather soft but far-carrying call *chik chirrik cheek chereek* with variations, preceded by a distinct *clik* at short range.

Habits: Inhabits reedbeds, irrigated cultivation, particularly sugarcane, tall grass and scrub near water. Feeds in pairs or coveys on ground on seeds, shoots and invertebrates. Shy and secretive but often calls from high perch. Rather crepuscular.

Distribution: Locally common breeding resident in northern plains and low hills from Pakistan east to northeast India including Nepal. Very rare in Bangladesh. Replaced in Deccan by next species.
Also occurs in W Asia.

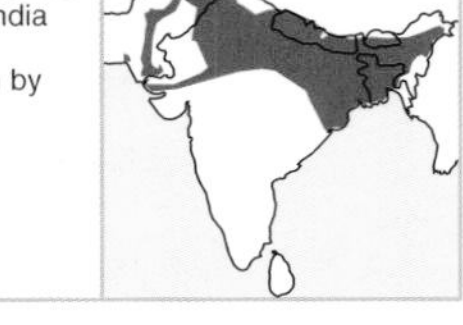

PAINTED FRANCOLIN (Painted Partridge)

Francolinus pictus 31 cm F: Phasianidae

Joanna Van Gruisen

Description: A heavily spotted, medium-sized game bird. Male has plain chestnut head. Upper-parts buff-edged rich brown with finely barred black and white rump and tail. Underparts and neck black heavily spotted with white. Rufous under tail. Female paler especially on head.

Voice: Noisy. Similar to **Black**, a grating *keek keek keeki* preceded by a quieter *kek*. Often calls in duet.

Habits: Inhabits thick cover including grassland, scrub, cultivation and forest edges. Feeds in pairs or coveys on ground on seeds, shoots and invertebrates. Very shy except when calling from exposed perch. Rarely flies, squatting if disturbed. Rather crepuscular.

Distribution: Locally common endemic breeding resident from southern edge of northern plains south through peninsula except parts of southern coastal strip. Rare in Sri Lanka.

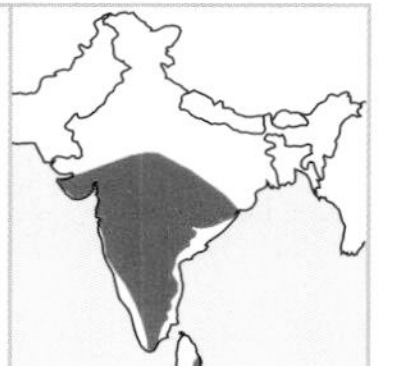

GREY FRANCOLIN (Grey Partridge)

Francolinus pondicerianus 34 cm F: Phasianidae

Otto Pfister

Description: Medium-sized grey and brown game bird with black-bordered, buffy-white throat. Barred buff, grey and rufous-brown above with orange supercilia and cheeks. Finely barred grey and white underparts and buff vent. Sexes alike. Young birds fly when only a third of adult size.

Voice: Very noisy throughout year. Loud coarse repeated *pat ee laa, kila kila kila, khirr khirr*.

Habits: Inhabits dry, open country including woodland, scrub, semi-desert and cultivation. Feeds on ground in pairs or coveys on seeds, shoots and invertebrates. Picks up grit and dust bathes on roads. Runs when disturbed. Flies low on whirring wings for short distances. Nests on ground.

Distribution: Very common breeding resident throughout lowlands except Northeast, Bangladesh and most of Nepal and Sri Lanka.

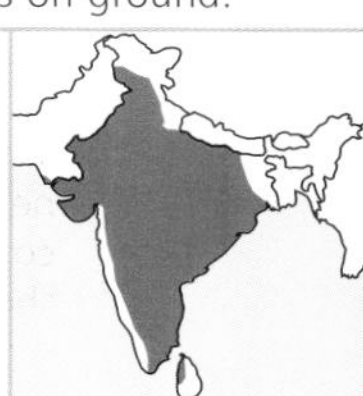

Otto Pfister

SWAMP FRANCOLIN (Swamp Partridge)

Francolinus gularis 37 cm F: Phasianidae

Description: A medium-sized, wet grassland, long-legged game bird with an orange throat. Finely buff barred brown above with chestnut flight feathers and outer tail. Heavily white-streaked brown below. Brown crown, buff supercilia, thin brown eyestripes extending to neck and unmarked dark orange throat. Long pale legs. Sexes similar.

Sharad Gaur

Voice: Noisy. Rythmn reminiscent of **Grey**. A loud *kyew kyew kyew kaa kaa* call and *chukeroo chukeroo chukeroo*. Alarm is *kaw care*.

Habits: Inhabits tall grassland and swamp. Also nearby sugar cane fields. Feeds on ground on seeds, roots, shoots and insects in pairs or coveys. Will wade in water and climb reeds. Nests in reeds.

Distribution: A globally threatened breeding endemic resident now restricted to protected areas in the terai of Nepal and northern India particularly the Brahmaputra Valley. Formerly more widespread in the north including Bangladesh.

TIBETAN PARTRIDGE

Perdix hodgsoniae 31 cm F: Phasianidae

Otto Pfister

Otto Pfister

Description: A medium-sized, high altitude game bird with black cheeks. Richly buff, barred brown above. Chestnut neck and flanks, heavily black barred and blotched on breast and belly. White throat, supercilia and upper neck contrasting with black eye stripes and cheek patches. Sexes similar. Northwestern race (shown) paler.

Voice: Call a rattling repeated *scherrrreck* and a high *chee chee chee*.

Habits: Inhabits high altitude semi-desert and rocky slopes with scattered low scrub. Feeds on ground in pairs or coveys on seeds, shoots and invertebrates. Runs rapidly uphill when disturbed or flies downhill.

Distribution: Locally common breeding resident in northern India, Nepal and Bhutan along border with Tibet.
Also occurs in Central Asia and China.

COMMON QUAIL

Coturnix coturnix 20 cm F: Phasianidae

Goren Ekstrom

Description: A small, plump game bird. Brown streaked buff above, buff below with streaked flanks. Head variable striped black and white with male usually showing black anchor mark on throat. Usually seen in fast, direct flight when very short tail and long-bowed wings obvious. Smaller **Rain Quail** *C. coromandelica* is heavily streaked black below in male but female very similar.

Voice: Far carrying *wet me lips*. Calls *wree*.

Habits: Inhabits grassland and crops. Secretive, feeding close to ground in pairs and small coveys on seeds and invertebrates. Usually only seen when flushed. Flies up to 200m before running to hide in cover. Difficult to flush again.

Distribution: Mainly a widespread but erratic winter visitor to lowlands in northern half of region but much overlooked. A few probably breed in the north. Formerly much more numerous on passage.
Also occurs in Europe, Africa and W and N Asia.

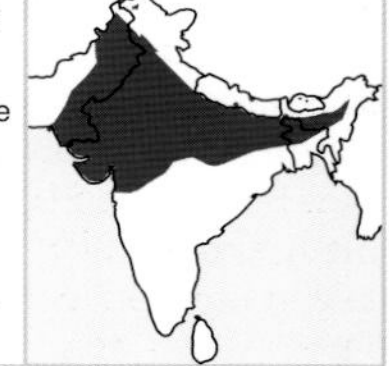

JUNGLE BUSH QUAIL

Perdicula asiatica 17 cm F: Phasianidae

Description: Tiny, barred game bird. Male heavily barred black and white below. Chestnut head with white supercilia and moustache. Upperparts brown barred black and white. Female pinky-buff below with barred wing coverts and buff supercilia. In similar but paler **Rock Bush Quail** *P. argoondah* male lacks moustache and female plainer with no supercilia or barring.

Voice: Loud *chee chee chee chuk*. Also musical whistling *whi whi whi* alarm.

Habits: Inhabits dry, often stony, country with scrub and grass. Feeds on ground on seeds and insects in pairs or coveys. Very secretive, squatting if disturbed. If flies, noisily scatters. Makes worn runs. Nests on ground.

Sharad Gaur

Distribution: Locally common endemic breeding resident throughout most of peninsula extending north to Himachal Pradesh but absent from most of the Gangetic plain and Northeast. Also found in Sri Lanka.

HILL PARTRIDGE (Common Hill Partridge)

Arborophila torqueola 28 cm F: Phasianidae

Description: A medium-sized, dark, forest game bird with a bright chestnut head. Black barred brown above. Grey below with chestnut and white flank streaks. Bright chestnut head with white-streaked black eye stripes, throat and neck. White neck collar. Female dull with brownish head.

Voice: A rather eerie, repeated whistle *po eer, po eer*. Also a rapid *kwikkwikkwik*.

Habits: Inhabits the ground of well-forested slopes and ravines feeding in pairs or small coveys on seeds and invertebrates. Very secretive, preferring to run if disturbed. Flies expertly through trees. Best located by sitting quietly and listening for sounds of feeding birds on leaf litter. Roosts in trees. Nests on ground.

Peter Morris

Distribution: Fairly common breeding resident in northern mountains from Himachal Pradesh to the Myanmar border.
Also occurs in Tibet and SE Asia.

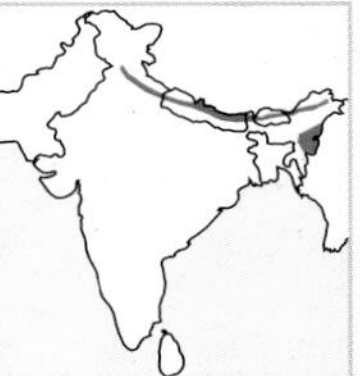

RED SPURFOWL

Galloperdix spadicea 36 cm F: Phasianidae

S Elamon

Description: A medium-sized, long-tailed game bird with red legs. Buff edged reddish-brown with grey head and neck and red eye patches. Sexes similar. Kerala race deep chestnut red.

Voice: Calls a cackling *kuk kuk kuk kuk karack* and a repeated crowing *ker kweck*.

Habits: Inhabits dense scrub and thickets, often on rocky hillsides. Also plantations and secondary growth. Extremely shy, running rapidly away if disturbed. Feeds on seeds, fruit and invertebrates on ground in pairs or small coveys. Feeds in clearings and on paths in early morning and evening. Roosts in trees. Nests on ground.

Distribution: Locally common endemic breeding resident throughout most of peninsular India from Uttar Pradesh southwards except extreme south and eastern India and Sri Lanka.

PAINTED SPURFOWL

Galloperdix lunulata 32 cm F: Phasianidae

Gertrud Denzau

Description: A medium-sized, stocky, dark game bird with grey legs. Male white-spotted chestnut above with blackish wings and tail. Finely white-barred black head and neck. Lightly black flecked yellow buff underparts. Female plain brown with paler throat and belly.

Voice: Calls not fully known but include a loud *chur chur chur*.

Habits: Inhabits rocky scrub and thickets on dry hillsides. Feeds on ground on seeds, fruit and invertebrates singly, in pairs or small coveys. Exceptionally shy, running away fast to hide in cover when disturbed. Best looked for in the shade of fruiting trees such as Banyans. Nests on ground.

Distribution: Locally common endemic breeding resident in inland peninsular India south from eastern Rajasthan and West Bengal.

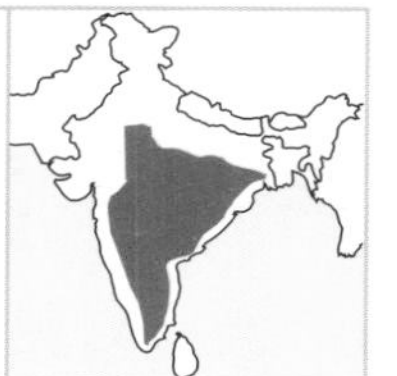

BLOOD PHEASANT

Ithaginus cruentus 38 cm F: Phasianidae

Description: Medium-sized, red and grey game bird with thick crest. Varies according to race. Males are white-streaked grey above and greyish-white with variable amounts of red below but always on vent and tail. Head and neck variably patterned red and black with grey or black crest Red skin round eyes. Female dark brown with grey nape and crest. Very short-tailed for a pheasant.

Voice: Squealing *kzeeuuk cheeu cheeu chee*, a repeated *chuck*.

Habits: Inhabits open forest and scrub near snowline. Feeds gregariously on ground scratching (often through deep snow) for shoots, moss and lichen. Usually very confiding. Runs strongly and rarely flies. Roosts in trees. Nests on ground.

Otto Pfister

Distribution: Locally common breeding resident in high northern mountains from western Nepal east to Arunachal Pradesh. Moves lower in winter. Also occurs in Tibet, Myanmar and China.

WESTERN TRAGOPAN

Tragopan melanocephalus Male 71 cm Female 60 cm F: Phasianidae

Description: A large, brightly coloured game bird with a golden breast. Body heavily spotted white on grey-brown above and black and red below. Black crown, bright red hind neck and cheeks and green throat. Female greyish-brown spotted white. Longish thick tail and long pink legs.

Voice: Loud bleating repeated *waah waah* call.

Habits: Inhabits montane forest undergrowth in pairs or singly. Feeds on ground on seeds, fruit, shoots, leaves and invertebrates often near snowmelt. Very skulking but comes into open early and late in day. Male displays horns and lappets on head when breeding. Roosts in trees. Nests on ground.

Thakur Dalip Singh

Distribution: Globally threatened and rare endemic breeding resident very locally in northwestern mountains from north Pakistan east to Himachal Pradesh. Formerly in Uttaranchal. Moves lower down in winter.

SATYR TRAGOPAN (Crimson Horned Pheasant)

Tragopan satyra Male 68 cm Female 59 cm F: Phasianidae

Sujan Chatterjee

Description: Large, brightly coloured game bird with blue face. Male's underparts mainly white-spotted crimson red. Black and white-spotted brown upperparts with red on wings and dark brown rump and tail. Blue face surrounded by black extending onto neck sides. Female white-spotted rufous-brown.

Voice: Loud wailing *guwaa guwaah guwaah* getting louder and more extended. Also a *wak wak* alarm.

Habits: Inhabits montane forest undergrowth particularly on slopes or in ravines. Feeds on ground on vegetable matter and invertebrates. Confiding where unmolested but tends to keep hidden. Feeds in open in morning and evening. Nests in tree.

Distribution: Scarce and very local breeding resident in northern mountains from Uttaranchal east to Arunachal Pradesh. Common only in Bhutan. Moves lower down in winter. Also occurs in Tibet.

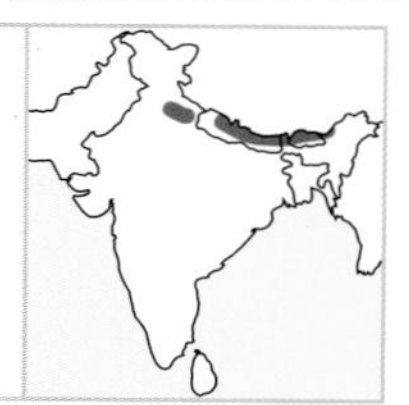

BLYTH'S TRAGOPAN

Tragopan blythii Male 68 cm Female 59 cm F: Phasianidae

Thakur Dalip Singh

Description: A large, brightly coloured game bird with a yellow face. Male has unspotted red head, neck and breast with black crown and cheek stripes. Flanks white spotted red and belly greyish-buff. Upperparts black and white spotted brown. Dark brown tail. Female white spotted brown and extremely similar to several other species.

Voice: A loud *hwaa ouwaa ouwaa*. Also *gock gock gock*.

Habits: Inhabits montane forest undergrowth on steep slopes. Food and behaviour as other tragopans but more often in small parties. Nesting poorly known but one nest found on cliff ledge.

Distribution: Globally threatened and rare breeding resident in northern mountains from Bhutan east to the Myanmar border. Moves lower down in winter.
Also occurs in Tibet and Myanmar.

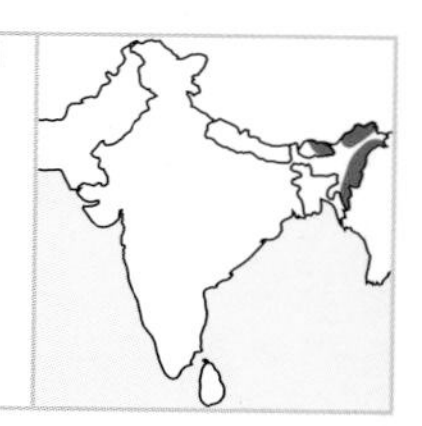

TEMMINCK'S TRAGOPAN

Tragopan temminckii Male 64 cm Female not known F: Phasianidae

Thakur Dalip Singh

Description: A large, brightly coloured game bird with a golden crown. Very similar to **Satyr** but male has bright blue face surrounded by black and upperparts red. Black and white-spotted brown wings and uppertail coverts and darker tail. Grey spots on red underparts. Female very similar to **Satyr** with large white spots below. This species is very little known in the wild.

Voice: A sequence of moaning *whoah* notes which get louder and longer and end in a nasal grumble.

Habits: Habitat, food and behaviour as other tragopans as far as is known but more arboreal. Nest not known.

Distribution: Rare resident in Arunachal Pradesh. Not currently classed as threatened but probably is in the region.
Also occurs in Tibet, Vietnam, Myanmar and China.

KOKLASS PHEASANT (Koklas Pheasant)

Pucrasia macrolopha Male 61 cm Female 53 cm F: Phasianidae

S Pandey

Description: Large, grey and brown game bird with long backswept crest. Races vary in amount of chestnut on underparts. Male has bottle green head with brown and golden crown and crest. Prominent white lower cheeks. Underparts chestnut. Flanks silvery-grey of variable extent. Upperparts similar. Dark pointed tail. Female paler with short crest and cheek spots.

Voice: Noisy. A crowing *khok kok kok kokha* and various chuckles.

Habits: Inhabits montane forested hillsides and ravines. Feeds in pairs on ground on vegetative matter and invertebrates, Very shy, running or flying rapidly when disturbed. Roosts in trees.

Distribution: Locally common breeding resident in northern mountains from north Pakistan east to central Nepal. Moves lower down in winter.
Also occurs in Central and N Asia and China.

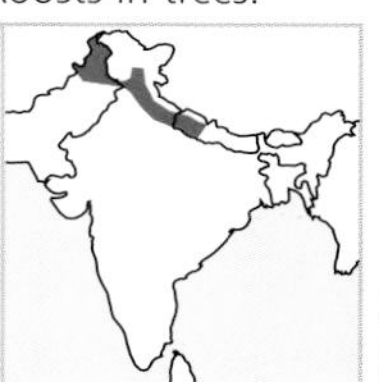

HIMALAYAN MONAL (Impeyan Pheasant)

Lophophorus impejanus Male 72 cm Female 63 cm F: Phasianidae

Otto Pfister

Description: Large, brightly coloured game bird with plumed crest. Male iridescent purple-blue above with chestnut tail and wings. Underparts black including foreneck and face. Blue round eye. Rear neck coppery-yellow becoming greener on mantle. White back patch. Female white flecked brown with white throat.

Voice: Loud whistle *whhee uu* like **Eurasian Curlew**.

Habits: Inhabits rocky and grassy slopes at or above treeline moving down into open forest in winter. Also active cultivation, particularly potatoes. Feeds on ground usually in small groups on roots, seeds, shoots, berries and invertebrates. Digs into ground, often through snow. More confiding than most game birds.

Distribution: Locally common breeding resident in northern mountains from north Pakistan east to Arunachal Pradesh. Moves lower down in winter. Also occurs in Tibet.

RED JUNGLEFOWL

Gallus gallus Male 65 cm Female 42 cm F: Phasianidae

Otto Pfister

Description: Origin of domestic chickens. Male has golden brown hackles covering neck and back, golden and green wings and black underparts. Thick, long curved black tail. White rump. Red comb and wattles. Female speckled reddish-brown, with golden neck and small red comb and wattles. Cocked tail. Beware hybrids with domestic fowl.

Voice: Shrill rushed crow *kuk ku rudi ru*, reminiscent of domestic chicken. Cackled alarm in flight.

Habits: Inhabits well-watered forest and secondary growth. Appears on roads in morning and evening. Feeds in groups, usually with one male, on ground on vegetative matter and insects. Very shy, running rapidly or flying when alarmed. Roosts in trees.

Distribution: Locally common breeding resident in northern foothills from Jammu and Kashmir east to Myanmar border and south into northeastern plains south to Orissa and Madhya Pradesh. Also Bangladesh. Associated with distribution of natural Sal forests. Also occurs in China and SE Asia.

GREY JUNGLEFOWL

Gallus sonneratii Male 80 cm Female 46 cm F: Phasianidae

Otto Pfister

Description: Long-tailed, silvery-grey game bird with pink legs. Male similar in shape to but longer than **Red** and overall plumage spangled silvery-grey with gold flush restricted to neck and wing coverts. Hackles shorter. Flight feathers black not golden and belly also silvery. Female reddish brown above and heavily white-streaked black below. Wings not barred as in **Red**.

Voice: Noisy. Loud confident crow *kuk kuk kur ra kuk* in morning.

Habits: Inhabits forest undergrowth, secondary growth and abandoned plantations. Food as **Red** but usually in pairs. Very shy but confiding where unharmed. Often seen on roads in morning and evening. Roosts in trees. Nests on ground.

Distribution: Locally common endemic breeding resident in low hills of western peninsular India from Rajasthan south to Kerala and east to Madhya Pradesh. Associated with distribution of natural Teak forests. Hybridises with Red in northeast of range.

WHITE-CRESTED KALIJ PHEASANT (KALEEJ)

Lophura leucomelanos hamiltonii 65 cm F: Phasianidae

Description: Large, dark game bird with white crest. Male mainly blackish with brown scaling on upperparts and silvery on rump. Long, thick black tail. Heavily feathered whitish flanks. Red face. Female has crest and red face but is dark brown with buff scaling. Race *melanota* (not illustrated) has black crest and is plain black above.

Otto Pfister

Voice: Male crows *kurr kurr kurchi kurr*. Various agitated squeaks, whistles and chuckles.

Habits: Inhabits forest undergrowth often in ravines and near water. Feeds on ground in pairs or small parties scratching for seeds, fruit, insects and small reptiles. Shy but emerges onto tracks in early morning and evening. Roosts in trees.

Distribution: Common endemic breeding resident in northwestern mountains from Pakistan to western Nepal. Moves lower down in winter. *Melanota* occurs from east Nepal east to Bhutan.

♀

RK Gaur

KALIJ PHEASANTS (Kaleej)

Lophura l. leucomelanos and *l. lathami* 65 cm F: Phasianidae

RK Gaur

RK Gaur

Descrption: A large, dark game bird with a black crest. *Leucomelanos* is similar to *hamiltoni* but with more limited scaling above and with a black crest. Can have whitish neck as shown. *Lathami* is all black with white scaling on rump. Black crest and red face. Race *moffitti* (not illustrated) is all black with no scaling. It also has a black crest and red face. Females very similar to female *hamiltoni*.

Voice: As *hamiltoni*.

Habits: As *hamiltoni*.

Distribution: Locally common breeding resident in northern mountains. *Leucomelanos* occurs in central and east Nepal, *lathami* in east Himalayas and northeast India and *moffitti* in central Bhutan. *Lathami* also occurs in Myanmar.

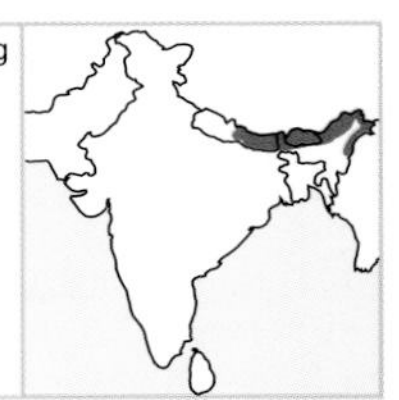

CHEER PHEASANT (Chir Pheasant)

Catreus wallichii Male 118 cm Female 76 cm F: Phasianidae

S Pandey

Description: Large, very long-tailed, pale game bird. Male black-barred grey above and on breast. Buffer on belly and browner on strongly barred tail and rump. Long back-swept brown crest and red face. Female similar but browner and shorter tailed.

Voice: Far-carrying crow *chir a pir cir a pir chi chir chirwa*. Also cackling *waa* and *tuk tuk* calls.

Habits: Inhabits secondary scrub and grass patches in or near forest. Often on steep rocky slopes. Feeds in pairs or parties on ground on vegetable matter and invertebrates. Extremely wary, running rapidly for cover if disturbed. If flushed hurtles downhill. Roosts in trees. Nests on ground.

Distribution: Globally threatened and very local endemic breeding resident in northern mountains from north Pakistan east to central Nepal. Very rare in Pakistan and Nepal.

INDIAN PEAFOWL (Common, Blue Peafowl, Peacock)

Pavo cristatus Male 110 cm (plus train) Female 86 cm F: Phasianidae

Description: Huge, very familiar game bird with tufted crest. Adult male has train of elongated green upper tail coverts up to 1.5m in length. Train covered by blue centred "eyes". Fanned high in shimmering display. Small head and long neck deep blue. Grey coverts and chestnut wings. Female brown, whitish below with black barred foreneck and breast and green hind neck.

Voice: Loud wailing *may yow*. Also cackles.

Habits: Inhabits forest undergrowth and, where feral, villages, cultivation and some towns. Feeds methodically on ground on vegetable matter, invertebrates and reptiles in small parties. Often confiding. Roosts high in trees. Nests on ground.

Kamal Sahai

Otto Pfister

Distribution: India's national bird. Locally common endemic breeding resident throughout lowlands of India, Sri Lanka and Nepal. More common in north. Large feral populations in northern, western and central areas. Very rare in Pakistan and gone from Bangladesh.
Also found ferally in Europe and elsewhere.

FULVOUS WHISTLING DUCK (Tree Duck)

Dendrocygna bicolor 51 cm F: Dendrocygnidae

Description: Larger than **Lesser** with buffish-white upper tail coverts, dark rear neck stripe and bright brown on crown. More obvious white flank feathers often hidden by wings. Otherwise very similar. Sexes alike.

Voice: Noisy. A high-pitched *pe weeo* call and a harsh *kee kee kee*. Wings make whistling sound.

Habits: Inhabits wetlands, flooded paddy fields and rivers with preference for thick aquatic vegetation and partly drowned trees. Usually in flocks feeding, mainly during night, on vegetation on surface and by diving. Wary. Nests mainly high in tree holes, sometimes in deserted nests of other birds.

Sumit Sen

Distribution: Now scarce and probably declining breeding resident in northeast India and Bangladesh. Rare wanderer elsewhere.
Also occurs in N and S America and Africa.

LESSER WHISTLING DUCK (Whistling-teal, Tree-duck)

Dendrocygna javanica 42 cm F: Dendrocygnidae

Otto Pfister

Otto Pfister

Description: Mainly orange-brown, medium-sized duck with scaly pattern on back, long legs, rather thin neck and large head. In flight, large rounded wings look black below and there are distinct maroon patches on the forewings and rump. Sexes alike.

Voice: A clear high-pitched double whistle; most un-duck like. Uttered frequently in usually rather low, clumsy flight.

Habits: Inhabits well vegetated freshwater jheels, flooded paddy fields and swamps where it feeds mainly at night on aquatic vegetation. Nests in reedbeds, tree holes and old bird nests. Frequently perches on open tree branches. Gregarious.

Distribution: Locally common breeding resident patchily throughout. Breeding monsoon visitor to the Northwest and Pakistan.
Also occurs in SE Asia.

MUTE SWAN

Cygnus olor 152 cm F: Anatidae

Otto Pfister

Description: Huge, long-necked, white waterbird. Pure white with dark grey legs and deep orange bill. Variably sized, fleshy, black knob on upper bill base with black extending to eyes and nostrils. Neck usually held in S shape and wings often arched. Sexes similar. Juveniles brownish-white. Similar **Whooper** *C. cygnus* and smaller **Tundra** *C. columbianus* **Swans** have yellow-based, black bills, erect necks and flat backs.

Voice: Usually silent but makes snorting sounds. Wings make loud throbbing in flight.

Habits: Large rivers, lakes and jheels. Swims most of time. Feeds on underwater vegetation by submerging neck or upending. Grazes grass on land. Shy.

Distribution: Winter vagrant mainly to Pakistan but also northwest India. Other two swan species have same rare status.
Also occurs in Europe east to Central Asia and south to N Africa.
In Europe and other temperate areas there are many feral populations in urban areas and it is common in bird collections.

GREATER WHITE-FRONTED GOOSE

Anser albifrons 70 cm F: Anatidae

Description: Large, greyish-brown goose with a white face. Size varies but usually smaller than similarly patterned **Grey-lag** with smaller bill and less contrast in wings when flying. Black barring on belly. Sexes similar. Immature lacks this and white face. Smaller **Lesser White-fronted** *A. erythrops* has shorter neck, stubby bill and obvious yellow eye rings. Larger **Bean** *A. fabalis* is browner with a darker neck and orange-banded dark bill.

Voice: Laughing *kew yew* in flight, like distant hounds.

Habits: Inhabits large wetlands and adjoining pasture and cultivation. Likely to be with commoner goose species grazing grass or roosting on water. Likely to be shy.

Tim Loseby

Distribution: Very rare winter visitor to Bangladesh, Pakistan and northern India. Scattered records elsewhere. Possibly overlooked in large Grey-lag flocks.
Also occurs in N America, Europe, N and E Asia.

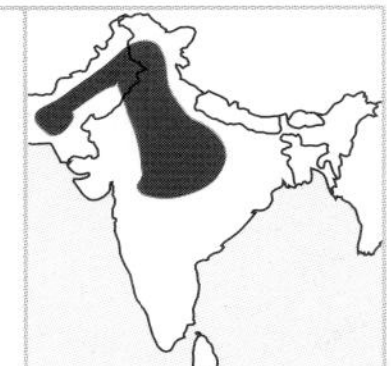

GREY-LAG GOOSE (Greylag)

Anser anser 75–90 cm F: Anatidae

Description: Large, brownish-grey goose, ancestor of most domestic geese. Whitish feather edges to upperparts and lower flanks. Tail coverts white, tail black with white borders. Strong bill and feet bright pink. In flight, shows a paler grey leading edge to the wings. Sexes alike but males larger.

Voice: Deep ringing honks, usually in flight or when alarmed. Conversational murmuring when feeding in large flocks.

Habits: Winter visitor favouring large, shallow freshwater bodies and large rivers during day. Feeds on aquatic vegetation while swimming. Also wet marshes where it crops grass. Flies out to young cereal crops to graze. Very gregarious.

Otto Pfister

Distribution: Locally common winter visitor mainly to northern lowlands. Also occurs in Europe, N Africa and N Asia.

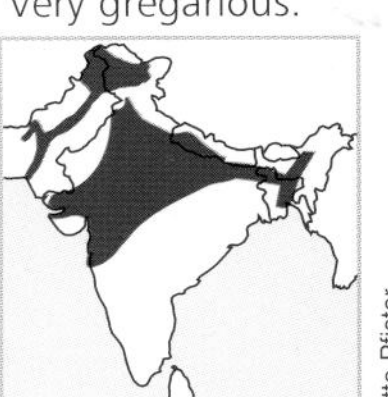

Otto Pfister

BAR-HEADED GOOSE

Anser indicus 75 cm F: Anatidae

Kamal Sahai

Otto Pfister

Description: Large, mainly silvery-grey goose with black-banded white head and white neck stripe down black neck. White tail coverts and grey tail. Juveniles lack the head stripes and have black caps. In flight silvery-grey forewing contrasts with dark flight feathers. Feet orange-yellow and black-tipped yellow bill.

Voice: A nasal, evocative, musical honking mainly in flight. Also a low murmuring when feeding.

Habits: Inhabits large freshwater bodies, wet, grassy fields and river sandbanks by day, often flying out to young cereals to graze at night. Very gregarious, sometimes mixing with **Grey-lag**.

Distribution: Breeds by high altitude lakes in Ladakh and migrates to lowlands throughout region in winter. Recorded flying at over 9300 m in Nepal. Locally common, particularly in the Northwest.
Also occurs in Central Asia and Myanmar.

RUDDY SHELDUCK (Brahminy Duck, Ruddy Sheldrake)

Tadorna ferruginea 65 cm F: Anatidae

Otto Pfister

Description: A rich orange-brown, goose-like duck with buffish head, black tail and a black neck ring in the male. In flight, white forewing contrasts with black flight feathers and green specula. Sexes similar.

Voice: A noisy bird, particularly when pairs are defending feeding territory with excited goose-like honking and trumpeting. Calls less in full flight.

Habits: Inhabits large open, usually fresh, water bodies favouring sand bars on large rivers. Feeds on animal and vegetable food, often grazing bankside grass. Less gregarious than other ducks but almost always in pairs or family parties during the winter. Wary in winter, quite confiding when breeding.

Distributiuon: Breeds by high altitude lakes in Ladakh. Locally common winter visitor throughout, most common in north.
Also occurs in SE Europe, Central Asia and S China.

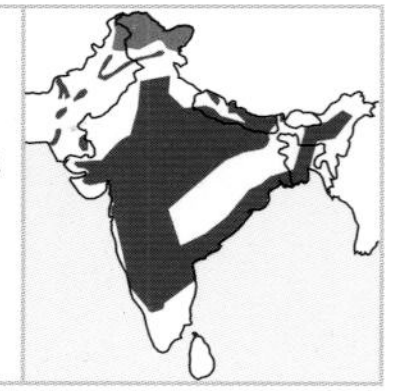

COMMON SHELDUCK

Tadorna tadorna 61 cm F: Anatidae

Description: A large, pied duck with a broad chestnut breast band. Size varies, female smaller but similar. Glossy green-black head and neck, black flight feathers, back and belly stripes. Green specula. Chestnut breast band continues round upper back and chestnut vent. Bright red bill (with large knob in breeding male) and dark pink legs.

Voice: Usually quiet in region. Sometimes a goose-like *a ang*.

Habits: Inhabits rivers, lakes, saltpans and the coast. Feeds on crustacea and aquatic plants by upending in water and walking on mud, sifting it carefully. Usually in pairs or small groups. Rather wary. Strong flier on long pointed wings.

Nikhil Devasar

Sujan Chatterjee

Distribution: Local and irregular winter visitor mainly to Pakistan and northwest India. Recorded less commonly east to Bangladesh and south to Andhra Pradesh.
Also occurs in N Africa and throughout Eurasia.

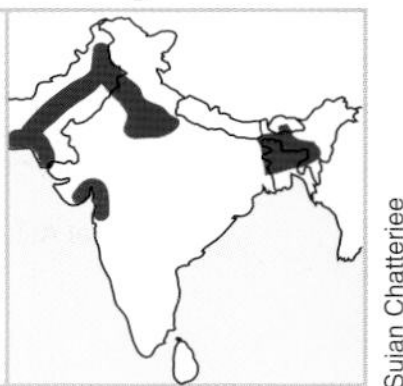

WHITE-WINGED DUCK (White-winged Wood Duck)

Cairina scutulata 65–80 cm F: Anatidae

Description: Probably India's rarest breeding duck. A large, dark duck, males larger than females and with whiter heads. Overall a greenish glossed blackish-brown with contrasting white leading edges to the wings in flight. Yellowish bill and feet.

Voice: Not noisy. An extended high-pitched honking in flight and various hisses.

Habits: Usually solitary or in pairs or family parties in mature, waterlogged forests. Favours small pools and streams with dead timber on which it roosts during the day. Nests in tree holes. Rarely wanders far from breeding areas and apparently declining rapidly with the loss of its habitat.

Thakur Dalip Singh

♀

Sharad Gaur

Distribution: Globally endangered breeding resident. Now restricted to a few areas in the Northeast.
Also occurs in SE Asia.

COMB DUCK (Nukta, Nakta, Knob-billed Goose)

Sarkidiornis melanotos 55–75 cm F: Anatidae

Otto Pfister

Description: Large, heavily built duck; male significantly larger and brighter. Glossy purple black above with greyish flanks, black bordered white breast and speckled black head. Bill and feet blackish. In breeding season male has large fleshy protuberance on bill (the comb) which is much reduced in winter. All dark wings contrast with white underparts. Immature pale brown below with dark cap and eyestripes.

Voice: Usually silent but occasionally croaks.

Habits: Inhabits well vegetated jheels and marshes with old timber. Perches on bare branches and usually nests in tree holes. Sometimes in small single-sex flocks. Feeds on aquatic animals and vegetation.

Distribution: Local resident throughout but spasmodic in appearance. Most frequent in the Northwest.
Also occurs in Africa, China and S America.

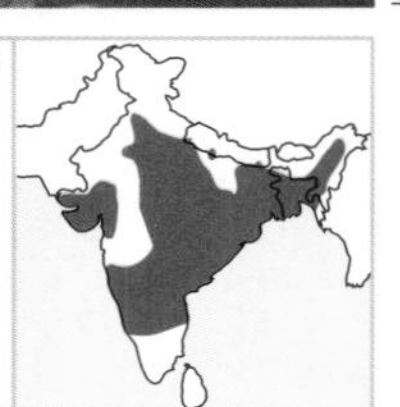

COTTON PYGMY-GOOSE (Cotton Teal, Quacky-duck)

Nettapus coromandelianus 32 cm F: Anatidae

Otto Pfister

Description: A tiny, pied duck with a very small bill. Breeding male has green-glossed black upperparts, white underparts with black cap and breast band and pale grey flanks. Shows striking white wingbars in flight. In non-breeding plumage more closely resembles dingier, browner female (also shown) but retains wingbars. Black bill and feet.

Voice: Male has rapid chattering quack in flight.

Habits: Inhabits well vegetated freshwater jheels and village ponds, often perching on bare waterside branches. Usually in pairs or small parties. Flight extremely swift and low. Engages in aerobatic courtship chases. Nests in tree holes. Feeds on aquatic vegetation and animals.

Distribution: Locally common breeding resident throughout but spasmodic in appearance. Mainly breeding monsoon visitor to the Northwest.
Also occurs in S China, Indonesia and Australia.

MANDARIN DUCK

Aix galericulata 48 cm F: Anatidae

Description: A strikingly coloured duck with a large, rounded head. Male in breeding plumage has red bill, brown crested crown, broad white supercilia and thick-feathered orange neck and cheeks. Usually two orange "sails" project up from wings. Breast black bordered white and flanks orange. Eclipse male and female grey with white throat, eye rings and eye stripes and dappled flanks. Male retains red bill.

Voice: Rather silent. Male may whistle *haweep* and female clucks.

Habits: Inhabits freshwater wetlands usually in flocks of commoner duck species. Dabbles and upends for aquatic vegetation. Regularly perches on tree branches.

Bikram Grewal

Distribution: Very rare but possibly increasing winter visitor to Bangladesh, eastern Nepal and northeast India. Also occurs in northeast Asia and, from feral stock, in Europe.

GADWALL

Anas strepera 50 cm F: Anatidae

Description: Medium-sized, dull grey-brown duck. Male has fine vermiculations on grey breast and flanks and subtle combination of grey and brown on upperparts contrasting with black stern. Bill grey and legs orange. Female has bright orange sides to bill and shares with eclipse male greyish head and neck contrasting with delicately scalloped brown body. In flight the black and white specula are obvious. Shows clear-cut white belly in flight and male has chestnut patch on forewing.

Voice: Rather quiet. Male has a soft whistle and female a hard quack.

Habits: Inhabits rivers, jheels and marshes. Gregarious, mixing with other ducks. Feeds on aquatic plants and animals.

Sumit Sen

Distribution: Common winter visitor throughout. Most numerous in north. Also occurs in N Africa, Eurasia, China and N America.

FALCATED DUCK (Falcated Teal)

Anas falcata 51 cm F: Anatidae

Sujan Chatterjee

Description: A striking, large-headed duck. Male has green head and high chestnut crown, black-bordered white throat and neck and black and white speckled breast. Rest of body grey with black rear and yellow vent patch. Elongated black and grey scapular plumes unique. Female barred brown with greyer head. Black bill and legs.

Voice: Quiet. Male has sharp whistle. Female quacks.

Habits: Inhabits larger wetlands often with emergent vegetation. Usually only seen singly or in small parties. Mixes with other ducks, particularly **Gadwall**. Feeds on surface vegetation and upends. Often wary.

Distribution: Very scarce but regular winter visitor mainly to Nepal, Bangladesh and northeast India. Rare elsewhere in north.
Also occurs in N and E Asia.

EURASIAN WIGEON (Wigeon, Widgeon)

Anas penelope 50 cm F: Anatidae

Nikhil Devasar

Description: Plump, large-headed, short-necked, medium-sized duck. Male distinctive with grey body, black and white stern and obvious white forewings. Head chestnut with yellow forehead and breast pinkish-grey. Female and eclipse male reddish-brown. Small black-tipped, goose-like bill. In flight, both sexes show white belly, long pointed wings and pointed tail.

Voice: Male has haunting *wheeo* whistle. Female a gruff bark.

Habits: Inhabits jheels, marshes and large rivers where there is access to short grass or cereals for grazing. Very gregarious often mixing with other ducks on water but as a grazer is more likely to be seen with geese when feeding.

Distribution: Common winter visitor throughout. Most numerous in north. Also occurs in N Africa and throughout Eurasia.

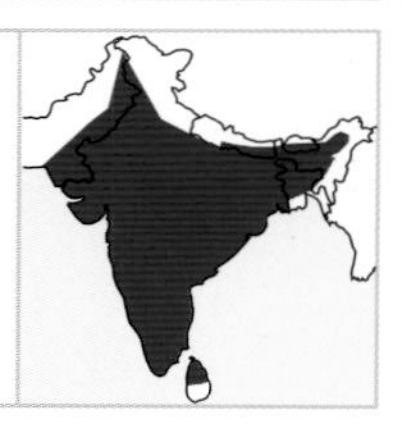

MALLARD (Wild Duck)

Anas platyrhynchos 60 cm F: Anatidae

Description: Large duck, ancestor of many domestic ducks. Male has distinctive glossy bottle green head and neck, white neck ring and dark purplish breast. Most of body grey and stern black. Tail has two curly feathers and bill yellowish. Female and eclipse male are streaked brown with more orange in bill. Both sexes have orange feet and white-edged bluish-purple specula.

Voice: Male has low rasp. Female makes loud, persistent quacking.

Habits: Frequents large rivers, jheels and swamps feeding both by upending for aquatic plants and animals and flying out to paddy and other cereal fields at night. Gregarious, almost always with other duck species.

Otto Pfister

Distribution: Scarce winter visitor mainly to northern areas and usually in small numbers. A few breed in the Himalayas.
Also occurs in N Africa, N America and Eurasia.

SPOT-BILLED DUCK (Spotbill)

Anas poecilorhyncha 60 cm F: Anatidae

Description: A large duck with sexes alike. Basically brown with whitish feather edgings on body, striking black and white-bordered green specula and white tertials. Breast spotted. Black crown and eyestripe create a masked appearance. Reddish spots in front of eyes and black bill boldly tipped yellow.

Voice: Quacks loudly.

Habits: The commonest breeding duck, inhabiting any sort of sizable vegetated freshwater, including park or garden ponds. Feeds on aquatic plants and animals. Nests on the ground in waterside vegetation. Fluffy young buff with dark patches and in broods of up to fifteen. Gregarious, mixes with other duck species but often on its own.

RK Gaur

Distribution: Locally common breeding resident throughout lowlands. Less common in east.
Also occurs in Myanmar, China and NE Asia.

NORTHERN SHOVELER (Shoveler, Shoveller)

Anas clypeata 50 cm F: Anatidae

Otto Pfister

Description: Medium-sized duck recognisable by unique spatulate bill. Male has bottle green head, white breast and rear and broad chestnut flank patch. Upperparts black and white. Iris yellow. Female patterned brown similar to female **Mallard**. Juvenile and eclipse male rather reddish brown. Both sexes show bluish forewings and green specula in flight. Swims low and front heavy.

Voice: Rather quiet. Male has muffled *tonk-tonk* call and female a hoarse quack.

Habits: Inhabits well-vegetated large rivers, jheels, coastal lagoons and marshes. Surface-sifts with large bill. One of the commonest wintering wildfowl and very gregarious, mixing readily with other species.

Distribution: Abundant winter visitor throughout lowlands. Less common in east.
Also occurs in Eurasia, N America, China and Africa.

NORTHERN PINTAIL (Common Pintail, Pintail)

Anas acuta 60 cm (excluding tail pins) F: Anatidae

Otto Pfister

Description: Large, slender, small-headed, duck with pointed tail. Grey bill. Adult male has dark brown head and upper neck with white side stripes extending to white breast. Flanks vermiculated grey. Stern and tertials black and yellow. Central tail pin feathers elongated by 10 cm. Female rather plain brown on head and neck. Buff-edged brown on rest of body. Specula brownish green with striking white rear borders. Streamlined and fast in flight.

Voice: Very quiet. Male whistles.

Habits: Inhabits well-vegetated large rivers, jheels, coastal lagoons and marshes. Feeds on aquatic plants and animals from surface and by upending but also grazes watersides. Very sociable, mixing readily with other species.

Distribution: Common winter visitor throughout lowlands.
Also occurs in Eurasia, N and Central America, Africa and China.

GARGANEY (Garganey Teal, Blue-winged Teal)

Anas querquedula 40 cm F: Anatidae

Description: Small duck with large grey bill. Male has striking white eyebrows on brown head and pale grey flanks contrasting with brown breast and spotted brown stern. In flight, shows silvery blue forewings. Female similar to **Common Teal**, mottled brownish but with distinct eyestripes and pale supercilia, loral and chin spots and larger bill.

Voice: Male has distinct dry cricket-like rattle. Female makes a quiet quack.

Habits: Inhabits thickly vegetated jheels, coastal lagoons, swamps and paddy. Feeds from surface on aquatic plants and animals. Gregarious, mixing with other species. Flight very fast and agile.

Otto Pfister

Distribution: Abundant winter visitor in south, mainly passage migrant in northern areas.
Also occurs in Africa, Eurasia and Indonesia.

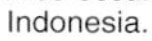

COMMON TEAL (Teal, Green-winged Teal)

Anas crecca 34 cm F: Anatidae

Description: Small, stubby-billed duck. Male has chestnut head and yellow-bordered green ear patch. Body finely vermiculated grey with white side stripes, brown-spotted buff breast and black-bordered bright yellow stern. Female buff-edged brown like small **Mallard**. Both sexes show white-edged green specula.

Voice: Male has high pitched piping whistle. Female quacks.

Habits: Inhabits well-vegetated large rivers, jheels, coastal lagoons and swamps. Also village ponds. Feeds on aquatic plants and animals from water surface and by upending. Grazes grass and cereals. Flight very fast and agile, with much twisting and turning. Very gregarious, mixing with other species.

Otto Pfister

Distribution: Abundant winter visitor throughout lowlands. Less common in east.
Also occurs in Africa, Eurasia and Indonesia.

RED-CRESTED POCHARD

Rhodonessa rufina 55 cm F: Anatidae

Otto Pfister

Description: Large, bulky, big-headed duck. Male has rich orange-crested head contrasting with red bill and black neck and breast. Back brown, flanks white and stern black. Female is coffee brown with obvious pale cheeks and dark bill. In flight, they both show broad white wing bars and male looks very black below. Swims high and proud.

Voice: Silent.

Habits: Inhabits well-vegetated large rivers and lakes. Feeds on aquatic plants and animals by diving and upending. Tends to keep to small flocks of its own species, sometimes near other diving ducks.

Distribution: Rather scarce winter visitor throughout lowlands. Commonest in north.
Also occurs in N Africa and Eurasia.

COMMON POCHARD

Aythya ferina 45 cm F: Anatidae

Otto Pfister

Description: A medium-sized, stocky, dome-headed diving duck. Male distinctive with brick red head and neck, grey body and black breast and stern. Quite large bill, black with broad grey band in male. Female brown, paler on cheeks and flanks. In fast, whirring flight both show black bordered grey wing bars.

Voice: Rather quiet. Female sometimes purrs in flight.

Habits: Inhabits all types of open water including village ponds. Feeds mainly by repeated diving for aquatic plants and animals. Gregarious, mixing mainly with other diving ducks.

Distribution: Common winter visitor throughout lowlands. Less common in east.
Also occurs in N Africa, Eurasia and S China.

FERRUGINOUS POCHARD (White-eyed Pochard)

Aythya nyroca 40 cm F: Anatidae

Description: Small, neat, dark, peak-headed diving duck. Mahogany brown, darker on the back and with bright white under the tail. Male brighter with bright white irises. Small bill grey with white tip and black nail. In fast whirring flight, long white wing bars and white belly obvious. Rarer **Baer's Pochard** *A. baeri* bulkier with green glossy head, white flank patches. Large bill.

Voice: Usually silent although female sometimes makes burring sound in flight.

Habits: Inhabits well-vegetated jheels, rivers, canals and ponds. Dives for aquatic plants and animals. Less gregarious than other diving ducks, usually keeping in small groups. Hides in emergent vegetation. Cautious.

Otto Pfister

Distribution: Scarce and local winter visitor throughout lowlands. Commonest in the Northwest but rarely in large numbers.
Also occurs in NE Africa, Eurasia and W Asia.

TUFTED DUCK (Tufted Pochard)

Aythya fuligula 45 cm F: Anatidae

Description: A medium-sized, round-bodied, round-headed diving duck usually showing a head tuft. Male is glossy black with broad white flanks. Female dark brown with paler flanks and sometimes white under the tail. In both sexes the irises are yellow and the bill black-tipped grey. In flight, shows white wing bars.

Voice: Rather quiet. Male has high bubbling note. Female croaks.

Habits: Inhabits all kinds of open water including reservoirs. Feeds by repeated diving for aquatic plants and animals. Very gregarious, often mixing with other diving ducks.

Otto Pfister

♀

Sharad Gaur

Distribution: Fairly common winter visitor throughout lowlands. Less common in east and south.
Also occurs in Eurasia, S China and SE Asia.

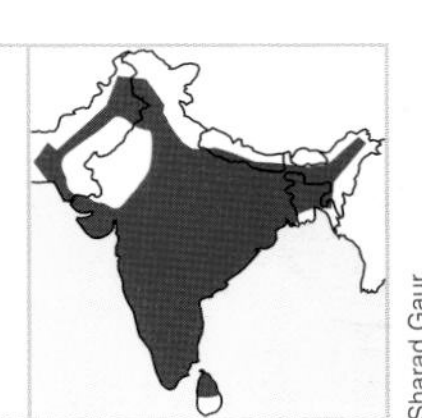

COMMON MERGANSER (Goosander, Merganser)

Mergus merganser 65 cm F: Anatidae

Otto Pfister

Description: Large, big-headed, long-billed and long-bodied sawbill duck. Male has domed glossy green head, pinkish-white body, pied back and grey stern. Female has distinct crest on back of brown head, grey body and whitish breast. Both have long, thin, hooked, red bill serrated along the sides to hold fish. In flight, male's wings are black and white, while female has grey forewings.

Voice: During courtship, male has far-carrying deep *kruu, kra* call. Female a shorter *praha* note.

Habits: Inhabits larger rivers in northern areas. Feeds solely on fish by repeated diving. Often swims with head submerged to locate prey. Rests on rocks. Flight path usually follows river course.

Distribution: Rather scarce. Breeds in Himalayas and northeastern mountains. Winters on northern rivers, most commonly in the Northeast. Also occurs in Europe, N, E and Central Asia and N America.

SMALL BUTTONQUAIL (Little Bustardquail)

Turnix sylvatica 13 cm F: Turnicidae

Balachandran

Description: Tiny, round-bodied, small-headed, quail-like bird with short wings. Basically brown with broad buff feather edgings above. Head paler and speckled, with distinct crown stripes. Orangey-buff below with black spotting on flanks. Small grey bill, yellowish legs and white irises. Female slightly larger. Three-toed.

Voice: Female has far-carrying, low-pitched, intermittent hoot at dusk and dawn.

Habits: Inhabits scrubby, dry, grassland. Exceptionally secretive. Fast, whirring but brief flight. Prefers to run or hide. Feeds on small seeds on the ground. Dust-bathes, offering best chance of seeing them. Female dominant and polyandrous. Male incubates eggs and rears young.

Distribution: Probably fairly common resident throughout lowlands but much overlooked. Also occurs in Africa, S Europe, China and SE Asia.

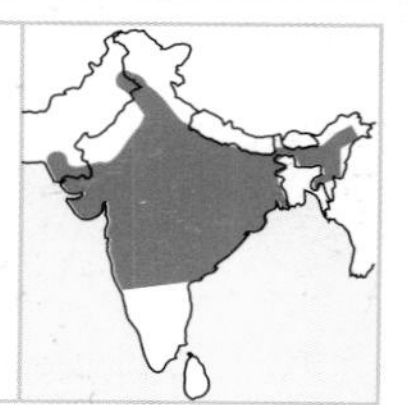

BARRED BUTTONQUAIL (Common Bustardquail)

Turnix suscitator 15 cm F: Turnicidae

Description: As **Small** but larger and darker. Both sexes have streaked greyish heads, dark brown upperparts with paler edgings and warm chestnut bellies and vents. Throat to belly barred black and white but larger female has broad blackish central area. Small bill and legs grey. Three-toed.

Voice: Female has a far carrying, burring note and an intermittent booming call.

Habits: As **Small** and often found in similar areas but tolerates damper ground and longer grass.

Sanctuary / Vivek Sinha

Distribution: Probably fairly common resident throughout but much overlooked.
Also occurs in China and SE Asia.

EURASIAN WRYNECK (Wryneck)

Jynx torquilla 17 cm F: Picidae

Description: Small, rather reptilian-looking woodpecker, beautifully coloured like tree bark. Basically grey and brown with dark eyestripes and very obvious dark stripe from crown to rump. Underparts barred and throat warm buff. Tail long and barred which can make it look rather shrike-like in flight. Bill short and pointed. Twists neck round and often raises crown feathers. Sexes alike.

Voice: High pitched *pee pee pee*. Hisses in nest.

Habits: Breeds in forest edges in northern mountains. Elsewhere in scrub and cultivation. Unobtrusively feeds with long tongue on ants on ground and tree branches (shown). Does not drill holes or use tail as support. Nests in tree hole.

Nikhil Devasar

Distribution: Breeds in Kashmir. Scarce but widespread winter visitor mainly to northern areas. Rare in south. Also occurs in Africa, Eurasia and SE Asia.

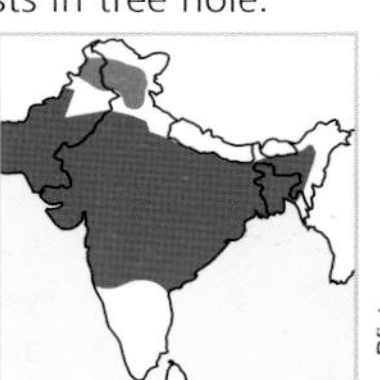

Otto Pfister

SPECKLED PICULET (Spotted Piculet)

Picumnus innominatus 10 cm F: Picidae

Otto Pfister

Description: Tiny, dumpy woodpecker with very short tail. Lime green above and black-spotted on yellowish-white below. Crown grey, head striped black and white. Male has orange forehead. **White-browed Piculet** *Sasia ochracea* of Northeast, rufous below with short white eyestripes.

Voice: Hard, rapid *tseep tseep* in flight. Drums on branches.

Habits: Inhabits highland forest, often in undergrowth and usually in pairs with mixed species flocks. Feeds on ants on small branches and bamboos, often hanging upside down or creeping along branches. Energetic tapping often first indication of presence. Perches across branches like passerine. Nests in self-made hole in bamboo stem or small branch.

Distribution: Local breeding resident in northern mountains, the Northeast, Eastern and Western Ghats.
Also occurs in SE Asia.

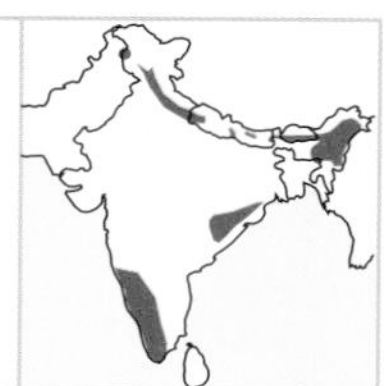

BROWN-CAPPED PYGMY WOODPECKER

Dendrocopos nanus 13 cm F: Picidae

♀

Otto Pfister

Description: A small brown and white woodpecker with distinctive pink-rimmed white irises. Barred brown and white above, lightly streaked dirty white below. Tail spotted white. Paler brown crown (edged red in male) and eyestripes contrasting with white supercilia and cheeks.

Voice: Rapid chatter *click rrrr* and quiet drumming.

Habits: Unobtrusive inhabitant of light woodland and village groves in lowlands. Feeds acrobatically on insects and grubs in tree bark, often close to ground and usually in pairs. Frequently joins mixed species flocks in canopy favouring outer branches. Uses tail as prop. Nests in a self-made hole in tree branch.

Distribution: Fairly common breeding resident throughout lowlands but absent from Pakistan and rare in Bangladesh.
Also occurs in SE Asia.

GREY-CAPPED PYGMY WOODPECKER

Dendrocopos canicapillus 14 cm F: Picidae

Description: A small, dark woodpecker with dark irises. Barred black and white above with, usually, unbarred central tail feathers. Dark buff below with prominent dark streaking. Grey crown (red nape on male) and black eyestripes contrasting with broad white supercilia and cheeks. Slight malar stripe.

Voice: A high rattling trill and a weak *pik* call.

Habits: Unobtrusive inhabitant of forest and open woodland in northern plains. Often in pairs with mixed species flocks feeding acrobatically high in canopy on smallest branches but also low on small trees. Feeds on insects and grubs in bark. Uses tail as prop. Nests in self-made hole.

Otto Pfister

Distribution: Fairly common breeding resident in northern foothills including parts of Bangladesh.
Also occurs in NE and SE Asia and China.

BROWN-FRONTED WOODPECKER

Dendrocopos auriceps 20 cm F: Picidae

Description: A medium-sized, pied woodpecker with yellow in crown. White-barred (rather than spotted) black upperparts, prominent black moustache extending to breast and black-streaked white underparts. Vent deep pink. In male forecrown brown, centre yellow, rear red with black rear neck. In female whole crown yellow.

Voice: A fast chattering *chik chik chik rrr* and a short *chik*.

Habits: Inhabits deciduous and coniferous hill forests feeding on invertebrates and tree seeds in trees and bushes. Often with mixed hunting groups. Uses tail as prop. Nests in self-made hole.

Distribution: Locally common endemic breeding resident in northern hills from Baluchistan, east to Nepal. Moves lower down in winter.

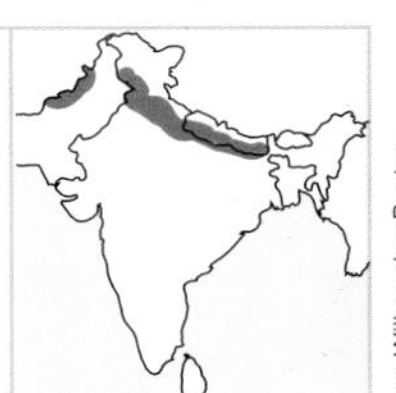

Jan Willem den Besten

FULVOUS-BREASTED WOODPECKER

Dendrocopos macei 18 cm F: Picidae

Toby Sinclair

Sumit Sen

Description: A medium-sized, pied woodpecker. Upperparts black, heavily barred white. Undertail red, breast and belly buff with light flank barring and slight side streaking. Whitish cheeks partly bordered by black line. Crown red in male with orange forehead, black in female.

Voice: A hard *tik*, sometimes extended into a chatter.

Habits: Inhabits open broad-leaved and coniferous forests and forest edges. Feeds on trunk and branch bark invertebrates often with mixed species flocks and sometimes in isolated trees in cultivation. Uses tail as prop. Nests in self-made tree hole.

Distribution: Common breeding resident in northern hills, the Northeast, Bangladesh and Eastern Ghats. Also occurs in SE Asia.

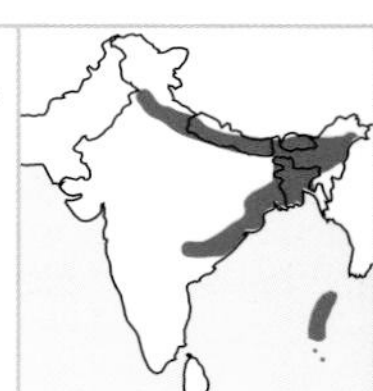

YELLOW-CROWNED WOODPECKER (Mahratta)

Dendrocopos mahrattensis 17 cm F: Picidae

Otto Pfister

Otto Pfister

Description: Medium-sized, pale-headed, pied woodpecker. Upperparts black, heavily spotted and barred white. Underparts dark, streaked dingy white with red belly patch. Irregular brown cheek and neck patches. Female has yellowish crown and nape. In male nape scarlet and fore-crown yellow.

Voice: A soft abrupt *click* often extended. Soft creaking contact call. Drums.

Habits: Commonest woodpecker. Inhabits dry, open woodland and scattered trees, gardens, villages and cultivation. Confiding. Feeds on tree trunk and branch invertebrates by digging into dead wood. Joins mixed hunting groups in canopies. Usually in pairs. Uses tail as prop.

Distribution: Common resident throughout lowlands except the Northeast and Bangladesh. Also occurs in Myanmar.

DARJEELING WOODPECKER (Darjeeling Pied)

Dendrocopos darjellensis 25 cm F: Picidae

Description: A medium-sized, pied woodpecker with yellow neck sides. Black upperparts with large white, scapular patches and white barred flight feathers and tail sides. Male has red nape patch. White cheeks and long black moustache extending to upper breast. Black-streaked, yellow-buff below with red vent. Long bill.

Voice: Short *tsik* calls and a rattling *di di di di dtttttt dit*.

Habits: Inhabits high forest feeding on invertebrates on and in upper branches and mossy trunks. Also feeds on dead wood on ground. Usually singly or in pairs, sometimes in mixed hunting groups. Nests in self-bored tree hole.

Sujan Chatterjee

Distribution: Locally common breeding resident in northern mountains from central Nepal east to Myanmar border. Also occurs in China and SE Asia.

GREAT SPOTTED WOODPECKER

Dendrocopos major 24 cm F: Picidae

Description: A medium-sized, pied woodpecker similar to three other pied species. Black upperparts with broad white patches from shoulder to lower back and white barring on flight feathers and tail edgings. Underparts buff with extensive bright red vent. Black mark like spread-eagled man on cheeks and neck. Male crown black and nape red, female's crown and nape black, juvenile male has red crown and black nape. Markedly undulating flight.

Voice: A hard *chek* sometimes trilled. Resonant creaking drumming.

Habits: Inhabits oak and pine forest. Feeds on invertebrates on trunks, branches and ground.

Distribution: Scarce breeding resident in the Northeast only.
Also occurs elsewhere in Eurasia and Myanmar.

Otto Pfister

HIMALAYAN WOODPECKER (Himalayan Pied)

Dendrocopos himalayensis 24 cm F: Picidae

Otto Pfister

Description: A medium-sized, pied woodpecker similar to three other pied species. Black above with broad white patches from shoulder to lower back, limited white barring on flight feathers and clean white tail edgings. Underparts and head plain pale buff with black Y-shaped mark on neck and cheeks, red crown in male and black in female. Black marks under eyes unique. Vent pink.

Voice: A short *kit*, sometimes extended excitedly.

Habits: Inhabits montane forests with preference for coniferous trees. Feeds on invertebrates in tree bark and also pine seeds from cones it has wedged into bark crevices. Uses tail as prop. Drums. Nests in self-made tree hole.

Distribution: Common breeding endemic resident in northwestern mountains.

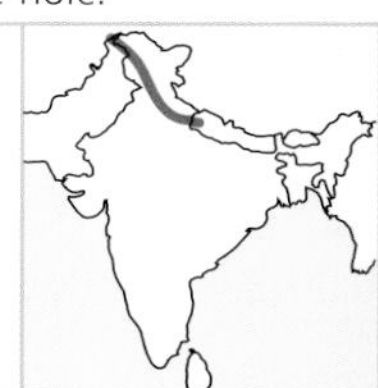

RUFOUS WOODPECKER (Brown Woodpecker)

Celeus brachyurus 25 cm F: Picidae

Sumit Sen

Description: A medium-sized, rufous-coloured woodpecker with a short crest and short black bill. The upperparts are finely barred black. Male has small red patches by eyes.

Voice: Loud, high-pitched *ke ke kre ke*. Also a distinctive staggered drumming.

Habits: Inhabits upland broadleaf forests. Feeds mainly on ball-like tree ant nests but also on ground. Also eats figs. Often in loose groups. Excavates nest chamber inside the active ant nests without apparently being bothered by the stinging ants.

Distribution: Scarce breeding resident in foothills of northern mountains and the Northeast, south to Bangladesh, Western and Eastern Ghats and Sri Lanka.
Also occurs in S China and SE Asia.

LESSER YELLOWNAPE (Small Yellow-naped)

Picus chlorolophus 27 cm F: Picidae

Description: A medium-sized, greenish woodpecker with a short-crested, yellow nape. Variable. Differs from **Greater** in smaller size, barred whitish underparts, white-barred outer and plain rufous inner primaries. Red eye and malar stripes. Smaller bill dark.

Voice: Noisy. Loud, sad *pee oo* and a descending *ke ke ke ke.*

Habits: Inhabits all types of open forest and plantations. Feeds on invertebrates at lower levels, including ground, and often in mixed hunting groups. Nests in self-made tree hole.

Morten Strange

Distribution: Locally common breeding resident in northern foothills from Himachal Pradesh to Myanmar border and peninsular hills, Sri Lanka and Bangladesh.
Also occurs in China and SE Asia.

GREATER YELLOWNAPE (Large Yellow-naped)

Picus flavinucha 33 cm F: Picidae

Description: Large, olive green woodpecker with prominent yellow-crested nape and throat. Dark olive green with grey underparts. Crown brownish and flight feathers chestnut barred with black. Bill often looks whitish. **Lesser** has smaller, dark bill, red markings on head, red stripe on whitish chin, smaller crest and barred underparts.

Voice: Loud, plaintive *keeyu* and hard *chep* note. Accelerating trill *quee, quee, quee*.

Habits: Inhabits upland broadleafed forests. Feeds on invertebrates from ground to canopies often with mixed species parties. Shy, flushes readily if disturbed. Uses tail as prop. Nests in self-made tree hole.

Sumit Sen

Distribution: Fairly common breeding resident of northern foothills, the Northeast, Bangladesh and Eastern Ghats.
Also occurs in S China and SE Asia.

STREAK-THROATED WOODPECKER (Scaly-bellied Green)

Picus xanthopygaeus 30 cm F: Picidae

Otto Pfister

Description: A medium-sized, green woodpecker with streaked throat and scaly whitish underparts. Green above with yellowish rump, white supercilia and white and black moustache. Crown red in male, blackish in female. Tail dark and plain. Small, dark bill.

Voice: Rather silent. A quiet *pik* occasionally. Also drums.

Habits: Inhabits broadleafed forest and plantations. Feeds on invertebrates on ground, tree trunks and branches, sometimes in mixed hunting groups. Uses tail as prop, working its way in spirals or directly upwards. Nests in self-made tree hole.

Distribution: Fairly common breeding resident throughout the lowlands except Pakistan and parts of the east. Also occurs in SE Asia.

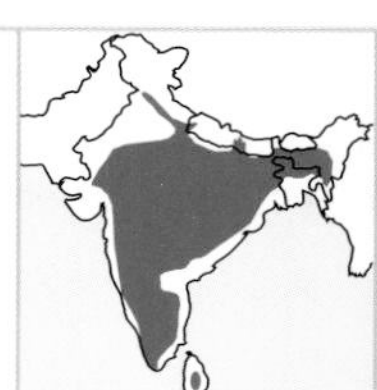

SCALY-BELLIED WOODPECKER (Large Scaly-bellied)

Picus squamatus 35 cm F: Picidae

Otto Pfister

Description: Large, green woodpecker with distinct scaling from breast to vent. Similar to **Streak-throated** but larger and with unstreaked throat and upper breast. Black moustache and black bordered white supercilia. Tail strongly barred. Crown red in male, blackish in female. Large pale bill.

Voice: Contact call a hard, repeated *quik*. Also a musical, repeated *peeko, peeko*.

Habits: Inhabits upland coniferous and broad-leafed forests; also open dry and cultivated country with scattered trees. Feeds mainly on ants on ground and tree trunks. Uses tail as prop. Nests in self-made tree hole.

Distribution: Fairly common breeding resident in northern mountain forests east to eastern Nepal.
Also occurs in W Asia to Afghanistan.

GREY-HEADED WOODPECKER (Black-naped Green)

Picus canus 32 cm F: Picidae

Description: A fairly large, green woodpecker with plain grey head. Crown in male red, nape black. Crown in female black. Black moustache and diagnostic plain grey-olive underparts. Rump yellow and tail dark. White-barred dark flight feathers.

Voice: High-pitched, repeated *keek, keek* call. Chattering alarm. Drums loudly in breeding season.

Habits: Inhabits upland moist woodlands. Feeds mainly on ants and beetles, often on ground but also on tree trunks. Digs into termite mounds and eats nectar and fruit. Uses tail as prop. Nests in self-made tree hole.

Otto Pfister

Distribution: Common breeding resident in northern foothills, Northeast, Bangladesh and Eastern Ghats. Also occurs in Eurasia and SE Asia.

COMMON FLAMEBACK (Golden-backed Three-toed)

Dinopium javanense 28 cm F: Picidae

Description: A medium-sized, golden-backed woodpecker with long and solid black moustachial stripes. Both sexes have black eyestripes joined to black rear neck stripe. Male has red, female black crown. Black-scaled white underparts and red rump contrasting with black tail. Rather small bill and only three toes.

Voice: Similar to but higher-pitched than **Black-rumped**.

Habits: Inhabits damp woodland and mangroves. Feeds on invertebrates at all levels but mainly rather low. Often in mixed hunting groups. Nests in self-made tree hole.

Morten Strange

Distribution: Locally common breeding resident in foothills of Western Ghats and northeastern hills. Also in the Bangladesh Sunderbans. Also occurs in SE Asia.

BLACK-RUMPED FLAMEBACK (Lesser Golden-backed)

Dinopium benghalense 30 cm F: Picidae

Otto Pfister

Description: A bright golden and black woodpecker with bold red crest. Black and yellow speckling on shoulders. Nape, rump and tail black; female has blackish forecrown. White-streaked black throat and eyestripes contrast with white cheeks. Dark scalloped whitish underparts.

Voice: Noisy bird with laughing cackle reminiscent of **White-throated Kingfisher**. Drums.

Habits: Most widespread of five similar species. Found where trees including road and canal avenues, gardens and plantations. Avoids thick forest. Feeds on ants on ground and at all levels in trees, often with mixed hunting groups. Confiding but flies off noisily. Uses tail as prop. Nests in self-made tree hole.

Distribution: Common breeding endemic resident throughout lowlands, except parts of the Northwest and the Northeast.

GREATER FLAMEBACK (Large Golden-backed)

Chrysocolaptes lucidus 32 cm F: Picidae

Sumit Sen

Description: Large golden, black, red and white woodpecker with pale irises and large bill. Golden above with red rump and black tail. Heavily black-scalloped, white shoulders and underparts. Broad black band from eyes to lower neck. Thin bifurcating lines from bill to lower throat. High-crested crown red in male, white-speckled black in female. White-spotted black flight feathers.

Voice: Noisy. High-pitched *tri tri tri* reminiscent of **Red-wattled Lapwing.**

Habits: Inhabits broad-leafed forests including mangroves. Feeds on ground on dead wood and in larger trees on invertebrates, often in mixed hunting groups. Uses tail as prop. Nests in self-made oval tree hole.

Distribution: Locally common breeding resident of northern foothills, Eastern and Western Ghats, Northeast, Bangladesh and Sri Lanka. Also occurs in SE Asia.

WHITE-NAPED WOODPECKER (Black-backed Woodpecker)

Chrysocolaptes festivus 29 cm F: Picidae

Description: Medium-sized, bright olive yellow and pied woodpecker with red crested crown in male, yellow in female. Back black with diagnostic, striking white central lozenge up to nape. Broad black lines extend up neck from nape to bill branching to rear crown. Contrasting white face. Two thin black lines extend across white throat to bill. Underparts white, broadly scalloped with black. Tail and rump black.

Voice: A thin, repeated *tee tee tee*.

Habits: Inhabits dry open woodland, plantations and scrub. Feeds on lower parts of tree trunks and on ground on invertebrates, mainly ants. Uses tail as prop. Nests in self-made tree hole.

RK Gaur

RK Gaur

Distribution: Scarce breeding endemic resident mainly restricted to peninsular India, dry zone Sri Lanka and far-western Nepal.

HEART-SPOTTED WOODPECKER

Hemicircus canente 16 cm F: Picidae

Description: A small, uniquely shaped, black and buff woodpecker with large, prominently recurved crest on head and very short tail. Basically black with heart-shaped black spots on white shoulders and broad white scapular patches and barring of flight feathers. Whitish throat and plain grey underparts. Female has white forecrown.

Voice: Noisy, especially in flight with rasping *kiruk* and repetitive trilling *trer trer trer*.

Habits: Inhabits damp hill forests, favouring bamboo and plantation shade trees. Feeds on arboreal ants in canopy, often in mixed hunting groups. Creeps along branches and perches across branch like a passerine. Nests in self-made branch hole.

Balachandran

Distribution: Scarce breeding resident of Western and Eastern Ghats and parts of northeastern India. Also occurs in SE Asia.

GREAT BARBET (Great Hill, Great Himalayan Barbet)

Megalaima virens 33 cm F: Megalaimidae

Otto Pfister

Otto Pfister

Description: A very large, bizarrely coloured barbet with a heavy yellow bill. Bluish-slate head, maroon-brown breast and mantle. Lower back, wings and tail green. Vent strikingly bright red and rest of underparts blue streaked pale yellow. Sexes alike.

Voice: Very noisy with far-carrying, oft-repeated rather raptor-like *pee-yo pee-yo*.

Habits: Inhabits highland forests. Feeds on fruit and flowers in treetops but will come low for favoured fruits. Often very difficult to spot. Usually solitary or in small groups. Nests in tree hole.

Distribution: Common breeding resident of northern and northeastern mountains. Moves lower down in winter.
Also occurs in China and SE Asia.

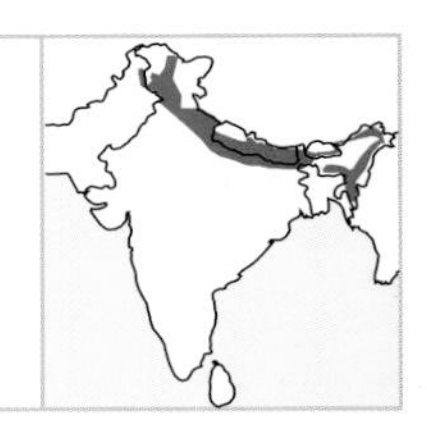

BROWN-HEADED BARBET (Green, Large Green Barbet)

Megalaima zeylanica 28 cm F: Megalaimidae

Nikhil Devasar

Description: Large-headed, brown and green barbet with bare orange patches round eyes which reach large reddish-orange bill. Head, neck, mantle and upper breast white streaked pale brown. White spots on wing coverts. Rest of body bright grass green, paler on belly. Sexes alike.

Voice: Loud, persistent *prutruk, prutruk* sometimes without terminal hard consonant. Usually beginning with long guttural churring note.

Habits: Inhabits lowland forests, groves, city parks and gardens. Feeds high in treetops on fruit especially *Ficus*. Often in groups in favoured trees. Has powerful, undulating flight. Very well camouflaged and rather secretive. Nests in tree hole.

Distribution: Common breeding endemic resident throughout lowlands except Pakistan, Bangladesh and northeast India.

LINEATED BARBET

Megalaima lineata 28 cm F: Megalaimidae

Description: Large green and brown barbet with bare, pale yellow patches round eyes and large yellowish bill. Similar to **Brown-headed** but slimmer and less top-heavy looking. Pale brownish head and breast. Rest of body bright green with unmarked wings. More extensive heavy white streaking on head, neck, mantle and underparts to upper belly. White throat. Bare eye patches do not reach bill. Sexes alike.

Voice: A persistent, rather subdued *kuruk, kuruk*, usually with long guttural start-up note.

Habits: Inhabits foothill forests, parks and gardens. Strictly arboreal. Feeds on fruit. Very well camouflaged and rather secretive. Nests in tree hole.

Otto Pfister

Distribution: Locally common breeding resident in the Northeast, Bangladesh and Himalayan foothills west to Himachal Pradesh.
Also occurs in SE Asia.

WHITE-CHEEKED BARBET (Small Green Barbet)

Megalaima viridis 23 cm F: Megalaimidae

Description: A medium-sized, green and brown barbet with short white stripes above and below eyes. Similar to last two species but smaller with no bare patches on face and yellow bill less massive. Brownish head and upper breast, darker on crown and nape and with white throat and heavy white streaking on breast. Rest of body unmarked grass green. Sexes alike.

Voice: A start-up burring followed by repeated, penetrating, rather high-pitched *putruk, putruk*.

Habits: Inhabits hill forests, groves and gardens. Strictly arboreal feeding on fruit. Well camouflaged and often difficult to spot. Nests in tree hole.

Clement M Francis

Distribution: Common endemic breeding resident in southwestern foothill forests north to Gujarat.

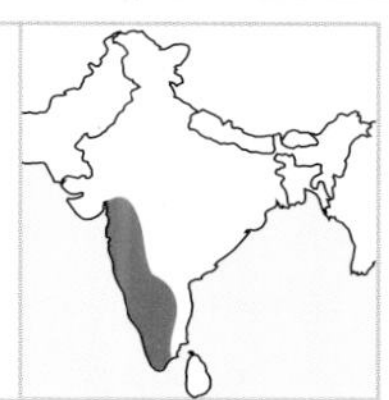

GOLDEN-THROATED BARBET

Megalaima franklinii 23 cm F: Megalaimidae

Morten Strange

Morten Strange

Description: A medium-sized, green barbet with a yellow throat. The yellow throat and centre to red crown, black eye stripes and whitish cheeks distinguish from similar **Blue-throated**. Bluish wing shoulders. Sexes alike.

Voice: Start-up call *krrr* followed by a monotonous *pukwok pukwok*.

Habits: Inhabits forest trees, particularly fruiting figs. Feeds mainly on fruit singly, in pairs or small parties. Difficult to see in foliage. Nests in tree hole.

Distribution: Locally common breeding resident in northern plains and hills from central Nepal east to the Myanmar border.
Also occurs in China and SE Asia.

BLUE-THROATED BARBET

Megalaima asiatica 23 cm F: Megalaimidae

Sumit Sen

Description: A medium-sized, grass green barbet with red-capped blue head. Blue of head includes chin and throat down to upper breast where there is a small red patch. Red crown interrupted by black and yellow patch in centre. Bill large and dark. Sexes alike. Young birds have green heads and blackish crowns.

Voice: A loud, persistent *kutt oo ruk, kutt oo ruk*, similar to **Brown-headed** but three notes in each phrase not two.

Habits: Inhabits forests and groves with fruiting trees. Strictly arboreal and usually solitary, though groups will gather in favoured trees, particularly *Ficus*. Rather secretive. Nests in tree hole.

Distribution: Locally common breeding in Himalayan foothills, the Northeast, Bangladesh and Nepal.
Also occurs in S China and SE Asia.

BLUE-EARED BARBET

Megalaima australis 17 cm F: Megalaimidae

Description: A small, green barbet with red-bordered, blue cheeks and a blue throat. Forecrown blackish and black spot in centre of breast. Larger **Blue-throated** has black and red crown. Sexes similar.

Voice: A rapidly repeated *koo trrk*.

Habits: Inhabits forest trees, particularly fruiting figs on which it feeds singly, in pairs or in small parties. Difficult to see in foliage. Nests in tree hole.

Morten Strange

Distribution: Scarce breeding resident in plains from eastern Nepal east to the Myanmar border and Bangladesh. Also occurs in SE Asia.

COPPERSMITH BARBET (Crimson-breasted Barbet)

Megalaima haemacephala 17 cm F: Megalaimidae

Description: Small, green barbet with brightly coloured head. Black-bordered yellow face with black malar and eye stripes, red forecrown and throat patch. Bill stout and dark. Upperparts grass green, underparts paler greenish, diffusely streaked with darker green. Young birds have no red. Sexes alike.

Voice: Oft repeated *tuk, tuk, tuk* like a small hammer on metal, hence its English name. One of India's most familiar sounds in the hot season.

Habits: Commonest barbet wherever there are trees. Avoids thick forests. Feeds on fruits and nectar. Perches on bare branches. Often confiding. Sociable and inquisitive. Nests in tree hole.

Otto Pfister

Distribution: Common breeding resident throughout lowlands. Also occurs in SE Asia.

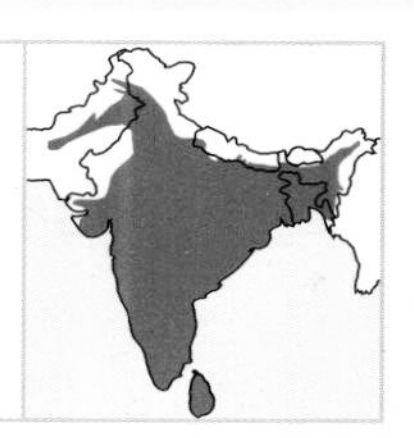

INDIAN GREY HORNBILL (Common Grey Hornbill)

Ocyceros birostris 60 cm F: Bucerotidae

Otto Pfister

Jan Willem den Besten

Description: Large, long-tailed, greyish bird with long, decurved bill. Sandy grey with black-and white-tipped tail, blackish eyestripes and black based yellow bill. Male's black casque longer and pointed. Bill less yellow. Whitish belly and wingtips.

Voice: Noisy, with strangely high-pitched squealing *ka ka ka keee*.

Habits: Inhabits open wooded country, parks and gardens. Often seen in sweeping flight between trees or perched on bare branches. Sociable and arboreal, feeding on fruit but also eats large insects, lizards and rodents, caught on the ground. Nests in tree hole; incubating female sealed in with mud by male. She moults during this period.

Distribution: Locally common endemic breeding resident throughout Indian lowlands and parts of Nepal. Rare in Pakistan, Bangladesh, the Northeast and parts of coast.

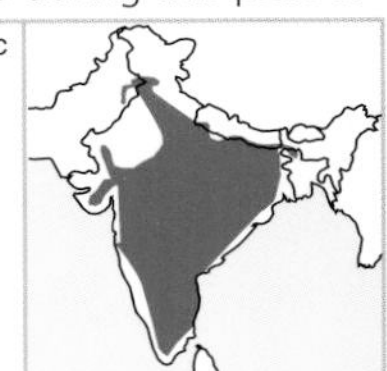

MALABAR GREY HORNBILL

Ocyceros griseus 58 cm F: Bucerotidae

Clement A. Francis

Description: Similar to **Indian Grey** but darker grey and with no casque on deep yellow bill. Dark grey, paler below and with distinct whitish supercilia and blackish crown and eyestripes. Tail tipped white. Female has black bill base, male reddish.

Voice: Noisy, with a squealing, laughing cry.

Habits: Restricted to open, broad-leafed Western Ghat forests where it feeds high in fruiting trees. Nests in tree hole as **Indian Grey**.

Distribution: Local endemic breeding resident in Western Ghats.

GREAT HORNBILL (Great Pied, Giant Hornbill)

Buceros bicornis 130 cm F: Bucerotidae

Otto Pfister

Description: Huge, pied hornbill with large casqued, yellow bill. Black upperparts with broad white wing bars, white wing tips and white tail with black central bar. White vent and thighs, black belly and lower breast and yellowish upper breast and neck. Face black. Large yellow casque extends from rear crown to where thick, yellow bill decurves. Sexes alike, although female smaller.

Voice: Noisy, with deep barking calls. Wings make whooshing noise in flight, audible at a kilometre.

Habits: Inhabits foothill forests. Feeds on treetop fruit, tossing it in the air before swallowing. Also eats insects, reptiles and rodents. Nesting habits as other hornbills.

Distribution: Scarce breeding resident in Himalayan foothills east from Uttaranchal to the Myanmar border and Western Ghats.
Also occurs in SE Asia.

Otto Pfister

RUFOUS-NECKED HORNBILL

Aceros nipalensis 100 cm F: Bucerotidae

Description: A large black and rufous hornbill with the terminal half of the tail white. Male and immature have head, neck and most of underparts rich orange-brown. Female has black head and underparts. Both have black upperparts with white wing and tail tips, bare blue skin around eyes and bill base and a bare, red chin pouch. Bill long, thick and yellow but no casque.

Voice: A deep, single, far-carrying staccato *wok*.

Habits: Inhabits primary forests. Arboreal, feeding mainly on fruit. Nests as other hornbills.

Morten Strange

Distribution: Globally threatened rare breeding resident in northeastern hill forests.
Also occurs in SE Asia.

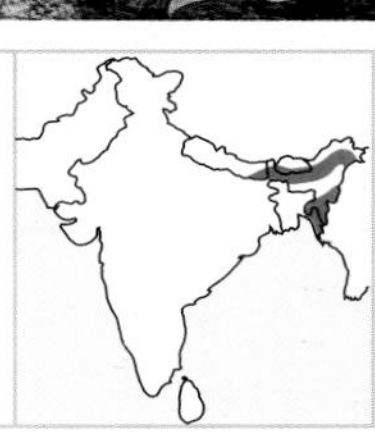

WREATHED HORNBILL

Aceros undulatus 110 cm F: Bucerotidae

Morten Strange

Description: A large, mainly black hornbill with a white tail. Larger male has whitish head with brown crown and nape. Red around eye and yellow throat pouch with a black stripe across it. Female's pouch is blue and head black. Corrugated bill.

Voice: A loud *kuk kwek*. Wings make whooshing sound in flight.

Habits: Inhabits forest with fruiting trees. Feeds mainly on fruit, travelling in, sometimes large, parties to favoured sites. Roosts communally when not breeding. Shy and best seen flying above canopy. Nests in high tree hole.

Distribution: Fairly common breeding resident in northeastern plains and foothills from West Bengal east to the Myanmar border.
Also occurs in SE Asia.

ORIENTAL PIED HORNBILL (Indian Pied, Pied Hornbill)

Anthracoceros albirostris 90 cm F: Bucerotidae

Morten Strange

Description: Medium-sized, pied hornbill with mainly yellow casqued yellow bill and white tips to outer tail feathers. Black all over with narrow white wing tips, belly and thighs. Black patch on underside of large casque and blue throat patch. Female has smaller, more rounded casque.

Voice: Various high-pitched cackles and squeals. Also a fast *ka ka ka ka*.

Habits: Inhabits open forest and groves. Sociable and arboreal, favouring fruiting trees but will feed on ground. Also eats young birds, large insects, reptiles and rodents. Nests as other hornbills.

Distribution: Scarce breeding resident in Himalayan foothills east from Haryana, the Northeast, Bangladesh and the Eastern Ghats south into northern Andhra Pradesh.
Also occurs in S China and SE Asia.

MALABAR PIED HORNBILL

Anthracoceros coronatus 92 cm F: Bucerotidae

Description: A medium-sized, pied hornbill very similar to **Oriental Pied** but with broad white outer tail feathers. White wing tips broader. Casque is also longer and largely black and throat patch pink. Female similar but less black on casque.

Voice: A loud *kek kek kek kek*.

Habits: Inhabits open forest and groves, particularly *Ficus* and other fruiting trees on hillsides. Food and nesting as **Oriental Pied**.

Otto Pfister

Distribution: Scarce endemic breeding resident in foothills of Western and Eastern Ghats and Sri Lanka, where commoner.
Probably overlaps in north-central India with the Oriental Pied.

Otto Pfister

COMMON HOOPOE (Hoopoe)

Upupa epops 31 cm F: Upupidae

Description: Striking, fan-crested pinkish-orange, black and white bird with long decurved bill. Orange or rufous pink (depending on race) with white-barred black tail and white bars on broad, black wings. Long crest black tipped, held flat on crown or raised in fan. Distinctive, halting flight with wings closed between beats.

Voice: Far-carrying, low *poo, poo, poo*.

Habits: Inhabits cultivation, open woodland, villages, parks and gardens. Feeds mainly on short grass, including lawns, probing ground with long beak. Often in pairs or loose groups. Feeds on large insects, worms, small reptiles and rodents. Confiding although can be inconspicuous.

Otto Pfister

Otto Pfister

Distribution: Common breeding resident throughout. Summer visitor to Himalayas.
Also occurs in Africa, Eurasia and SE Asia.

MALABAR TROGON (Ceylon Trogon)

Harpactes fasciatus 30 cm F: Trogonidae

S Elamon

Description: A striking black, fawn and crimson bird with a long square ended, white-edged tail. Sooty-black head and neck with white gorget on upper breast. Underparts rich crimson. Wings black-edged white. Back and tail fawn-brown, the latter with black tip. Female brown with blackish wings. Stout, dark bill.

Voice: Diagnostic. Somewhat whistling *cue, cue, cue* extending to up to eight notes.

Habits: Inhabits forest. The only trogon in its range. Sits for long periods high in trees often with back to observer. Thus easily overlooked unless calling, bobbing or tail-flicking. Fly-catches in canopy. Also eats fruit. Solitary or in pairs.

Distribution: Rather scarce endemic breeding resident in Western and Eastern Ghats and Sri Lanka, where commoner.

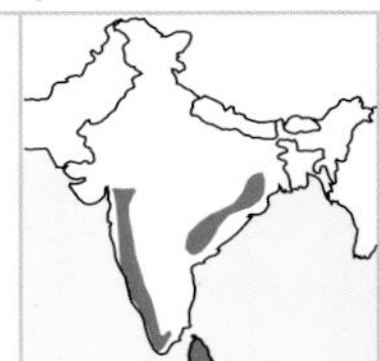

RED-HEADED TROGON

Harpactes erythrocephalus 35 cm F: Trogonidae

Morten Strange

Description: A mainly red and fawn-brown bird with long, square-ended tail edged with black and white. Head and neck bright crimson, white gorget and pinker underparts. Wings black closely barred with white and rest of upperparts fawn-brown. Tail broadly edged black with narrow white outer borders to sides.

Voice: A scaled sequence of *chaup chaup chaup* notes.

Habits: Inhabits dense broadleafed forests. Habits as **Malabar Trogon**. Equally difficult to see.

Distribution: Scarce breeding resident in the Northeast including parts of Nepal and Bangladesh.
Also occurs in S China and SE Asia.

EUROPEAN ROLLER (Roller)

Coracias garrulus 31 cm F: Coraciidae

Description: A large-headed, mainly blue bird. Whole body powdery-blue with short, black eyestripes and black apical tips to tail. Mantle cinnamon-brown. Wing shoulders and rump darker blue and flight feathers blackish. Bill dark and crow-like. Sexes alike but juveniles much greyer and duller with faint streaking on breast.

Voice: Loud, harsh *rack rack* call.

Habits: Inhabits open, dry country and cultivation, frequently perching on wires and posts. Feeds on large insects, reptiles and rodents, usually catching them on the ground in a shrike-like pounce. Solitary.

Otto Pfister

Distribution: Breeding resident in northern Pakistan and Kashmir. Scarce winter visitor further south, mainly in Pakistan and northwest India but scattered reports from elsewhere. Probably mainly a passage migrant enroute to Eastern Africa. Also occurs in Africa and Eurasia.

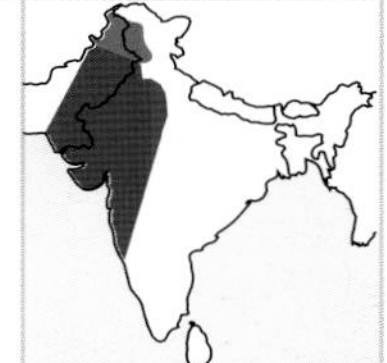

INDIAN ROLLER (Blue Jay)

Coracias benghalensis 31 cm F: Coraciidae

Description: Large-headed, dark pink bird with bright wings in two shades of blue. Crown, inner wing, outer tail and belly turquoise, tail centre and tip and most of flight feathers darker blue. Neck and breast dark pink with white streaking on throat. Mantle brownish-pink with greenish gloss. Large dark crow-like bill. Sexes alike. Juvenile dull.

Voice: Occasional harsh *kak kak* calls. Very noisy in rolling display flight.

Habits: Inhabits open country, light woodland and cultivation. Perches prominently, pouncing on insects, rodents and reptiles. Also chases insects in flight. Solitary or in pairs but groups sometimes gather. Nests in tree hole.

Distribution: Common breeding resident throughout lowlands, subject to local movements. Also occurs in W and SE Asia.

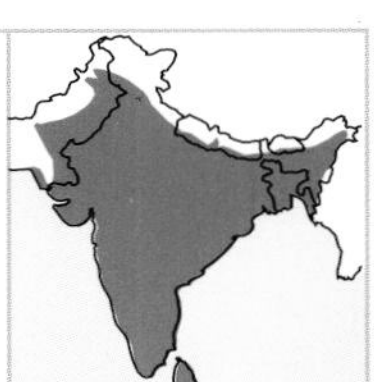

Otto Pfister

Otto Pfister

DOLLARBIRD (Broad-billed, Dark, Asian Broad-billed Roller)

Eurystomus orientalis 31 cm F: Coraciidae

Morten Strange

Description: A dark greenish bird with a very stout red bill. Looks black in distance but closer examination reveals bluish on throat and wings. Also a silvery blue patch in primaries. Legs and eye rings also red. Sexes alike. Juvenile duller.

Voice: A rasping *drak drak*.

Habits: Inhabits forest edges and clearings, also nearby cultivation. Perches high in trees on bare branches from where it launches itself to catch large flying insects. Crepuscular and sociable, groups often hunting after dusk. Frequently overlooked. Nests in tree hole.

Distribution: Scarce breeding resident and partial migrant in the Northeast and Western Ghats. Summer visitor to Nepal, winter visitor to Bangladesh. Rare in Sri Lanka.
Also occurs in E and SE Asia and Australasia.

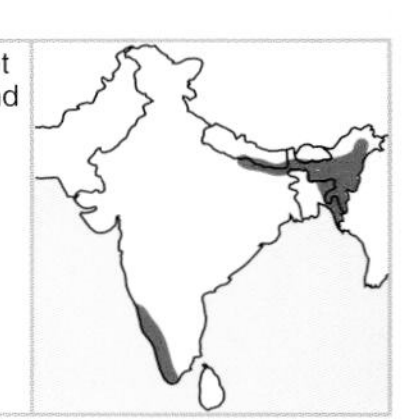

COMMON KINGFISHER (Small Blue, Eurasian Kingfisher)

Alcedo atthis 18 cm F: Alcedinidae

Otto Pfister

Otto Pfister

Description: Small, blue and orange kingfisher with long, dark bill and very short tail. Bright sky blue above with greenish-blue wings and finely black-barred, blue crown. Bluer in southern form (also shown). Underparts orange-brown. Bill black and dagger-like, females sometimes showing red on base. Tiny feet waxy red. Sexes similar.

Voice: A short piercing *chee,* mainly in flight.

Habits: Inhabits fresh and coastal waters. Feeds on small fish by diving from overhanging perch or after hovering. Favours small canals, streams and fish ponds, avoiding larger water bodies and forests. Flies low and fast over water. Nests in self-made burrow in sand bank.

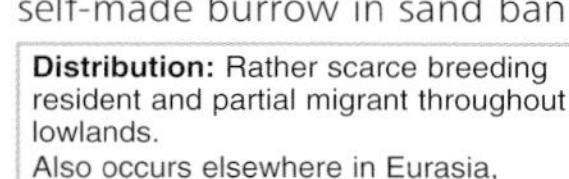

Distribution: Rather scarce breeding resident and partial migrant throughout lowlands.
Also occurs elsewhere in Eurasia, Indonesia and the SW Pacific.

ORIENTAL DWARF KINGFISHER (Three-toed)

Ceyx erithacus 13 cm F: Alcedinidae

Description: A tiny, mainly violet and orange kingfisher with a long, waxy red bill and very short tail. Head, tail and underparts orange-chestnut with violet on crown, back and rump. Wings blackish-purple, blue and white cheek spots. Tiny red feet.

Voice: A squeaky *chichee*.

Habits: Inhabits small pools and streams in hill forest. Feeds on small fish and insects in forest streams and the forest understorey. Shy and easily overlooked. Solitary. Sometimes attracted to lights during nocturnal migration.

Joanna Van Gruisen

Distribution: Scarce breeding resident subject to local movements in the Northeast, west coast and Sri Lanka. Also occurs in S China and SE Asia.

BROWN-WINGED KINGFISHER

Halycon amauroptera 36 cm F: Halcyonidae

Description: Very large, brown and orange kingfisher with a huge red bill. Similar to **Stork-billed** but brown above with blue restricted to rump and lower back. Lacks brown cap and is generally warmer orange-buff on head and underparts. Sexes alike.

Voice: Laughing, descendent *cha cha cha* and a low, whistling *chow chow chow*.

Habits: Inhabits coasts with mangroves and wooded estuaries. Very rarely on inland wetlands. Feeds on fish and other water animals by plunging into water or onto mud from favoured perch. Will feed in sea-surf, often resting within breaking waves. Rather sluggish but has powerful high flight. Nests in self-excavated bank tunnels but rarely observed.

Sujan Chatterjee

Distribution: Near-threatened globally. Scarce breeding resident and winter visitor to northeastern Indian and Bangladesh coasts, occasionally inland in Bangladesh and Assam.
Also occurs in SE Asia.

STORK-BILLED KINGFISHER

Halcyon capensis 38 cm F: Halcyonidae

Sumit Sen

Description: A very large, mainly blue and buff kingfisher with a large head and huge red bill, making it appear top-heavy. Dull blue upperparts, brown crown and rich yellow-buff collar and underparts. Tail fairly long. Feet small and red. Sexes alike.

Voice: Very noisy. A loud, ringing, oft-repeated *kee kee kee*.

Habits: Inhabits larger, well-wooded lakes, rivers and forest streams. Also wooded coasts and mangroves. Stays hidden in vegetation for long periods. Feeds on fish by diving but also eats crustacea, reptiles, rodents and young birds. Nests in self-made burrow in bank.

Distribution: Fairly common breeding resident in most of region except the Northwest and Pakistan.
Also occurs in SE Asia.

WHITE-THROATED KINGFISHER (White-breasted)

Halcyon smyrnensis 28 cm F: Halcyonidae

Otto Pfister

Otto Pfister

Description: Large, blue, brown and white kingfisher with powerful red bill and long tail. Head and underparts brown with white throat and central breast. Upperparts turquoise with broad brown and black wing shoulders. Black flight feathers with white patches. Sexes alike.

Voice: Noisy. A very loud, laughing *kila kila kila,* often in flight. Adapted as song when breeding.

Habits: Commonest and most catholic kingfisher. Found in cultivation, parks, gardens, open forest and fresh and coastal wetland. Frequently perches on wires, posts and in trees. Feeds by pouncing on insects, reptiles, amphibians and rodents on ground. Occasionally dives for fish. Nests in banks.

Distribution: Abundant breeding resident throughout lowland parts of region.
Also occurs elsewhere in tropical and sub-tropical Asia.

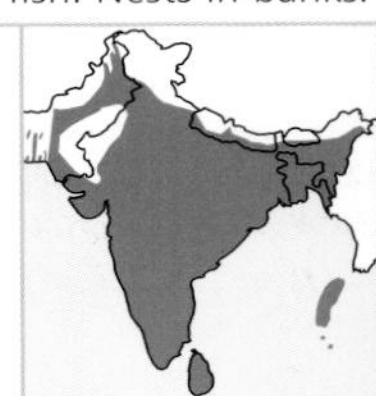

BLACK-CAPPED KINGFISHER (Black-capped Purple)

Halcyon pileata 30 cm F: Halcyonidae

Description: A large, black, purplish and white kingfisher with a powerful red bill. Black crown to below eyes contrasts with broad white collar and breast. Pale rufous underparts. Purple-blue above with bluer rump and blackish wing shoulders. Sexes alike.

Voice: A shrill, loud cackle, similar to, but distinct from, **White-throated**.

Habits: Inhabits mangrove creeks and wooded estuaries. Sometimes penetrates well inland up rivers. Feeds mainly by skim-diving for fish but also takes crustacea, insects and amphibians from the ground. Usually solitary. Nests in hole in bank.

Kamal Sahai

Distribution: Local resident and winter visitor to coasts from Maharashtra round to Bangladesh. Also Sri Lanka, the Andamans and the Nicobars. Rare inland. Breeding status unknown in region.
Also occurs in China and SE Asia.

COLLARED KINGFISHER (White-collared Kingfisher)

Todiramphus chloris 24 cm F: Halcyonidae

Description: A medium-sized, blue and white kingfisher. Head turquoise with variable dark and white markings depending on race. Throat, broad collar and underparts white, sometimes with a buff wash. Upperparts bright greenish-turquoise, bluer on wings. Can look surprisingly dark above against the light. Bill largely black with whitish on lower mandible. Sexes alike.

Voice: A hard *krerk krerk krerk*.

Habits: Inhabits tidal creeks, mudflats and mangroves where it feeds mainly on small crabs and skippers. Solitary or in pairs. Excavates own nest hole in tree trunk, ant or termite nest.

Otto Pfister

Distribution: Scarce breeding resident, sporadically along coast from Bangladesh round to Goa. Commoner in the Northeast and on Andaman and Nicobar Islands.
Also occurs in W Asia, SE Asia, Pacific Islands and Australia.

CRESTED KINGFISHER (Greater Pied, Large Pied)

Megaceryle lugubris 41 cm F: Cerylidae

Nikhil Devasar

Description: A huge, crested, dark grey and white kingfisher. Heavily barred greyish-black and white above with shaggy, white-streaked black crest and white collar. Face white with black malar streak turning to orange neck patch on male, black-speckled white on female. Underparts white with grey flank barring. Long tail, strongly barred black and white. Powerful dark bill.

Voice: A quiet bird, giving the occasional *kik* note.

Habits: Inhabits fast-flowing rivers. Perches on overhanging branches and rocks, diving from perch for fish. Solitary and very loyal to favoured perches.

Distribution: Scarce breeding resident of northern and northeastern uplands. Also occurs in China, Japan, Korea and SE Asia.

PIED KINGFISHER (Lesser Pied, Small Pied)

Ceryle rudis 31 cm F: Cerylidae

Kamal Sahai

♀

Otto Pfister

Description: Large, pied kingfisher frequently hovering. Black above with white bars and patches, white head with black-crested crown and broad black eye stripes. White below with broad black breast band. Male has second band below. Long tail and long black bill.

Voice: An excited *chirruk chirruk* usually in flight and when pairs meet up.

Habits: Inhabits larger rivers, lakes and the coast. Feeds almost entirely on fish mainly by plunge-diving from extended hovering. Batters them on perch before swallowing. Sits on wires and other high perches over water but also sits low. Sociable, nesting in tunnels in banks in groups.

Distribution: Common breeding resident throughout lowlands. Also occurs in Africa, S China, W and SE Asia.

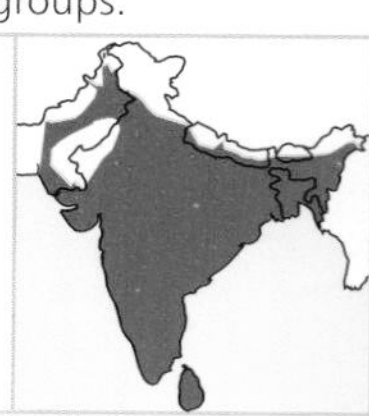

BLUE-BEARDED BEE-EATER

Nyctyornis athertoni 34 cm F: Meropidae

Description: A large, green bee-eater with a blue throat and long square-ended tail. Overall rather pale green with bluish wash to face, shaggy throat and upper breast. Green-streaked buff underparts. Pale orange undertail. Powerful, decurved dark bill. Often sits in shade when can look very dark.

Voice: Deep *korrr korrr* and various chuckling notes.

Habits: Inhabits wet forest edges. Sluggish and shy, keeping to treetops; often along forest roads and in clearings, and frequently in small groups. Catches aerial insects from perch and batters them before swallowing. Excavates deep tunnel for nest in bank, often in road cuttings.

Otto Pfister

Distribution: Scarce breeding resident of Western and Eastern Ghats, the Himalayan foothills and the Northeast. Also occurs in S China and SE Asia.

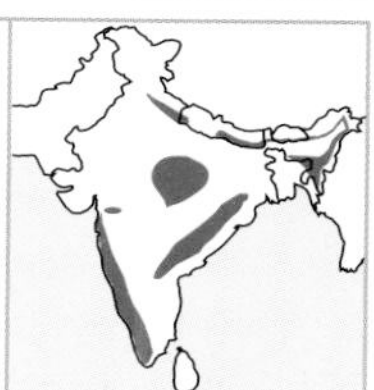

GREEN BEE-EATER (Little Green Bee-eater)

Merops orientalis 21 cm including tail pins F: Meropidae

Description: Small, active, green bee-eater with black gorget and eyestripes. Green with slight buff wash on crown, bluish cheeks and throat. Darker chestnut on crown and nape in eastern race *ferrugeiceps*. Long central tail pins on adult, not present in duller juvenile. Slender, dark, decurved bill. Sexes alike.

Voice: Noisy. Repeated trilling, usually in flight.

Habits: Inhabits open country, cultivation and light woodland. Feeds acrobatically on flying insects including bees. Robs beehives. Sociable, perching on wires and bare branches or on sandy ground. Frequently dust-bathes. Nests in self-excavated tunnels in banks or obliquely into flat ground.

Otto Pfister

Distribution: Common breeding resident throughout lowlands. Also occurs in Africa, W. China, W and SE Asia.

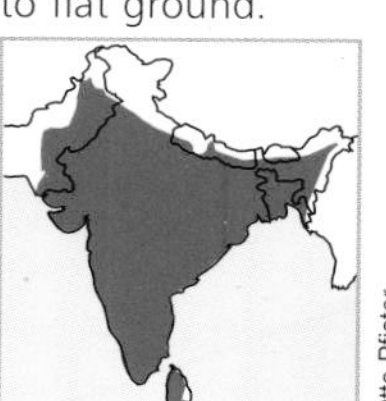

Otto Pfister

BLUE-CHEEKED BEE-EATER

Merops persicus 30 cm F: Meropidae

Otto Pfister

Description: Large, bright green bee-eater with chestnut underwings and brightly coloured face. Bright grass green all over with black eyestripes with silvery-blue above and below them and chestnut-edged yellow throat. Adults have long central tail pins. Juveniles are duller with pointed or square ended tails. Sexes alike.

Voice: Noisy. A liquid, disyllabic *priit priit* mainly in high flight or when birds meet on wires.

Habits: Inhabits dry, open country near water. Feeds largely on dragonflies in flight. Sociable, often perching on wires, bare branches and open ground. Often flies high, almost out of sight, after prey; frequent contact calls alone alerting the observer. Nests in self-excavated burrows.

Distribution: Locally common breeding summer visitor to Pakistan and northwest India where it overlaps with and mixes with Blue-tailed. Also occurs in Africa, W and Central Asia.

BLUE-TAILED BEE-EATER

Merops philippinus 30 cm F: Meropidae

Otto Pfister

imm

Otto Pfister

Description: Large, dull green bee-eater with blue rump and tail. Duller grass green than **Blue-cheeked** with brownish wash and blue rump and tail diagnostic but not always obvious. Wings can also show blue wash. Black eyestripes with slight blue edging above. Throat extensively chestnut, chin and cheek stripe yellow. Chestnut underwings. Adults have long tail pins. Duller juveniles have pointed or square ended tails. Sexes alike.

Voice: Noisy. Similar to **Blue-cheeked** but harsher.

Habits: Inhabits open, often wooded, country near water bodies. Feeds aerially mainly on dragonflies. Very sociable. Perches on wires, bare branches and ground. Nests in self-excavated burrows in banks.

Distribution: Summer visitor to northern lowlands, wintering further south down to Sri Lanka. Also occurs in S China, Indonesia, SW Pacific and Papua New Guinea.

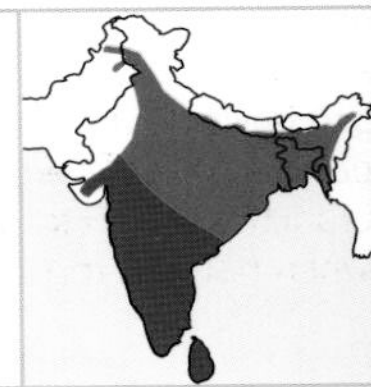

EUROPEAN BEE-EATER

Merops apiaster 27 cm F: Meropidae

Description: A large, richly coloured bee-eater with a black-bordered yellow throat. Crown to mantle rich chestnut merging into golden further down back. Yellow forehead. Wings green and chestnut, tail and underparts emerald green. Short tail pins on adults. Juveniles much duller with square-ended tails. Pale orange underwings. Sexes alike.

Voice: Noisy. A liquid monosyllabic *prut prut,* given mostly in flight.

Habits: Inhabits dry, open country and cultivation. Behaviour as **Blue-tailed** and **Blue-Cheeked** species but feeds mainly on bees and therefore does not need to be near water.

Otto Pfister

Distribution: Common breeding summer visitor to Pakistan and Kashmir. Infrequent winter visitor to northern and peninsular India. Most birds presumably on passage to E Africa.
Also occurs in Africa and Eurasia.

CHESTNUT-HEADED BEE-EATER

Merops leschenaulti 21 cm F: Meropidae

Description: A small chestnut, yellow and green bee-eater, with a square-ended tail which usually shows a slight notch. Chestnut crown to mantle contrasts with bright green upperparts and paler green below. Throat bright yellow, edged by black and chestnut gorget and black eyestripes. Sexes alike. Juveniles duller.

Voice: A musical, repeated *perrup perrup*.

Habits: Inhabits stream and pool margins in forests. Feeds by acrobatic sallies from exposed perches on flying insects, particularly utilising bare tree branches. Often in small parties, when noisy at roosts.

Otto Pfister

Distribution: Common breeding resident of Himalayan foothills, the Northeast, southern peninsula hills and Sri Lanka.
Also occurs in SE Asia.

PIED CUCKOO (Pied Crested, Jacobin Cuckoo)

Clamator jacobinus 33 cm F: Cuculidae

Otto Pfister

Description: Medium-sized, black and white cuckoo with distinct rear crown tuft. Black with white wing patch and white-tipped tail. White below. Small, decurved bill. In flight, very long, graduated tail and white markings on black striking. Juvenile sooty-brown above and yellowish below. Often near babbler foster parents. Sexes alike.

Voice: Noisy; loud penetrating *pew, pew, pew* in various combinations given in flight and when perched.

Habits: Inhabits open and lightly wooded country, including marshes. Brood-parasatises babblers. Feeds on insects, particularly caterpillars, on leaves and on ground. Singly or in small groups.

Distribution: A widespread monsoon breeding visitor to the lowlands. Thought to winter in E Africa. Also occurs throughout Africa and in Myanmar.

CHESTNUT-WINGED CUCKOO (Red-winged Crested)

Clamator coromandus 45 cm F: Cuculidae

Otto Pfister

Description: A large, chestnut and black cuckoo with a backswept crest. Black crown, mantle rump and tail contrast with chestnut wings. Underparts white with warm orange wash to throat and upper breast. White collar. Juveniles have rufous feather edges above. Sexes alike.

Voice: In breeding season a whistling *peep peep*. Also a hoarse screech.

Habits: Inhabits forest and thick scrub. Usually singly in canopy but descends to bushes to feed. Inconspicuous. Feeds on invertebrates, particularly caterpillars. Parasitises mainly laughing-thrushes.

Distribution: Scarce and local breeding summer visitor to northern foothills from Uttaranchal east to Arunachal Pradesh, Nepal, NE India and Bangladesh. Winters sparsely in extreme south India and Sri Lanka but possible anywhere in peninsula on passage. Also occurs in China and SE Asia.

LARGE HAWK CUCKOO

Hierococcyx sparverioides 38 cm F: Cuculidae

Description: Large, hawk-like, brownish-grey cuckoo with long, broad barred tail and strongly barred underparts. Four very broad tail bars, including tip and obvious barring from belly to breast, latter streaked chestnut. Throat whitish. Irises yellow and strong, decurved, dark bill. Juvenile much browner. Sexes alike.

Voice: A measured *kee-keeah kee-keeah* usually without a crescendo. Heard at night as well as day.

Habits: Inhabits broadleafed forests. Keeps well hidden in foliage where it feeds on caterpillars and other insects. Flap and glide hawk-like flight. Heard much more than seen. Brood-parasitises smaller birds.

Otto Pfister

Distribution: Fairly common breeding summer visitor to Himalayas and northeast mountains. Winters sparsely in lowlands.
Also occurs in SE Asia.

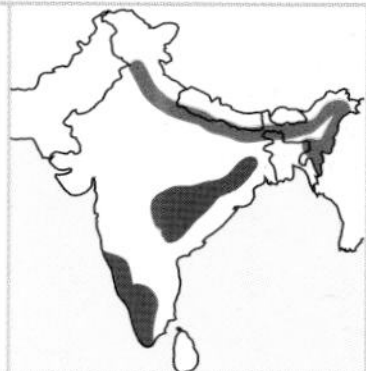

COMMON HAWK CUCKOO (Brain-fever Bird)

Hierococcyx varius 34 cm F: Cuculidae

Description: Medium-sized cuckoo with clear grey upperparts, long, barred tail and plain underparts. Resembles male **Shikra** in flight. Upperparts pale grey, four narrow black tail bars. Underparts greyish-white with pale chestnut chest and light barred belly and flanks. Dark yellow-based bill. Juvenile browner, streaked below.

Voice: Classic *brain fever* or *dee dee dit,* oft-repeated and rising to manic crescendo. Uttered all day and night, especially during the hot season.

Habits: Inhabits all types of wooded areas, plantations, parks and gardens. Feeds mainly on caterpillars. Brood-parasitises babblers.

Kamal Sahai

Sumit Sen

Distribution: Common breeding endemic resident in lowlands throughout; monsoon breeding visitor in northern areas.

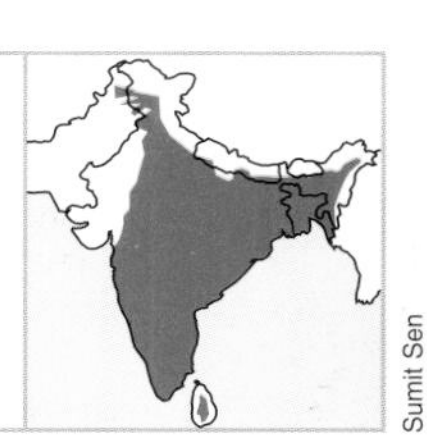

INDIAN CUCKOO

Cuculus micropterus 33 cm F: Cuculidae

Sharad Gaur

Sharad Gaur

Description: Medium-sized, grey cuckoo with distinctly brownish cast to upperparts. Head and throat greyer, underparts white with broad black banding. Long tail, with extensive black subterminal band and obvious white barring on outer and central feathers. Brownish irises. Female browner on throat and upper breast. Juvenile barred brown.

Voice: The best id feature. Four noted mellow whistle *bo ko tako* or *I'm a cuckoo,* trailing off at end. Very noisy in summer, even at night.

Habits: Inhabits wooded areas. Inconspicuous except when calling, hiding in upper foliage where it eats mainly caterpillars. Brood-parasitises orioles.

Distribution: Scarce breeding resident mainly of northern, eastern and southern lowlands and Sri Lanka. Subject to local movements.
Also occurs in E Asia.

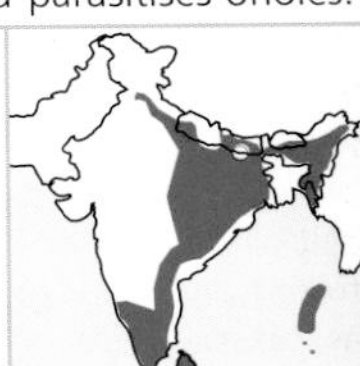

EURASIAN CUCKOO (Cuckoo, Common Cuckoo)

Cuculus canorus 33 cm F: Cuculidae

Otto Pfister

Otto Pfister

Description: Medium-sized, grey cuckoo with finely barred underparts. Pale blue-grey above, including head and throat to upper breast. Darker on flight feathers and black and white bordered tail. From lower breast white, closely barred with darker grey. Irises and eye rings yellow. Female browner with barred breast. Hepatic reddish-brown form illustrated. Juvenile barred brown with white nape.

Voice: Well-known, repetitive *cuc koo*. Also rapid bubbling, mainly by female.

Habits: Inhabits woodlands and scrub, where it brood-parasitises small birds. Winters in wooded country and cultivation. Typically flies with wings held low and rather humped back.

Distribution: Common breeding summer visitor to northern mountains. Scarce winter visitor elsewhere in lowlands.
Also occurs throughout Eurasia and winters in Africa.

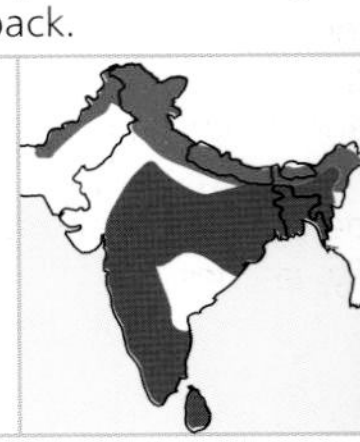

ORIENTAL CUCKOO (Himalayan Cuckoo)

Cuculus saturatus 31 cm F: Cuculidae

Description: Medium-sized, typical grey cuckoo almost identical to **Eurasian**. Separated by slightly smaller size and darker back, tendency to bolder barring below. Often with plain vent and buffish wash to under tail-coverts. Underwing less barred. Irises and eye rings yellow or brown. Female and juvenile differences similar to **Eurasian,** including hepatic form.

Voice: Best id feature. Distinctive repetitive *po po po po* in series of four to eight notes. Longer than very similar call of **Hoopoe**. Often calls at night.

Habits: Inhabits hill forests. More arboreal than **Eurasian** and difficult to see except when calling from treetops. Flight and shape as Eurasian. Brood-parasitises warblers.

Toby Sinclair

Distribution: Breeding summer visitor to northern hills. Scarce but overlooked winter visitor to lowlands.
Also occurs in Central, E and SE Asia.

GREY-BELLIED CUCKOO (Indian Plaintive)

Cacomantis passerinus 23 cm F: Cuculidae

Description: A small, grey cuckoo. All grey, slightly darker above and whitish undertail. White bars on outer tail feathers and at bend of wing below. Bill all dark. Irises reddish. Juvenile and some females barred brownish above in hepatic form.

Voice: Noisy in monsoon. A range of calls including a mournful *piteeer* and a shrill, repeated and rising *pee pip pe pe*.

Habits: Inhabits wooded areas including parks and large gardens. Feeds on caterpillars and brood-parasitises warblers. Although easily overlooked, can be very active and flighty. Often fly-catches.

Balachandran

Distribution: Rather scarce monsoon endemic breeding visitor to northern areas. Scarce resident further south.

PLAINTIVE CUCKOO (Rufous-bellied Plaintive Cuckoo)

Cacomantis merulinus 23 cm F: Cuculidae

Morten Strange

Description: A small, grey and orange cuckoo. Head and throat paler grey than upperparts. Breast to vent warm orange. Hepatic female of this and **Grey-bellied** very difficult to separate, but latter more strongly barred especially on tail and underparts have buff, not white, ground colour.

Voice: A mournful, rising *play to tee* often in a trailed-off crescendo.

Habits: Inhabits wooded country including parks and gardens. Secretive but active often calling in rapid restless flights. Feeds on caterpillars and brood-parasitises warblers.

Distribution: Generally scarce breeding resident in the Northeast, most frequent in Bangladesh. Also occurs in S China and SE Asia.

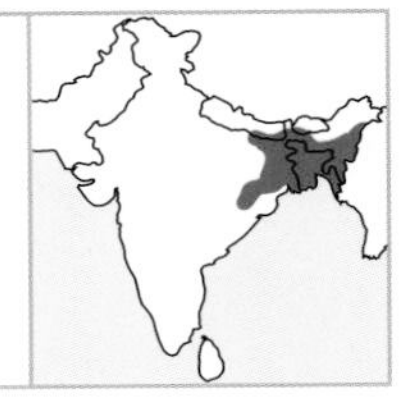

DRONGO CUCKOO

Surniculus lugubris 25 cm F: Cuculidae

Morten Strange

Description: A small, all black cuckoo with long, white-barred undertail coverts. Very similar to juvenile **Black Drongo** but tail only has slight notch. Undertail coverts best id feature but also has white thighs and nape patch. Juveniles duller black with white spots. Slender, dark, decurved bill.

Voice: Noisy in monsoon. A loud, whistled and rising *pee pee pee pee pee pee* abruptly ended, then soon restarted.

Habits: Inhabits open forests and plantations where keeps to foliage to feed on insects. Often sings from bare branch. Rather sluggish and inconspicuous. Brood-parasitises small babblers, and perhaps drongos and minivets, but breeding still imperfectly known.

Distribution: Scarce breeding resident in southern hills and Sri Lanka. Locally common monsoon breeding visitor to Himalayas and northeastern hills. Also occurs in S China and SE Asia.

ASIAN KOEL (Koel, Indian Koel, Koel Cuckoo)

Eudynamys scolopacea 43 cm F: Cuculidae

Description: Large, striped or glossy black long-tailed cuckoo with white bill and ruby red eyes. Male all black. Female and juveniles dark grey-brown, above barred and spotted whitish. Dark-barred white on underparts, spotted on throat.

Voice: Familiar pre-monsoon sound starting before dawn. Loud, rising and sometimes manic *ko-el ko-el ko-el*. Also steady bubbling note and various gurgles and shrieks. Young caw like crows.

Habits: Inhabits all types of wooded country, city parks and gardens. Adults eat mainly fruit but robs birds' nests. Brood-parasitises **House** and **Large-billed** Crows. Young accompany foster parents for several weeks, noisily demanding food.

Gertrud Denzau

♀

Kamal Sahai

Distribution: Common breeding resident throughout though mainly monsoon breeding visitor to the Northwest.
Also occurs in China, SE Asia, SW Pacific Islands and Papua New Guinea.

GREEN-BILLED MALKOHA (Large Green-billed)

Phaenicophaeus tristis 51 cm F: Cuculidae

Description: A large, pale-billed, grey cuckoo with a very long, black and white tail. Whole body greenish-grey with paler head and underparts and graduated tail, black with white feather tips. Bare red skin round eyes and heavy, curved bill pale apple green.Sexes alike.

Voice: Rather silent but will sometimes croak quietly.

Habits: A very shy inhabitant of forest and scrub undergrowth. Feeds, creeping with surprising manoeuvrability, low down in vegetation tangles. Eats insects and reptiles. Flies poorly and usually only for short distances. Builds own nest.

Morten Strange

Distribution: A scarce breeding resident of the northern foothills, the Northeast and the Eastern Ghats.
Also occurs in S China and SE Asia.

BLUE-FACED MALKOHA (Small Green-billed)

Phaenicophaeus viridirostris 39 cm F: Cuculidae

Balachandran

Description: A large, dark greenish-grey cuckoo with a very long black and white tail. Darker than the above with bluish streaked throat, blue bare skin round eyes and broader white tips to tail feathers. Bill apple green. Sexes alike.

Voice: Rather silent but will sometimes croak quietly.

Habits: A very shy inhabitant of thick thorn scrub and open forest undergrowth. Habits very similar to **Green-billed**. Builds own nest.

Distribution: Scarce but overlooked endemic breeding resident of southern India and Sri Lanka.

SIRKEER MALKOHA (Sirkeer Cuckoo, Sirkeer)

Phaenicophaeus leschenaultii 43 cm F: Cuculidae

Joanna Van Gruisen

Description: A large, pale brown cuckoo with a hooked, yellow-tipped red bill and a very long tail. Overall a rather nondescript tawny-brown with paler throat, light-streaked gorget and distinct white-lined, black eye patches. Underside of black tail looks white because of the width of the white feather tips. Raptor-like hook on colourful bill unique among cuckoos. Sexes alike.

Voice: Rather quiet but has a high pitched raptor-like *kek kek kek*.

Habits: Feeds mainly on reptiles on the ground, running low and rat-like when disturbed. Very shy and given to hiding in deep shade. Occasionally perches in open on low shrubs. Flies low and poorly on laboured wings. Builds own nest.

Distribution: Scarce breeding endemic resident of thorn scrub and semi-desert throughout. Rare or absent from Pakistan, Bangladesh and the Northeast.

GREATER COUCAL (Crow-pheasant)

Centropus sinensis 50 cm F: Centropodidae

Description: Very large, black and chestnut cuckoo with long, broad black tail. All glossy black with chestnut mantle and wings. Irises red. Strong bill. Female similar but larger. Juvenile barred white. Often looks shaggy headed.

Voice: Resounding and repeated *poop poop poop* call, often fading away. Croaks and chuckles.

Habits: Inhabits grassland, cultivation, scrub, open woodland, forest edges and reed beds. Feeds on reptiles and large insects which it locates by walking rather grandly over the ground with tail held horizontally. Clambers about in thick cover. Robs birds' nests. Flight low and laboured, rarely sustained. Often climbs to a height in order to glide to destination.

Toby Sinclair

Distribution: Common breeding resident throughout.
Also occurs in S China and SE Asia.

LESSER COUCAL (Small Coucal)

Centropus bengalensis 33 cm F: Centropodidae

Description: A black and chestnut cuckoo. Similar to, but smaller than, **Greater** with duller brown upperparts and chestnut under wings. Has buff-streaked non-breeding plumage although wings remain chestnut and tail black. Young birds similar but have barred wings and tail.

Voice: A deep, rapid *oop oop oop* that varies in tone. Also *kutook kutook* and variants.

Habits: Inhabits grass, reeds and thick scrub. Feeds mainly on grasshoppers. Usually singly or in pairs. Sits quietly for long periods and can be very unobtrusive. Nests in thick cover.

Clement M Francis

Distribution: Local breeding resident and summer visitor to northern foothills from Uttaranchal east to Myanmar border, throughout the Northeast south to Orissa and including Bangladesh. Also southwest India. Some local movements in non-breeding season. Also occurs in S China and SE Asia.

VERNAL HANGING PARROT (Indian Lorikeet)

Loriculus vernalis 15 cm F: Psittacidae

Otto Pfister

Description: A tiny, grass green, gymnastic parrot with a short, square tail. Bright crimson rump patch and blue throat on male. Female similar but lacks blue. Red parrot-shaped bill. Yellowish irises.

Voice: A quiet but distinctive and rapid *zzit zzit* in flight.

Habits: Inhabits forests. Very arboreal. Feeds usually in pairs on nectar, flowers and small fruit but occasionally gathers in flocks. Feeds very actively climbing swiftly around high branches and often hanging upside down to reach food. Unique among birds, lorikeets sleep in this bat-like position. Nests in small tree holes. Rapid, undulating finch-like flight between trees. Not shy but restless and thus difficult to see well.

Distribution: Scarce breeding resident of lower hills of southwestern and eastern India and Bangladesh, subject to local movements.
Also occurs in Vietnam.

ALEXANDRINE PARAKEET (Large Indian)

Psittacula eupatria 53 cm F: Psittacidae

Otto Pfister

Description: A large, boisterous, grass green parakeet with red shoulder patches. Male has black malar stripes merging into pink collar round lavender grey cheek. Long tail bluish with yellow tip and undertail. Smaller female lacks face markings but both have huge red bills.

Voice: Noisy. A loud but abrupt strangulated squawk in flight and when perched.

Habits: Inhabits wooded areas and plantations, including city parks and gardens. Eats fruit, cereals and vegetables, often raiding farms and orchards. Rather shy, keeping high in trees. Fast but rather heavy flight. A popular, if illegal, cage-bird. Pairs very affectionate. Excavates tree hole for nesting.

Distribution: Fairly common breeding resident throughout lowlands. Rare in extreme south.
Also occurs in Afghanistan and east to Vietnam through SE Asia.

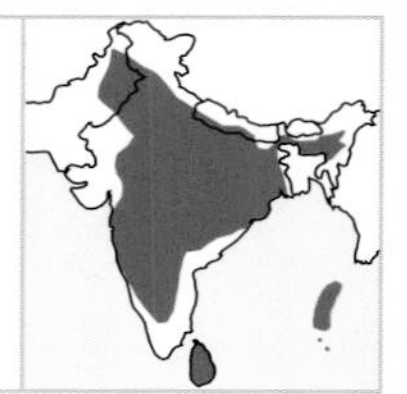

ROSE-RINGED PARAKEET (Green Parrot)

Psittacula krameri 42 cm F: Psittacidae

Description: The most well-known parakeet. Grass green with long bluish tail and red bill. Male has black and pink malar stripes and collar after third year. Female lacks these. No shoulder patches.

Voice: Very noisy in flight and at rest. A loud shrill *keeak,* often repeated. Also conversational murmuring between pairs.

Habits: Occurs wherever there are trees. Particularly favours city parks and gardens and cultivation where huge flocks can seriously damage fruit, vegetable and grain crops. Thousands gather at grain silos and at favoured city roosts. Although the most popular cage-bird in the region, it remains abundant. Pairs very affectionate. Nests in holes in trees and walls, often enlarging an existing site.

Otto Pfister

Distribution: Abundant breeding resident throughout lowlands.
Also occurs in Africa and Myanmar.
Feral in Europe.

SLATY-HEADED PARAKEET (Himalayan Parakeet)

Psittacula himalayana 40 cm F: Psittacidae

Description: A slender, green parakeet with a dark grey head. Grass green with small red shoulder patches and extensive yellow tip to very pointed, green tail. Black chin and collar, bluish hindneck. Upper mandible red, lower yellow. Female alike but without red shoulders. Smaller, eastern **Grey-headed** *P. finschii*, has a paler grey head.

Voice: Noisy. Distinctive high-pitched, musical *tooi tooi* deeper than **Plum-headed**.

Habits: Inhabits hill forests and cultivation, descending much lower in winter. Usually very arboreal but a pest to fruit and cereal growers. Very rapid, acrobatic flight through trees usually in synchronised flocks. Nests in tree hole.

Vivek Menon

Distribution: Common breeding resident of northern hills.
Also occurs in Afghanistan, S China and Vietnam.

PLUM-HEADED PARAKEET (Blossom-headed)

Psittacula cyanocephala 36 cm F: Psittacidae

Nikhil Devasar

Description: A slender, green parakeet with purple-red head in male, dark lavender in female. Male and most females have red shoulder patches, both have white tipped blue tail, and yellow upper and black lower mandibles. Male has black and blue collar, female yellow. Looks small-bodied in very rapid flight.

Voice: Noisy. High, musical questioning *tooi? tooi?* notes, usually in flight.

Habits: Inhabits wooded areas and plantations. Occurs in city parks and gardens although tends to be more rural. Does raid fruit and cereal crops but rarely in large numbers. Pairs are very affectionate. Nests in tree hole.

Distribution: Fairly common endemic breeding resident in lowlands. Very local in Pakistan and the Northeast.

MALABAR PARAKEET (Blue-winged Parakeet)

Psittacula columboides 38 cm F: Psittacidae

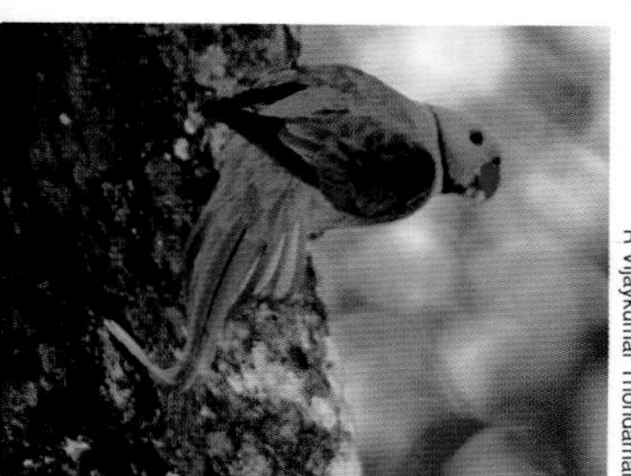

R Vijaykumar Thondaman

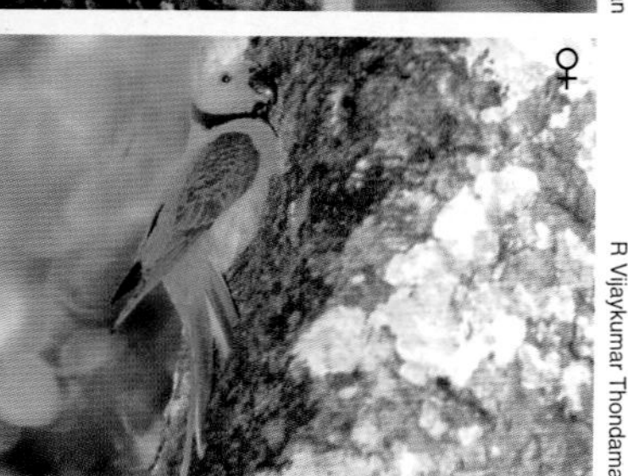

♀

R Vijaykumar Thondaman

Description: A long-tailed, mainly blue and grey parakeet. Blue-grey head, breast and mantle, blue flight feathers and yellow-edged blue tail. Both sexes have black collar but male has it edged blue-green. Irises pale.

Voice: A hard squeaky *screet screet*, harsher than **Plum-headed**.

Habits: Inhabits mainly evergreen hill forest feeding high in canopy on fruit and flowers. Also feeds on grain on ground. Very sociable often mixing with **Plum-headed** in foothills. Nests in tree hole.

Distribution: Fairly common endemic breeding resident restricted to the Western Ghats.

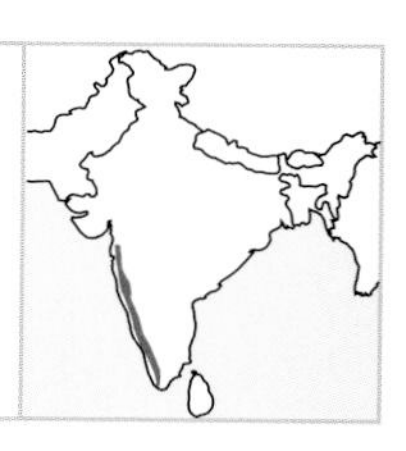

RED-BREASTED PARAKEET (Moustached, Rose-breasted)

Psittacula alexandri 38 cm F: Psittacidae

Description: A stocky, grass green parakeet with a pinky-red breast. Both have bluish-grey head with broad black eye and malar stripes and collar. Yellowish wing patches. Male's bill red, female's black. Overall female a shade paler. Long tail bluish.

Voice: Rather quiet. A sharp *quawnk*.

Habits: Inhabits foothill forests. Usually in small groups. Raids orchards and hill farms. Flight less dashing than other parakeets. Nests in tree hole.

Otto Pfister

Distribution: Scarce breeding resident in northern foothills east from Uttaranchal.
Also occurs in S China and SE Asia.

LONG-TAILED PARAKEET

Pscittacula longicauda 47 cm F: Psittacidae

Description: A typical parakeet with pinkish cheeks. Crown green and broad black chin stripe. Back often with bluish wash and very long tail blue (in moult in photo). Lower mandibles black, upper red in male, black in female.

Voice: Similar to **Rose-ringed** but less screeching.

Habits: Inhabits forests and plantations. Raids fruit and cereal crops. Acrobatic and very fast flight. Nests in tree hole.

Otto Pfister

Distribution: Common breeding resident in, and restricted to, Andaman and Nicobar Islands.
Also occurs in SE Asia.

GLOSSY SWIFTLET (White-bellied Swiftlet)

Collocalia esculenta 10 cm F: Apodidae

Morten Strange

Description: A tiny, dark, fast-flying swift. Glossy blue-black above. Dark grey throat and breast, darker vent and white belly. Only swiftlet with a square-ended tail. White belly distinguishes from **Edible-nest, Indian** *C. unicolor* and significantly larger **Himalayan** *C. brevirostris* which anyway only occurs in the Himalayas.

Voice: A hard, twittering call.

Habits: Nests in buildings. All swiftlets feed high in the sky on tiny insects. Often over forests as well as towns. Difficult to observe closely away from nest.

Distribution: Common breeding resident in Andaman and Nicobar Islands.
Also occurs in SE Asia, New Guinea and S Pacific islands.

EDIBLE-NEST SWIFTLET (Andaman Grey-rumped)

Collocalia fuciphaga 12 cm F: Apodidae

Morten Strange

Description: A small, dark-brown swift. Duller blackish-brown above than **Glossy** with underparts greyish brown and a narrow, greyish rump band. Tail notched. Very similar to larger **Himalayan**. **Indian** (also known as **Indian Edible-nest Swiftlet**) is paler with no rump contrast.

Voice: Loud, metallic *zwing*.

Habits: Feeds aerially on insects over mangroves, the coast and towns. More coastal than **Glossy**. Roosts and nests in caves which it navigates by echolocation. Large numbers cling to walls like bats. Enter and leave roost in rushing cascades. White nest made of saliva affixed to cave wall. Much prized for birds' nest soup and some colonies are frequently and illegally raided for nests.

Distribution: Common breeding resident on Andaman Islands, less common on Nicobars.
Also occurs in Indonesia.
Indian Swiftlet is an endemic breeding resident of the W Indian coast, the Western Ghats and Sri Lanka.

BROWN-BACKED NEEDLETAIL (Spinetail)

Hirundapus giganteus 23 cm F: Apodidae

Description: A large, dark, extremely fast-flying swift with white vent crescent and pointed tail. All sooty-brown with paler back, distinct white vent extending as crescent up flanks and white loral spots. Needles in tail feathers visible and long enough to form pointed tail. Wings often audible in low flight.

Voice: Rather quiet. Various soft squeaks and trills.

Habits: Inhabits forested areas. Hunts aerial insects low over forest in loose parties. Roosts and breeds in large hollow trees. Flight exceptionally powerful and flexible; quite difficult to follow with binoculars!

Morten Strange

Distribution: Scarce breeding resident of Western Ghats, Sri Lanka and the Northeast.
Also occurs in SE Asia.

ASIAN PALM SWIFT (Palm Swift)

Cypsiurus balasiensis 13 cm F: Apodidae

Description: Very small, slender, long-tailed, brown swift. Pale brown, paler on throat and belly. Narrow back-swept wings and very long, forked tail usually held closed and appears as a point. Resembles **Tree Swift** in shape.

Voice: Trills in flight, especially near nest.

Habits: Strictly associated with palmyra and other palms. Hunts in loose parties over cultivation and open woodland. Frequently near villages. Rather fluttery flight. Unique nest is glutinous pad of feathers and other material collected in air. It is stuck to the underside of a palm leaf or sometimes a thatched roof. The two eggs are stuck to it and the hatched young cling by their claws until fledging.

Morten Strange

Distribution: Rather scarce breeding resident of lowlands. Most common in south and east.
Also occurs in SE Asia.

ALPINE SWIFT

Tachymarptis melba 23 cm F: Apodidae

Goren Ekstrom

Description: A large, powerful, falcon-like swift with a striking white belly. All dull brown, including under-tail coverts and breast band. White throat often difficult to see (as in photo) but white belly obvious. Wings sickle-shaped and short tail forked. Often flies very high. When low, wings make audible "swoosh".

Voice: A distinct, shrill trill, usually in flight when flocks are feeding.

Habits: Feeds high aerially on insects, over forests, wetlands and cultivation. May come low during thunderstorms and passing weather fronts. Drinks by skimming water surfaces. Nests on tall buildings and cliffs.

Distribution: Scarce breeding resident of western hills and Sri Lanka. Breeding summer visitor to northern hills. Widespread records from elsewhere as it wanders far for feeding. Also occurs in S Europe, Africa and W Asia.

FORK-TAILED SWIFT (Pacific, Large White-rumped Swift)

Apus pacificus 18 cm F: Apodidae

Toby Sinclair

Description: A medium-sized, long winged and fork-tailed, blackish swift with a white rump band. Basically blackish-brown (looks black) with a small white throat patch; the narrow white rump band is the most striking feature. The tail fork is deep and obvious and wings sickle-shaped.

Voice: A hard *shkree* in flight.

Habits: High aerial feeder (often out of sight) in loose flocks or singly with other aerial feeders. Best chance of seeing it is when it flies low before a weather front or a passing thunderstorm. Indeed in such conditions the prudent observer will search the skies for swifts. Very fast, lucid flight without fluttering, quickly distinguishes it from **House Swift**, before shape is noted.

Distribution: Locally common breeding summer visitor to northern mountains. Wintering areas not known but widespread scattered records. Also occurs in NE and SE Asia and Australia.

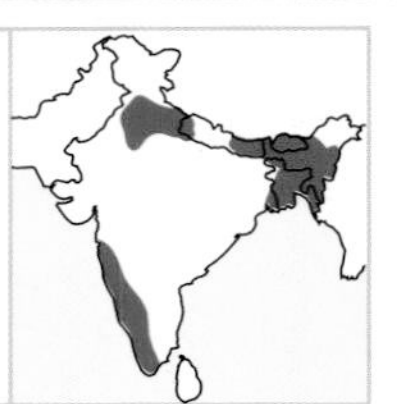

HOUSE SWIFT (Little Swift)

Apus affinis 15 cm F: Apodidae

Description: Small, broad-winged, blackish swift with almost square tail. Very dark, with large and obvious white throat patch and square white rump extending down sides. Wings pointed but broad-based, producing fluttering bat-like flight. Tail short and square-ended; notched in *nipalensis*.

Voice: Very noisy, especially when breeding. A high-pitched screeching and insistent trill.

Habits: Commonest swift. Particularly favours urban areas and ancient buildings where it builds untidy nests colonially under bridges, the eaves of buildings and on old walls. Feeds in nesting area, at all levels, on aerial insects. Engages in spectacular acrobatic displays when breeding.

Morten Strange

Sharad Gaur

Distribution: Abundant breeding resident throughout.
Also occurs in Africa, W and SE Asia, and S China.

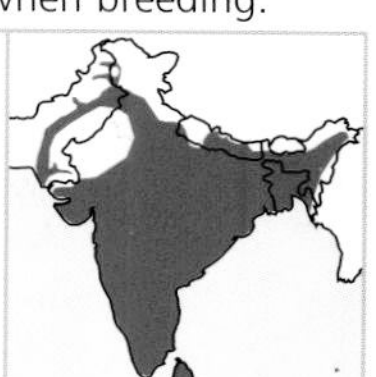

CRESTED TREESWIFT (Crested Swift)

Hemiprocne coronata 23 cm F: Hemiprocnidae

Description: A slender, medium-sized, very long-tailed, grey swift with a pronounced back-swept, head tuft. Basically bluish-grey with darker wings and tail and whitish underparts. The male has pale chestnut cheeks and the juvenile is very scaly looking. Forked tail very long and usually held in a point so it looks like a large **Asian Palm Swift** in flight. Supremely elegant.

Voice: A hard *whik whik* call in flight.

Habits: Inhabits forest and open woodlands, usually on hills. Easy, relaxed flight in small loose groups over trees in search of aerial insects. Frequently perches on high, bare boughs. Makes tiny saucer-shaped nest which it affixes to a bare branch.

Otto Pfister

Distribution: Scarce resident of the peninsula and Sri Lanka north to the Himalayan foothills. Subject to local movements.
Also occurs in SE Asia.

BARN OWL

Tyto alba 35 cm F: Tytonidae

Thakur Dalip Singh

Description: A medium-sized, golden and white owl, appearing ghostly at night. Long, heart-shaped face with small black eyes. Upperparts golden-brown with black and white speckles. Whiter below and particularly on underwings. Wavering low flight on long wings with long feathered legs often dangling and tail rather short. Can look completely white in artificial light.

Voice: Vocal. An extended and wild shriek, plus hissing and snoring noises near nest.

Habits: Commensual with man but nevertheless often overlooked. Frequents urban areas and cultivation, nesting in building or tree hole. Nocturnal and crepuscular, feeding mainly on small rodents.

Distribution: Scarce and local breeding resident throughout lowlands. Worldwide distribution throughout Europe, Asia, Australasia, Africa and the Americas, including many oceanic islands.

PALLID SCOPS OWL (Striated Scops Owl)

Otus brucei 21 cm F: Strigidae

Otto Pfister

Description: A small, secretive, cryptically plumaged fawn-grey owl with small ear tufts. Plainer and greyer than the similar but more widespread **Oriental Scops** and **Mountain Scops** *O. spilocephalus* and with distinct underpart streaking and indistinct scapular lines. Narrow but well-defined black border to pale facial discs. Yellow irises. Ear tufts often depressed. Sexes alike.

Voice: A resonant, extended *hoop hoop hoop*. **Mountain** calls a liquid *toop toop*.

Habits: Inhabits dry mountain ranges. Highly nocturnal, roosting by day in foliage, tree hole or rock crevice. Feeds on insects.

Distribution: Little known resident of northern Pakistan mountains east to Ladakh. Some winter records from northwest India.
Also occurs in W Asia and Socotra Island.

ORIENTAL SCOPS OWL (Little Scops Owl)

Otus sunia 19 cm F: Strigidae

Description: A tiny, eared owl with yellow irises. Cryptic, mottled plumage in three morphs, rufous (shown), brown and grey. Latter two not readily separable from rarer **Eurasian** *O. scops* and call only reliable id factor. Well-streaked crown and underparts distinguish from **Mountain** and **Pallid**. Prominent white scapular lines. Large vertical ear tufts are however often suppressed.

Tim Loseby

Voice: Spaced *uk kuk kruk*. **Eurasian** calls a plaintive *peo*.

Habits: Inhabits forests, groves and orchards. Strictly nocturnal, roosting in day close to trunk in thickly foliaged tree. Stretches against trunk if disturbed. Feeds mainly on invertebrates caught in flight or by pouncing from perch. Nests in hole.

Distribution: Locally common breeding resident in plains and lower hills mainly in north from northern Pakistan east to the Myanmar border, Bangladesh, Sri Lanka, the Western Ghats and southern peninsular India. Scattered records elsewhere and probably under-recorded.
Also occurs in SE Asia.

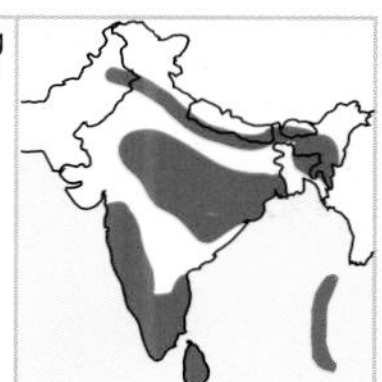

COLLARED SCOPS OWL

Otus bakkamoena 23 cm F: Strigidae

Description: A stocky, eared owl, either brownish or greyish (both illustrated). Buff collar, finely streaked underparts and dark irises. Indistinct scapular lines on cryptically patterned upperparts which may have a rufous, brown or grey cast. Sexes alike.

Voice: Irregular but persistent and questioning *wut wut?*.

Habits: Inhabits well-wooded country including gardens and cultivation. Nocturnal, roosting on well-vegetated branches and in tree holes. With care can be found in favoured roosting sites and then quite tolerant of close, daytime observation. Feeds on insects, reptiles, small birds and rodents.

Otto Pfister

Distribution: Widespread and common breeding resident throughout lowlands. Also occurs in W Asia, Myanmar and Thailand.

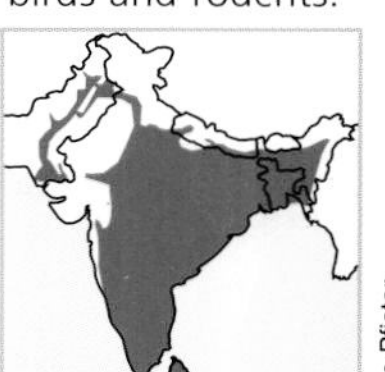

Otto Pfister

EURASIAN EAGLE OWL (Eagle, Great Horned)

Bubo bubo hemachalana 60 cm F: Strigidae

Otto Pfister

Otto Pfister

Description: A huge, streaked, yellowish-brown long-eared owl with orange irises. Upperparts dark yellowishbrown, with dark streaks and chevrons and barred wings. Underparts dark streaked on upper breast, faintly barred below. Dark-rimmed, pale face with fierce expression enhanced by vertical, cat-like ear tufts. Long, very broad wings with dark U-shaped carpal mark. Sexes alike.

Voice: Very deep, booming *hu hooo*.

Habits: Inhabits mountainous and desert areas. Mainly nocturnal, roosting by day in cliffs and old ruins. Sometimes perches in open, where it has a good view. Feeds on rodents, reptiles and birds. Nests in hole in cliff or building.

Distribution: Scarce breeding resident in northern mountains from Pakistan east to Arunachal Pradesh. Also the deserts of western Pakistan.
Also occurs in Europe, W and N Asia and China,

ROCK EAGLE OWL (Bengal Eagle Owl)

Bubo (bubo) bengalensis 56 cm F: Strigidae

Otto Pfister

Otto Pfister

Description: Huge streaked, dark brown, long-eared owl with orange irises. Could be conspecific with **Eurasian** but richer brown, very heavily streaked above and with underpart streaking extending to belly. Voice and habitats differ. Long, very broad wings with U-shaped carpal marks.

Voice: Higher pitched than **Eurasian**, *hu hu hu*.

Habits: Inhabits well-wooded areas, particularly with ravines, large canals and other wetlands, forts, ruins and rocky, bushy hill-sides. Nocturnal. Roosts in large trees, rocks and on ground under cover. Often perches in open at dusk and dawn. If flushed, flies long distances in open. Feeds on rodents, birds and reptiles.

Distribution: Scarce endemic breeding resident throughout lowlands.

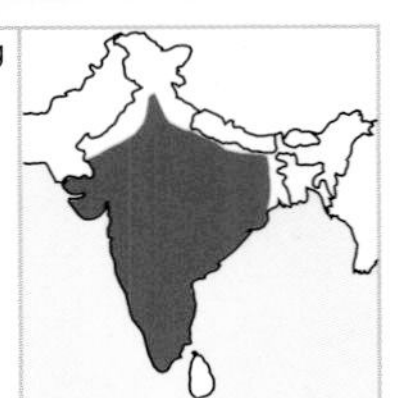

DUSKY EAGLE OWL (Dusky Horned Owl)

Bubo coromandus 58 cm F: Strigidae

Description: Huge, grey, long-eared owl with yellow irises. Finely streaked and barred above and pale with fine streaking below and strongly barred wings. Dark-rimmed, pale facial discs and thick vertical ear tufts, often depressed. Sexes alike.

Otto Pfister

Voice: A long, booming but fading *woo woo woo oo oo oo,* often started in the afternoon.

Habits: Inhabits well-wooded and watered areas, often near habitation. Mainly nocturnal but more active on cloudy days than most night owls. Will perch in small groups on exposed bare branches before dusk and after dawn. Roosts in foliage and feeds on mammals, reptiles and birds. Usually nests in old raptor's nest.

Distribution: Scarce breeding resident throughout lowlands.
Also occurs in Myanmar and Malaysia.

Otto Pfister

BROWN FISH OWL

Ketupa zeylonensis 56 cm F: Strigidae

Description: Huge, flat-headed, streaked brown owl with horizontally held ear tufts and yellow irises. Dark brown above with buff feather edges. Dark streaking on paler underparts. Yellow irises. Very short tail and broad barred wings. Scarcer **Tawny Fish Owl** *K. flavipes* is larger, richer brown with heavier streaking below and whitish forehead. Sexes alike.

Otto Pfister

Voice: A deep *whom, whom*. Also an undulating *cur-ree*.

Habits: Inhabits well-wooded areas with water, including village groves near fish-rich ponds. Mainly nocturnal, roosting in leafy trees. Hunts fish, crabs and frogs either by pouncing from perch or flying low over surface. Sits for hours waiting for prey. Nests in large, deserted bird's nest.

Distribution: Scarce breeding resident throughout lowlands.
Also occurs in W and SE Asia and S China.

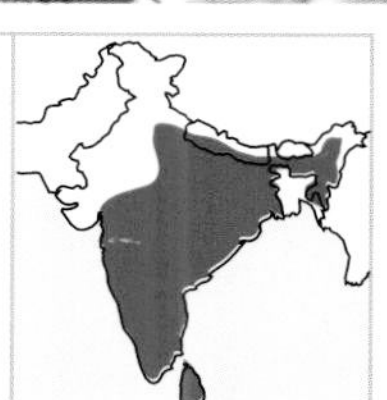

MOTTLED WOOD OWL

Strix ocellata 48 cm F: Strigidae

Nanda Rana

Description: A medium-sized, dark owl with dark-rimmed, pale, barred facial discs around dark eyes. Can look like a very dark **Barn Owl.** Brown above and quite reddish-brown, barred underparts. White partial collar above breast. High-browed, somewhat startled expression at rest. Fairly long tail. Sexes alike.

Voice: Quavering *whooaa*.

Habits: Scarce and local resident in much of the lowlands in wooded country in cultivation and near villages. Sometimes hunts round lights. Nocturnal, roosting in thickly foliaged trees. Feeds on large insects, reptiles, mammals and birds. Difficult to observe so locating the distinctive call is helpful. Usually singly or in pairs. Nests in tree hole.

Distribution: Rare endemic breeding resident through most of peninsula and Gangetic plains, except in east.

TAWNY OWL (Tawny Wood Owl)

Strix aluco 46 cm F: Strigidae

Otto Pfister

Description: Medium-sized, stocky owl with large, rounded head. Ranges from greyish to rufous-brown. Dark streaking on upper and underparts and well barred wings and tail. Whitish scapular lines and V mark on forehead. Large, pale, dark-rimmed facial disc and dark eyes create benign expression. Sexes alike.

Voice: Wavering *tu who oo* and a sharp *ker wik*. Sometimes combined to sound like *tu wit to woo*.

Habits: Inhabits mountain broadleaf and coniferous forests. Nocturnal, roosting high and close to tree trunk when its discovery by diurnal birds offers best chance of locating. Hunts mammals and birds mainly from perch. Nests in tree hole.

Distribution: Fairly common breeding resident of northern and northeastern mountains.
Also occurs in Europe, W, N and Central Asia and China.

COLLARED OWLET (Collared Pygmy Owlet)

Glaucidium brodiei 17 cm F: Strigidae

Description: Tiny, short-tailed, dark, barred owlet with distinctive, cobra-like "face" marks on hind neck. Heavily barred grey or rufous-brown with white throat, centre breast and belly. No contrasting rufous in wings. Streaked flanks and spotted crown. Size, "face" pattern on hind neck (often shown by perched bird) and whiter belly best distinction from larger **Jungle**. Sexes alike.

Voice: Noisy when breeding. Liquid *poop po poop poop*.

Habits: Inhabits thick forest at lower levels. Feeds on insects, birds and other animals, hunting from perch near track or clearing, by day as well as night. Often mobbed by diurnal birds but rarely flies far. Nests in tree hole.

Morten Strange

Distribution: Locally common breeding resident in northern mountains from north Pakistan east to Myanmar border. Moves lower down in winter.
Also occurs in China and SE Asia.

ASIAN BARRED OWLET (Barred Owlet)

Glaucidium cuculoides 23 cm F: Strigidae

Description: Small, heavily barred owl with relatively long tail. Larger than **Jungle** and **Collared**. Heavily barred brown on buff upperparts with no contrasting rufous in wings. Heavily and broadly barred on breast and flanks with streaking on lower flanks. Usually shows white on lower throat and central underparts. Rounded head with yellow irises. Surprised, rather than fierce, look. Yellow feet. Sexes alike. Eastern race more rufous.

Voice: Bubbling *wowowo-wowowo* which gets deeper and louder. Also abrupt *kao kuk*.

Habits: Inhabits hill forest often perching conspicuously and hunting in day. Singly or in pairs. Eats insects, birds and rodents. Waves tail. Undulating flight.

Otto Pfister

Distribution: Locally common breeding resident in northern hills from north Pakistan east to Myanmar border and parts of Bangladesh.
Also occurs in China and SE Asia.

JUNGLE OWLET (Barred Owlet)

Glaucidium radiatum 20 cm F: Strigidae

Otto Pfister

Description: A small, round-headed owl with yellow irises. Dark brown, heavily barred rufous and white above with contrasting rufous and black wing barring. Below barred rufous and white with white moustache, front collar, central breast patch and vent. Sexes alike.

Voice: Noisy. An accelerating, musical *kuo kak kuo kak* which ends abruptly.

Habits: Inhabits forests and woodland, favouring teak and sal forests. Nocturnal but not strictly so, sometimes hunting after dawn or on cloudy days. Roosts in foliage but readily flies if discovered, showing very short tail and rounded wings. Feeds on insects, rodents, reptiles and small birds. Nests in tree hole.

Distribution: Common endemic breeding resident in most of lowlands except the Northwest and Pakistan.

LITTLE OWL

Athene noctua 23 cm F: Strigidae

Otto Pfister

Otto Pfister

Description: Small, flat-crowned white-spotted brown owl. Two subspecies (both illustrated). In the west, east to Ladakh is smaller, paler **Ludlow's Owl** *A. n. ludlowi*. From Ladakh eastwards is the darker **Hutton's Owl** *A.n.bactriana*. Both are brown spotted white, the spots forming lines on the crown. Underparts streaked. Wings and tail broadly barred. White collar and eyebrows, pale facial disc and yellow irises give fierce expression. No ear tufts. Sexes alike.

Voice: A repeated *kee u* and a low bark.

Habits: Inhabits rocky desert areas with cliffs and ruins. Often diurnal, hunting small mammals and birds. Undulating flight. Bobs and stretches.

Distribution: Scarce and very local breeding resident in northwestern mountains east to Nepal.
Also occurs in Europe, N and E Africa, W and N Asia and China.

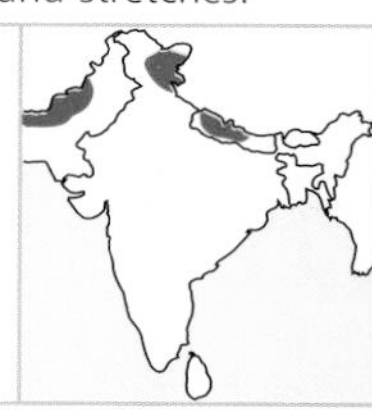

SPOTTED OWLET

Athene brama 21 cm F: Strigidae

Description: Small, greyish brown owl with yellow irises. White spotting on brown crown and white markings on greyish-brown upperparts. Wings and tail narrowly barred. Underparts white-barred pale brown. White eyebrows and rounded crown give startled expression. Broken white collar. Sexes alike.

Voice: Harsh, repeated *churew* often developing into cackling. Calls by day as well as night.

Habits: Commonest owl, inhabiting open woods, cultivation and near habitation. Often diurnal, waking to bob and stare before flying short distance. Not shy. Family parties stay close together. Feeds on reptiles, insects, small birds and mammals. Nests in hole in tree, wall or cliff.

Otto Pfister

Distribution: Common breeding resident throughout the lowlands. Absent from Sri Lanka. Also occurs in Iran and SE Asia.

FOREST OWLET (Forest Spotted Owlet)

Athene blewitti 23 cm F: Strigidae

Description: A small, round-headed, dark brown owl with yellow irises. Similar to **Spotted** but much darker brown with whitish spotting with broad light wing and tail barring. Brown crown unspotted and contrasts with rather white face. Underparts white with broad brown breast band and flank barring. Large claws.

Voice: Not well known but *oh hu* and *kweek kweek* calls reported.

Habits: Very little known. Found to be diurnal in open, dry deciduous forest on low hills. May still occur elsewhere in north-central India where all old records come from.

Distribution: Rare endemic breeding rediscovered in extreme northwestern Maharastra in 1997 after no sightings for over a century. Also now known from sites in Madhya Pradesh.

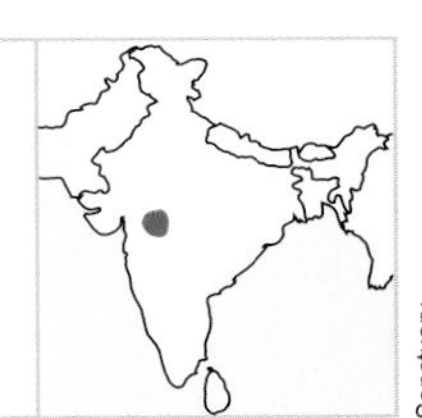

Sanctuary

BROWN HAWK OWL (Brown Boobook)

Ninox scutulata 32 cm F: Strigidae

Kushal Mookherji

Description: Medium-sized, slim, very dark owl with long banded tail. Chocolate brown with some white on back and wings and heavily-streaked underparts. Andaman and Nicobar race all dark. Face mainly dark. No ear tufts. Irises yellow. Sexes alike.

Voice: Extended, liquid *oowup oowup oowup*.

Habits: Inhabits well-wooded country, usually near water. Hunts, singly or in pairs, insects, birds and other animals from favoured perch. Sweeping hawk-like flight with flaps and long glides. Usually crepuscular or nocturnal, roosting in thick canopy. Will also hunt on cloudy days. Nests in tree hole.

Distribution: Locally common breeding resident mainly in lower hills from Uttarranchal east to the Myanmar border, Bangladesh, peninsular India and Sri Lanka.
Also occurs in N, E and SE Asia.

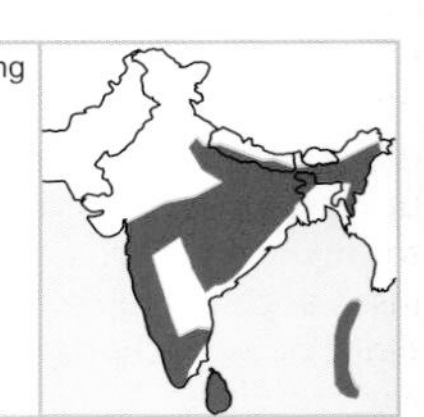

SHORT-EARED OWL

Asio flammeus 38 cm F: Strigidae

Otto Pfister

Nikhil Devasar

Description: Medium-sized, heavily-streaked buff, terrestrial owl with long wings. Upperparts and breast streaked blackish on buff ground colour. Upper primaries yellowish with black tips and white trailing edge. Under-wing whitish with black tips and carpal crescents. Small head with short ear tufts and bold facial disc. Black mask round small yellow irises. Cat-like face. Sexes alike.

Voice: Usually silent.

Habits: Inhabits dry, open country with scattered bushes. Diurnal hunter covering ground buoyantly. Rests in grass clumps or in shade of thorn bush. If flushed often flies high when may be mobbed by raptors or corvids. Will perch on posts.

Distribution: Scarce but widespread winter visitor throughout lowlands.
Also occurs in Europe, N Africa, N Asia and the Americas.

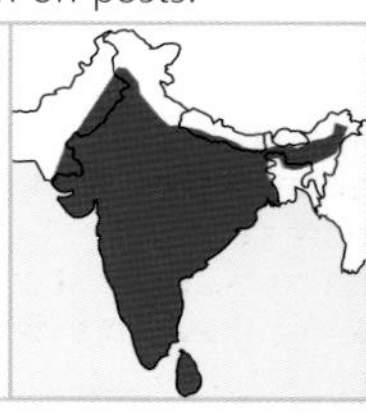

SRI LANKA FROGMOUTH (Ceylon Frogmouth)

Batrachostomus moniliger 24 cm F: Batrachostomidae

Description: A secretive, nocturnal, nightjar-like bird that perches vertically and has a large head and huge gape. Male cryptic greyish-brown mottled white, buff and black with white neck bar. Female plainer orange-rufous with some black and white spotting. Thick, hooked bill almost covered with feathers.

Voice: A repeated, decelerating and chuckling *whuo whuo*.

Habits: Inhabits thick, wet evergreen forests. Nocturnal, roosting in thick undergrowth, often perched vertically on the end of a tree stump. Points bill upwards when disturbed. Loyal to favoured roost. Catches flying insects, particularly moths, in flight or by ground pounces. Nests on branch.

S Kartikeyan

Distribution: Scarce endemic breeding resident of Western Ghats and Sri Lanka, where it is the only frogmouth.

GREY NIGHTJAR (Indian Jungle, Jungle, Highland Nightjar)

Caprimulgus indicus 30 cm F: Caprimulgidae

Description: Large, dark greyish nightjar. Grey with cryptic mottling and barring of black, buff and white. Small, whitish throat patch and scapular line and orange patches on neck sides. Male has small white patches at base of primaries and white subterminal band to tail. Female has no white. Long wings and tail, wide gape and small bill.

Voice: Extended knocking *tuckoo tuckoo* or *tuck tuck* through night. Also a hoarser *kukroo*.

Habits: Inhabits open woodland and wooded rocky areas. Roosts on exposed tree branch or ground. Difficult to spot unless roost site known. Hunts flying insects mainly just after dusk and before dawn in twisting, wheeling flight. Lays eggs on bare ground.

Otto Pfister

Distribution: Widespread and fairly common breeding resident throughout lowlands. Very local in Pakistan and Bangladesh.
Also occurs in NE and SE Asia and China.

EURASIAN NIGHTJAR (European Nightjar)

Caprimulgus europaeus 25 cm F: Caprimulgidae

Tim Loseby

Description: A medium-sized, long-winged nightjar with only a very small, white throat patch. Black-streaked brownish-grey with prominent buff scapular and covert spots. Male has white spots in wing and tail tips. Female has no white. Larger **Grey** darker and **Savannah** less streaked. Some migrants from Central Asia much paler sandy-grey.

Voice: A rhythmic churring like a distant motor-cycle. Soft *quoit quoit* flight call.

Habits: Inhabits rocky and bushy hillsides. Crepuscular and nocturnal, spending day, sleeping on ground or along branch. Well-camouflaged and difficult to find in day unless inadvertently flushed. Feeds on flying insects. Nests on ground.

Distribution: Locally common breeding summer visitor to hills of western and northern Pakistan. Also scarce passage migrant probably en route to Eastern Africa. Mainly in Indus Valley and western Gujarat. Very rare elsewhere. Also occurs in Europe, Africa, W, Central and N Asia.

LARGE-TAILED NIGHTJAR (Long-tailed Nightjar)

Caprimulgus macrurus 30 cm F: Caprimulgidae

Toby Sinclair

Description: Large, brightly-hued and broad-tailed nightjar. Richly but cryptically marked with buff, and black on brown. Bright buff-edged black scapular lines and white moustache and breast patch noticeable. Male has broad white (female buff) apical tips to tail and primary patches. Long wings and noticeably long, broad tail. Wide gape and small bill.

Voice: A resounding, rather slow *chunk chunk chunk*.

Habits: Inhabits lowland forests and woodland. Roosts on ground well-camouflaged by dead leaves. Hunts flying insects at night with wheeling flight, often resting on tracks. Nests on ground.

Distribution: Fairly common resident along Gangetic Plain and throughout northeast south to Andhra Pradesh. Similar but smaller Jerdon's Nightjar *C. atripennis* replaces it in the south. Also occurs in SE Asia. Australia and South Pacific Islands.

INDIAN NIGHTJAR (Common Indian, Little)

Caprimulgus asiaticus 24 cm F: Caprimulgidae

Description: A small, rather bright nightjar with a rufous collar. Similar to **Large-tailed** though scapular line less striking and collar obvious. Has similar white breast patch and moustache. Both sexes have white apical tips to tail and white primary patches. Slightly shorter winged and tailed than other nightjars. Wide gape and small bill.

Voice: A loud steady *chuck chuck chuck kerrrr* accelerating as it fades away.

Habits: Inhabits dry, open scrub country with few trees but including fallow and sparse plantations. Habits as other nightjars but roosts exclusively on ground. Often rests on tracks after dark. Lays eggs on bare ground.

Otto Pfister

Distribution: Fairly common breeding resident throughout lowlands and Sri Lanka. Very local in Pakistan and Bangladesh.
Also occurs in SE Asia, Madagascar and Aldabra Island.

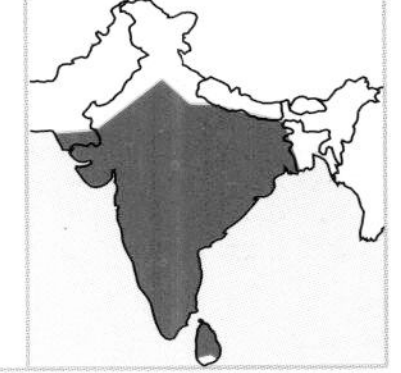

SAVANNAH NIGHTJAR (Franklin's, Allied Nightjar)

Caprimulgus affinis 23 cm F: Caprimulgidae

Description: A small, rather mottled, grey-brown nightjar. Mottled buff, grey and brown with no dark markings. Striking buff V-marking from shoulders to lower back. Small, white throat patch. Male has broad white outer tail and large white primary patches. Female, buff wing patches. Long wings and tail. Wide gape and small bill.

Voice: A distinctive, parakeet-like screeching *shreek*, abrupt and repeated, usually in flight.

Habits: Inhabits open forest particularly with rocky areas and dry stream beds. Habits as other nightjars and roosts on ground. In parts of Indonesia has adapted to urban areas, utilising flat roof tops and building plots.

Distribution: Locally common breeding resident throughout lowlands but local in Bangladesh and absent from Sri Lanka.
Also occurs in Indonesia and the Philippines.

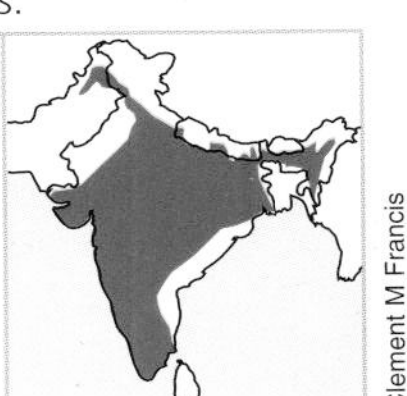

Otto Pfister

Clement M Francis

ROCK PIGEON (Rock Dove, Feral, Blue Rock Pigeon)

Columba livia 33 cm F: Columbidae

Otto Pfister

Description: Medium-sized, blue-grey pigeon with darker head and neck and black wing bars but it interbreeds with domestic pigeons so various colour varieties possible. Basically blue-grey with green-purple sheen across neck. Wings darker and tail black sub-terminal band. Some races have white lower back. Always has two long black bars across wing coverts. Small head, stocky body and fast direct flight. Sexes alike.

Voice: A deep *tru troo tru*.

Habits: Inhabits urban areas and cultivation. Gregarious, mixing with other doves. Feeds in flocks on spilt grain and sprouting cereals. Also in remote rocky places with cliffs and ruins. Nests colonially in cavities and on ledges in buildings and cliffs.

Distribution: Abundant feral and wild breeding resident throughout. Also occurs in Europe, Africa, W and E Asia, often with domestic pigeons living ferally.

HILL PIGEON

Columba rupestris 33 cm F: Columbidae

Otto Pfister

Description: A medium-sized, pale grey pigeon with a dark tail with a striking, broad, white sub-terminal band. Very similar to, but paler than, **Rock Pigeon**. Usually shows white lower back and shorter black wing bars. Tail pattern is diagnostic. Sexes alike.

Voice: Rather high-pitched *guk guk guk*.

Habits: Inhabits high rocky country and village cultivation, frequently confiding and mixing with **Rock Pigeons**. Usually in flocks. Nests in crevice in cliff or building.

RK Gaur

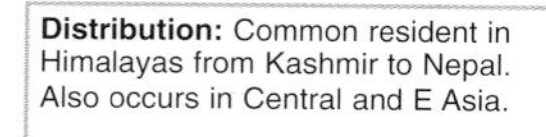

Distribution: Common resident in Himalayas from Kashmir to Nepal. Also occurs in Central and E Asia.

SNOW PIGEON

Columba leuconota 34 cm F: Columbidae

Description: A medium-sized, very distinctive, pale pigeon with a dark grey head. White collar, underparts and rump, pale brownish back and grey wings with three dark wing bars. Dark grey tail with broad white chevron forming a very obvious bar, particularly in flight. Sexes alike.

Voice: A high *coo coo* and various croaks.

Habits: Inhabits cliffs, gorges, steep slopes and cultivation. Forages in flocks on seeds, including cereal gleanings, and bulbs. Often feeds round snow meltwater. Strong wheeling flight. Nests colonially on cliffs.

Otto Pfister

Distribution: Locally common breeding resident of high Himalayas. Also occurs in Tibet and China.

COMMON WOOD PIGEON (Wood Pigeon)

Columba palumbus 43 cm F: Columbidae

Description: A large, stocky, grey dove with white wing flashes. Small head, long tail and plump body. Grey above with a broadly black-tipped tail. Grey head with green and whitish neck marks. Deep vinous-pink breast. Small red and yellow bill and pink legs. Pale irises. Sexes alike.

Voice: A deep crooning *dooh doo daw daw doo*. Wings clatter on take off.

Habits: Inhabits hill scrub and woodland. Also cultivation. Often in flocks. Feeds mainly on seeds, grain, acorns and buds. Climbs about in foliage. On ground crouches low when feeding. Wary especially where hunted. Nests in tree.

Jan Willem den Besten

Distribution: Local breeding resident in northern hills of Pakistan and Kashmir. Erratic but sometimes numerous winter visitor further south in hills of northwest India east to Nepal. Also occurs in Europe, Russia, N Africa and Iran.

Jan Willem den Besten

NILGIRI WOOD PIGEON

Columba elphinstonii 42 cm F: Columbidae

Otto Pfister

Description: A large, dark pigeon with diagnostic, chessboard patch on rear neck. Rather small head and bulky body. Head and underparts pale grey with white throat. Upperparts dark purplish-brown, glossed green. Tail plain dark grey. Bill red with pale tip. Yellow irises, red eye rings. Sexes alike.

Voice: Langur-like *who* followed by a deep, eerie *who whu who* of three to five notes.

Habits: Inhabits wet, evergreen and shola forests and plantations. Arboreal, eating mainly fruit and fresh buds. Will descend to ground for fallen fruit. Solitary or in small parties. Flies strongly and skilfully through trees. Sits quietly for long periods. Makes flimsy stick nest in tree branches.

Distribution: Scarce endemic breeding resident of Western Ghats

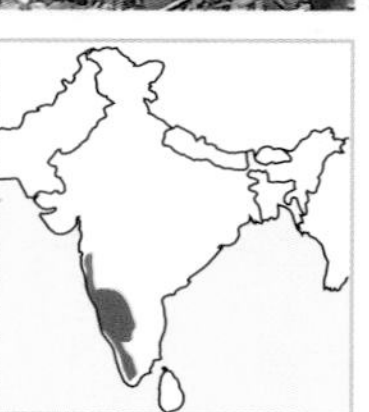

ORIENTAL TURTLE DOVE (Rufous, Eastern Rufous)

Stretopelia orientalis 33 cm F: Columbidae

Otto Pfister

Nikhil Devasar

Description: A medium-sized, stocky, pink, rufous and black dove. Head and underparts dull pink with grey forehead and blue and black-barred patch on neck. Upperparts look scaly being black broadly edged rufous. Grey on wing edges and rump. Long, fanned tail blackish with broad greyish-white feather tips. Can look rather dusky. Sexes alike.

Voice: Deep, grating *ghur ghroo goo*.

Habits: Inhabits open forest and cultivation. Mainly singly or in small groups. Rather shy although will perch on bare branches and wires. Feeds on ground on seeds including cereal gleanings. Nests in low tree branches.

Distribution: Locally common resident of Himalayas, the Northeast and east-central India. Widespread throughout in winter.
Also occurs in Myanmar and E Asia.

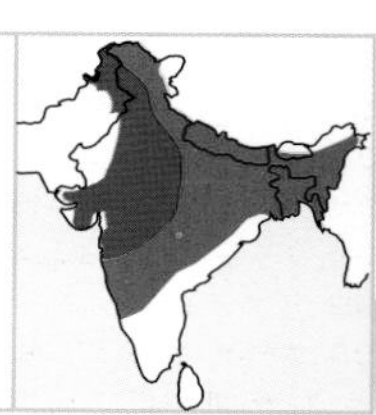

LAUGHING DOVE (Little Brown, Palm Dove)

Streptopelia senegalensis 27 cm F: Columbidae

Description: A small, slender, long-tailed, brown and grey dove. Pinkish-brown head and underparts with extensive black speckling on upper breast. Upperparts fawn-brown with extensive grey wing patches and grey tail with black basal and apical patches. Sexes alike.

Voice: A hurried, soft *cru do do do do*.

Habits: Inhabits dry, rocky and bushy country, cultivation, villages, parks, gardens and urban areas. Very confiding although unobtrusive. Feeds on seeds on ground usually in pairs, though large flocks congregate at harvest or round grain silos. Rather territorial when nesting. Slight twig nest usually located low in creeper or bush.

Otto Pfister

Distribution: Abundant breeding resident throughout lowlands. Absent from Sri Lanka, Bangladesh and most of the Northeast.
Also occurs in Africa, W and Central Asia.

SPOTTED DOVE

Streptopelia chinensis 30 cm F: Columbidae

Description: Medium-sized, pink and spotted brown dove with large white-spotted black, hind neck patch. Deep pink head (sometimes grey on fore crown) and underparts, light brown upperparts with numerous small buff spots and some grey in wings and rump. Tail dark with broad white borders. Sexes alike.

Voice: A rather doleful *cro cro cro* or *cruk cru croo*.

Habits: Inhabits open deciduous forest, cultivation with groves, villages, parks and gardens. Quite confiding, flying noisily a short distance when disturbed. Often found eating grit on roads and tracks. Drinks frequently. Feeds, usually in pairs or small groups, on seeds on ground. Nests low in tree or bush.

Otto Pfister

Distribution: Common resident in much of lowlands but very local in the Northwest and Pakistan.
Also occurs in China and SE Asia.

RED COLLARED DOVE (Red Turtle Dove)

Streptopelia tranquebarica 23 cm F: Columbidae

Otto Pfister

Description: A small, stocky, deep pink dove in which, unique among *Streptopelia*, the sexes differ. Both shown. Male a striking wine pink above with contrasting bluish-grey head and small black hind collar. Rump grey and tail black with broad white tips. Underparts paler pink. Female very like a small, stocky **Collared Dove** but is darker fawn. Rather short wings and tail.

Voice: Rolling *gru gur goo*.

Habits: Inhabits open woodland and scrub to breed spreading into cultivation and villages afterwards. Usually in pairs or small parties. Joins large dove flocks at harvest and in cold weather. Feeds on seeds on ground. Often perches on bare branches and wires. Nests high on tree branch.

Distribution: Locally common breeding resident throughout lowlands except extreme south and Sri Lanka. Some local movements for breeding. Also occurs in Tibet and SE Asia.

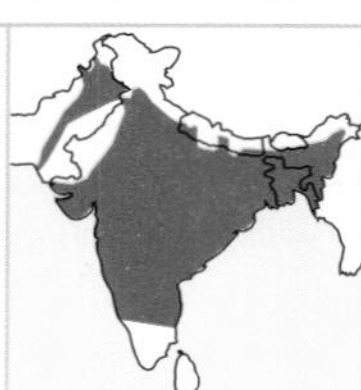

EURASIAN COLLARED DOVE (Indian Ring, Collared)

Streptopelia decaocto 32 cm F: Columbidae

Otto Pfister

Description: Medium-sized, fawn brown dove with black hind collar. Greyish brown, paler below with grey wing coverts bordering blackish flight feathers. Tail dusky with broad white tips. Underneath looks blacker on base. Beady black eyes. Sexes alike.

Voice: Soothing, repetitive *doo dooo doo*. Also a harsh *kreer* on landing and in display.

Habits: Inhabits open woodland and scrub, cultivation, villages and urban areas. Avoids thick forest. Feeds on seeds often in flocks on ground, usually forming the bulk of harvest and winter dove flocks. Frequently perches on wires. Nests low down in tree, bush or building. Territorial and defensive of nest, driving off corvids and even **Shikra**.

Distribution: Abundant breeding resident throughout lowlands, including deserts. Also occurs throughout Europe, W, N and E Asia. Spread west rapidly in 20th century.

EMERALD DOVE (Green-winged Pigeon, Green-winged)

Chalcophaps indica 26 cm F: Columbidae

Description: A medium-sized, very stocky, dark dove with waxy red bill. Both sexes are basically deep vinaceous-pink with glossy dark green upperparts. Two white bars on black rump and all blackish tail and flight feathers. Male has pale grey crown and whitish shoulder patches.

Voice: A deep, mournful and repetitive *hoo oon*.

Habits: Secretive inhabitant of evergreen broadleafed and wet deciduous forest. Keeps to ground under forest trees, often in semi-darkness. Will feed on paths and in clearings. Usually singly or in pairs and rarely seen before flushed. Rapid, low, skilful flight through trees. Feeds on seeds and fruits on ground. Nests low in forest undergrowth.

Sumit Sen

Distribution: Scarce breeding resident throughout much of region except the Northwest and Pakistan.
Also occurs in SE Asia, Australia and South Pacific.

NICOBAR PIGEON

Caloenas nicobarica 41 cm F: Columbidae

Description: Unmistakable. A large, portly, slate blue-grey and green pigeon, the male with neck and back hackles. Fat body, small head on long neck and short white tail. Most of body and long neck hackles slate blue-grey. Glossy green and bronze on back and wing coverts with shorter hackles. Juvenile has no hackles and black tail. Dark eyes, bill and feet.

Voice: Usually silent but will occasionally croak.

Habits: Inhabits undisturbed wet evergreen forests. Often flies between islands. Feeds on fruit and seeds on forest floor. Flies noisily up into treetops when disturbed. Nests in forest trees, sometimes colonially.

Morten Strange

Distribution: Rare endemic breeding resident of Nicobar Islands, sometimes visiting Andamans.

POMPADOUR GREEN PIGEON (Grey-fronted Green)

Treron pompadora 28 cm F: Columbidae

Kamal Sahai

Description: A medium-sized, green pigeon with a chestnut-maroon back (in male) and grey crown. Female has green back but both sexes have yellow and black wing edgings and grey terminal band on tail. Male has maroon undertail coverts and orangey wash on breast. Pale grey bill and red-rimmed pale irises. Feet reddish.

Voice: High-pitched wavering whistles, *wooweeyooweeyoo*.

Habits: Inhabits moist forests and plantations. Feeds inconspicuously on fruit in treetops, sometimes in large flocks. Coloration can make them very difficult to spot. Fast, direct flight. Nests on tree branch.

Distribution: Locally common breeding resident of Sri Lanka, Western Ghats, Nepal and the Northeast.
Also occurs in SE Asia.

THICK-BILLED GREEN PIGEON (Thick-billed Pigeon)

Treron curvirostra 27 cm F: Columbidae

Morten Strange

Description: A medium-sized. green pigeon with a thick red, based bill and bluish eye patch. Similar plumages to **Pompadour** but without orange wash on breast of male and with paler brown and barred undertail coverts. Dark eyes. Reddish feet.

Voice: Musical whistles and a throaty *kloo kloo*.

Habits: Inhabits forests and groves. Feeds on fruit often hidden in foliage and sometimes in flocks. Fast, direct flight. Nests on tree branch.

Distribution: Scarce resident of the Northeast and parts of Bangladesh and Nepal.
Also occurs in SE Asia.

YELLOW-FOOTED GREEN PIGEON (Bengal Green)

Treron phoenicoptera 33 cm F: Columbidae

Description: Medium-sized, yellowish-green pigeon with diagnostic yellow feet. Green, yellower below and round neck. Grey head, shawl and tail and lilac shoulders. Yellow-edged blackish flight feathers. Barred undertail coverts, yellow-tipped maroon in adults. Undertail has basal half black, rest white. Bill pale greenish. Small head on long neck and bulky body. Sexes alike.

Voice: Conversational, lilting *wheet wa hoo* and *who who*.

Habits: Commonest green pigeon. Inhabits open forest, groves, parks and gardens. Flies far for fruiting trees. Acrobatic in search of food. Frequently in large flocks, conspicuous in fast, direct flight. Will sun themselves on exposed bare branches.

Otto Pfister

Distribution: Locally common breeding resident throughout lowlands. Also occurs in SE Asia.

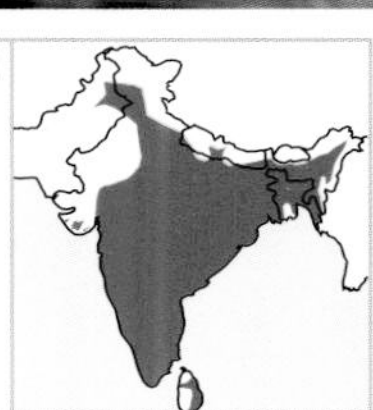

GREEN IMPERIAL PIGEON

Ducula aenea 45 cm F: Columbidae

Description: A very large, green-backed pigeon with pale grey head and underparts. Metallic sheen to green wings and back and plain green tail. Undertail coverts chestnut. Eyes dark. Pinkish feet. Sexes alike.

Voice: A loud *kurr hoo* with only the second syllable audible at a distance.

Habits: Inhabits all types of forest. Usually solitary or in pairs feeding on fruit in treetops. Sometimes small groups gather, often mixing with other species. Flies high above canopy and very powerfully. Nests in small tree.

Morten Strange

Distribution: Fairly common breeding resident of Western and Eastern Ghats, east and northeast India and Sri Lanka. Also occurs in SE Asia.

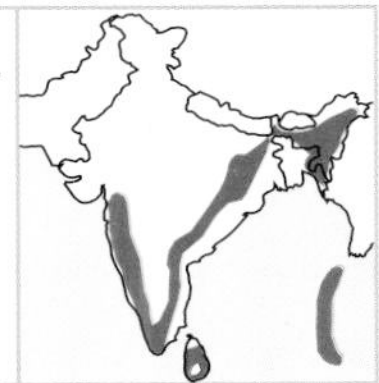

MOUNTAIN IMPERIAL PIGEON (Imperial Pigeon)

Ducula badia 45 cm F: Columbidae

Clement M Francis

Description: A very large, brown-backed pigeon with a dark-based tail. Upperparts dark brown, head and underparts pale pinkish-grey with whiter throat and under tail coverts. Red-based bill and red eye rings emphasising pale irises. Pinkish feet. Sexes alike.

Voice: A booming *oomp, oomp*.

Habits: Inhabits closed forests. Behaves as **Green Imperial Pigeon**, feeding high in fruiting trees and flies powerfully above canopy. Nests in small tree.

Distribution: Fairly common breeding resident of Western Ghats, the Northeast and parts of Nepal. Also occurs in SE Asia.

PIED IMPERIAL PIGEON

Ducula bicolor 41 cm F: Columbidae

Morten Strange

Description: A very large, white pigeon with black wings and tail. White has creamy look. Eyes dark and bill and feet bluish. Sexes alike.

Voice: A deep *kru kroo* and a chuckling *hu hu hu*.

Habits: Inhabits forests, particularly mangroves. Feeds high in fruiting trees in flocks, moving from island to island with high powerful flight in search of the best food sources. Nests in trees.

Distribution: Common breeding resident on Nicobar Islands. Visits the Andamans rarely. Also occurs on small Indonesian islands.

INDIAN BUSTARD (Great Indian Bustard)

Ardeotis nigriceps Male 120 cm Female 92 cm F: Otididae

Description: Huge, stocky terrestrial bird with long, thick neck and long, yellow legs. Slow, stately walk. Pale grey neck, black crown, black breast band and belly. Brown upperparts, white-spotted black coverts and grey wings. Female smaller with greyer neck and incomplete breast band. Irises yellow.

Voice: Usually silent but can make a loud, far-carrying *whonk*.

Habits: Inhabits extensive dry grassland with scattered scrub. Shy, running fast to escape intruders. Flies ponderously on broad wings. Feeds on seeds, fruit and termites. Often in small parties, the males gathering to display in traditional leks; swelling throats, drooping wings and cocking tails. Nests on ground.

Joanna Van Gruisen

Distribution: A globally threatened endemic breeding resident now very local in parts of northwest India and the central Deccan.

MACQUEEN'S BUSTARD (Houbara)

Chlamydotis macqueeni 65 cm F: Otididae

Description: A large, slim, sandy bustard with a grey neck and upper breast. Male has small crest and long black neck line. Female is smaller, has much less black on neck and less crest. Also less white in wings. Immature is paler with no black. Head brownish. Underparts white. Upper parts sandy brown vermiculated with black. Extensive white patch in black wings. Iris yellow.

Voice: Nothing recorded.

Habits: Inhabits semi-desert and associated cultivation, particularly mustard fields. Extremely shy, walking rapidly away from danger but also lurking under low cover with neck outstretched. A main quarry for falconers and declining seriously as a result.

Otto Pfister

Distribution: Now mainly a rare winter visitor to semi-desert areas of Pakistan and northwest India.
Also occurs in W and Central Asia.

BENGAL FLORICAN

Houbaropsis bengalensis 66 cm F: Otididae

Joanna Van Gruisen

Description: Large, stocky bustard with black head, neck and underparts. Almost completely white wings. Back warm brown with black feather tips. Larger female and first-year male buff on head, neck and breast and white on belly. Crown is black with buff central stripe. Wings buff. Yellow irises.

Voice: A clicking sound in display and when flushed.

Habits: Inhabits wet grasslands. Most obvious in breeding season when males leap above grassland in exuberant display. Feeds on seeds and plant shoots, locusts and other insects. Ventures into short grass and burnt-over areas at beginning and end of day. Usually solitary but males do gather to lek. Nests on ground.

Distribution: A globally threatened breeding resident now restricted to a few sites in the terai and Assam. Also occurs in Cambodia.

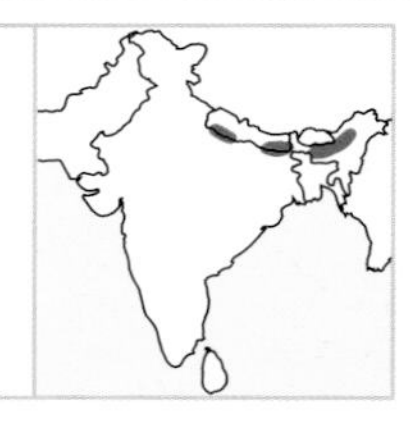

LESSER FLORICAN (Likh)

Sypheotides indica 45 cm F: Otididae

Joanna Van Gruisen

Description: Small, slim bustard with black head spatulas, head, neck and underparts and white shawl and inner wing patch. Black-tipped greyish-brown upperparts. Flight feathers black-edged brown contrasting with white coverts. Larger female and immature buff with black streaking on head, neck and underparts and more rufous in wings.

Voice: A rattling click in display and a whistle when flushed.

Habits: Inhabits dry, scrubby grasslands and tall crops. Fairly tolerant of humans. If flushed, flight surprisingly fast. Males leap into aerial display in breeding season. Usually solitary or in pairs. Feeds on locusts and other insects, seeds and plant shoots. Nests on ground.

Distribution: A globally threatened endemic breeding visitor to a few sites in north central India. Reportedly moves southeast after breeding but rarely recorded there.

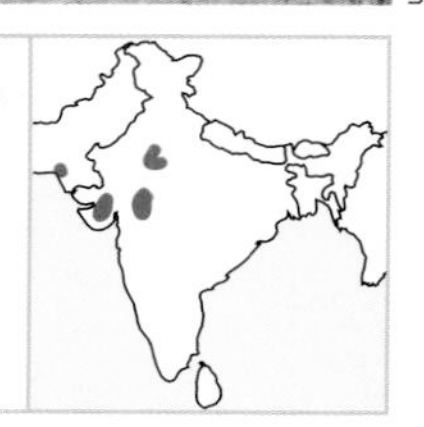

SIBERIAN CRANE

Grus leucogeranus 140 cm F: Gruidae

Oto Pfister

Description: A large, white crane with bare, red face mask. Black primaries hidden by long, curved tertials when on ground. Long bill, reddish-brown and decurved. Long legs dark pink. Juvenile has no mask and has rather scruffy, pale brownish patches on the white plumage. Sexes alike.

Voice: A soft *koonk koonk*.

Habits: Probably India's rarest bird. Beware the use of "Siberian" to describe any striking migrants and "crane" to describe any large long-legged, long-billed and long-necked bird. Also the distant possibility of confusion with **White Stork**. Feeds mainly on grass tubers in shallow jheels, often exploiting areas "snouted-up" by wild boar.

Distribution: Globally threatened winter visitor from western Siberia. Less than a handful of birds arrive each autumn at Keoladeo Ghana National Park in southeastern Rajasthan. Sightings elsewhere should be carefully checked and reported.
Also occurs in Afghanistan, Iran, Siberia, and China.

Bikram Grewal

SARUS CRANE

Grus antigone 165 cm F: Gruidae

Oto Pfister

Description: Huge, pale grey crane with a bare, red head and upper neck. Whiter drooping scapulars and upper breast. Black primaries often hidden. Crown grey. Powerful grey bill. Legs pinkish. Juvenile has brownish cast, rusty head and no red. Sexes alike but female slightly shorter.

Voice: Far-carrying bugling particularly in greeting and display.

Habits: The world's tallest flying bird and important symbol of fidelity to Indians as they pair for life. Often quite approachable. Inhabits marshes, jheels, irrigated cultivation; sometimes breeding in small swamps. In pairs or small flocks. Dancing courtship. Feeds on grain, tubers, large insects, fish and amphibians. Makes large built-up nest on ground.

Distribution: A globally threatened declining breeding resident of northern and central India.
Also occurs in SE Asia and N Australia.

DEMOISELLE CRANE

Grus virgo 95 cm F: Gruidae

Otto Pfister

Otto Pfister

Description: A rather small, grey and black crane with very long tertials and lower neck feathers. Head, foreneck and breast black with white ear tufts and grey crown. Rest of body grey with blackish flight feathers. Legs grey and bill pink and grey. Sexes alike. Immature greyer and lacks the elongated feathers.

Voice: Rather high-pitched *krook krook* in flight. Guttural courtship calls.

Habits: Very sociable. Feeds mainly in cultivation, consuming large quantities of grain and pulses. Also shallow jheels and marshes where they often roost.

Distribution: A locally numerous winter visitor mainly to Gujarat and Rajasthan. Huge concentration in Kitchen village (as shown), where fed by local people. More sporadically almost anywhere in the Northwest and the Deccan.
Also occurs in E Europe, Africa, NE and Central Asia and China.

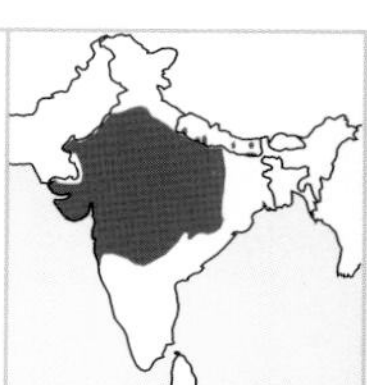

COMMON CRANE

Grus grus 120 cm F: Gruidae

Otto Pfister

Description: A large, grey crane with long, bushy tertials drooping over tail and white striped black neck. The white rear neck extends from the eyes to half way down the neck and the crown is red. Tertial bush has black tips and flight feathers are black. Bill and legs grey. Sexes alike. Immature (also shown) has rusty brown head and neck and brownish cast to the grey upperparts.

Voice: A far-carrying strident *kerr ohk,* often from high in the sky but also when settled.

Habits: Feeds on insects, grain, beans and other crops in cultivation but spending much time on shallow jheels or large rivers, particularly to roost. Occurs in large flocks. Often indulges in bouncing display dance. Wary.

Distribution: A fairly common winter visitor to northern India and parts of Pakistan.
Also occurs in Europe, E and NE Africa, N and Central Asia and China.

BLACK-NECKED CRANE

Grus nigricollis 140 cm F: Gruidae

Description: A large, grey crane with black head and neck and black, drooping tertials. Flight feathers black. Crown red. Legs blackish, bill grey. Sexes alike. Immature rusty brown.

Voice: A loud, rather high-pitched *ker kra krew*.

Habits: Breeds by high altitude lakes and winters on wet fallow, stubble and marshes. Often very tame as revered by Buddhists. Feeds on grain, tubers and large insects. Nests on lake islands.

Otto Pfister

Distribution: A globally endangered endemic breeding resident that breeds in small numbers in the Himalayas and winters mainly in Bhutan.

SLATY-BREASTED RAIL (Blue-breasted Banded Rail)

Gallirallus striatus 27 cm F: Rallidae

Description: A medium-sized, heavily-barred crake with chestnut crown and rear neck and slaty-blue foreneck and breast. Upperparts brown vermiculated with white. Lower breast to vent white vermiculated with black. Fairly long, rather heavy and slightly decurved red bill. Grey legs. Sexes alike though female has unmarked white central belly. Juvenile is streaked brown.

Voice: A harsh, repeated *gelek*.

Habits: Extremely shy and somewhat crepuscular. Inhabits well-vegetated freshwater marshes and adjacent scrub. Feeds on insects and seeds. Nest well-hidden on ground.

Morten Strange

Distribution: Scarce but overlooked resident throughout peninsula, the Northeast, Bangladesh and Sri Lanka. Also occurs in China and SE Asia.

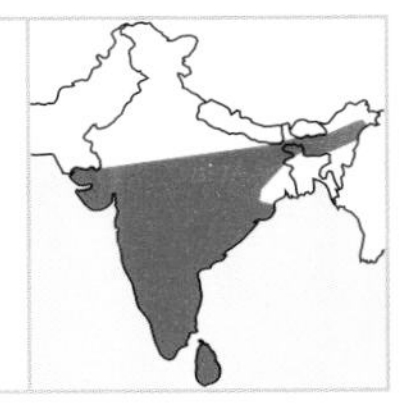

WATER RAIL

Rallus aquaticus 28 cm F: Rallidae

Kamal Sahai

Description: A medium-sized, grey and brown crake with white-barred black flanks. Crown, rear neck and upperparts warm brown heavily streaked with black. Rest of head, neck down to belly blue-grey. Undertail coverts bright buff. Long, slender, decurved red bill and pink legs. Sexes alike. Juvenile dull buff-brown below in place of blue-grey.

Voice: A startled pig-like squeal which fades away. Also a regular *kip kip kip* song.

Habits: Inhabits freshwater reed-beds where difficult to see. Will come out cautiously onto mud and paths to feed on invertebrates and plant matter. Also known to eat carrion. Solitary or in pairs. Runs fast but if flushed flies rapidly with dangling legs.

Distribution: Breeds in Kashmir. Winters mainly in the Indus and north Indian river valleys.
Also occurs in Europe, N Africa, N and W Asia and China.

CORN CRAKE

Crex crex 25 cm F: Rallidae

Sanctuary

Description: A medium-sized, stout-billed crake with white-barred chestnut flanks and a large chestnut wing patch. Head, neck and breast pale grey with brown crown, eyestripe and hind neck. Upperparts warm brown heavily patterned with black. Bill and long legs pinkish. Sexes alike.

Voice: Unlikely to be heard in our region but it is a rasping *crex crex* like rubbing a comb.

Habits: Inhabits lush, often damp grasslands. Somewhat crepuscular and very secretive except when calling. Usually creeps away if disturbed but, if flushed, rather laboured flight with dangling legs.

Distribution: Globally threatened and very rare vagrant to Kashmir but secretive and probably overlooked. Also occurs in Europe, Africa and Central Asia.

BROWN CRAKE

Amaurornis akool 28 cm F: Rallidae

Toby Sinclair

Description: A fairly large, plain crake with dark pink legs and short, greenish bill. Overall dark olive-brown, paler on undertail coverts and grey from face to belly. Throat whitish. Sexes alike. Beware juvenile **White-breasted Waterhen** which has rufous undertail coverts and juvenile **Moorhen** which is warmer brown and has white undertail coverts and, usually, flank stripes.

Voice: Rather quiet. A vibrating trill, reminiscent of **Little Grebe**.

Habits: Crepuscular, secretive and solitary or in pairs. Inhabits freshwater marshes and water courses with luxuriant vegetation. Favours narrow flowing streams, even when polluted. Feeds on invertebrates and seeds on mud margins constantly flicking tail.

Distribution: Common breeding resident and local migrant mainly in northern and central India. Also occurs in China and SE Asia.

WHITE-BREASTED WATERHEN

Amaurornis phoenicurus 32 cm F: Rallidae

Otto Pfister

Description: Large, long-necked grey and white crake. Dark slaty-grey from rear crown to tail, including flanks. Forecrown to belly white. Lower belly to vent rufous orange. Long yellowish legs. Bill yellowish-green with red basal patch. Sexes alike. Juvenile brown above and grey below.

Voice: Noisy when breeding. Grunts and chuckles. Song a repeated *krrr kuk kuk*.

Habits: Commonest crake. Inhabits marshes, village ponds, streams and thick vegetation. Can be secretive but often confiding. Striding walk often jerking tail. Flies with dangling legs. Frequently climbs up into trees and bushes. Solitary or in small groups. Feeds on invertebrates and plants. Nests on ground.

Distribution: Common breeding resident throughout region. Also occurs in SE Asia.

LITTLE CRAKE

Porzana parva 20 cm F: Rallidae

Description: Very small, short-billed and long-winged crake. Male deep grey on head, neck and underparts with white barring from thighs back to black and white-barred undertail coverts. Upperparts brown with blackish lines and sparse white spots. Female has whitish underparts, fawn on flanks and belly. Green legs and red-based green bill. Immature similar to female but whiter on head, neck and breast. Faintly red bill base.

Voice: Unlikely to be heard but song far-carrying rapid yapping *kua kua kua* that fades.

Habits: Inhabits well-vegetated marshes and jheels with floating vegetation. Secretive but confiding if undisturbed. Feeds on invertebrates and plant matter.

Distribution: Rare winter visitor mainly to Pakistan. Also occurs in Europe, Africa, W and Central Asia.

BAILLON'S CRAKE

Porzana pusilla 19 cm F: Rallidae

imm

Otto Pfister

Description: Smallest crake, spotted white above and well-barred below. Adults alike with stubby wings and tail. Brown upperparts, well-streaked black and spotted white. Underparts paler grey than **Little** with black and white barring from belly to tail. Legs and bill usually greenish with no red. Juvenile has no grey and whitish barring starts from breast. Legs and bill pale brown.

Voice: Unlikely to be heard unless breeding but it is a frog or grasshopper-like dry rattling *trrr trrr*.

Habits: Much overlooked. Inhabits marshes, reedy jheel edges, small ponds and wet paddy. Will venture out to feed if undisturbed and then quite confiding. Feeds on invertebrates.

Distribution: Common breeding visitor to Kashmir. Scarce passage migrant and winter visitor throughout region. Also occurs in Europe, Africa, W, Central and SE Asia and Australia.

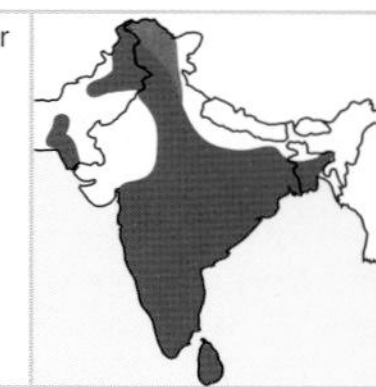

SPOTTED CRAKE

Porzana porzana 22 cm F: Rallidae

Description: A stocky, short-billed crake with plain buff undertail coverts. Black-flecked brown with dense white spotting on head, wings and breast and white barred flanks. Grey supercilia. Red-based yellow bill and green legs. White leading edge to wings in flight. Sexes similar. Larger than **Baillon's** and **Little** and lacking grey on underparts.

Voice: A far-carrying, whipping *huit huit* heard mainly at night.

Habits: Inhabits reedbeds and swamps. Also muddy jheel, canal and stream margins with good cover nearby. Crepuscular and secretive but will come into open to feed mainly on invertebrates and seeds.

Tim Loseby

Distribution: Scarce winter visitor mainly to lowland Pakistan but also, more rarely, in northwest and western India east to Nepal and south to Karnataka. Probably overlooked. Also occurs in Europe, N Africa, W and Central Asia.

RUDDY-BREASTED CRAKE (Ruddy Crake)

Porzana fusca 22 cm F: Rallidae

Description: A small, reddish-brown crake with bright pink legs. Adults alike with brown upperparts, pale throat, dark red underparts and white-barred black undertail coverts. Juveniles browner with faint white barring below.

Voice: Rather quiet but makes soft *crake* calls, sometimes accelerated into a trill.

Habits: Secretive and crepuscular and thus easily overlooked. Inhabits well-vegetated marshes, jheels and river banks. Also paddy fields and quite dry low scrub. Feeds on invertebrates.

Morten Strange

Distribution: Scarce breeding resident and local migrant most often recorded from northern Pakistan, Nepal, Bangladesh, Sri Lanka and the Western Ghats. Also occurs in China, Japan, and SE Asia.

WATERCOCK

Gallicrex cinerea Male 43 cm Female 36 cm F: Rallidae

Morten Strange

Bikram Grewal

Description: A huge, chicken-like crake. In breeding plumage the male is mainly blackish with brownish feather edgings to upperparts. Legs red and bill orange with upstanding red shield. Otherwise resembles immature and smaller female which is variably barred brown, darker above with greenish legs and yellowish bill.

Voice: A booming call when breeding.

Habits: Inhabits extensive reedbeds, marshes and wet paddy. Also tidal estuaries when migrating. Crepuscular and secretive but sometimes seen in flight. Aggressive when breeding and still used in Bangladesh for "cock-fighting".

Distribution: Generally a scarce local migrant throughout lowlands, breeding mainly in the Northeast and Bangladesh and in Sri Lanka. Also occurs in E and SE Asia.

PURPLE SWAMPHEN (Purple Moorhen, Purple Gallinule)

Porphyrio porphyrio 45 cm F: Rallidae

Otto Pfister

Description: Huge, fat-bodied, purplish-blue crake with a small head and massive red bill and shield. Mainly bright purplish-blue, becoming progressively paler towards head. White under stumpy tail and wings greenish-blue. Long legs and toes pinkish. Sexes alike. Juvenile greyer.

Voice: Quite noisy with various deep hoots and cackles. Most frequent is a sudden, thick *quark*.

Habits: Inhabits reedbeds, marshes and well-vegetated jheels, particularly favouring water hyacinth which it eats. Also eats other plant food, invertebrates and amphibians. Holds plant food with toes like a parrot. Often in large flocks. Flies reluctantly but strongly. Nests on ground.

Distribution: Common breeding resident throughout lowlands. Also occurs in S Europe, Africa, W and SE Asia, Australia and S Pacific, but possibly more than one species involved.

COMMON MOORHEN (Waterhen, Indian Gallinule)

Gallinula chloropus 32 cm F: Rallidae

Description: Medium-sized, dark aquatic crake with red shield and bill with yellow tip. Head, neck and breast deep slaty-grey, paler on belly. White flank stripe and undertail coverts, the latter with dark central stripe. Back and wings browner. Long legs and toes greenish with red garters. Sexes alike. Juvenile browner.

Voice: An abrupt *perruk*. When breeding *kereck kereck* calls.

Habits: Inhabits well-vegetated wetlands. Feeds on water plants, small fish and invertebrates. Swims with tail jerking and head bobbing. Walks on floating vegetation, mud margins and on waterside grass. Frequently climbs up into waterside bushes. Often in scattered parties. Aggressive in breeding season. Nests in reeds.

Otto Pfister

Distribution: Common breeding resident throughout lowlands.
Also occurs in Europe, Africa, throughout Asia, Pacific Islands and the Americas.

COMMON COOT (Coot, Eurasian Coot, Black Coot)

Fulica atra 40 cm F: Rallidae

Description: Large, blackish aquatic crake with prominent white bill and frontal shield. Looks tailless. Sooty black head and neck with greyer black body and whitish trailing edge to wings. Sexes alike. Juvenile has whitish throat and foreneck, grey bill, no shield. Greenish feet.

Voice: Quite noisy. An abrupt *towk*, a sharper *pseek* and various conversational chucklings.

Habits: Large swimming flocks on open water of jheels and rivers, often mixing with duck. Walks on floating vegetation. Feeds on vegetable matter. Flocks may panic and splatter clumsily across the water. Flies low but strongly, with lobed toes projecting. Builds anchored, floating reed nest.

Distribution: Common winter visitor throughout lowlands but rather sporadic breeder.
Also ocurs in Europe, N Africa, W, SE and Central Asia and Australia.

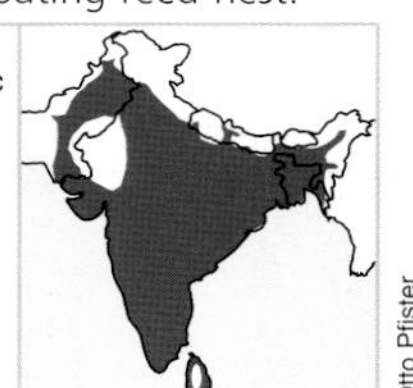

Otto Pfister

Otto Pfister

TIBETAN SANDGROUSE

Syrrhaptes tibetanus 48 cm F: Pteroclidae

Otto Pfister

Otto Pfister

Description: A large, high-altitude, pintailed sandgrouse. Pigeon-shaped with small head and bill and feathered legs. Long, pointed wings and fast flight. Male rather pale sandy-brown, especially below. Orange face, fine black barring on breast and spots on shoulders. White belly and feet feathering. Underside of wings, flight feathers and trailing edge contrasting black. Female heavily vermiculated black from crown to rump.

Voice: Deep musical *guk guk*.

Habits: Inhabits high altitude, stony pastures and semi-desert where it is the only sandgrouse. Tame and approachable. Seldom drinks. Feeds on plant seeds. Lays eggs on bare ground. Young precocial.

Distribution: Locally common breeding resident in Ladakh, Himachal Pradesh and Sikkim.
Also occurs in Central Asia.

CHESTNUT-BELLIED SANDGROUSE (Indian)

Pterocles exustus 28 cm including pin tail F: Pteroclidae

Otto Pfister

Description: Medium-sized, pin-tailed sandgrouse with dark belly and underwing. Breeding male sandy with soft grey wash on neck, diffuse pale patches on mantle and fine black spotting on wings. Face orange. Thin black breast bar and deep chestnut belly patch. Underwings and flight feathers black. Trailing edge white. Female heavily streaked blackish on upperparts.

Voice: Clucking *kut tu ro* in flight.

Habits: Inhabits semi-desert, dry fallow, pastures and open thorn scrub. Flocks fly long distances to drink in mid-morning, wetting belly feathers if incubating eggs. Feeds on seeds shuffling close to ground. Flushes with noisy clutter. Flies very fast. Nests as **Tibetan**.

Distribution: Fairly common but declining breeding resident throughout peninsula and Indo-Gangetic Plain.
Also occurs in Africa and W Asia.

BLACK-BELLIED SANDGROUSE (Imperial Sandgrouse)

Pterocles orientalis 39 cm F: Pteroclidae

Description: A large, black-bellied sandgrouse with white underwings and black wing tips. Shape as **Tibetan** but shorter pointed tail. Male has grey head and neck and orange face. Upperparts rather yellowish, mottled and barred darker. Female heavily marked with black on head, neck and upperparts. Extensive black bellies and thin black breast bands. Underside diagnostic.

Voice: Noisy at drinking places. A chuckling *kerr rer rer*.

Habits: Inhabits semi-desert and fallow. Formerly sometimes in enormous flocks when slaughtered by "sportsmen". Very shy. Rises vertically if flushed. Flies very fast and long distances to drink at waterholes, mid morning and afternoon. Feeds on plant seeds.

Goren Ekstrom

Distribution: Winter visitor in variable numbers now mainly to Rajasthan and Baluchistan. Straggler further east and south.
Also occurs in S Europe, N Africa, W and Central Asia.

PAINTED SANDGROUSE (Close-barred Sandgrouse)

Pterocles indicus 28 cm F: Pteroclidae

Description: Small, plump sandgrouse with blunt tail. Shape as **Tibetan**. Male has striking black and white crown pattern and plain sandy head and neck. Black and chestnut-edged yellow breast band. Rest of underparts closely barred black and white. Upperparts sandy broadly barred black with plainer sandy wing coverts. Underwings plain. Female sandy, heavily barred black with plain greyish throat.

Voice: A thick, clucking *wuko wuko* in flight.

Habits: Inhabits dry, open woodland and thorn scrub usually on hill sides; more vegetated habitat than other sandgrouse. Less sociable and usually flies to drink after dusk. Sits tight but often runs to escape. Nests as **Tibetan**.

Toby Sinclair

Distribution: Scarce breeding endemic resident mainly in dry parts of northern and Central India.

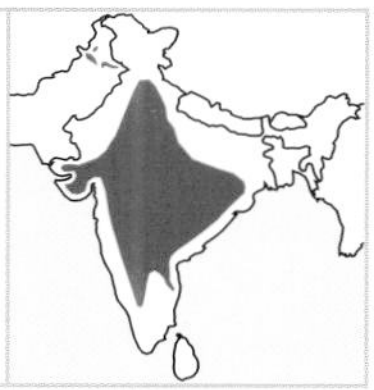

SOLITARY SNIPE

Gallinago solitaria 30 cm F: Scolopacidae

Joanna Van Gruisen

Description: Plump, cryptically plumaged, very long-billed wader. Rather dull with indistinct head stripes and orange wash on breast. Upperparts dark brown with whitish stripes. Yellowish legs. Slow, heavy flight on pointed wings, with white wing bars and heavily barred beneath. Sexes alike. Endangered **Wood Snipe** *G. nemoricola* is darker and heavier looking with slower flight.

Voice: Rather quiet. A loud, deep *pench* on flushing.

Habits: Inhabits alpine bogs and streams with thick cover. Solitary and unobtrusive. If flushed zig-zags but settles quickly. Feeds on worms. **Wood Snipe** also in high altitude bogs. Winters more widely, particularly in southwest and favours forest edge swamps.

Distribution: Rare breeding resident of northern mountains moving down to foothills in winter. Very rarely seen elsewhere.
Also occurs in Central and E Asia and Myanmar.

SWINHOE'S SNIPE

Gallinago megala 28 cm F: Scolopacidae

Balachandran

Description: Plump, cryptically plumaged, very long-billed wader. Very similar to **Pintail** *G. stenura* and not always separable in field. Usually noticeably longer-billed and longer-tailed with toes only just projecting in flight. Yellower, thicker legs. Grey trailing edge to wings. Shares with **Pintail** rather plain wing panels, barred underwings and pin-like feathers in outer tail. Sexes alike.

Voice: In flight a sudden *chrek* with abrupt ending. **Pintail** has a more subdued *tesch*.

Habits: Inhabits marshes and wet paddy, favouring saline, coastal grasslands. Also on drier rice stubble and fallow. Comes into open but their reaction to intrusion is to crouch low in grass. Feeds mainly on worms in mud.

Distribution: Probably fairly common but overlooked winter visitor mainly to east and south of region. Pintail is more widespread though rare in the Northwest.
Also occurs in Central, E and SE Asia and Australia.

COMMON SNIPE (Fantail Snipe, Snipe)

Gallinago gallinago 28 cm F: Scolopacidae

Description: A plump, cryptically plumaged, very long-billed wader. Similar to **Swinhoe's** but more brightly marked and with white trailing wing edges and white stripes on underwings. Whitish back stripes on brown back noticeable. Sexes alike. **Jack Snipe** *G. minutus* is tiny and short-billed, with broad yellow back stripes and a dark crown.

Otto Pfister

Voice: A sudden, rather frenetic, rasping *schkarp* in flight.

Habits: Inhabits wet paddy, marshes and jheel edges. Usually in small, scattered parties (wisps). Feeds by probing mud for worms. Zig-zags noisily off on flushing but more direct in high flight. Shy, preferring to crouch than fly. Much scarcer **Jack** rarely calls and only flies a short distance.

Distribution: Common winter visitor throughout lowlands. Less common than Pintail in south.
Also occurs in Europe, Africa, Asia and the Americas.

BLACK-TAILED GODWIT

Limosa limosa 45 cm F: Scolopacidae

Description: A very large, long-necked, long-billed and long-legged wader with striking white wing bars and rump and black tail. Size varies, females larger and eastern race smaller. Non-breeding plumage plain brownish grey, palest on head and neck. Becomes orange in breeding plumage, breast black-barred and back darker. Short white supercilia. Long straight pinkish bill with black tip.

Nikhil Devasar

Voice: Rather quiet but flocks call *keka keka*.

Habits: Inhabits freshwater marshes, jheels and rivers. Less frequent on coastal mudflats, although eastern race more coastal. Feeds on aquatic invertebrates in large flocks, often up to belly in deep water. Flies high between feeding grounds.

Distribution: Common winter visitor throughout lowlands. Some stay over in summer.
Also occurs in Europe, Africa and W Asia.

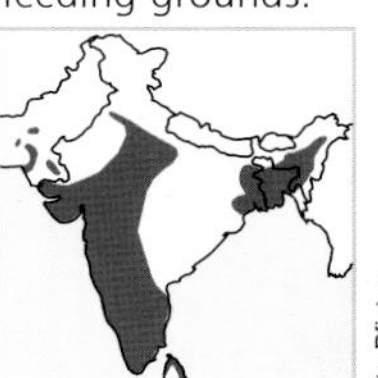

Otto Pfister

BAR-TAILED GODWIT

Limosa lapponica 40 cm F: Scolopacidae

Balachandran

Description: A large, long-necked, long-billed and long-legged wader with a wedge-shaped white rump patch and no wing bars. Stockier than **Black-tailed** with shorter, slightly upturned bill and shorter legs. Size varies; females and eastern race larger. In non-breeding plumage, finely streaked brown above and on head and neck. Latter become rich orange, as do all underparts, in breeding plumage. Back becomes darker and wing coverts silvery-grey.

Voice: Rather quiet. A soft *kik kik kik* in flight.

Habits: Coastal. Favours open mudflats. Feeds on invertebrates in mud or shallow water in scattered flocks.

Distribution: Rather scarce winter visitor. Commoner on northwestern coasts. Rare inland.
Also occurs in Europe, Africa, N, E and W Asia, Australia, Pacific Islands and N America.

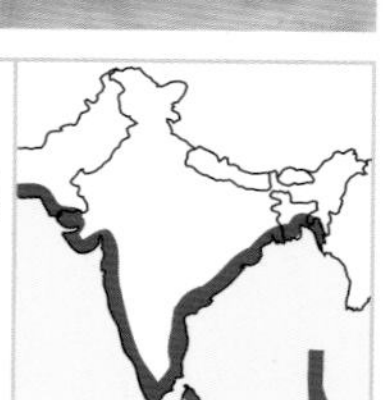

WHIMBREL

Numenius phaeopus 43 cm F: Scolopacidae

Otto Pfister

Description: A large, brown-streaked, long-necked and long-legged wader with a white rump wedge and no wing bars. Bill fairly long and decurved. Markedly striped crown pattern. White belly. Sexes alike.

Voice: Noisy. A trilling, usually seven note, whistle *tee tee tee tee tee tee tee*.

Habits: Coastal. Inhabits muddy and rocky shores, particularly favouring mangroves on which it perchs at high tide. Feeds mainly on crabs on mud and sand flats. Also feeds on short grass near coast, including golf courses and large lawns. Rather shy.

Distribution: Fairly common winter visitor to most coasts. Rare inland.
Also occurs in Europe, N, E, W and Central Asia, Indonesia, Australia and the Americas.

EURASIAN CURLEW (Curlew)

Numenius arquata 58 cm F: Scolopacidae

Description: A large, brown-streaked wader with a very long, down-curved bill and a white rump wedge but no wing bars. Head plain and body streaking paler than **Whimbrel**. Bill length varies, females being longer. Sexes alike in plumage. Note that the even longer-billed but dark-rumped **Eastern Curlew** *N. madagascariensis* occurs rarely in Bangladesh and could occur on other eastern coasts.

Voice: Noisy. A loud wild and musical *cour lee*.

Habits: Inhabits coastal mud-flats, jheels and rivers. Feeds in scattered parties on crustaceans and worms in mud. Also takes mudskippers. Will feed on damp pasture and arable. Very shy.

Kamal Sahai

Distribution: Fairly common winter visitor to most coasts and inland in north in small numbers.
Also occurs in Europe, Africa and N, W, Central and SE Asia.

SPOTTED REDSHANK (Dusky Redshank)

Tringa erythropus 33 cm F: Scolopacidae

Description: A fairly large, long-necked, long-legged and long-billed wader, with a thin white back stripe and no wing bars. In non-breeding plumage. rather pale grey upperparts with fine black speckles. Head and neck paler with distinct but short white supercilia. Underparts white. Long bill dark with red lower mandible and slight downcurve at tip. Legs bright red. In breeding plumage, strikingly black on head, neck and underparts and white-spotted dark grey upperparts. Sexes alike.

Voice: Noisy. An explosive *chew it*.

Habits: Inhabits marshes, jheels and rivers. Feeds on aquatic invertebrates in deep water, often swimming and sometimes up-ending like a duck. Fast, direct flight. Mixes with other waders.

Sumit Sen

Distribution: Fairly common winter visitor throughout lowlands. Scarce on coast and in the Northeast.
Also occurs in Europe, Africa, N, W and Central Asia and China.

COMMON REDSHANK (Redshank)

Tringa totanus 28 cm F: Scolopacidae

Otto Pfister

Description: A medium-sized brownish wader with broad white trailing wing edges and a white rump edge. Shorter and straighter billed than **Spotted** with red bill base, but can be very similar in non-breeding plumage. Legs more orange. Upperparts browner grey becoming darker and more streaked especially on head and neck in breeding plumage. Sexes alike.

Voice: Noisy. A three note frenetic *tew lee uu* and a hard *kip kip*.

Habits: Inhabits marshes, jheels, rivers, muddy coasts and estuaries. Feeds on aquatic invertebrates on and in mud and shallow water. Sociable, mixing with other waders. Wary, often the first to warn of intrusion. Bobs when alarmed.

Distribution: Breeding summer visitor to Kashmir. Fairly common winter visitor throughout lowlands and most coasts.
Also occurs in Europe, throughout Asia, Africa and Australia.

MARSH SANDPIPER

Tringa stagnatilis 23 cm F: Scolopacidae

Otto Pfister

Description: Medium-sized, pale wader with very fine bill. Like small **Greenshank** but all dark, needle-thin bill. Upperparts neatly marked grey with darker wing shoulder. Neat grey cap and grey hind neck. Rest of underparts all white. In breeding season darker and more strongly marked above with heavy streaking on head and neck. Long, narrow white back stripe and no wing bars. Legs yellowish-green. Sexes alike.

Voice: Rather quiet. A briefly repeated *klew,* rather like first note of **Common Redshank**.

Habits: Inhabits shallow fresh-water jheels, village ponds, wet paddy and river shoals. Feeds on aquatic invertebrates, stepping delicately through water. Mixes with other waders. Fairly tame.

Distribution: Fairly common winter visitor throughout lowlands. Less common on coasts.
Also occurs in Europe, Africa N, W, Central and SE Asia and Australia.

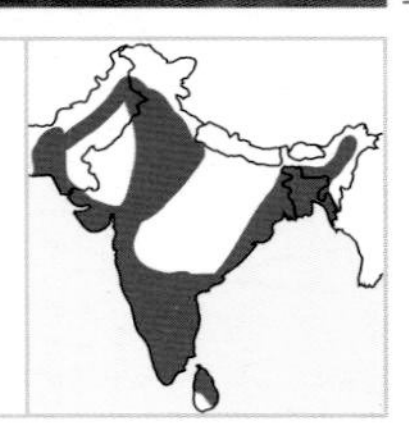

COMMON GREENSHANK (Greenshank)

Tringa nebularia 35 cm F: Scolopacidae

Description: A fairly large, grey wader with a long, powerful, upturned bill and a broad white back stripe and no wing bars. Grey brown above with darker wing shoulder. Head lightly streaked and looks very pale at any distance. More heavily streaked and darker in breeding plumage. Greenish legs. Sexes alike.

Voice: Noisy in flight. A ringing, musical *teu teu teu*.

Habits: Inhabits both fresh and coastal waters including village ponds and mangroves. Often solitary or in small groups but will mix with other waders. Wary with powerful flight. Feeds on aquatic invertebrates and small fish, often chasing them with bill under water.

Otto Pfister

Distribution: Fairly common winter visitor throughout lowlands.
Also occurs in Europe, Africa, W, N, Central, E and SE Asia.

GREEN SANDPIPER

Tringa ochropus 23 cm F: Scolopacidae

Description: A medium-sized, dark brown and white wader with diagnostic blackish under-wings. Upperparts dark olive brown lightly spoted with white. Tail blackish with white rump square. Head, neck and breast heavily streaked, stopping abruptly to contrast with white belly. White eye-ring and patch before eye. Legs greenish. Black and white in rapid, towering flight. Sexes alike.

Voice: A ringing *tlu eet weet*.

Habits: Usually seen singly and away from other waders in small, well-vegetated pools, streams, mangroves and village ponds. Often only seen when flushed or flying over calling. Feeds along water margins on insects, sometimes bobbing rear. Wary.

Bikram Grewal

Distribution: Widespread but rather scarce winter visitor.
Also occurs in Europe, Africa and W, N, Central, E and SE Asia.

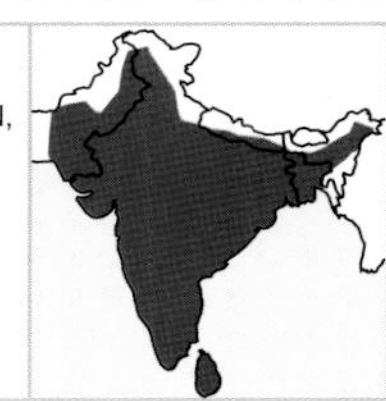

WOOD SANDPIPER (Spotted Sandpiper)

Tringa glareola 21 cm F: Scolopacidae

Otto Pfister

Description: Medium-sized, rather slender wader with square white rump, no wing bars and whitish underwings. Above brown, strongly spotted white. Some head and neck streaking and white supercilia. Streaking extends beyond breast and onto flanks with no clear-cut divide. Tail barred, so less contrast with white rump than **Green**. Darker with more spots in breeding plumage. Legs yellowish. Toes project in flight. Sexes alike.

Voice: Noisy. A rather soft but insistent *chiff chiff chiff*.

Habits: Inhabits marshes, jheels, riversides, wet paddy and muddy coasts. Often in large flocks. Bobs when alarmed. Feeds in shallow water or muddy margins on aquatic invertebrates. Not shy.

Distribution: Abundant winter visitor throughout lowlands.
Also occurs in Europe, Africa, Asia and Australasia.

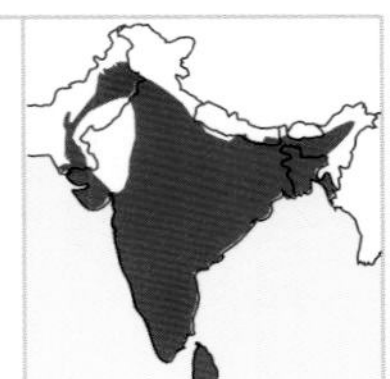

TEREK SANDPIPER

Xenus cinereus 24 cm F: Scolopacidae

Morten Strange

Description: A medium-sized, short-legged wader with a horizontal stance and long, upcurved bill. Grey-brown above, darker in breeding plumage when pale marked head and neck becomes more streaked. Dark primaries and white secondary tips contrast with pale grey tail and back in flight. Long, dark bill looks out of proportion and has reddish base. Legs bright yellow. Sexes alike.

Voice: Ringing *twoot wee wee*.

Habits: Inhabits muddy coasts and mangroves. When inland mostly on sandy rivers or muddy jheels. Usually feeds very actively, running rapidly then stopping, with a few bobs, to pick up invertebrates from surface of mud. Scattered flocks, often with other waders.

Distribution: Rather scarce winter visitor to most coasts. Most numerous on Arabian Sea coast. Rare inland.
Also occurs in Europe, Africa, N, E and SE Asia and Australasia.

COMMON SANDPIPER

Actitis hypoleucos 20 cm F: Scolopacidae

Otto Pfister

Description: Small, short-legged wader with horizontal stance and constant bobbing action. Brown on head, neck and upperparts demarcated from white lower breast and with white lobe between wing shoulder and breast. White eye ring and supercilia. Dark eye stripes. Legs usually greenish-brown. Broad white wing bar and brown rump. Unique low, hesitant flight with fast, shallow flaps interspersed with short glides. Sexes alike.

Voice: Sibilant *twee see see see swee,* usually when flushed.

Habits: Inhabits village ponds, rivers, mangroves, rocky coasts and estuaries. Usually singly or in small scattered parties. Feeds along water margins on insects. Nests by rocky streams.

Distribution: Breeds in Himalayas. Fairly common winter visitor throughout lowlands.
Also occurs in Europe, Africa, Asia and Australia.

RUDDY TURNSTONE (Turnstone)

Arenaria interpres 22 cm F: Scolopacidae

Description: A medium-sized, short-billed, stocky wader with distinctive pied upperpart pattern in flight. In non-breeding plumage, mottled brown above with blackish breast patch. Wedge-shaped bill and short orange legs. In breeding plumage upperparts rich chestnut and black and head is patterned black, grey and white. In flight, white wing and scapular bars, white mantle and bar across black tail. Sexes alike.

Voice: A rapid *tuka tuk tuk* in flight.

Habits: Inhabits mainly rocky coasts. Also sandy flats and by large rivers. In small groups hurriedly turning over stones and seaweed for invertebrates. Not shy but easily overlooked.

Distribution: Scarce winter visitor to most coasts. Commonest in the Northwest. Rare inland.
Also occurs in Europe, Africa, Asia, Australasia and the Americas.

Otto Pfister

Joanna Van Gruisen

ASIAN DOWITCHER (Snipe-billed Godwit)

Limnodromus semipalmatus 34 cm F: Scolopacidae

Morten Strange

Description: Fairly large, plump wader with long, straight bill. Like a stocky **Bar-tailed Godwit** except for straight, all-black bill. Greyish above in non-breeding plumage, with distinct white supercilia and barred flanks. In breeding plumage, brick orange from crown to belly. Barring on lower flanks and vent. Upperparts darker with more brown. In flight rather featureless above with only slightly paler, barred back and rump. Sexes alike.

Voice: Usually silent. Sometimes a soft *maou*.

Habits: Inhabits mudflats and if inland, jheels and flooded fields. Mixes with other waders. Has characteristic vertical "sewing machine" action when feeding in shallow water on invertebrates.

Distribution: Rare passage migrant to eastern and southern coasts. Rarer inland.
Also occurs in N, E and SE Asia.

GREAT KNOT (Eastern Knot)

Calidris tenuirostris 28 cm F: Scolopacidae

Balachandran

Description: Shown on right with **Red Knot**. A rather large, elongated but deep-chested wader with a short, slightly decurved bill. Mainly grey upperparts and head in non-breeding plumage with sharp demarcation between streaked breast and white belly. Develops black streaking from head to breast and rufous on back for breeding plumage, which bird shown is acquiring. Slight white wingbars and small whitish rump patch in flight. Sexes alike.

Voice: Rather quiet. A low *chukka chukka*.

Habits: Inhabits coastal mudflats and salt pans. In small numbers only, usually with other waders. Feeds by picking invertebrates from exposed tidal mud.

Distribution: Rare winter visitor that could occur on any coast and even inland in the Northeast.
Also occurs in N, E and SE Asia and Australasia.

RED KNOT (Knot)

Calidris canutus 24 cm F: Scolopacidae

Description: A medium-sized, chunky wader with short legs and a rather short, straight bill and small head. In non-breeding plumage, very similar to **Great Knot** except for size and shape. In breeding plumage (see previous species), head, neck and underparts become brick orange and upperparts become darker. In flight, narrow white wing bars and pale grey rump. Sexes alike.

Voice: Rather quiet. Sometimes a soft *chuk chuk*.

Habits: Inhabits coastal mudflats. Feeds with other waders by picking invertebrates from the surface of the mud or by wading in shallow water.

Tim Loseby

Distribution: Rare winter visitor to east coast and particularly Sri Lanka. Also occurs in Europe, Africa, N and E Asia, Australia and the Americas.

SANDERLING

Calidris alba 20 cm F: Scolopacidae

Description: A small, very pale coastal wader. In non-breeding plumage, very pale scalloped grey above and white below. Often shows blackish shoulder. Rather short, straight and stout black bill and black legs with no hind toe. In breeding plumage becomes rufous on head and chest and mottled rufous, black and grey above. In flight, shows broad black-edged white wing bars and a black central stripe to white rump and grey tail. Sexes alike.

Voice: A metallic *plit*.

Habits: Almost entirely restricted to sandy coasts where small parties run rapidly along the waves' edge picking up surface invertebrates. Also sometimes on mudflats.

Otto Pfister

Distribution: Local winter visitor to most coasts. Very rare inland. Also occurs in Europe, Africa, Asia, Australasia and the Americas.

LITTLE STINT

Calidris minuta 14 cm F: Scolopacidae

Otto Pfister

Description: Tiny, short-billed wader with black legs. Non-breeding adult is finely scalloped grey above with diffuse grey breast sides. When breeding, grey becomes rufous with black feather centres on wings and fine white mantle lines. Throat white. Juvenile plumage similar to adult breeding but less rufous. Mantle lines more prominent, dark-capped with split white supercilia. In flight, thin wing bars and black line on white rump and grey tail.

Voice: Piercing repeated *stit-it*.

Habits: Inhabits coastal mudflats, and salt pans, shallow open freshwater, muddy lake and river margins. Normally in flocks. Very active, running and picking small invertebrates from surface. Flocks rise and wheel in unison.

Distribution: Abundant winter visitor throughout lowlands and all coasts. Also occurs in Europe, Africa and W and N Asia.

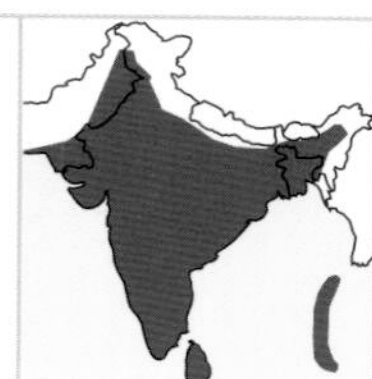

RED-NECKED STINT (Eastern Little, Rufous-necked Stint)

Calidris ruficollis 15 cm F: Scolopacidae

Morten Strange

Description: A tiny, rather elongated wader with black legs. Very similar to **Little** but looks shorter-legged and shorter-billed. Has a more attenuated rear end. Bill thicker and blunter. Almost identical in non-breeding plumage (shown) but plainer. In breeding plumage easier when whole head rich orange with streaked brown crown and breast gorget. No mantle lines. Juvenile plumage plainer than **Little** with rather grey wing coverts. Flight pattern identical. Sexes alike

Voice: A hoarse *chrit it it*.

Habits: Inhabits coastal mudflats and saltpans but also inland in Bangladesh. Behaviour as **Little** but in much smaller numbers.

Distribution: Scarce but overlooked winter visitor, most regular in Bangladesh, Tamil Nadu and Sri Lanka. Also occurs in N and E Asia, Indonesia and Australia.

TEMMINCK'S STINT

Calidris temminckii 14 cm F: Scolopacidae

Description: A tiny, plain wader with greenish or yellowish legs. In non-breeding and juvenile plumages, olive-brown above finely scalloped darker. Complete greyish breast band with white side lobes so recalls **Common Sandpiper.** In breeding plumage patterned with black and grey above. Tail extends beyond wings. In flight tail sides white. Beware legs darkened by mud (as shown). Sexes alike.

Voice: A dry, high-pitched trilling in flight.

Habits: Inhabits vegetated fresh-water habitats favouring flooded short grass and paddy. Feeds singly or in small scattered parties, often crouching when disturbed. Towers when flushed. In spring may perform song flight.

Otto Pfister

Distribution: Common winter visitor throughout lowlands. Scarce on coast. Also occurs in Europe, Africa and N and E Asia.

LONG-TOED STINT

Calidris subminuta 14 cm F: Scolopacidae

Description: Tiny, long-toed and long-necked wader with long, yellowish legs. Recalls **Wood Sandpiper** in pattern and stance. Warm brown, rufous and black above with white feather edgings and lines. Streaked breast band. Distinct brown cap and white supercilia with brown cheek spots and forehead. Shows thin wing bars and black line through white rump and grey tail. Short, dark, slightly down-curved bill has pale base to lower mandible.

Voice: A soft *chrup*.

Habits: Inhabits both fresh and brackish marshes, well-vegetated jheels and mudflats. Usually in scattered parties. Wades or walks on floating vegetation picking up invertebrates. Stretches neck when alarmed.

Morten Strange

Distribution: Local winter visitor mainly to south and east. Also occurs in N, E and SE Asia.

DUNLIN

Calidris alpina 19 cm F: Scolopacidae

Otto Pfister

Description: A variable, small wader with slightly decurved, black bill and black legs. Size, plumage and bill length varies. In non-breeding plumage, plain grey above. Grey breast sides. In breeding plumage mixture of rufous, black and grey above with rufous-streaked crown and unique black belly patch below black-streaked breast. Juvenile similar but belly spotted black. Shows thin wing bars and black line through white rump and grey tail. Sexes similar.

Voice: A trilling *treep* in flight.

Habits: Inhabits coastal mudflats. Inland, occurs on sandy rivers and muddy jheels. Sociable, mixing with other waders. Feeds by probing or picking invertebrates off mud or in shallow water.

Distribution: Locally common winter visitor to most coasts. Most common in Northwest. Decidedly scarcer inland. Also occurs in Europe, Africa, W, N and E Asia and N America.

CURLEW SANDPIPER

Calidris ferruginea 21 cm F: Scolopacidae

Tim Loseby

Description: An elegant, small wader with long, decurved, black bill and long black legs. Similar to **Dunlin** but longer necked, billed and legged. In non-breeding plumage grey with distinct white supercilia. In breeding plumage, deep brick red below with white chin and black, grey and rufous upperparts. Juvenile scaly-brown and grey above with peachy breast. Shows thin wing bars and white rump square though the rump can be barred. Sexes alike.

Voice: Trilling *chirrup* in flight.

Habits: Inhabits coastal mud and sandflats and saltpans. Inland on floods, muddy jheels and sandy rivers. Usually in small flocks often with other waders. Often wades up to belly in water, submerging head to probe for invertebrates.

Distribution: Common winter visitor to most coasts, particularly in south and east. Scarce passage migrant inland. Also occurs in Europe, Africa, N, E and SE Asia and Australia.

BROAD-BILLED SANDPIPER

Limicola falcinellus 18 cm F: Scolopacidae

Description: A small, stripe-headed wader with a kink at end of black bill. In all plumages shows distinct split supercilia contrasting with dark crown. In non-breeding plumage, lightly marked grey above with breast streaking. In breeding and juvenile plumages, rufous, black and white above with heavier breast streaking. Short legs greenish-brown. Dark wings show only slight wing bars. Black stripe through white rump and grey tail. Sexes similar.

Voice: A buzzing, metallic *brr eet* flight call.

Habits: Inhabits mud and sandflats and saltpans. Feeds on invertebrates singly or in small parties often with other waders. Can appear rather furtive.

Tim Loseby

Distribution: Scarce winter visitor to most coasts, most regular in Tamil Nadu. Rare on passage inland.
Also occurs in Europe, Africa, N, E, W and SE Asia and Australia.

RUFF (Reeve for female)

Philomachus pugnax Male 30 cm Female 23 cm F: Scolopacidae

Description: Large-bodied, small-headed, erect wader, very variable in plumage and size. In non-breeding plumage, scalloped brownish-grey above with greyish, brownish or occasionally, in males, white head, neck and breast. Male acquires crown tippets and neck ruff of varying colours in spring. Legs usually orange, greenish when young. Bill rather short and slightly down-curved. In characteristic, rather easy, flight shows thin wing bars and white ovals on dark rump.

Voice: Silent. No call recorded.

Habits: Inhabits marshes, jheels, stubble, cultivation, mudflats and saltpans. Feeds in large, often single sex flocks. Eats insects and grain.

Distribution: Abundant winter visitor mainly to western and southern parts of region.
Also occurs in Europe, Africa and throughout Asia.

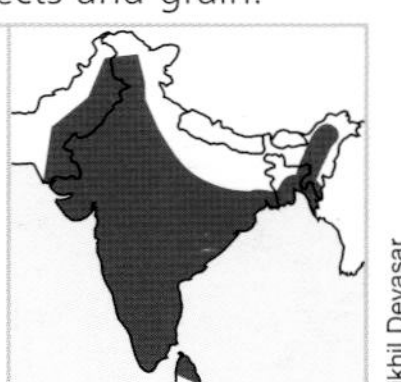

Otto Pfister

Nikhil Devasar

RED-NECKED PHALAROPE

Phalaropus lobatus 20 cm F: Scolopacidae

Tim Loseby

Description: A small, finely built wader with a needle-like bill. Most likely to be seen swimming buoyantly on sea. In non-breeding plumage, grey above and white below with dark rear crown and hindneck and black eyestripes. Juvenile darker and browner. In breeding plumage, buff and brown above, grey below with brick red neck and white throat. Female brighter. Feet lobed like **Coot.** In flight shows narrow wingbars.

Voice: A husky *chet*.

Habits: Marine. Inland, prefers salt lakes. Feeds swimming rapidly. Eats aquatic surface insects. Sometimes wades or walks on mud. In small flocks at sea, singly inland. Duller male incubates eggs and rears young.

Distribution: Scarce winter visitor to Arabian Sea and off Sri Lanka coast. Occasionally inland.
Also occurs in Europe, N America and throughout Asia and on all oceans.

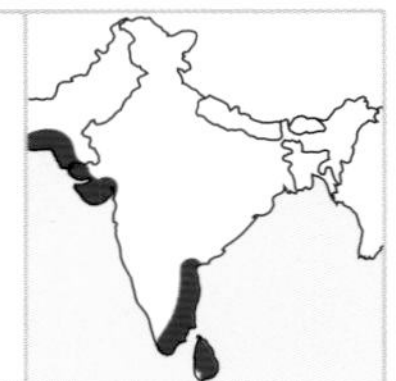

GREATER PAINTED-SNIPE (Painted Snipe)

Rostratula benghalensis 25 cm F: Rostratulidae

RK Gaur

Description: Medium-sized, plump wader with down-curved yellow bill. Female brighter with metallic green upperparts and yellow braces. Neck maroon. Crown and breast band black. White eyestripes. Belly white. Male and juvenile mixed buff and black above with pale brown neck and whitish throat.

Voice: Normally silent but polyandrous female has soft, booming *phom phom* in lek display.

Habits: Secretive and crepuscular, usually in pairs or singly. Inhabits marshes and wet paddy, feeding on invertebrates. Crouches when disturbed and difficult to flush. Flies low on broad wings with dangling legs; pitching, then running to hide. Male incubates eggs and rears young.

Distribution: Locally common breeding resident throughout lowlands but only monsoon breeding visitor to Northwest.
Also occurs in Africa, SE Asia and Australasia.

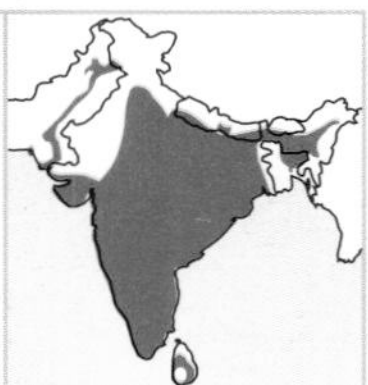

PHEASANT-TAILED JACANA

Hydrophasianus chirurgus 30 cm (without streamers) F: Jacanidae

Description: Fairly large, long-legged waterbird with 15-cm black tail streamers in breeding plumage. Black-edged yellow hindneck, black underparts and brown back. In all plumages outer wing strikingly white with black wing tips. Non-breeding plumage, paler brown above and white below as is juvenile, which has no yellow in neck. Very long toes and short bill. Sexes alike.

Voice: Startling mewing *mee ow* call.

Habits: Inhabits well-vegetated jheels, marshes, village ponds and rivers with water hyacinth or lotus. Sociable. Swims well. Walks on floating vegetation, eating seeds and other plant matter. Usually flies low, dangling legs. Will fly high.

Otto Pfister

Bikram Grewal

Distribution: Common breeding resident throughout lowlands though mainly monsoon breeding visitor to the north.
Also occurs in S China and SE Asia.

BRONZE-WINGED JACANA

Metopidius indicus 30 cm F: Jacanidae

Description: A fairly large, very dark waterbird with long legs and toes. Short chestnut tail. Glossy black below with white eyestripes. Bronze above. Underwings also dark. In flight and in poor light resembles **Moorhen**. Bill short and yellow with bluish shield. Juvenile brown above and on hind-neck and crown, white below with orange across breast. Sexes alike.

Voice: In breeding season an insistent raptor-like *keek keek keek*.

Habits: Habitat as commoner **Pheasant-tailed** with which it is often seen. Habits similar but flies less and is much shyer, often half-burying itself in floating vegetation if disturbed. Will fly high.

Kamal Sahai

Otto Pfister

Distribution: Common breeding resident throughout lowlands except parts of Northwest and Pakistan. Mainly monsoon breeding visitor to north.
Also occurs in SE Asia.

EURASIAN THICK-KNEE (Stone-curlew, Stone-Plover)

Burhinus oedicnemus 41 cm F: Burhinidae

Otto Pfister

Description: Large, streaked, sandy, dry country wader with big eyes and long yellow legs. Large rounded head has white surround to eyes. Short, thick, yellow-based black bill. Irises pale yellow and owl-like. Black and white bars on coverts, white patches in black wing tips. Belly white. Sexes alike.

Voice: Loud, rising, mournful *kur lee* after dusk. Alarm *pik pik*.

Habits: Inhabits dry woodland, rocky and sandy scrub, dry riverbeds, urban parks, waste ground and flat rooftops. Crepuscular and nocturnal. Furtive, spending day standing or sitting motionless in shade. Solitary or in pairs. If disturbed, prefers to squat low or run, rather than fly. Feeds on small reptiles and invertebrates.

Distribution: Common breeding resident throughout most of lowlands. Also occurs in Europe, Africa and W, Central and SE Asia.

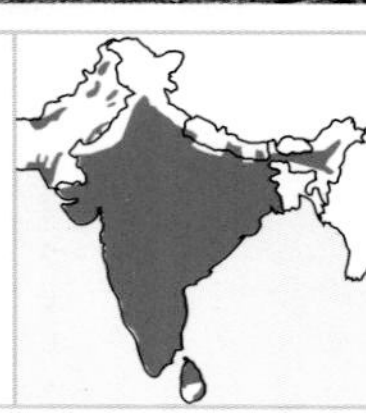

GREAT THICK-KNEE (Great Stone Plover)

Esacus recurvirostris 51 cm F: Burhinidae

Otto Pfister

Description: A very large, pale wader with enormous chisel-shaped, black bill. Rather plain sandy-grey with a dark bar on wing coverts and white patches in black flight feathers. Head has three black stripes on white back-ground. Bill base and irises yellow. Eyes look huge. Sexes alike.

Voice: Harsh *kwak kwak* alarm call. High rising whistle *tsee tsee tsee* in same rhythm as **Eurasian** but thinner.

Habits: Almost always near water, favouring river islands, rocks and sandbanks, open jheels, salt pans and flats. Crepuscular and nocturnal, spending day resting in open, regardless of sun. Can be very difficult to spot. Solitary or in pairs. Feeds on invertebrates and reptiles. Lays eggs on ground.

Distribution: Scarce and local breeding resident mainly in north and west lowlands and parts of south and the Northeast. Also occurs in Myanmar.

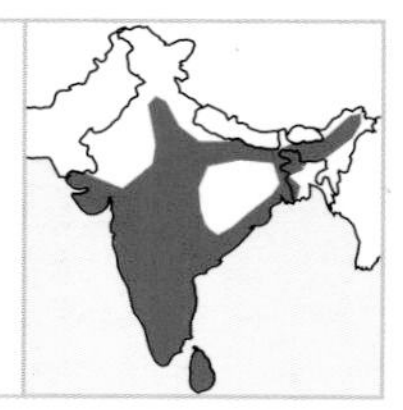

EURASIAN OYSTERCATCHER (Common Oystercatcher)

Haematopus ostralegus 42 cm F: Charadriidae

Description: A large, plump, black and white wader with a long, straight orange bill and rather short, pink legs. Head, neck, breast and upperparts black. White below and bold white wing bars. Mantle to tail white with black terminal tail bar. In non-breeding plumage (shown), white cheek stripes. Sexes alike.

Voice: Noisy. A high piping *peep,* often hurriedly repeated and made on ground as well as in flight.

Habits: Feeds in scattered flocks on open mud and sandflats and among rocks, probing for worms and molluscs. Mixes with other waders, particularly at roosts.

Sharad Gaur

Distribution: Scarce winter visitor mainly to western and southern coasts including Sri Lanka.
Also occurs in Europe, Africa, W and N Asia and China.

IBISBILL

Ibidorhyncha struthersii 41 cm F: Charadriidae

Description: Large, plump, grey wader with black face and long, down-curved, dark red bill. Upperparts sandy grey with short white wing bars. Head, neck and breast lavender grey with white-edged black crown and face and white and black breast bands. Underparts white. Short legs orange. Sexes alike. Juvenile has grey bill and pale face.

Voice: A high *ti ti ti ti ti, klu klu; wicka tik tik* near nest.

Habits: Inhabits pebble beds in fast-flowing rivers, blending well among stones. Usually singly or in small groups. Wary, but loyal to favoured spots. Bobs when alarmed. Feeds on invertebrates by probing and turning over stones often wading belly deep. Lays eggs among stones.

Otto Pfister

Distribution: Rare breeding resident in northern mountains moving lower in winter.
Also occurs in Central Asia.

BLACK-WINGED STILT

Himantopus himantopus 35 cm F: Charadriidae

Otto Pfister

Description: Slender, pied wader with extremely long, pink legs. Black above, with brownish wash on female and juvenile. Head, neck and all underparts white with variable amounts of dark grey on crown, face and hind neck. Bill very fine, straight and black. In flight, black wings and white tail and rump wedge. Juvenile often dusky on head with white trailing edges. Legs extend equivalent to body in flight.

Voice: Noisy. A high *keep keep*.

Habits: Inhabits jheels, sewage seepage, rivers, village ponds, saltpans and mudflats. Graceful high-stepping gait. Gregarious. Wades and swims, picking or probing for insects and small fish. Nests, often colonially, on islands and floating vegetation.

Distribution: Abundant breeding resident and winter visitor throughout lowlands.
Also occurs in Europe, Africa, W, N and Central Asia and China.

PIED AVOCET (Avocet)

Recurvirostra avosetta 45 cm F: Charadriidae

Otto Pfister

Otto Pfister

Description: A large, supremely elegant, black and white wader with long, bluish legs and a long, fine, upcurved bill. Basically white with black cap and wing patches. In flight black wing tips, covert bars and back patches. Legs extend beyond tail. Sexes alike. Juvenile has some brownish mottling.

Voice: Noisy. A loud piping *klute klute klute*.

Habits: Found on coast and inland favouring alkaline water of saltpans and the coast. Also large rivers, muddy jheels. floods and village ponds. Gregarious, often with stilts. Wary. Feeds mainly by wading or swimming in quite deep water, sweeping bill from side to side in search of invertebrates and small fish. Will also upend.

Distribution: Breeding resident in Gujarat and Pakistan. Fairly common winter visitor in the Northwest and southeast including Sri Lanka. Local and sporadic elsewhere.
Also occurs in Europe, Africa, W, N and Central Asia and China.

EUROPEAN GOLDEN PLOVER (Eurasian, Greater)

Pluvialis apricaria 27 cm F: Charadriidae

Description: A stocky, short-billed wader with spangled yellowish upperparts. White supercilia and belly. In breeding plumage, white extends from forehead to vent in broad band bordering black face and underparts. Stockier, shorter legged and billed than **Pacific Golden** with white not grey underwings. **Grey** is larger, lacks yellow and has black armpits. Sexes similar.

Voice: A plaintive whistle *tu lee*.

Habits: Inhabits damp grasslands, jheel margins and muddy coasts. Most likely to be seen in flocks of **Pacific Golden** or with other waders. Feeds on invertebrates on ground in typical plover fashion.

Tim Loseby

Distribution: Very rare winter visitor or passage migrant recorded from Pakistan and northern India east to Assam. Possibly overlooked.
Also occurs in Europe, N Africa, W and N Asia.

PACIFIC GOLDEN PLOVER (Lesser, Asiatic Golden Plover)

Pluvialis fulva 42 cm F: Charadriidae

Description: Medium-sized, short-billed wader with golden-brown upperparts. Typical plover with rounded head, large eye. and long grey legs. In non-breeding plumage (shown), brown-edged and/or spangled with gold above, streaked brown on buff head, neck and breast. Darker cap and cheek spots and whitish forehead and supercilia. In breeding plumage, black from face to belly with broad white border from supercilia to flanks. Smoky underwings. Sexes alike.

Voice: A distinct *tu leep* like a soft **Spotted Redshank**.

Habits: Inhabits coastal mudflats saltpans and grasslands. Inland on marshes and jheels. Feeds on surface invertebrates by running and picking from mud or grass.

Otto Pfister

Distribution: Locally common winter visitor to all coasts and inland in Bangladesh and the Northeast. Most common in southeast. Scarce but regular passage migrant inland elsewhere.
Also occurs in N and SE Asia, Australasia and Alaska.

GREY PLOVER (Black-bellied Plover)

Pluvialis squatarola 41 cm F: Charadriidae

Otto Pfister

Jan Willem den Besten

Description: A fairly large, dumpy wader with spangled grey-brown upperparts and black armpits. Non-breeding plumage similar to a large **Pacific Golden** but grey above. In breeding plumage, black below to thighs and broad white border ends on breast sides. Upperparts become more spangled silver, black and white. White wingbars and rump square. In flight diagnostic black armpits (axilliaries). Sexes alike.

Voice: A loud, penetrating *pee-ou ee*.

Habits: Coastal favouring open mudflats where it feeds on surface invertebrates in typical plover fashion in widely scattered flocks. Inland on open jheels, floods and rivers.

Distribution: Uncommon winter visitor to most coasts.Rare inland on passage. Also occurs in Europe, Africa, N, W and SE Asia, Australasia and N America.

COMMON RINGED PLOVER (Ringed Plover)

Charadrius hiaticula 19 cm F: Charadriidae

Ron Saldino

Description: Small, stocky wader with broad, black breast band extending right round neck. Muddy-brown upperparts and crown. Broad white wing bars and tail sides. Bright orange legs. In breeding plumage black face with white forehead and cheek patch and black-tipped orange bill. In non-breeding plumage black duller, breast band narrower, less black on face and dusky bill. Sexes similar. Juvenile similar but often has broken breast band and yellower legs. Wing bars, legs and lack of eye rings best distinctions from **Little Ringed**.

Voice: Rising disyllabic *poo eep*.

Habits: Inhabits muddy and sandy margins of lakes, rivers and the coast. Mixes with other waders.

Distribution: A rare winter visitor to Pakistan, Sri Lanka and Indian east coast. Probably mainly passage migrant inland in Northwest. Possibly overlooked. Also occurs in Europe, Africa, W and N Asia and N America.

LONG-BILLED PLOVER (Long-billed Ringed Plover)

Charadrius placidus 22 cm F: Charadriidae

Description: A medium-sized wader with black breast band and brown cheeks. Largest ringed plover with a longish, blackish bill and yellowish-brown legs. Muddy-brown above including crown. White forehead and rear eye-stripe. Black fore-crown bar and narrow breast band, the latter may appear broken as shown. Brown cheeks in all plumages. In flight shows thin white wingbar and trailing edges and black sub-terminal tail band. Sexes alike. Bird shown is in head moult.

Otto Pfister

Voice: A high *pee wee*.

Habits: Inhabits shingle banks on rivers and muddy banks. Solitary and unobtrusive. Feeds in typical plover fashion on invertebrates.

Distribution: Rare winter visitor to northern rivers and Gujarat coast. Probably overlooked in winter. Also occurs in China, N, SE and Central Asia.

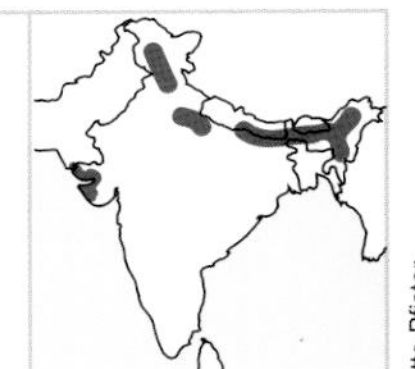

Otto Pfister

LITTLE RINGED PLOVER (Little Plover)

Charadrius dubius 16 cm F: Charadriidae

Description: Small, slim, muddy-brown wader with black head and breast markings. Long-bodied and small-headed. In breeding plumage, white fore-crown band, black borders to forehead and black cheeks and breast band. No obvious wing-bars. Legs yellowish-grey, bill mostly black. Prominent yellow eyering. In non-breeding plumage, black becomes brown and whole head suffused buff; juvenile similar. Sexes alike.

Otto Pfister

Voice: A sad, lilting *pee ou*.

Habits: Inhabits muddy jheels, grass fields, rivers, coastal saltpans and mudflats. Feeds on invertebrates, sometimes vibrating them to surface with foot. Often in scattered parties, camouflaged on mud. Nests on ground.

Distribution: Common resident and winter visitor throughout lowlands. Also occurs in Europe, N, W, Central and SE Asia, China, Australasia and Pacific Islands.

KENTISH PLOVER

Charadrius alexandrinus 17 cm F: Charadriidae

Tim Loseby

Description: A small, sandy-brown wader with blackish legs and black bill. Typical small plover with sandy-brown upperparts, crown and cheeks and small dark breast patches. White collar. Breeding plumage male has chestnut rear crown and black forehead, cheeks and breast-sides. Strong white wingbars and edges to tail and rump.

Voice: A soft *dri ip* and a *whi it* alarm on ground. In flight an abrupt *tit tit*.

Habits: Inhabits coastal sandflats, salt pans but inland on river sand bars, open lakeshores and drying floods. Behaviour as **Little Ringed** but more sociable. Often quite tame. Lays eggs on sand or mud.

Distribution: Fairly common winter visitor and local breeding resident to most coasts and throughout lowlands. Commonest in the Northwest and Pakistan.
Also occurs in Europe, Africa and W, Central, E and SE Asia.

LESSER SAND PLOVER (Mongolian)

Charadrius mongolus 19 cm F: Charadriidae

Otto Pfister

♀

Otto Pfister

Description: A small, compact, muddy-brown wader with dark legs. Round-headed, gentle expression. In non-breeding plumage, muddy-brown on crown, cheeks, breast sides and whole upperparts including collar (see photo of **Greater**). Breeding female has peachy breast band. Breeding male has black mask, broad chestnut breast band and neck sides to forecrown. Stubby black bill. In flight white wing bar and tail sides.

Voice: Hard, repeated *tri ip*. Also a trilling *trrp*.

Habits: Coastal sand and mudflats. River and jheel margins and flooded fields inland. Nests on ground in high-altitude semi-desert. Highly sociable, feeding in scattered flocks.

Distribution: Race *atrifrons* breeds in Ladakh and elsewhere in northern mountains. Common winter visitor to most coasts. Scarce on passage inland.
Also occurs in N, W, Central and SE Asia and Australia.

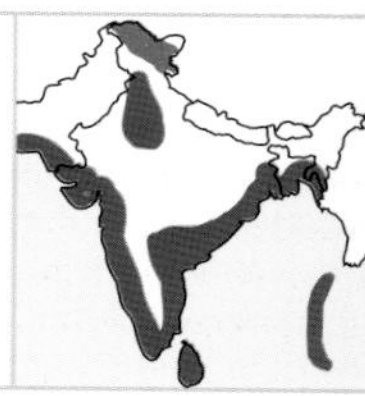

GREATER SAND PLOVER (Large Sand Plover)

Charadrius leschenaultii 22 cm F: Charadriidae

Description: A medium-sized, rather gangly, muddy-brown wader with greenish or yellowish legs. Very similar plumages to **Lesser Sand** (shown to right in non-breeding plumage; other three are **Lesser Sand**) but larger with longer, paler legs and longer, often disproportionate-looking, bill with marked thickening. Head looks flatter and eyes larger so has more gaunt expression. Shows longer wing bars and toes projecting beyond tail in flight. In breeding plumage as **Lesser** but male has white forehead patch and narrower, paler breast band.

Voice: A trilling *trrr t*.

Habits: Habitat and behaviour as **Lesser**, with which it often mixes.

Otto Pfister

Distribution: Rather scarce winter visitor to most coasts. Commonest in northwest. Very rare inland.
Also occurs in Africa, W, Central, E, and SE Asia and Australasia.

NORTHERN LAPWING (Eurasian Lapwing, Green Plover)

Vanellus vanellus 31 cm F: Charadriidae

Description: A large, stocky, black and white wader with long up-swept crest. Upperparts have purple and green sheen. Undertail coverts orange and breast black. Crown, face, cheek stripes and wispy crest black. Legs pinkish and short bill dark. Rounded wings have white tips and under-wing coverts. Tail white with black tip. Sexes similar but female has white patches in black areas. Juveniles rather scaly above.

Voice: An abrupt, rising *pee wit* (hence old English name Peewit).

Habits: Inhabits wet grassland, jheel and river margins and fallow, rarely mixing with other waders. Usually in parties, feeding by running and stopping to pick up surface invertebrates. Relaxed, low, flapping flight. Shy in region.

Nikhil Devasar

Distribution: Rather scarce winter visitor mainly to northern lowlands.
Also occurs in Europe,and W, N and E Asia.

YELLOW-WATTLED LAPWING

Vanellus malarbaricus 27 cm F: Charadriidae

Otto Pfister

Description: Large, sandy, dry country wader with black skull gap and yellow bill wattles. Sandy brown upperparts including clearly demarcated breast, neck and head. Black throat and thin breast band. White rear eyestripe below velvet black cap. Yellow legs, eye-rings and fleshy wattles hanging down from base of short beak. In flight narrow white-bordered black outer wing and black-banded tail. Sexes alike.

Voice: Rather quiet. A plaintive *tee ee* and a hard *tik tik tik*.

Habits: Inhabits dry, short grass plains, fallow and bare wasteland. Still occurs in some urban areas. Unobtrusive but confiding. Prefers to run from disturbance. In pairs or scattered parties feeding on insects in typical plover manner.

Distribution: A declining and local endemic breeding resident throughout lowlands except Pakistan, Bangladesh and parts of the Northwest and Northeast.

RIVER LAPWING (Spur-winged Plover)

Vanellus duvaucelii 30 cm F: Charadriidae

Otto Pfister

Description: A large, sandy-grey riverine wader, black from throat to crested crown. Sandy-grey above with white-bordered black wing tips and shoulder. White below with small black belly patch and sandy breast band. Spurs on wing shoulders rarely visible. Sexes alike. Recently separated from western **Spur-winged Plover** *V. spinosus.*

Voice: A hard *dee dit dee dit*.

Habits: Inhabits river sand bars and margins usually in pairs or scattered flocks. Sometimes on coasts, riverside fields and jheel margins. Spends much time sitting hunched watching the world go by. Feeds as typical plover on surface invertebrates. Unobtrusive and quite confiding.

Distribution: Fairly common resident in the Northeast from Himachal Pradesh.
Also occurs in SE Asia.

GREY-HEADED LAPWING (Grey-headed Plover)

Vanellus cinereus 37 cm F: Charadriidae

Sumit Sen

Description: An obviously large, pale wader with long yellow legs. Sandy-brown with pale grey head, neck and breast. Black breast band and black-tipped yellow bill. In flight, extensive black wing-tips, white secondaries and black-banded white tail. Red irises and bare yellow lores. Sexes alike. Beware smaller and aquatic **White-tailed**.

Voice: Rather quiet. A high *chee e it*, recalls **Red-wattled**.

Habits: Inhabits open, usually, wet grassland, jheel and river edges and fallow. Seeks out recently burnt, damp grasslands. Feeds in typical plover fashion on invertebrates on ground, usually in scattered parties and sometimes with other waders.

Distribution: Local winter visitor to lowland Nepal, northeast India and Bangladesh. Rare elsewhere in north India.
Also occurs in N, E and SE Asia.

RED-WATTLED LAPWING (Did-ye-do-it? Bird)

Vanellus indicus 35 cm F: Charadriidae

Description: Large, brown, black and white wader with short red bill wattles and eye rings. Warm brown above with black head, neck and breast broken by broad white side-neck stripe. White below, broad white border to black wing tips. Black-tipped red bill and yellow legs. Sexes alike.

Voice: Very noisy. Well-known call is *deed ye do it?* usually preceded by *de de de*. Calls at night.

Habits: Inhabits any sort of open country but preferring the vicinity of water. Breeds in villages, parks, even large gardens. Extremely vigilant to intrusion when breeding but tolerant of close approach. In pairs or small parties, crepuscular and nocturnal. Lays eggs on ground or flat roofs.

Distribution: Abundant breeding resident throughout region.
Also occurs in W and SE Asia.

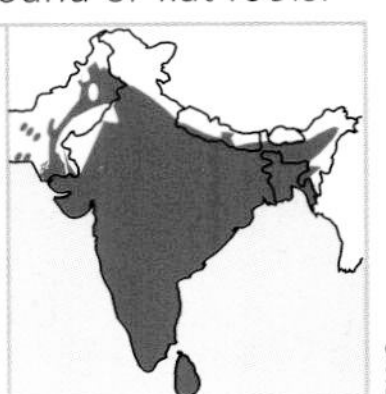

Otto Pfister

RK Gaur

WHITE-TAILED LAPWING

Vanellus leucurus 28 cm F: Charadriidae

Otto Pfister

Description: A large, pale, wetland wader with long yellow legs. Rather plain sandy-brown head, neck, breast and upperparts. White below. Black eye and bill provide only head contrast. In flight, striking white tail and rump and broad white borders to black flight feathers. Sexes alike.

Voice: A short emphatic *pee wik* usually in flight.

Habits: Inhabits well-vegetated freshwater wetlands. Always by water in which it wades in a most un-ploverlike fashion to pick up surface invertebrates. Also feeds on water hyacinth mats and damp cultivation. Usually in scattered groups. Rather unobtrusive but quite confiding.

Distribution: Common winter visitor to Northwest. Rare elsewhere.
Also occurs in NE Africa and W and Central Asia. Slowly spreading west into Europe.

CRAB-PLOVER

Dromas ardeola 41 cm F: Glareolidae

Balachandran

Description: Very large, black and white coastal wader with long, bluish legs and large, heavy, black bill. Disproportionately large, white head with black-edged eye standing out. Sometimes blackish rear crown. Primarily white with black mantle and scapulars and mainly black flight feathers and outer coverts. Looks ungainly and flies low with extended legs. Sexes alike.

Voice: A goose-like honking *qurk qurk qurk*.

Habits: Inhabits sandy coasts and reefs. Feeds in plover-like manner, scattered over flats or among exposed reefs, mainly on crabs but also mudskippers. Nests in deep burrows in sand banks. Feeds at low tide. Often confiding.

Distribution: Mainly a very local winter visitor to most coasts. Has bred in northern Sri Lanka. Most regular in Gujarat and Tamil Nadu.
Also occurs in Africa, W Asia and Indian Ocean Islands.

CREAM-COLOURED COURSER

Cursorius cursor 23 cm F: Glareolidae

Description: A pale, arid country wader. Much paler sandy-buff than **Indian** with little variation between upper and underparts. Pale lores, pale brown forecrown and striking lavender-grey hind crown. No white on rump. Underwings blacker. Sexes alike. Often confused with pale **Indian**.

Voice: A liquid *whit whit* and *whek whek*.

Habits: Inhabits desert and semi-desert including dry grass flats. Sociable, sometimes mixing with **Indian**. Feeds in typical courser manner on surface invertebrates, running and picking at prey. Also digs into dung or soft ground. Stands tall or runs in preference to flying when disturbed. Fast, direct, often high, flight. Fairly confiding.

Distribution: Scarce breeding resident and winter visitor to Pakistan and the Thar Desert in India. Rare winter visitor further south and east in Rajasthan, east to Haryana.
Also occurs in N Africa, W Asia and Atlantic Islands.

Nikhil Devasar

Nikhil Devasar

INDIAN COURSER

Cursorius coromandelicus 23 cm F: Glareolidae

Description: Medium-sized, upright, dry country wader with rich chestnut crown, white supercilia and black eyestripes. Plover-like with short, down-curved bill and startlingly white legs. Upperparts warm muddy-brown. Neck buff, breast to belly warm orange, black belly patch and white vent. In flight, rather rounded wings appear all dark below. Outer wings, black with white edges. Sexes alike.

Voice: Rather silent; a gruff cluck when flushed.

Habits: Inhabits dry sandy flats, grassland and fallow land. Prefers flat country with minimal scrub to maximise view. Feeds as plovers on invertebrates. Runs by preference but if flushed flies strongly. Usually in pairs. Nests on ground.

Distribution: Local and declining endemic resident mainly in the Northwest and south. Local migrant turning up to breed then disappearing.

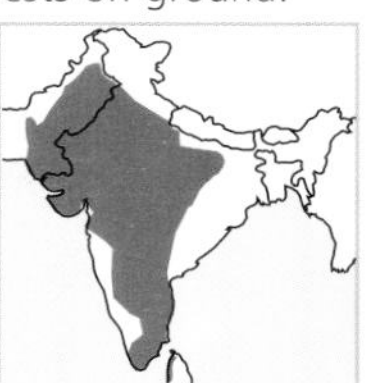

Otto Pfister

Otto Pfister

COLLARED PRATINCOLE (Common, European)

Glareola pratincola 25 cm F: Glareolidae

Otto Pfister

Description: Medium-sized, short-legged wader with long, narrow wings and long forked tail. Sandy brown with sharp demarcation from white lower breast. Short, black, hooked bill with red base. Black lores and necklace, white partial eye rings. Darker brown outerwings, white trailing edges, black forked tail, extending to or beyond wingtips. White outer tail feathers, uppertail coverts and rump. Chestnut underwing difficult to see. Sexes alike.

Voice: A tern-like *kirrik* in flight.

Habits: Inhabits open, bare mud usually near drying jheels, along rivers or near coast. Crouches in flocks, flying up to catch swarming aerial insects, which it hunts in graceful swallow-like manner Crepuscular. Nests on ground.

Distribution: Breeds in lower Pakistan but rare winter visitor elsewhere in region.
Also occurs in Europe, Africa and W Asia.

ORIENTAL PRATINCOLE (Large Indian Pratincole)

Glareola maldivarum 25 cm F: Glareolidae

Otto Pfister

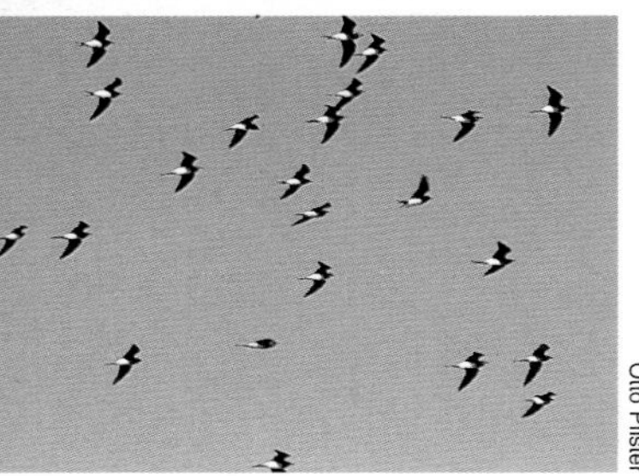

Otto Pfister

Description: A medium-sized, short-legged wader with long, narrow wings and a short, forked tail. Almost identical to **Collared** except brown on breast extends down to belly and merges into the white and the tail tips are always obviously shorter than wing tips. In flight, much less contrast above and no white trailing edges. Sexes alike.

Voice: A tern-like *krek krek* in flight.

Habits: Habitat and behaviour as **Collared** but perhaps more tolerant of dry fallow away from water. Favours river sand banks. Nests on bare ground but adventitiously, not always returning to previous sites. Can be remarkably tame when on ground but also very well camouflaged.

Distribution: Locally common breeding resident and local migrant. Most frequent in north and extreme south, including Sri Lanka.
Also occurs in SE Asia.

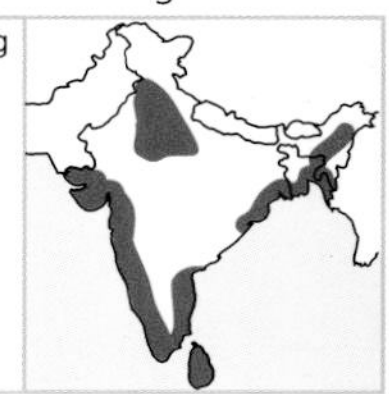

SMALL PRATINCOLE (Small Indian, Milky, Little Pratincole)

Glareola lactea 17 cm F: Glareolidae

Description: A small, stumpy wader, the colour of milky tea. Shape as other pratincoles, but short square-ended or shallow forked tail and shorter wings. Distinctly pale grey-brown, paler below becoming white on belly. Red-based black bill and black lores. Striking wing pattern with black outer wings, above and below, and broad white bars. White rump and tail with broad black terminal band. Sexes alike.

Otto Pfister

Voice: High-pitched *territ, territ*.

Habits: As other pratincoles but more dependent on water, preferring river sand banks, muddy jheel margins and mudflats on coast. Hunts swarming insects in wheeling flocks, often very high over water. Lays eggs on sand. Confiding but easily bypassed.

Distribution: Widespread but local resident to most of lowlands but rare in south and only monsoon breeding visitor to the Northwest and Pakistan. Also occurs in SE Asia.

INDIAN SKIMMER

Rynchops albicollis 40 cm F: Laridae

Description: A large, black and white, tern-like bird with huge red and yellow bill. Black upperparts with white trailing edges , white collar and black crown. White forehead and underparts. Lower mandible red with yellow tip and bent downwards. Shorter, red upper mandible. Red legs. Diagnostic feeding method. Sexes alike. Immatures browner with black-tipped orange bill and legs.

Nikhil Devasar

Voice: Usually quiet. A deep yapping call.

Habits: Inhabits large, slow rivers and lakes. Sits in groups on sandbars and islands. Crepuscular. Feeds on small fish in still water by dragging lower mandible through surface in powerful flight, often in groups. Lays eggs on sand.

Distribution: Globally threatened and very local resident found sparingly in the northern lowlands. Rare on coast. Probably still also occurs in Myanmar.

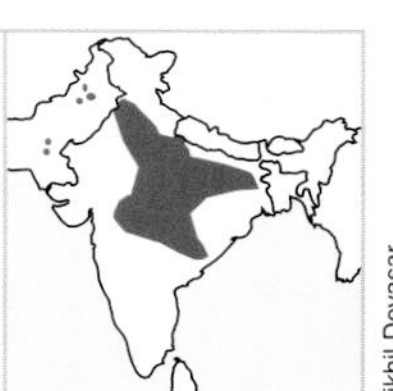

Nikhil Devasar

ARMENIAN GULL (Herring Gull)

Larus armenicus 56 cm F: Laridae

Nikhil Devasar

Description: Large, grey and white gull with strikingly patterned bill. Size varies. Recently given specific status. Similar to **Baraba Gull** *L.(c. or h.) barabensis* which darker, longer billed, usually pale irises and grey in wing tips. In our view **Armenian** occurs in northwest India at least. Adult darker grey than *L. cachinnans,* with extensive black wing tips with no grey and only one or two white mirrors. Bill shorter with white tip and black and red subterminal band. Eyes dark and head rounded, giving benign expression. Immatures very similar to **Yellow-legged**.

Voice: As **Yellow-legged**.

Habits: As **Yellow-legged**. So far only recorded on northwestern Indian rivers and lakes.

Distribution: Status uncertain due to identification problems. Probably a rare or scarce winter visitor to Pakistan and northwest India. Baraba Gull may also occur, but no documented sightings. Also occurs in south-central Asia and winters in W Asia east at least to the Persian Gulf. Baraba occurs in Central Asia and winters in W Asia east at least to the Persian Gulf.

YELLOW-LEGGED GULL (Caspian, Herring Gull)

Larus cachinnans 60 cm F: Laridae

Nikhil Devasar

Description: A large, grey and white gull with yellow legs. Size varies by 5 cm either way. Note the large gulls of Eurasia are still subject to debate about their speciation. Adult white with grey upperparts and black wing tips with white mirrors. Yellow bill with red spot. Pale irises and rather mean expression. Mature after four years. Immatures start streaked brown becoming greyer and whiter. Bill dark in juvenile.

Voice: Nasal *kyow* and *ag ag*.

Habits: Inhabits coastal waters especially near fishing villages and also inland lakes and large rivers. Bouyant, powerful flight. Scavenges on fish and other waste. Robs smaller birds. Sociable, often with other large gulls.

Distribution: Scarce winter visitor to most of region, common in Pakistan. Scarcer in south and inland but regular on northwestern rivers.
Also found in E Europe, N and E Africa and W and Central Asia.

HEUGLIN'S GULL (Herring, Lesser Black-backed Gull)

Larus heuglini 65 cm F: Laridae

R Vijaykumar Thodamam

Description: Large, dark, grey and white gull, often with pink legs. Size varies. Recently given specific status but much debate about relationships. Very similar to **Yellow-legged** but averages larger, much darker grey above (can appear almost black as in photo) and with more powerful bill. Often has pink legs in winter and in immatures. Immature structurally large and with more contrasting tail band. Note similar, but smaller, **Lesser Black-backed Gull** *L. fuscus* not now accepted for region.

Voice: As **Yellow-legged**.

Habits: As **Yellow-legged** but probably even more coastal.

Distribution: Status still uncertain due to identification problems. Probably scarce winter visitor to western and southern coasts from Pakistan to Andhra Pradesh. Also Sri Lanka. Regular in small numbers on north-western rivers and lakes. Baraba Gull *L.(c. or h.) barabensis* may be a small sub-species. Also occurs in N Asia, migrating to W Asia and E Africa.

Nikhil Devasar

PALLAS'S GULL (Great Black-headed Gull)

Larus ichthyaetus 69 cm F: Laridae

Description: A strikingly large, pale, deep-chested, long-winged gull with long-sloping forehead and powerful yellow bill with black and orange tip. Breeding plumage adult has black head and white eye rings. In non-breeding plumage head becomes dusky. Very pale grey above with whiter outer wing and isolated black wing spots near tips. Immature browner but still pale and shows grey mantle and wing panel. Dark bill and white below on juvenile. Sexes alike.

Voice: Rather quiet. Nasal *argh*.

Habits: Inhabits coasts and large rivers and lakes. Often solitary, but readily mixes with smaller gulls and terns. Scavenges fish waste and robs others for fish. Heavy buoyant heron-like flight.

Nikhil Devasar

Distribution: A scarce winter visitor to most coasts and inland.
Also occurs in E Africa and W, N and Central Asia.

BROWN-HEADED GULL (Indian Black-headed)

Larus brunnicephalus 42 cm F: Laridae

Otto Pfister

Otto Pfister

Description: A medium-sized, pale gull with large white spots in black wing tips. Adult is pearly-grey above, white below, with chocolate brown partial hood in breeding plumage, reducing to earmark when not breeding. First years have brownish in wings and black tail bars. Has very pale irises and white eye rings. Sexes alike.

Voice: A deep *kraaa*.

Habits: Inhabits coasts, rivers and lakes. Very gregarious, often with **Black-headed**. Scavenges, especially fish waste from humans, robs terns but also catches flying and surface insects. Light, easy flight. Swims buoyantly, often among ducks and **Coots**. Breeds colonially in high-altitude bogs.

Distribution: Breeds in Ladakh. Common winter visitor to most coasts and inland in north. Also occurs in Central Asia.

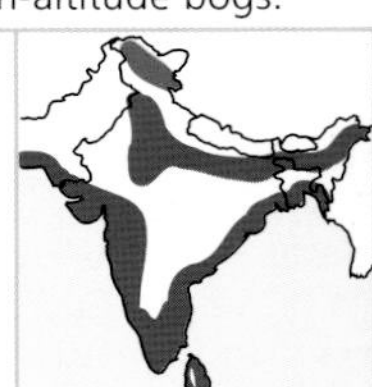

BLACK-HEADED GULL (Common Black-headed Gull)

Larus ridibundus 38 cm F: Laridae

Otto Pfister

Sharad Gaur

Description: A medium-sized, pale gull with mainly white outer flight feathers. Slighter than **Brown-headed** with more pointed, whiter wings. In breeding plumage shows a darker brown partial hood and white eye rings but dark irises. In non-breeding plumage (shown with a single **Brown-headed**) dark ear patch. Immature is browner but has more grey in wings than **Brown-headed**. Sexes alike.

Voice: A higher *keerraa* than **Brown-headed**.

Habits: Inhabits coasts, large rivers and lakes. Often mixing with **Brown-headed**. Similar behaviour but even more prone to long and high roosting flights in disciplined skeins.

Distribution: Common winter visitor to most coasts and inland. Also occurs in Europe, Africa and Asia.

GULL-BILLED TERN

Gelochelidon nilotica 38 cm F: Laridae

Description: Large, pale tern with a thick, black bill and long, black legs. Bull-necked, gull-like mien. In breeding plumage has sleek, black cap but this reduces to black patch behind eye in non-breeding plumage. Pale grey above including rump and tail. Dark trailing edge to primaries noticeable in flight. Similar, but rarer and coastal, **Sandwich Tern** *Sterna sandvicensis* has long,thin, black bill often with yellow tip, shaggy hind crown, white rump and tail. Sexes alike.

Voice: Normally rather quiet but a guttural *ger erk*.

Habits: Inhabits rivers, jheels and coasts. Hawks for flying insects, often over dry land but also picks fish from water surface. Nests on ground.

Mohit Aggarwal

Distribution: Common winter visitor to coasts and inland. Local breeder. Also occurs in Europe, Africa, W, Central, E and SE Asia and the Americas.

CASPIAN TERN

Sterna caspia 53 cm F: Laridae

Description: A huge, pale tern with a powerful, scarlet bill. The size of a large gull with a large head and dagger-like bill with black band near yellow tip. In breeding plumage has a black crown becoming speckled white in non-breeding season. Upper-parts pale grey but noticeable blackish outer primaries in flight. Sexes alike.

Voice: A deep heron-like *kaarrh* in flight.

Habits: Inhabits coasts, large rivers and lakes. Feeds by plunge diving for fish. A heavy, strong and determined flyer. Tends to be solitary but small groups gather on sandbanks. Lays eggs on sand.

Nikhil Devasar

Distribution: Mainly scarce winter visitor to some coasts and inland. Most frequent in west and south. Breeds in Gujarat and Pakistan. Also occurs in Europe, Africa, Central Asia, Australasia and N America.

RIVER TERN (Indian River Tern)

Sterna aurantia 42 cm F: Laridae

Nikhil Devasar

Description: A large, silvery tern with a striking yellow bill and red legs. Silvery-grey above but shows contrasting white flight feathers which are a useful id aid at a distance. Long, white, forked tail. Underparts white with black cap in breeding season, speckled white when not breeding and has black bill tip. Sexes alike.

Voice: A rather high *kiuck* call.

Habits: Inhabits large rivers, lakes and estuaries. Patrols over water, diving for small fish. Usually in small parties, often with other terns. Rests and lays eggs on sandbanks, usually in colonies.

Distribution: Common breeding resident throughout, less so in south. Also occurs in SE Asia.

GREAT CRESTED TERN (Swift, Large Crested Tern)

Sterna bergii 47 cm F: Laridae

Morten Strange

Description: A robust, dark grey-backed, marine tern with a powerful, greenish-yellow bill. Heavily built with dark grey upperparts and blackish wing tips. Always shows a white forehead. Shaggy black cap in breeding plumage. In non-breeding plumage (shown), black cap reduced. Dark legs. Sexes alike. Smaller and slimmer **Lesser Crested Tern** *S. bengalensis* is paler grey above and has a thinner orange bill.

Voice: A coarse *kerrick*.

Habits: Seen mainly off-shore or around the archipelagos, where it nests on islets.

Distribution: Local resident. Most regular off Pakistan, Sri Lanka, western and southern Indian coasts.
Also occurs in Africa, W and SE Asia, Pacific and Indian Ocean Islands and Australasia.

BLACK-NAPED TERN

Sterna sumatrana 33 cm F: Laridae

Description: A slim, startling white tern with narrow black lines from eyes to black nape. Deep tail fork. Very pale grey wash above and dark shafts to outer primary only. May show pink flush below when breeding. Black legs and bill. Sexes alike. Juvenile is black flecked above. Appears even whiter than **Roseate** *S. dougallii*, which is otherwise similar in non-breeding plumage. **White** is stockier with white head, shallow tail fork and huge black eyes.

Voice: A sharp *kik*.

Habits: Strictly maritime but often feeds close inshore in flocks. Surface picks or plunge-dives for small fish. Perches on posts, buoys and fish traps. Nests on rocky islets.

Morten Strange

Distribution: Local breeding resident restricted to Maldives (where it is the commonest tern), Andaman and Nicobar Islands. Very rarely off coasts of Bangladesh, south India and Sri Lanka but should be looked for off east coast.
Also occurs in SE Asia, Pacific Islands and Australasia.

COMMON TERN

Sterna hirundo 35 cm F: Laridae

Description: A medium-sized, pale tern with greyish wash to underparts and a black-tipped, red bill. Mantle and wings, pale grey contrasting with white, rump and white deeply-forked tail. In flight shows darker tips to primaries. Black cap reduces to hind crown in non-breeding plumage and underparts whiter. Sexes alike. Marine **Roseate Tern** is whiter, often with pink flush and has much longer tail streamers and long black bill.

Voice: Loud, harsh *kirrah kirrah*.

Habits: Inhabits coasts and, rarely, inland lakes and rivers. Feeds by plunge-diving after fish. Usually in scattered flocks, often with other terns. Breeds on ground by high-altitude lakes.

Otto Pfister

Distribution: Breeds in Ladakh. Local winter visitor to most coasts and sometimes inland. Most common off Pakistan, south India and Sri Lanka. Also occurs in Europe, Africa, Asia, New Guinea and the Americas.

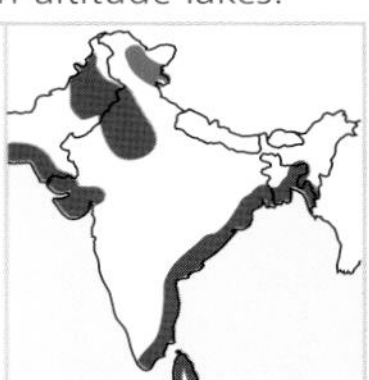

Otto Pfister

LITTLE TERN

Sterna albifrons 23 cm F: Laridae

Morten Strange

Description: A very small tern with black-tipped yellow bill and white forehead. Pale grey back and wings with dark outer primary. Forehead patch extends as point above eye. Yellow legs. In non-breeding plumage, bill black and black crown reduced. Sexes alike. Similar but marine **Saunder's Tern** *S. saundersi* has square-ended forehead patch, more black in primaries, browner legs.

Voice: A high, hurried *kirik kirik*.

Habits: Inhabits sandy coasts, large rivers and lakes. Disperses widely but chiefly to coast in non-breeding season. Distinct, fluttering flight interspersed by hovering from which it makes sudden plunge dives for small fish. Lays eggs in sand on ground.

Distribution: Local breeder in north, southern coasts and Sri Lanka. Disperses widely, mainly to coasts, in non-breeding season.
Also occurs in Africa, W, E and SE Asia and Australasia.

BLACK-BELLIED TERN

Sterna acuticauda 33 cm F: Laridae

Morten Strange

Description: A small, but very long-tailed, tern with black underparts. Slender build and long, deep orange bill distinguish from **River**. Black cap, rather dark grey upperparts, white face, neck and upperparts. Underparts, black from lower breast to vent (unlike **Whiskered** which has a whitish vent at all times). Sexes alike.

Voice: Shrill *krek krek* and a fast *kek kek kek*.

Habits: Inhabits inland rivers and lakes. Feeds on small fish by plunge-diving and surface picking. Also flies high after aerial insects. Nests on sand bars in dry season but much affected by barrage construction and consequent changes in water levels. As a result, becoming very rare.

Distribution: Globally threatened and now very local breeding resident throughout most of region. Most regular in parts of northwest India.
Also occurs in SE Asia.

WHISKERED TERN

Childonias hybridus 25 cm F: Laridae

Kamal Sahai

Description: Small, stocky tern with shallow tail fork and slender, black bill. Mostly seen in non-breeding plumage when pale grey above, including rump and tail. White below with blackish speckled rear crown. In breeding plumage deep grey on underparts, white cheeks and vent and ull black cap. Immatures have dark saddles.

Voice: Rather quiet. A shrill *kerk kerk*.

Habits: The commonest tern. Inhabits mainly large rivers, jheels, wet paddy and the coast. Nests on floating vegetation. Usually seen in large synchronised flocks, dipping the water surface for insects. Occasionally dives and wheels high for swarming insects. Perches on wires.

Distribution: Local breeding resident in Kashmir and Assam, occasionally elsewhere in north India. Abundant winter visitor throughout.
Also occurs in Europe, Africa, throughout Asia and in Australasia.

Otto Pfister

WHITE-WINGED TERN (White-winged Black Tern)

Childonias leucopterus 23 cm F: Laridae

Morten Strange

Description: Small, stocky tern with short, black bill. Non-breeding plumage like **Whiskered**, but more paddle-shaped wings. Has black ear, muffs and very restricted, speckled, rear crown. In breeding plumage, very distinctive. Black all over, including striking underwing coverts (which are the first and last feathers to moult), with silvery grey wings and mantle, white forewings, rump, tail and vent. Sexes alike. Immature has dark saddle like **Whiskered**. Rarer **Black Tern** *C. niger* always shows black breast patches.

Voice: Quiet. A rasping *chree*.

Habits: Inhabits rivers, lakes and estuaries. Feeds by surface-dipping. Does not dive.

Distribution: Scarce passage migrant and winter visitor throughout. Most regular in Gujarat, Tamil Nadu and Sri Lanka but probably overlooked among Whiskered Terns.
Also occurs in Europe, Africa, Central, E and SE Asia and Australasia.

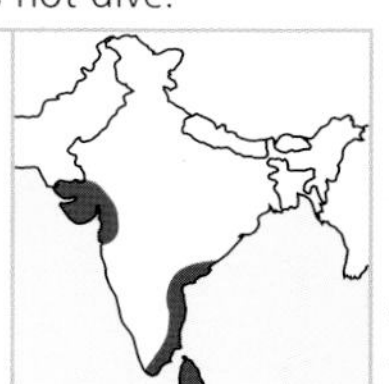

Morten Strange

WHITE TERN (Indian Ocean White, Fairy Tern)

Gygis alba 29 cm F: Laridae

Tim Loseby

Description: A medium-sized, pure white, marine tern with large, black eyes. Very pointed, slightly upturned black bill. Blue legs. Immature has light brown barring. Sexes alike.

Voice: A harsh *grich grich*.

Habits: Extremely confiding at nest. It lays a single egg in a tree hollow or the base of a palm frond, often quite low and usually looking most precarious. Feeds offshore on small fish shoals, usually catching them in the air as they jump to escape predators.

Distribution: Common resident only in Maldives where breeds. In non-breeding seasons wanders Indian Ocean and could be storm-blown to mainland coasts.
Also occurs on the Pacific, Indian and Atlantic Ocean Islands.

OSPREY (Fish-hawk)

Pandion haliaetus 56 cm F: Accipitridae

Otto Pfister

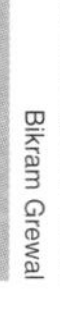

Bikram Grewal

Description: A large, brown and white raptor with long, angled wings. Dark brown above with white-crested head and brown stripe from eye to brown neck and brownish crown. Below white with brown-streaked breast band. Underwing pale with large, dark carpal patches, dark tips and crossbar. Rather short tail, faintly barred. Flies with wings angled down at carpal joints and tips raised.

Voice: Usually silent. A *queep* contact call.

Habits: Inhabits large rivers, lakes and the coast. Nests in tree. Feeds on fish, which it catches with its feet in an impressive dive, often submerging. Perches on bare tree branches and posts near water. Rarely soars high. Solitary.

Distribution: Scarce breeding summer visitor to northern mountains. Scarce winter visitor throughout region.
Also occurs throughout Europe, Africa, Asia, Australasia and the Americas.

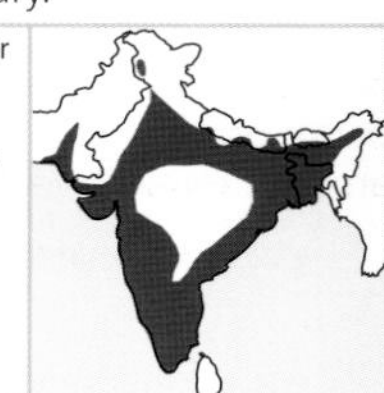

JERDON'S BAZA (Blyth's, Brown Baza)

Aviceda jerdoni 48 cm F: Accipitridae

Description: A medium-sized, hawk-like raptor with an erect, white-tipped crest. Pale head, brown upperparts. Banded tail and rufous-barred underparts. Wings nearly reach tail tip. In flap and glide flight, broad paddle-shaped wings, pinched in near body, and rather pale below. Can appear similar to **Crested Goshawk**. Sexes alike.

Voice: During breeding calls *kip kip kip* and *keeya keeya*. Also *pee ow*.

Habits: Inhabits evergreen forest and plantations. Crepuscular and sluggish, spending hours on often concealed perch looking for reptiles, amphibians and insects, which it catches on ground. Often in small parties but easily overlooked. Nests in tree.

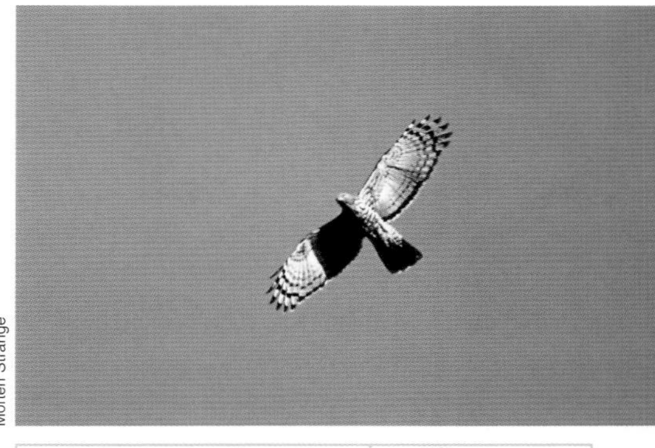

Morten Strange

Distribution: Rare breeding resident mainly in northeastern hills and Bangladesh. Also Sri Lanka and south peninsular hills.
Also occurs in SE Asia.

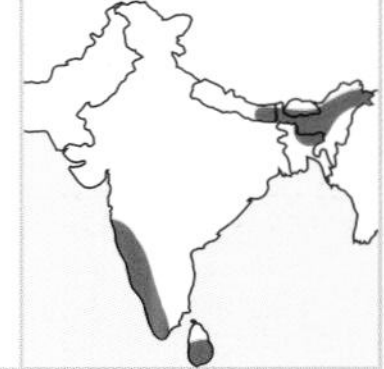

BLACK BAZA (Indian Black Crested, Black-crested Baza)

Aviceda leuphotes 32 cm F: Accipitridae

Description: A small, black and white raptor with a long, vertical crest. Black head, upperparts and thighs with some white and chestnut markings. White upper-breast band, followed by black and then chestnut bands. Chestnut and buff barring below. Rounded wings have distinctive grey and black pattern below. Tail grey. Sexes alike.

Voice: A high, squealing *pee aa*.

Habits: Inhabits forests. Rather crepuscular, sitting motionless for long periods, high in canopy or circling above it in small parties. Mainly eats insects and arboreal reptiles. Nests in tree.

Sujan Chatterjee

Distribution: Scarce breeding resident and local migrant mainly in Nepal, the Northeast and the southwest. Also rare passage migrant and winter visitor to Himalayan foothills, the Tamil Nadu coast and Sri Lanka.
Also occurs in China.and SE Asia.

ORIENTAL HONEY-BUZZARD (Crested, Eurasian)

Pernis ptilorhyncus 65 cm F: Accipitridae

Otto Pfister

Nikhil Devasar

Description: Large, pigeon-headed raptor with two broad tail bands. Variable, usually dark brown, often with grey head. Closely barred or whitish below. Barred underwings with dark trailing edges (broad in male) and three bars across flight feathers. Bare yellow legs. Adults have yellow irises, juveniles dark. Soaring wings held flat; long neck and small head protrude.

Voice: Gull-like *peee ou* when breeding.

Habits: Inhabits open forest, woodland, cultivation, villages and urban parks. Usually singly or in pairs. Feeds on grubs of bees and wasps, unconcernedly tearing up combs in trees or on ground. Confiding. Also eats reptiles and nestlings.

Distribution: Common and widespread breeding resident throughout lowlands. Also occurs in N, E and SE Asia.

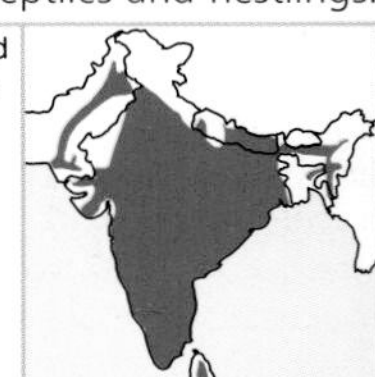

BLACK-SHOULDERED KITE (Black-winged Kite)

Elanus caeruleus 35 cm F: Accipitridae

Kamal Sahai

Kamal Sahai

Decscription: A small, pale grey raptor with striking ruby eyes. Pale grey above with broad black shoulders and black wingtips. Head and underparts white. Dark around large, red eyes. Shallow-forked, white tail. Buoyant flight on narrow wings. Sexes alike. Immatures duskier.

Voice: A high *peeya*.

Habits: Inhabits cultivation, open grassland with scrub and open woodland. Usually solitary or in pairs. Often sits on wires, poles or dead branches. Quite confiding. Graceful tern-like flight. Often hovers before floating down to pounce on reptiles and large insects. Also pounces from high perch. Nests in small, usually isolated, trees.

Distribution: A widespread but local breeding resident in the lowlands subject to local movements. Also occurs in S Europe, Africa, W and SE Asia and New Guinea.

RED KITE (Kite)

Milvus milvus 70 cm F: Accipitridae

Description: A medium-sized, slender raptor with a long, deeply forked rufous tail which is often twisted to aid steering. Adults rufous-brown with contrasting whitish-streaked head. Underwing in flight shows obvious whitish base to primaries contrasting with black wing tips. Sexes alike. Immatures browner.

Voice: Usually quiet but sometimes mews *wee u wee u*.

Habits: Inhabits sparsely wooded country and semi-desert. Soars searching for prey, which includes carrion, small mammals and birds. Also scavenges at garbage dumps. A very adept flier, more graceful than **Black Kite** but great care needed with well-marked immatures of **Black-eared** race of **Black**.

Otto Pfister

Distribution: Rare winter visitor mainly to northern areas.
Also occurs in Europe, N Africa, W Asia and Atlantic Islands.

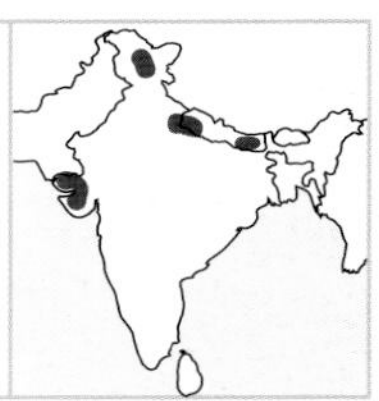

BLACK KITE (Pariah, Dark, Black-eared Kite)

Milvus migrans 61 cm F: Accipitridae

Description: Brown raptor with shallow-forked tail. Dark brown with paler head, Long tail normally shows fork unless spread or in moult. Sexes alike. Juvenile more streaked. Larger migrant **Black-eared** race, *M. m. lineatus,* shows larger pale base to primaries and dark ear patch. Juvenile can be very pale.

Voice: A rippling *qweeee*.

Habits: Most abundant raptor. Inhabits towns, large villages and along rivers and lake-shores. Soars in scattered groups in search of food, twisting tail to steer. Gathers on pylons and other structures near food sources. Scavenges garbage and dead animals, particularly fish. Also feeds on worms on fresh plough. Nests high in tree.

Otto Pfister

Otto Pfister

Distribution: Abundant breeding resident. The Black-eared race is a common winter visitor mainly in north and breeds in Himalayas.
Also occurs in Europe, Africa, throughout Asia and Australasia.

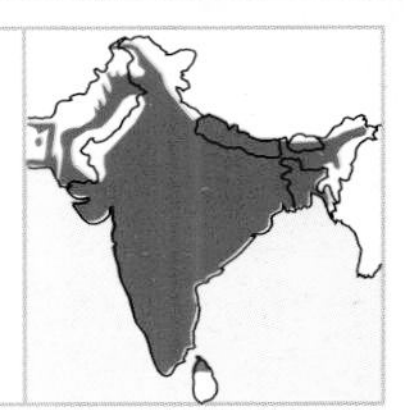

BRAHMINY KITE

Haliastur indus 50 cm F: Accipitridae

Otto Pfister

Kamal Sahai

Description: A medium-sized, rich chestnut raptor with a white head, neck and breast. Kite-like but with broader wings angled back at carpals and rounded tail. Black wing tips. Sexes alike. Juvenile browner with buzzard-like pale base to primaries.

Voice: A nasal, wavering *kee ah*.

Habits: Inhabits coasts especially near fishing villages and estuaries. Also inland by large rivers and lakes. Soars low, hanging in wind. Scavenges mainly fish and fish waste but also takes small animal prey. Often confiding. Sometimes in large gatherings but usually solitary. Will mix with **Black Kites**. Nests in isolated trees near water, often village palms.

Distribution: Rather local but widespread breeding resident, commonest on coast and decidedly scarce in the Northwest. Also occurs in China, SE Asia and Australasia.

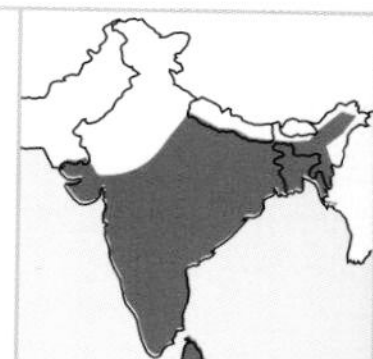

WHITE-BELLIED SEA EAGLE (White-bellied Fish-eagle)

Haliaeetus leucogaster 70 cm F: Accipitridae

Joanna Van Gruisen

Description: A large grey and white raptor with a black-based, white tail. Grey upperparts with black flight feathers. Smallish head, neck and underparts white. Wings broad and bulging. Very short wedge-shaped tail. Bluish bill and yellow legs. Sexes alike. Immatures whitish with pale bases to primaries, dark secondaries and tail band.

Voice: A resonant honking *ahnk ahnk ahnk,* particularly when breeding.

Habits: Inhabits estuaries, mangroves, coastal lagoons and offshore islets. Rare inland. Solitary or in pairs. Catches sea and water snakes by plunge-diving from lookout or sky and grasping with feet. Makes huge nest on isolated tree or coastal rock.

Distribution: Local resident on most coasts from Mumbai round to Bangladesh. Also Sri Lanka, the Andaman and Nicobar Islands. Also occurs in China, SE Asia and Australasia.

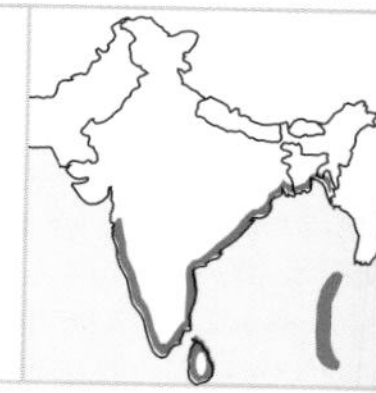

PALLAS'S FISH EAGLE (Pallas's, Ring-tailed Fishing Eagle)

Haliaeetus leucoryphus 80 cm F: Accipitridae

Otto Pfister

Description: Very large, brown raptor with black-banded white tail. Buff head and neck, more rufous on mantle, breast and darker belly. Dark, long, narrow wings held level. Contrasting tail often spread. Protruding head and long neck. Sexes alike. Immature, patchy brown with pale wedges in inner primaries and dark tail.

Voice: A high yapping *kay ou kay ou,* especially when breeding.

Habits: Inhabits large rivers, lakes and marshes. Feeds mainly on fish, which it catches from the surface, finds dead or pirates. Also robs nesting colonies of waterbirds and takes wildfowl. Solitary or in pairs. Nests in isolated tree near water.

Distribution: Globally threatened and very local breeding resident now mainly in the northern river valleys and the Northeast.
Also occurs in W and Central Asia and Myanmar.

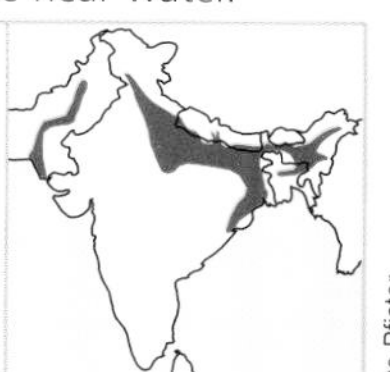

imm
Otto Pfister

WHITE-TAILED EAGLE (White-tailed Sea Eagle)

Haliaeetus albicilla 77 cm F: Accipitridae

Description: Huge, broad-winged eagle with rather short, white, wedge-shaped tail. Powerful, yellow, deeply-hooked bill. Adult dark brown with paler head and neck and pure white tail. Sexes alike. Immatures have brown tail and much darker head at first, taking five years to mature. Well protruding head and neck in flight but striking feature is very broad "barn-door" wings.

Voice: Yapping *klee klee klee*.

Habits: Inhabits large rivers and lakes, only rarely on the coast in this region. Spends time perched, often on low eminence. Feeds on fish by snatching from surface. Also waterbirds, small mammals and carrion. Hunts low with deep, easy wing-beats but also soars high.

Jan Willem den Besten

Distribution: Rare winter visitor mainly to north India and Pakistan but records throughout region south to Kerala and east to Arunachal Pradesh.
Also occurs in Greenland and throughout northern Eurasia east to Japan and China.

GREY-HEADED FISH EAGLE (Himalayan Fishing Eagle)

Ichthyophaga ichthyaetus 75 cm F: Accipitridae

Otto Pfister

Description: Large, brown eagle with grey head and neck. Black-banded white tail. Brown above with darker flight feathers. Warm brown breast and underwings. White belly and vent. Sexes alike. Immature streaked pale brown with whitish flight feathers. Broad, level wings and small, protruding head. Rarer and smaller **Lesser Fish Eagle** *I. humilis* has plain grey-brown tail.

Voice: A ringing *kerrah kerrah,* especially when breeding.

Habits: Inhabits well-wooded rivers and swamps. Sometimes in mangroves. Solitary or in pairs. Feeds mainly by grasping surface fish which it watches for, in upright mode, most of day from high perches. Builds large nest in isolated tree near water.

Distribution: Scarce breeding resident in the Northeast. Now rare elsewhere including Sri Lanka, Nepal, Bangladesh and a few sites in the peninsula and the Himalayan foothills.
Also occurs in SE Asia.

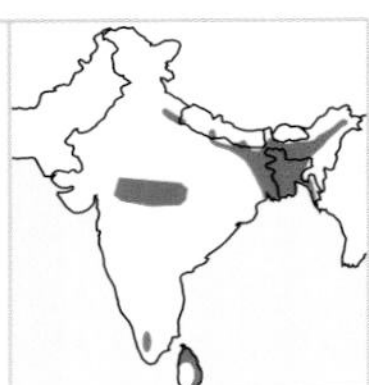

LAMMERGEIER (Bearded Vulture)

Gypaetus barbatus 125 cm F: Accipitridae

Otto Pfister

Description: Huge, grey and orange raptor with long wedge-shaped tail. Dark grey above and orange-buff below, darkest on breast. Black round eyes extends as hanging tuft over bill. Under-wing coverts blackish. Sexes alike. Immature same shape but dark brown. **Egyptian** is much smaller.

Voice: Usually silent. Sometimes a deep croak.

Habits: Inhabits high mountain ranges. Usually solitary. A master flier, effortlessly cruising mountain slopes often close to ground. Scavenges dead animals, feeding on the bones which it breaks by dropping from great height. Raids garbage dumps but avoids mixing with other vultures. Perches and nests on cliff ledges.

Distribution: Common breeding resident in high northern mountains. Also occurs in S Europe, Africa and W, Central and E Asia.

EGYPTIAN VULTURE (Scavenger, Small White, Neophron)

Neophron percnopterus 65 cm F: Accipitridae

Description: A medium-sized, dirty-white raptor with a wedge-shaped tail. White plumage often stained rusty. All flight feathers black. Small, shaggy head with yellow face and thin yellow or grey bill. Sexes alike. Immature blackish brown with grey face.

Voice: Usually silent.

Habits: Probably now the commonest vulture in the region. Usually near villages and towns, where it feeds on garbage and haunts dumps, usually in small parties. Will also feed, cautiously, on carrion with other vultures. Soars over wide areas in search of food. Often perches on ruins, high roofs and tombs. Nests on buildings, cliffs or trees.

Distribution: Fairly common breeding resident throughout region but rare or unknown in the Northeast, Bangladesh and Sri Lanka.
Also occurs in S Europe, Africa and W and Central Asia.

Kamal Sahai

Kamal Sahai

WHITE-RUMPED VULTURE (Oriental White-backed)

Gyps bengalensis 85 cm F: Accipitridae

Description: A large, dark, broad-winged raptor with large white back patch. Adult blackish-brown with whitish ruff and underwing coverts and brownish-grey, naked head and neck. Broad wings, small head and short tail. Bill blackish. Sexes similar. Immature lighter streaked brown with paler neck and no underwing contrast.

Voice: Squabbling grunts and hisses when feeding.

Habits: Where it still occurs in towns it scavenges at garbage dumps, tanneries and slaughterhouses. In rural areas feeds on dead livestock and predator kills. Spends much time soaring, covering huge areas. Waddles on ground and sits hunched. Very sociable where it can be. Nests high in trees.

Distribution: Formerly very common breeding resident throughout lowlands except Sri Lanka. Now globally threatened for unknown reasons and very local. Most frequent in northern wildlife sanctuaries and now very scarce in towns.
Also still occurs rarely in SE Asia.

Otto Pfister

Otto Pfister

INDIAN VULTURE (Long-billed, Indian Griffon)

Gyps indicus 90 cm F: Accipitridae

Otto Pfister

Description: Large, pale brown raptor with yellow bill. Mid-brown, paler below with white thighs and ruff and bare dark brown head and neck with scattered white fluff. Striking contrast between pale underparts and dark flight feathers and tail. Broad wings, small head, long neck and short tail.Sexes alike. Immature even paler with pale head and neck and grey bill. This is a peninsular vulture, recently separated from the northern **Slender-billed** *G. tenuirostris* which has a plain, dark neck and head and mainly dark, slender bill.

Voice: Quieter than **White-rumped**.

Habits: As **White-rumped**. Nests colonially exclusively on cliff ledges and buildings.

Distribution: Formerly common breeding resident except in extreme south and Sri Lanka. Now globally threatened for unknown reasons and very local. Most frequent in large wildlife sanctuaries.
Also still occurs rarely in SE Asia.

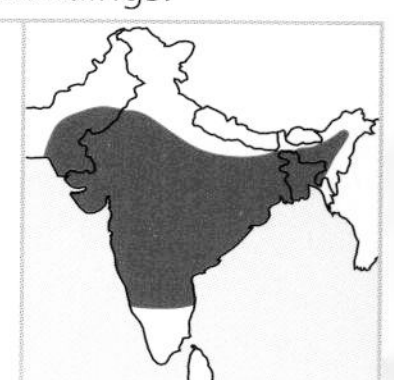

HIMALAYAN GRIFFON (Himalayan Griffon Vulture)

Gyps himalayensis 125 cm F: Accipitridae

Otto Pfister

Otto Pfister

Description: A huge, pale raptor with whitish underparts contrasting with black flight feathers. Upperparts sandy-buff with bare buff head and neck and shaggy ruff. Pale bill and cere. Feet pinkish. Sexes alike. Immature darker brown above streaked white. Soars on huge, flat wings with upturned fingers. Tail short, head looks small.

Voice: Grunts and hisses when feeding.

Habits: Usually seen soaring high over mountains singly or in small groups. Covers wide areas in search of food. Follows cattle graziers on their altitudinal movements to find carcasses on which it feeds. Rests and nests colonially on cliffs.

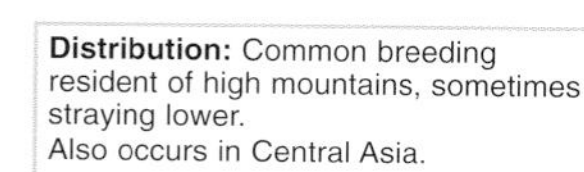

Distribution: Common breeding resident of high mountains, sometimes straying lower.
Also occurs in Central Asia.

EURASIAN GRIFFON (Griffon Vulture)

Gyps fulvus 100 cm F: Accipitridae

Description: A huge, rufous-brown vulture with a thick white neck ruff on adults. Darker and smaller than **Himalayan**. White head and neck. Bill yellow in adult, grey in immature. In flight shows dark, rather streaked, body, dark flight feathers and tail and pale underwing coverts with some white lines. Sexes similar.

Voice: Quiet. Grunts and hisses when feeding.

Habits: Inhabits mountains and semi-desert, wandering into nearby dry plains in winter. Usually in pairs or small groups, often with other vultures. Spends much time soaring looking for carcasses. Nests in small groups on cliff ledges.

Otto Pfister

Distribution: Common breeding resident in hills of western Pakistan and northern India and Nepal east to Sikkim. Winters widely in lowlands of Pakistan and northwest India south to Gujarat and Madhya Pradesh. Scattered records elsewhere.
Also occurs in S Europe, N Africa, W and Central Asia.

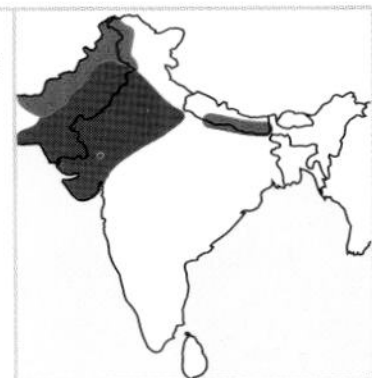

CINEREOUS VULTURE (Black, Monk Vulture)

Aegypius monachus 115 cm F: Accipitridae

Description: A huge, chocolate brown raptor with a blackish face mark. Rather angular head, dark eyes, whitish feet and powerful, hooked bill. Has a paler brown ruff extending up neck. Sexes alike. Immature even darker, especially on head. Broad, parallel wings. Short, often wedge-shaped, tail. Small head and habit of giving occasional deep flap when soaring separate it from large, dark eagles. Holds tail up when landing.

Voice: Usually silent.

Habits: Inhabits open country, ranging widely and usually singly looking for carcasses, often near rivers. Dominant over other species at a carcass. Nests on small trees growing out of cliffs.

Joanna Van Gruisen

Distribution: Mainly scarce winter visitor to northern mountains and river valleys. A few pairs probably breed in Pakistan.
Also occurs in S Europe, N Africa and Central and E Asia.

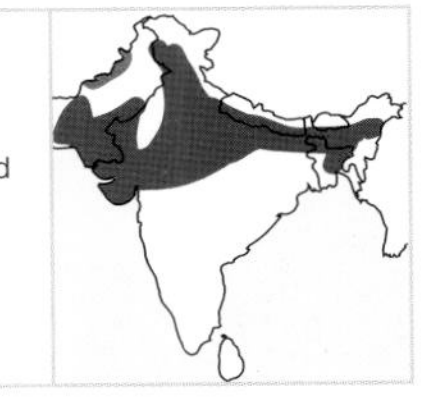

RED-HEADED VULTURE (King, Black Vulture)

Sarcogyps calvus 85 cm F: Accipitridae

Kamal Sahai

Kamal Sahai

Description: A large, blackish raptor with white breast and thighs. Adults have bare, red head and neck and reddish feet. Eyes and cere yellow, sharply hooked bill, black. Sexes alike. Immature browner with white down on head. In flight, white thighs, underwing stripe and breast patch and, often visible, red head make identification easy. Immature has white belly and pink head.

Voice: Some grunts and hisses when feeding.

Habits: Usually singly or in pairs. It joins other vultures at carcasses but is rather timid and, in spite of bill size, often has to wait its turn. Inhabits cultivation, semi-desert and open woodland. Nests high in a lone tree, often near a village.

Distribution: A scarce but widespread breeding resident throughout the lowlands.
Also occurs in China and SE Asia.

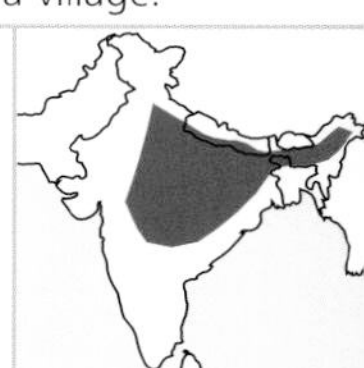

SHORT-TOED SNAKE EAGLE (Short-toed Eagle)

Circaetus gallicus 65 cm F: Acciptridae

RK Gaur

Sharad Gaur

Description: A medium-sized, but bulky, raptor which is usually rather pale below. Variable but most have closely barred, white underparts and underwings and barred, square-ended tail. No carpal patches and head to upper breast often brown as upperparts. Large owl-like head, prominent yellow irises and grey legs. Some individuals very pale. Sexes alike.

Voice: A loud *kee yo*.

Habits: Inhabits open country with a preference for scrubby grasslands. Hunts largely snakes and other reptiles, which it spots from rather clumsy hovering, soaring or from a high perch. Makes a steep dive to capture. Usually solitary.

Distribution: Scarce breeding resident in lowlands and more common and widespread winter visitor except in the Northeast, Bangladesh and Sri Lanka. Also occurs in S Europe, N Africa and W, Central and E Asia.

CRESTED SERPENT EAGLE

Spilornis cheela 75 cm F: Accipitridae

Description: Large, big-headed raptor with striking black and white pattern on tail and wings. Dark purplish-brown above with fine white speckling. Crown and nape have bulky but short, black and white crest. Underparts rufous vermiculated white and grey. Black tail with broad white central band. Broad wings have black trailing edges and black bands. Sexes alike. Immature paler, more speckled. Yellow face, cere and legs.

Voice: Calls frequently in flight, *kee kee ke*.

Habits: Inhabits well-wooded lowlands, including forest and often near water. Soars high, often in calling pairs, but also patrols low over canopy looking for arboreal reptiles and mammals.

Otto Pfister

Sumit Sen

Distribution: Scarce but widespread breeding resident. Rare in the Northwest.
Also occurs in SE Asia and China.

EURASIAN MARSH HARRIER (Western Marsh Harrier)

Circus aeruginosus 55 cm F: Accipitridae

Description: Medium-sized, dark brown raptor with characteristic low, quartering flight. Females and immatures chocolate brown usually with variable amounts of white or yellow on crown, throat and leading edges of wings. Rarer adult male, pale grey in wings and tail and paler, more streaked body. Dark eyes, long yellow legs.

Voice: Usually silent but a rapid, squealing *shee shee shee shee* when alarmed.

Habits: Wetlands, favouring reedbeds. Flies low with broad wings in shallow V, dropping long legs to catch mainly passerines and smaller waterbirds, reptiles and amphibians. Perches close to ground for long periods. Will also scavenge kills of eagles.

Otto Pfister

Otto Pfister

Distribution: Very common winter visitor throughout lowlands replaced by quite different Eastern race or species *C. spilontotus* in the extreme Northeast.
Also occurs in Europe, N Africa and W and Central Asia.

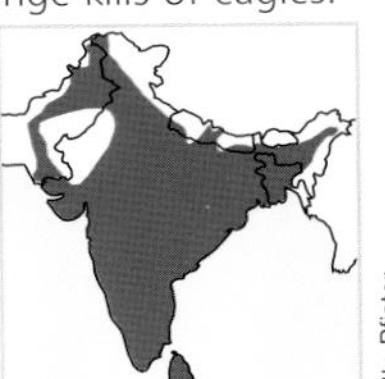

PIED HARRIER

Circus melanoleucos 48 cm F: Accipitridae

Goren Ekstrom

Description: A slender, medium-sized raptor; the adult male is strikingly black and white with black upperparts, head and breast. White forewings and underparts and grey flight feathers and tail. The female is well-streaked brown like other harriers but much greyer in wings and tail. Immatures are plain rufous below.

Voice: Usually silent.

Habits: Inhabits open country including cultivation. Hunts in typical harrier fashion by quartering ground and dropping on rodents, small birds, reptiles and insects. Roosts in groups, often mixed with other harriers, in long grass.

Distribution: Scarce winter visitor mainly to the Northeast, Nepal and extreme south including Sri Lanka. Breeds in Nepal.
Also occurs in Central, N, E and SE Asia.

MONTAGU'S HARRIER

Circus pygargus 48 cm F: Accipitridae

Otto Pfister

Description: Slender, medium-sized raptor. Dark grey head, breast, back and inner wing, paler grey on rest of wing and tail with black wing tips and narrow white rump. Black secondary bar and chestnut streaking on belly and underwing. Similar **Pallid Harrier** *C. macrourus* is pale grey with narrow, black wing tips and larger **Hen Harrier** *C. cyaneus* is grey with broad, black wing tips. Females and immatures of all three are streaked brown with owl-like faces and white rumps.

Voice: Usually silent but alarm is a high-pitched chatter.

Habits: Inhabits open country including cultivation. Often in scattered groups. Roosts communally in long grass. Hunts birds, rodents and large insects.

Distribution: Rather scarce winter visitor throughout peninsula. Passage migrant through Northwest. Rare in east.
Also occurs in Europe, Africa, Central Asia and China.

CRESTED GOSHAWK

Accipiter trivirgatus 42 cm F: Accipitridae

Otto Pfister

Description: Medium-sized, dark grey raptor with broad tail bands. Usually shows a short crest on hind crown and black throat stripe. Underparts strongly marked deep chestnut, the breast streaked and the flanks and belly barred. White undertail coverts often flared out. Very rounded wings and long tail. Orange irises. Female larger.

Voice: A high pitched scream *ke ke ke ke*.

Habits: Inhabits forest. Has favourite perches in thick foliage, often near forest clearings, from which it pounces on prey. Eats medium-sized birds and rodents. Flies low over canopy with fast wing beats, interspersed with long glides. Frequently soars. Nests high in tree.

Distribution: Fairly common breeding resident in the Himalayan foothills, Nepal, the Northeast, the Western Ghats and Sri Lanka.
Also occurs in China and SE Asia.

SHIKRA

Accipiter badius 35 cm F: Accipitridae

Mohit Aggarwal

Otto Pfister

Description: Small, pale raptor with few tail bars. Smaller adult male, pale grey above, finely-barred, pale chestnut below. Larger female darker and browner. Immature has black throat stripe and brown drop-like streaking below. Irises yellow.

Voice: Noisy. Mostly a loud, piercing **Black Drongo**-like *kitou kitou*.

Habits: Inhabits trees, including urban parks and gardens. Not shy. Very aggressive hunter, often following a daily beat. Eats reptiles and birds. Dashes after prey through branches or close to ground. Frequently soars in breeding display. Hurried flight with fast wingbeats and long glides. Usually singly or in pairs. Nests high in tree.

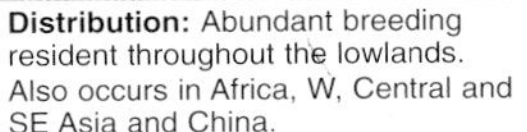

Distribution: Abundant breeding resident throughout the lowlands.
Also occurs in Africa, W, Central and SE Asia and China.

BESRA (Besra Sparrow-hawk)

Accipiter virgatus 32 cm F: Accipitridae

Otto Pfister

Nikhil Devasar

Description: A small, dark raptor with three broad bands on tail. Upperparts, very dark grey. Underparts, dark orange, streaked on breast, barred on belly. Strongly barred underwings, unlike **Shikra**. Black throat stripe and orange irises. Sexes similar. Immature browner above and with brown markings on white underparts.

Voice: Quite noisy. A high-pitched *chew chew chew,* when perched and in flight.

Habits: Inhabits forest and open woodland. Flies rapidly and expertly through trees in pursuit of prey which is mainly small birds. Usually singly or in pairs. Noisy display. Nests high in trees, often in old nest of another species.

Distribution: Scarce breeding resident in Himalayan foothills, Nepal, the Northeast, the Western Ghats and Sri Lanka. Winters in northern plains. Also occurs in Central, E and SE Asia and Australasia.

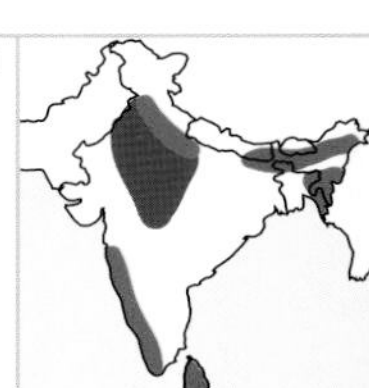

EURASIAN SPARROWHAWK (Northern Sparrowhawk)

Accipiter nisus 35 cm F: Accipitridae

Otto Pfister

Otto Pfister

Description: A small, grey raptor with well-barred tail and plain throat. Male dark grey above with white throat, dark orange-barred underparts and well-barred under wings. Larger female browner with whitish supercilia and no orange. Immature browner still with underparts streaked as well as barred. Irises yellow. Himalayan resident *A. n. melaschistos* (shown) is blackish grey above.

Voice: Rather quiet. A rapid *kew kew kew kew* when nesting.

Habits: Inhabits open wooded country including cultivation. Probably overlooked. Habits as **Shikra**.

Distribution: Fairly common breeding resident of northern mountains, wintering mainly in northern plains particularly in Pakistan. Rare further south.
Also occurs in Europe, N Africa and W, Central, N and E Asia.

NORTHERN GOSHAWK (Goshawk)

Accipiter gentilis Male 50 cm Female 61 cm F: Accipitridae

Description: Large, grey deep-chested raptor with very obvious white supercilia and undertail coverts. Blue-grey above and finely barred grey below. Immature browner and streaked brown below. Long tail and wings barred. Flight shape characteristic with bulging secondaries, thick-based tail and well-protruding neck with large head. Large female **Sparrowhawks** near male **Goshawks** in size.

Mohit Aggarwal

Voice: Usually quiet but pairs call chattering *yek yek yek* and female has high *kee aw* call.

Habits: Usually singly or in pairs, often soaring high over oak and coniferous forest. Also open country. Pursues prey relentlessly, taking birds, hares and large rodents. Shy and rather secretive.

Distribution: Scarce winter visitor mainly in north where it may breed in the Himalayas.
Also occurs in Europe, N Africa, W, Central, N and E Asia and N America.

WHITE-EYED BUZZARD (White-eyed Buzzard Eagle)

Butastur teesa 45 cm F: Accipitridae

Description: A medium-sized, slender, brown raptor with long, bare legs. Adult, which has whitish irises, is brownish above with pale wing patches. Throat has three dark stripes and nape has a whitish patch. Tail rufous and underparts barred rufous. Immature is paler with dark irises and indistinct throat markings. Rounded wings and long tail give hawk-like impression in flight.

Toby Sinclair

Voice: Noisy. A shrill, repeated *te twee*.

Habits: Inhabits open country including cultivation. Usually singly or in pairs. Sluggish, sitting on bare branches, wires and posts for hours, occasionally dropping on prey of reptiles, rodents and insects. Searches burnt ground for prey. Nests high in tree.

Distribution: Widespread but scarce near-endemic breeding resident in lowlands subject to local movements. May appear in an area for quite short periods.
Also occurs in Myanmar.

Otto Pfister

COMMON BUZZARD (Buzzard, Eurasian Buteo)

Buteo buteo 55 cm F: Accipitridae

Otto Pfister

Morten Strange

Description: A very variable, medium-sized, stocky raptor. Three races occur. Usually quite dark brown but can have pale or rufous head and underparts. Tail usually well-barred but can be plain orange. Shows black carpal patches on underwing. Wings and tail broad, neck thick and head small. Buzzard species are not always separable in field.

Voice: Usually silent but has haunting *pee ow* call.

Habits: Inhabits open country including cultivation. Spends much time in high soaring flight or sitting on high perches looking for prey. Normal flight rather heavy with deep wing beats. Rather sluggish. May hover clumsily. Preys by pouncing on rodents and reptiles.

Distrbution: Mainly scarce winter visitor to northern areas. Scarce in south and Sri Lanka. A resident race breeds in Himalayas.
Also occurs in Europe, Africa and W, Central, N and E Asia.

LONG-LEGGED BUZZARD (Long-legged Buteo)

Buteo rufinus 61 cm F: Accipitridae

Otto Pfister

Otto Pfister

Description: Medium-sized, long-winged raptor with, usually unbarred, orange tail. Overall streaked sandy-brown. Larger and paler than Common particularly on head and breast which contrast with darker rufous belly. Very prominent black carpal patches, trailing edges and wingtips contrast with lightly barred, white flight feathers. Long bare yellow legs. Rangy looking in flight.

Voice: Rather quiet. A short *peeu* occasionally.

Habits: Inhabits dry, open country, including cultivation. Usually seen singly, often soaring with wings in a shallow V. Pounces on mainly reptile and rodent prey from flight or a low perch. Often sits on ground. Nests in tree or on cliff-face.

Distribution: Breeds in a few northern sites. Winters commonly in the Northwest, particularly Pakistan.
Rare elsewhere.
Also occurs in E Europe, N Africa and W and Central Asia.

UPLAND BUZZARD (Upland Buteo)

Buteo hemilasius 70 cm F: Accipitridae

Description: A large, usually pale-headed, raptor with blackish thighs. Variable but can usually be told from other buzzards by large size and thickly feathered legs. Pale form has heavily-streaked, whitish head and breast and dull brown upperparts. Dark form has deep chocolate body plumage. Both show very white bases to flight feathers, contrasting with dark carpals and trailing edges and lightly-barred, greyish tail. Eagle-like shape and wings held in deep V when soaring. Not always identifiable in field.

Otto Pfister

Voice: A nasal mewing *pee ou*.

Habits: Inhabits open often rocky country, frequently soaring. Feeds mainly on rodents.

Distribution: Scarce, probably breeding, resident of northern mountains but status unclear due to identification difficulties. Winters rarely in northern plains.
Also occurs in Central and E Asia.

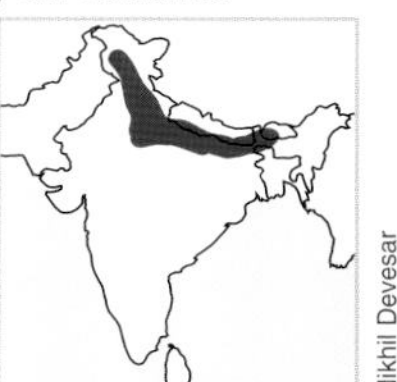

Nikhil Devesar

BLACK EAGLE

Ictinaetus malayensis 75 cm F: Accipitridae

Description: A large, very dark eagle with striking, yellow cere and legs. Adult blackish-brown with faint barring on tail and flight feathers. Immature has streaked buff head and underparts. Long wings reach tail tip at rest. In flight noticeably pinched in wings which bulge at inner primaries and long, broad and square-ended tail. Wings held in shallow V with long, upturned fingers when gliding. Small head.

Jan Willem den Besten

Voice: Rather quiet. A loud *kee kee kee* when courting.

Habits: Inhabits hill and mangrove forest. Normally seen soaring alone or in pairs over canopy. Feeds on birds' eggs and nestlings, birds, arboreal mammals, amphibians and reptiles. Nests in tall forest tree.

Distribution: Local breeding resident in well-watered northern hills from Punjab east to Myanmar border, Bangladesh, and the hills of the Eastern and Western Ghats and Sri Lanka. Scattered records elsewhere presumably stragglers.
Also occurs in S China and SE Asia.

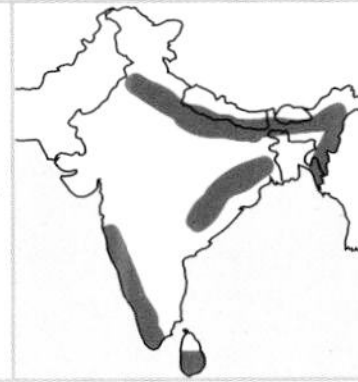

LESSER SPOTTED EAGLE

Aquila pomarina 62 cm F: Accipitridae

Sanctuary/Vivek Sinha

imm

Nikhil Devasar

Description: Large, dark raptor with broad wings and short tail. All dark brown though immature plumage has variable amounts of white spotting in wings and, usually, rufous nape patch. Very like **Greater** but slender with narrower wings and smaller head. Thinner legs. Best distinctions are dark flight feathers contrasting with paler coverts; two white crescents on carpals. Circular nostrils.

Voice: A yapping *kyeep kyeep*.

Habits: Inhabits open wooded country and fields. Less tied to water than **Greater**. Soars frequently with wings held below horizontal. Hunts rodents, reptiles and birds from perch or by low, steady quartering of ground. Nests high in tree.

Distribution: Very local breeding resident mainly in northern and eastern lowlands. Rare further south. Not confirmed from Pakistan and Sri Lanka. Race *A.p.hastata* may be separate species. Also occurs in Myanmar. Nominate race occurs in Europe, Africa and W and Central Asia.

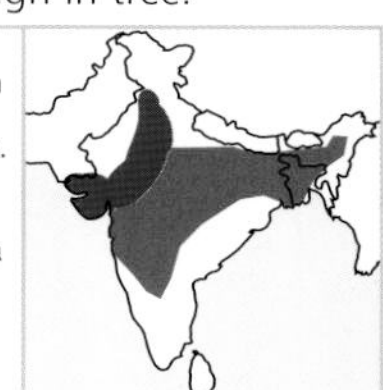

GREATER SPOTTED EAGLE (Spotted Eagle)

Aquila clanga 67 cm F: Accipitridae

Otto Pfister

imm

Otto Pfister

Description: Large, dark raptor with broad wings and short tail. All dark brown though immature has white spotting on upperparts and thighs and white rump crescent. Also a rare pale buff juvenile phase. See **Lesser Spotted** for distinctions. Wings reach end of tail at rest. Legs fully feathered. Circular nostrils. Head often appears large.

Voice: Rather noisy. Ringing *kaek kaek kaek* from perch or in flight. Often several call in unison.

Habits: Inhabits larger wetlands, including mangroves. Flocks where prey abundant. Spends much time perching, often on ground. Soars with wings below horizontal and tail spread. Feeds on waterfowl and carrion, including fish.

Distribution: Globally threatened but locally common winter visitor mainly to northern plains, including Pakistan where it breeds in very small numbers. Scarcer and more local further south. Not recorded from Sri Lanka. Also occurs in Europe, Africa and W, Central, N and E Asia.

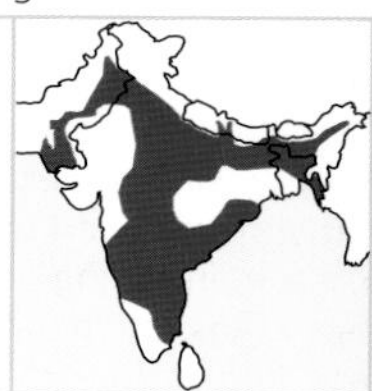

TAWNY EAGLE (Eurasian Tawny Eagle)

Aquila rapax 70 cm F: Accipitridae

Description: A large, variably coloured raptor with yellow irises. Similar to **Steppe** in dark brown phases but smaller with less barring, larger head, shorter tail and less broad wing tips. Gape extends only to middle of eye. Pale phases have creamy buff body contrasting with dark flight feathers and tail. Immatures lack white underwing bars. Thick feathered legs and oval nostrils.

Voice: Rather quiet. A cackling *ke ke ke ke*.

Habits: Inhabits mainly semi-desert and dry rocky scrub country but also cultivation in winter. Spends much time perched on bare branches and poles. Often robs other raptors of their prey and feeds on carrion. Sluggish with slow, heavy flight.

RK Gaur

Distribution: Widespread but scarce breeding resident throughout lowlands; common only in the Northwest. Wanders in winter.
Also occurs in Africa, W Asia and Myanmar.

STEPPE EAGLE

Aquila nipalensis 80 cm F: Accipitridae

Description: Large, dark eagle with dark irises. Larger and more fierce looking than **Tawny** but very similar to dark phases. Adults dark brown, barred on flight feathers and tail and often yellowish nape patch. Always has pale throat. Immatures paler with broad white under-wing bars, trailing edges, tail tip and rump crescent. Gape extends to rear of eye. Thick feathered legs and oval nostrils.

Voice: Usually silent.

Habits: Inhabits all types of lightly wooded and open country, with a preference for wetlands, unlike **Tawny**. Pirates animal prey from other raptors and also eats carrion, including fish. Soars on flat wings, often with tips raised and fingers spread.

Otto Pfister

Distribution: Fairly common winter visitor mainly to northern lowlands. Scarcer further south.
Also occurs in E Europe, Africa and W, Central and E Asia.

IMPERIAL EAGLE (Eastern Imperial Eagle)

Aquila heliaca 85 cm F: Accipitridae

Otto Pfister

Nikhil Devesar

Description: Very large eagle with long, broad wings, square-ended tail and protruding head and neck. Adult, dark brown with white scapular patches and light golden rear neck. Immature streaked sandy-brown with streaked breast. Flight feathers dark with pale wedge on inner primaries and white lines on upperwings only. No white rump patch. Oval nostrils. Soars on flat wings.

Voice: Usually silent. Gruff barking occasionally.

Habits: Inhabits wetlands and other open country. Spends much time in high soaring or sitting on ground or bare tree, from which it pursues other raptors to rob them of their animal prey. Also kills waterbirds. Dominates other eagles in lowlands.

Distribution: Globally threatened and scarce winter visitor mainly to the Northwest. Very rare elsewhere. Also occurs in E Europe, Africa and Central and N Asia.

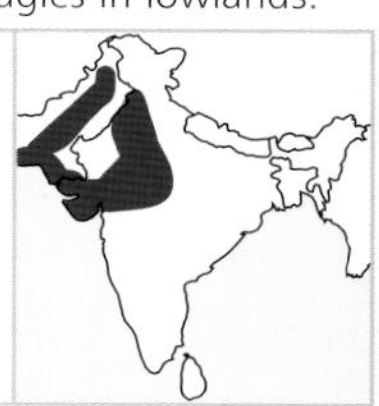

GOLDEN EAGLE

Aquila chrysaetos 90 cm F: Accipitridae

Otto Pfister

Otto Pfister

Description: Huge raptor with broad, long wings, long tail, protruding head and neck. Adults brown with paler coverts, golden crown and nape. Tail tip is darker. Immatures have different amounts of white but, most typically, broad white wing flashes and tail base. Heavily feathered legs. Flies with wings in pronounced V with wing tips raised. Trailing edges markedly curved.

Voice: Usually silent.

Habits: Inhabits mountains above treeline and near steep cliffs. Open country in plains. Usually solitary or in pairs. Extremely powerful hunter, locating animal prey in soaring flight and pursuing it close to ground. Nests on cliff ledge or tree.

Distribution: Scarce breeding resident of northern mountains, rarely wandering to plains in winter. Also occurs in Europe, N Africa, W, N and E Asia and N America.

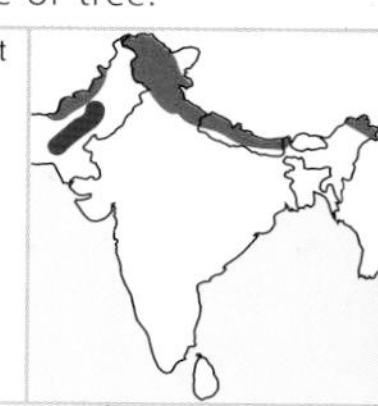

BONELLI'S EAGLE (Bonelli's Hawk Eagle)

Hieraaetus fasciatus 70 cm F: Accipitridae

Description: Fairly large raptor with long wings and tail and protruding neck. Adults dark brown with streaked white underparts, pale mantle and dark-ended grey tail. Underwings show contrasting blackish coverts. Immature gingery-brown below with barred tail and pale rump crescent. Long, feathered legs. Strong, dark-tipped, pale bill.

Voice: A high *kee kee kee* when breeding.

Habits: Inhabits well-wooded country, particularly near water. Fierce hunter of medium-sized birds, mammals and reptiles, often pursuing prey in pairs on regular beat. Robs nesting waterbirds of fish. Soars much less than other eagles, preferring to stay close to prey sources. Nests high in large tree.

Nikhil Devasar

Nikhil Devasar

Distribution: Scarce breeding resident throughout most of lowlands, very rare in the Northeast, Bangladesh and Sri Lanka. Wanders in winter.
Also occurs in S Europe, N Africa and W, Central and E Asia.

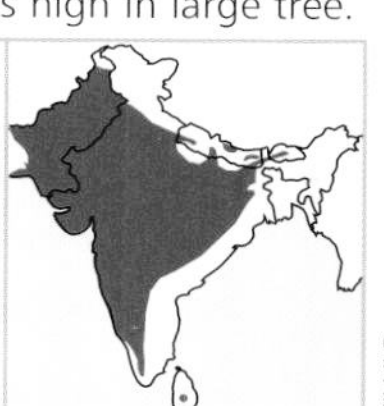

BOOTED EAGLE (Booted Hawk Eagle)

Hieraaetus pennatus 52 cm F: Accipitridae

Description: Medium-sized (smallest eagle), stocky raptor with square-cut, long tail and two colour phases (both shown). Dark phase is dark brown with distinct, pale wedge on inner primaries, pale upperwing patches and small white "landing lights" at the wing junctions. Beware square-tailed **Black Kite**. Pale phase is streaked, rather creamy-brown and buffish white below. Wing markings are the same. Sexes similar. Soars with flat wings.

Voice: Noisy when breeding. A rapid *kwe kwe*.

Habits: Inhabits wooded and open country including cultivation. Locates animal prey from soaring flight, then pounces on it in a spectacular dive. Pairs often hunt in unison. Nests in tall trees.

Otto Pfister

Goren Ekstrom

Distribution: Fairly common winter visitor throughout lowlands. Scarce in the Northeast, Nepal, Bangladesh and Sri Lanka. Breeds in northern mountains and occasionally in peninsula.
Also occurs in S Europe, N Africa and W and Central Asia.

RUFOUS-BELLIED EAGLE (Rufous-bellied Hawk Eagle)

Hieraaetus kienerii 60 cm F: Accipitridae

Nikhil Devasar

Description: Medium-sized, forest raptor with black crown. Adult blackish above and dark chestnut below, lightly streaked white from throat to breast. Immature is white below with diagnostic black armpits, black wing tips and underwing bar. Both have grey-barred flight feathers and tail, with black terminal bar. Short crest on rear crown. Soars with narrow, rounded wings held flat. Secondaries bulge. Tail looks short.

Voice: Normally silent, but has a plaintive scream when breeding.

Habits: Inhabits humid forests. Normally seen soaring over canopy. Stoops rapidly on, mainly bird, prey. Also pounces from concealed perch in foliage. Nests high in forest tree.

Distribution: Rare breeding resident of Western Ghats, Sri Lanka, Nepal, Bangladesh, the Northeast and the Himalayan foothills.
Also occurs in SE Asia.

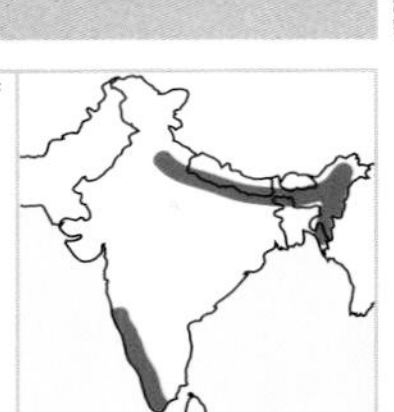

CHANGEABLE HAWK EAGLE (Crested Hawk Eagle)

Spizaetus cirrhatus 70 cm F: Accipitridae

Nanda Rana

Otto Pfister

Description: Large, crested raptor with variable plumage. Usually dark brown above with well, barred tail and prominent crest (slight in Himalayan race *S. c. limnaetus*). Below can be almost pure white or heavily streaked brown, vent often dark. Can be all black. Long, parallel-sided wings pinched in at bodyline, and long tail.

Voice: A drawn out whistling *ke ke ke ke kee*.

Habits: Inhabits well-wooded country. Sometimes in more open country. Usually seen singly or in pairs. Soars infrequently. Hides in well-foliaged tree, often on clearing edge, to locate mammal, bird and reptile prey. Nests high in forest tree.

Distribution: Scarce breeding resident throughout lowlands and foothills. Absent from Pakistan and very rare in most of Northwest.
Also occurs in SE Asia.

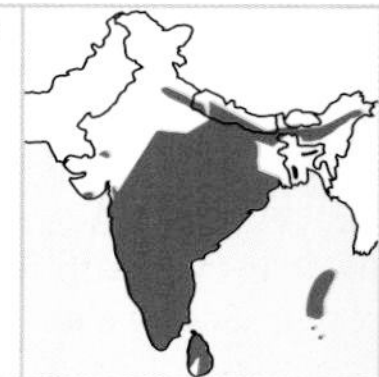

⊙ MOUNTAIN HAWK EAGLE (Hodgson's, Legge's)

Spizaetus nipalensis 72 cm F: Accipitridae

Description: A large, crested, forest raptor with well-barred underparts. Not always easily separable from **Changeable** but less variable, tail shorter and more prominently barred. Belly heavily barred, not streaked. Wing shape different with distinct bulge to secondaries. Southern race has rufous barring below. Immature, pale buff below with well-streaked head. Always has long crest though this often laid flat.

Voice: In breeding season, a high whistling *peeo peeo*.

Habits: Inhabits upland forests. Frequently soars over canopy, often in pairs, but otherwise behaves as **Changeable**.

Otto Pfister

Distribution: Locally common breeding resident in northern mountains, Western Ghats and Sri Lanka.
Also occurs in Myanmar, China and Japan.

⊙ COLLARED FALCONET (Red-breasted, Red-thighed)

Microhierax caerulescens 18 cm F: Falconidae

Description: A tiny, black, rufous and white raptor. Looks rather shrike-like and is barely the size of a **Bay-backed Shrike.** Slaty-grey above, on crown and cheeks with a white collar and neck. Buff below with variable amounts of chestnut from vent to breast. Rather long tail, barred white below. Underwings barred black and white. Sexes alike. The more northeasterly **Pied Falconet** *M. melanoleucos* is larger, black and white and has no collar.

Voice: A shrill whistle *killi killi*.

Habits: Inhabits clearings in, and the edges of, broadleafed tropical forest and plantations. Perches high in bare branches catching flying insects with feet in regular sorties. Bobs head and wags tail. Sometimes in small parties.

Mohit Aggarwal

Distribution: Scarce breeding resident of Himalayan foothills and Nepal.
Also occurs in SE Asia.

COMMON KESTREL (Kestrel, Eurasian Kestrel, Windhover)

Falco tinnunculus 35 cm F: Falconidae

Otto Pfister

♀

Otto Pfister

Description: Small, long-winged and long-tailed raptor. Male, grey-headed with black moustache, grey tail with broad black band. Lightly-spotted, chestnut upperparts. Spotted buff below. Barred underwings. Female and immature streaked brown with barred tail. Rarer **Lesser Kestrel** *F. naumanni* male, unspotted with no moustachial stripes and blue shoulders. White under wings.

Voice: A high *kee kee kee kee*.

Habits: Inhabits mountains, semi-desert and cultivation. Often near ruins and cliffs, where nests on ledges. Perches on wires and poles. Steady flight. Characteristic hovering with tail open, followed by gradual drop on, mainly rodent, prey.

Distribution: Common resident of Western Ghats and northern mountains including Nepal and Pakistan. Common and widespread winter visitor throughout lowlands. Also occurs in Europe, Africa and throughout Asia.

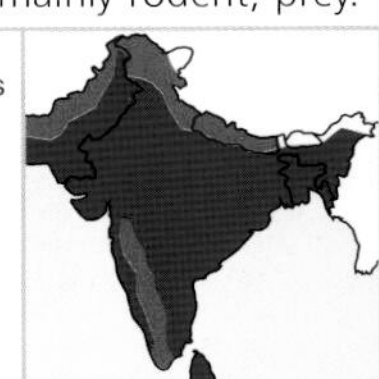

RED-NECKED FALCON (Red-headed Falcon/Merlin)

Falco chicquera 35 cm F: Falconidae

Mohit Aggarwal

Sunita Choudhry

Description: A small, powerful, long-winged falcon with pale grey upperparts, chestnut crown and nape. Chestnut moustachial stripes and black barred rather short looking tail. Upperparts silvery in bright light and contrast sharply with black wing tips and tail tip. Face and underparts white with fine black barring. Female larger.

Voice: A shrieking *ki ki ki ki*, rather parakeet-like.

Habits: Inhabits open country, including cultivation, with scattered trees. Dashing low flight. Pairs hunt as team, one bird flushing birds for other to catch. Crepuscular, often hunting around passerine roosts in reed beds. Nests in old crow's nest or date palm.

Distribution: Rare but widespread breeding resident throughout lowlands, except Sri Lanka. Also occurs in Africa.

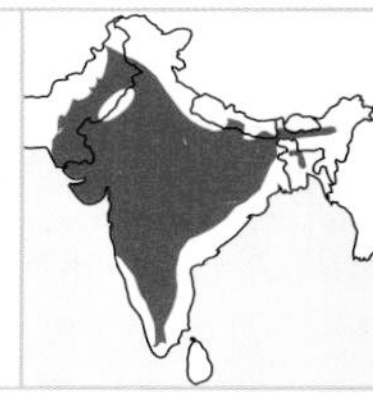

AMUR FALCON (Eastern Red-footed Falcon)

Falco amurensis 30 cm F: Falconidae

Description: Small, dark falcon with red legs and cere. Adult male, grey with rusty vent and lower belly, black moustache and striking white underwing coverts. Adult female grey above with dark buff vent, black-streaked, creamy underparts and a white face with black moustache. Immature has dark-edged, brown crown, buff-edged, browner upperparts and finely barred tail.

Voice: A high-pitched *kew kew kew* at roost.

Habits: Inhabits open country, often near water. Very sociable, feeding mainly on aerial insects which it holds in foot to eat. Also hovers and catches insects on ground. Rather crepuscular. Nests in old crow's nest.

Jan Willem den Besten

Jan Willem den Besten

Distribution: Scarce passage migrant (mainly in autumn) between NE Asia and south-central Africa. Mainly recorded in NE India, Bangladesh and Nepal but scattered records in NW India, throughout the peninsula and Sri Lanka. A few winter in NE India where there are isolated breeding records.
Also occurs in Africa and E Asia.

MERLIN

Falco columbarius 29 cm F: Falconidae

Description: Small, stocky raptor with extremely dashing flight. Rather short, pointed wings fall well short of relatively short tail when perched. Male is blue-grey above with dark crown and moustache. Rich orange-buff below with dark streaking and plain buff thighs. Wing tips and tail end black. Larger female and immature streaked brown above.

Voice: Usually silent.

Habits: Inhabits open country including cultivation and semi-desert. Frequently perches on ground and rarely in trees. Usually hunts low in pursuit of flocks of small birds which it may try to surprise by approaching from behind cover. Extremely agile in air, sometimes towering high and twisting steeply down in pursuit.

Otto Pfister

Distribution: Scarce winter visitor to northern lowlands.
Also occurs in Europe, throughout Asia and in N America.

EURASIAN HOBBY (Hobby, Northern Hobby)

Falco subbuteo 33 cm F: Falconidae

Otto Pfister

Description: Small, slender, long-winged raptor. Thin, pointed wings and rather short tail give swift-like silhouette. Adults are dark grey above with black crown and broad black moustache, contrasting with white face. Buff below with dense black streaking, red thighs and vent. Immature browner, scaly above and lack red. Rare northern resident **Oriental Hobby** *F. severus,* even darker with chestnut underparts.

Voice: A hurried *ki ki ki ki*.

Habits: Inhabits open, wooded country often over water. Aerial feeder, pursuing bats and small birds. Catches dragonflies, which it eats from feet in mid-air. Rather crepuscular. Extremely fluent flier, given to aerobatics when in groups. Nests in old crow's nests.

Distribution: Breeds in Himalayas including Nepal. Scarce passage migrant mainly in the Northwest but scattered records throughout the lowlands. Probably en route to E Africa. Rarely winters.
Also occurs in Europe, Africa and throughout Asia.

LAGGAR FALCON (Lanner Falcon)

Falco jugger 45 cm F: Falconidae

Joanna Van Gruisen

Description: Medium-sized, dark brown raptor with broad, pointed wings. Very dark brown including, diagnostically, thighs and armpits; sometimes all underparts in immature birds. Brown crown and narrow moustache. Whitish face, throat and breast with variable streaking. Tail appears unbarred. Female larger.

Voice: A high *kee he he*.

Habits: Inhabits all types of open, often rocky, country including deserts. Often hunts in pairs in low, fast flight after flying birds, rarely stooping from height like **Peregrine**. Perches on poles and bare branches and soars in search of prey, which also includes reptiles and rodents. Nests mainly in old crow's nest, but also on cliff ledges.

Distribution: Scarce and apparently declining near-endemic breeding resident throughout lowlands. Rare in south and east. Unknown in Sri Lanka.
Also occurs in Central Asia.

SAKER FALCON (Cherrug, Lanner)

Falco cherrug 55 cm F: Falconidae

Description: Large, long-winged and long-tailed, brownish falcon. Variable and difficult to identify. Usually pale-edged brown above but there is a greyer morph (shown). Head rather pale with long, thin moustache. Variably blotched dark brown on whitish underparts, usually most dense on lower flanks and upper thighs. Long tail clearly extends beyond long wings at rest. Female larger. **Laggar** is smaller and much darker.

Voice: Usually silent. A harsh *kerk kerk kerk* occasionally.

Habits: Inhabits dry, open country in mountains and plains. Hunts, usually singly, reptiles, mammals and birds mainly in low pursuit. Also stoops. Perches for long periods on ground or low objects.

Jan Willem den Besten

Distribution: This race is now a rare winter visitor to western Pakistan, northwestern India as far south as Delhi, Rajasthan and Gujarat. Also occurs in E Europe, NE Africa and W and Central Asia. Note that the Lanner *F. biarmicus* is a quite separate, more westerly, species of Europe and Africa, which has not been recorded in our region.

SHANGAR FALCON (Eastern Saker Falcon)

Falco cherrug milvipes 55 cm F: Falconidae

Description: A very variable, large falcon, currently considered a subspecies in the **Saker** group. Basically as **Saker** but northern birds are broadly barred buff above (not dark as in female **Common Kestrel** as often cited) and more heavily dark-blotched below. Southern birds (shown) resemble **Barbary (Peregrine) Falcon** below with fine barring on flanks and very limited blotching on chest (but obviously larger, and a **Saker** rather than a **Peregrine** in shape). Upperparts as northern form, except crown more obviously dark-edged rufous and moustache rather indistinct.

Voice: As **Saker**.

Habits: As **Saker**; no differences known but might use higher perches, more like **Peregrine**.

Otto Pfister

Distribution: Northern birds are rare winter visitors to the same areas as Saker plus Nepal. Southern birds may be rare breeding residents at least in Ladakh. Note that this form is the subject of current research and this photograph has been confidently identified by the researchers. Some southern birds at least may well turn out to be a separate species.

PEREGRINE FALCON (Peregrine)

Falco peregrinus 43 cm F: Falconidae

Asad Rahmani

Description: A powerful, medium-sized, mainly grey raptor. Rather broad but pointed wings and short tail. Resident **Shaheen** *F. p. peregrinator,* blackish-grey above with chestnut underparts, black-barred belly and thighs. Migrant race, *F. p. calidus* (shown) grey with black crown and broad black moustache on white face. Breast white but underparts finely barred black, as are underwings and tail. Sexes alike.

Voice: A high *keek keek keek*.

Habits: **Shaheen** inhabits open, usually rocky, country and some towns. Migrant race winters close to water, including the coast. Both usually seen singly or in pairs. Seeks bird prey from soaring, then stoops rapidly to kill it in mid air.

Distribution: Shaheen is scarce breeding resident throughout region. Migrant races are scarce winter visitors throughout. Smaller, pale Barbary race *P. p. babylonicus* breeds sparsely in N and W Pakistan perhaps east to Ladakh; winters rarely in the Northwest. Also occurs in Europe, Africa, throughout Asia, in Australasia and the Americas.

LITTLE GREBE (Dabchick)

Tachybaptus ruficollis 25 cm F: Podicipedidae

Otto Pfister

Description: Small, very dumpy, waterbird with distinctly fluffy rear and short yellow-based bill. In breeding plumage (shown), dark brown above, including breast with chestnut cheeks and fore-neck, buff flanks and whitish rear. In non-breeding plumage loses chestnut, and all buff below. White secondaries. Lobed feet set well back on body. Sexes alike. Young have stripey necks.

Voice: A high whinnying trill.

Habits: Inhabits well vegetated wetlands including village ponds. Often in scattered flocks on rivers in winter. Usually shy, diving when disturbed and hiding in emergent vegetation. Loathe to fly, which it does after pattering over surface. Feeds on small fish. Builds anchored, floating nest.

Distribution: Common breeding resident throughout region. Makes local movements to temporary water bodies. Also occurs in Europe, Africa and throughout Asia.

GREAT CRESTED GREBE

Podiceps cristatus 50 cm F: Podicipedidae

Otto Pfister

Description: A fairly large, slender waterbird with long neck and bill and silky white underparts. In breeding plumage acquires black tippets and black-edged chestnut ruffs. Dark brown above, buffish on flanks. In non-breeding plumage less ornate with dark crown and rear neck. Shows white leading and trailing edges to wings in flight. Lobed feet set well back. Sexes alike.

Voice: Loud guttural croaks when breeding.

Habits: Inhabits lakes, rivers and coasts. Requires quite deep open water where it dives expertly for fish. Swims low. Sometimes in scattered flocks. Rises after pattering surface. Flies low and in a laboured way. Makes floating, but anchored, nest.

Distribution: Scarce breeder in northern mountains and Gujarat. Scarce winter visitor to northern lowlands.
Also occurs in Europe, Africa, Central and N Asia, China and Australasia.

Nikhil Devasar

BLACK-NECKED GREBE (Eared Grebe)

Podiceps nigricollis 30 cm F: Podicipedidae

Description: A small, dumpy, dark-backed waterbird with a short uptilted bill. Fluffy rear end. In breeding plumage (shown), black upperparts, head and neck with rufous flanks and feathery yellow cheek spray, highlighting red irises. In non-breeding plumage, black crown and cheeks, rear neck and upperparts, dusky foreneck and white underparts. Sexes alike. Black upperparts, upturned bill and dusky neck separate from **Little** in winter.

Otto Pfister

Voice: Trills and whistles when breeding.

Habits: Inhabits well-vegetated wetlands and, in winter, the coast. Wanders more widely in north in winter. Behaves as **Little Grebe** but more often on open water.

Distribution: Scarce and very local breeding resident in parts of the Northwest. Wanders more widely in winter but rare except in Pakistan.
Also occurs in Europe, Africa, Russia and the Americas.

DARTER (Indian, Oriental Darter, Snakebird)

Anhinga melanogaster 90 cm F: Anhingidae

Otto Pfister

imm

Otto Pfister

Description: Large, dark waterbird with very slender neck and bill and long tail. Adults black above and below with silvery feathers on wings, dark brown head and neck with white throat and side stripes. Nestlings very white looking and immatures are brown with white from throat to belly.

Voice: Silent, but croaks and squeals near nest.

Habits: Usually singly or in small parties away from breeding colonies which are in trees with other waterbirds. Inhabits larger freshwater bodies, including rivers. Expert diver for fish which it spears. Usually swims with only thin neck exposed, sinking slowly to dive. Frequently sits on ground, posts or dead trees with wings open to dry.

Distribution: Globally threatened but locally common breeding resident throughout lowlands, especially in Northwest.
Also occurs in SE Asia.

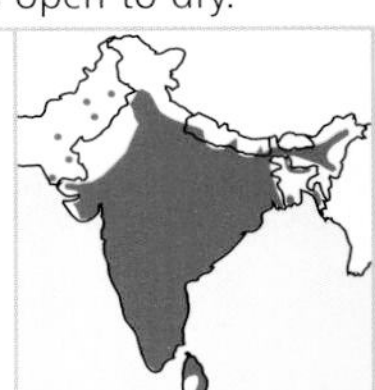

LITTLE CORMORANT

Phalacrocorax niger 50 cm F: Phalacrocoracidae

Otto Pfister

Description: Black, duck-sized waterbird with long tail and short, thick neck. When breeding, all black with greenish-blue sheen. White flecks on head and back and no throat patch. Throat white when not breeding. Rounded short head and stubby bill. Sexes alike. Immature browner.

Voice: Usually silent but croaks near nest.

Habits: Inhabits water bodies of all sizes, including village ponds and drains. Also the coast. Usually quite gregarious and confiding, often mixing with **Indian Cormorant** to fish. Dives with jump for fish and tadpoles. Swims low. Flies rapidly often close to water. Holds wings open to dry. Nests in tree colonies with other waterbirds.

Distribution: Common breeding resident throughout region, scarce in most of Pakistan and Nepal.
Also occurs in SE Asia.

INDIAN CORMORANT (Indian Shag)

Phalacrocorax fuscicollis 63 cm F: Phalacrocoracidae

Description: A medium-sized, black waterbird with a yellowish throat patch. Slender neck and bill and long, flat head. In breeding plumage, glossed bronze above and has white ear tufts. Browner in non-breeding plumage with whitish throat. Sexes alike. Immature paler with mottled white underparts. Tail appears shorter than **Little** and flight action recalls **Great.**

Voice: Usually silent but croaks near nest.

Habits: Inhabits large water bodies and the coast in large flocks. Dives with jump to pursue fish, often in a concerted drive by flocks. Swims low. Flies strongly and often high. Holds wings open to dry. Nests in tree colonies with other waterbirds.

Otto Pfister

Distribution: Locally common breeding resident throughout lowlands, subject to local movements.
Also occurs in Myanmar.

GREAT CORMORANT (Cormorant, Large Cormorant)

Phalacrocorax carbo 80 cm F: Phalacrocoracidae

Description: Goose-sized, mainly black, waterbird with yellow face patch. Long, thick neck and large head. Long, hooked, grey bill. All black with green sheen. In breeding plumage white feathers on head and neck and white thigh patches. In non-breeding plumage loses the white and the sheen but has whitish throat. Immature browner and usually largely white from chin to vent.

Voice: Usually silent. Deep croaks near nest.

Habits: Inhabits large water bodies, rivers and the coast. Usually breeds in colonies in tall trees and away from other birds. Roosts communally on high perches such as pylons. Dives with jump to catch large fish. Holds wings open to dry.

Kamal Sahai

Kamal Sahai

Distribution: Locally common breeding resident and winter visitor throughout.
Also occurs in Europe, Africa, Central, N and E Asia, Australasia and N America.

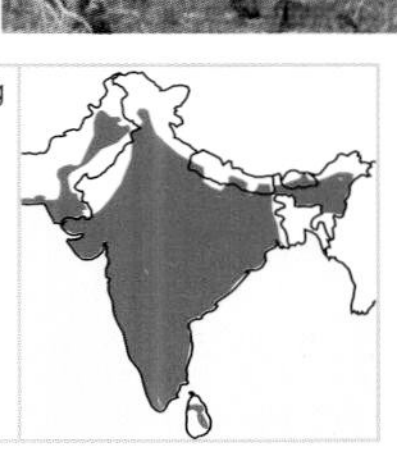

LITTLE EGRET

Egretta garzetta 65 cm F: Ardeidae

Otto Pfister

Description: Medium-sized, snow white heron with black legs and yellow feet. Slender build and long, thin neck. Bill thin and all black. In breeding season acquires two plumes on nape and others on back and breast. Bluish lores turn briefly red. Flies with neck pulled in and legs projecting.

Voice: A short croak. Most noisy at nest.

Habits: Inhabits all types of mainly fresh, shallow water. Also occurs in estuaries and mangroves. Catches fish, frogs, and other aquatic animals, either by patient stalking or by rather energetic chasing and stabbing. Usually in groups, particularly for roosting. Nests in tree colonies with other waterbirds.

Distribution: Fairly common breeding resident throughout, subject to local movements.
Also occurs in Europe, Africa, throughout Asia and Australasia.

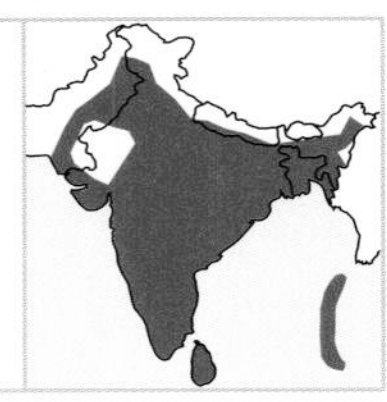

WESTERN REEF EGRET (Indian Reef Heron)

Egretta gularis 65 cm F: Ardeidae

James Hancock

Description: Medium-sized, dimorphic heron. Difficult to tell from **Little** in white phase breeding plumage. Bill stouter with curved mandibles and brownish-yellow or yellow. Legs appear stouter and brownish with more greenish feet, but in breeding plumage they are close to **Little**. Neck longer and more S-shaped. Dark phase is slaty-grey with prominent white throat. Intermediates, with grey upper and white underparts, do occur.

Voice: Occasionally croaks.

Habits: Inhabits coastline. Solitary and crepuscular although flocks gather to roost and breed. Feeds by patient stalking over mud or shallow water feeding mainly on mudskippers and crabs.

Distribution: Scarce breeding resident from Pakistan round coast to the Sundarbans in Bangladesh so overlaps with the Pacific Reef. Commonest in extreme Northwest, rare in south and inland.
Also occurs in E Africa and W Asia.

PACIFIC REEF EGRET (Eastern Reef Heron)

Egretta sacra 60 cm F: Ardeidae

Description: A medium-sized, exclusively coastal, dimorphic heron. As **Western** in terms of colour phases and confusion with **Little** but obviously stockier, thicker-necked bird with shorter legs. Only toes project beyond tail in flight. Colour ranges identical but grey phase lacks obvious white throat. In breeding plumage, acquires a short, bushy crest.

Voice: Usually silent. A harsh *squak* call when disturbed. Also grunts.

Habits: Inhabits rocky and muddy coasts, nesting colonially in mangroves or among rocks. Solitary, often sits or paces in hunched up mode. Will run after prey but prefers to wait and see. Feeds on fish and other marine animals.

Otto Pfister

Distribution: Common breeding resident on Andaman and Nicobar Islands. Rare visitor to Bangladesh and northeastern Indian coast so overlaps with the Western.
Also occurs in SE Asia and Australasia.

GREY HERON (Heron)

Ardea cinerea 100 cm F: Ardeidae

Description: Large, grey and black heron with powerful, yellow bill. Grey above with black flight feathers. Below white, usually with greyish wash. Black crown sides, nape plumes and foreneck. In flight shows white leading edges. Sexes alike. Immature greyer below. Crown grey. Bowed wings in stately flight with neck retracted and feet extended.

Voice: A harsh *kar aark*.

Habits: Inhabits all types of inland and coastal water. Normally singly or in scattered groups but roosts and breeds colonially high in trees. Feeds mainly standing hunched and motionless, waiting to spear prey. Will take incautious waterbirds and mammals as well as fish and amphibians. Flies high between feeding grounds.

Otto Pfister

Distribution: Locally common breeding resident and winter visitor throughout. Also occurs in Europe, Africa and throughout Asia.

GOLIATH HERON (Giant Heron)

Ardea goliath 147 cm F: Ardeidae

Sujan Chatterjee

Description: Huge, grey and orange heron; the largest in the world. Superficially like much slighter **Purple**. Chestnut-orange on head, neck, underwings and wing shoulders. No head plumes. White throat to upper breast with black markings. Deep blue-grey above. Huge, black bill with yellow lower mandible. Black legs. Sexes alike. Immature more rufous. Equally rare **White-bellied** *A. insignis,* dark grey above and white below. Pale grey head and neck, short crest and no orange.

Voice: Deep raucous *kworgh kworgh*.

Habits: Inhabits lakes, rivers, estuaries and mangroves. Solitary and shy. Feeds on fish and other water animals, waiting to pounce or wading after prey.

Distribution: A bit of a mystery species! Either very rare, presumed breeding, resident or an occasional visitor from Africa, where it is common. Only reliably recorded in northeast India, Bangladesh and Sri Lanka. Also occurs throughout Africa but nowhere else in Asia.

PURPLE HERON

Ardea purpurea 90 cm F: Ardeidae

Kamal Sahai

Description: Large, slender-necked, rufous and grey heron. Upperparts, purplish-grey with rufous shoulders and forewings. Serpentine head and neck is rufous with black lines and black-plumed crown. Underparts largely rufous. Can look very dark. Sexes alike. Immature plainer, sandy-brown. In flight, large, bunched, extended toes and deeper neck bulge separate from **Grey**.

Voice: A croaking, sudden *kruck*.

Habits: Inhabits well-vegetated wetlands of all types, including mangroves. Crepuscular and secretive, waiting with out-stretched neck to locate prey in dense cover. Eats fish and other aquatic animals. Nests in water-bird colonies in trees and in scattered groups in reed beds.

Distribution: Locally common breeding resident throughout lowlands. Also occurs in Europe, Africa, W and SE Asia.

GREAT EGRET (Large Egret, Great White Heron)

Casmerodius albus 90 cm F: Ardeidae

Description: Large, snow white heron with long, thin neck and very long legs. In non-breeding plumage, bill all yellow with dusky tip and legs all black. In breeding plumage, bill gradually becomes black, lores blue and tibia red. Also develops long back and short breast plumes. Typically stands with pronounced kink in neck or with neck completely straight when hunting. Legs protrude well beyond tail in flight. Deep neck bulge. Sexes alike.

Voice: A very deep *krark*.

Habits: Inhabits larger wetlands and coast. Feeds singly or in scattered groups in shallow water or short vegetation, waiting for fish or aquatic animal prey to approach. Breeds in tree colonies.

Distribution: Locally common breeding resident throughout lowlands and on coast.
Also occurs in Europe, Africa, throughout Asia, in Australasia and the Americas.

Otto Pfister

Sumit Sen

INTERMEDIATE EGRET (Smaller, Median, Yellow-billed)

Mesophoyx intermedia 80 cm F: Ardeidae

Description: Large, snow white heron with thick neck and short, thick bill. Separated from **Great** by shape, and gape-line halting below eye. In non-breeding plumage, bill all yellow, often with black tip, yellow lores and brownish legs with greenish tibia. In breeding plumage, bill becomes gradually black and lores greenish. Develops longer breast and shorter back plumes than **Great**. Neck has gentle curve. Legs do not protrude so far in flight.

Voice: Usually silent.

Habits: Inhabits wetlands and the coast. Habits as **Great**, with which it often mixes in feeding grounds, often in quite large parties. Stalks more and less commonly stands with neck straight.

Distribution: Locally common breeding resident throughout lowlands and the coast.
Also occurs in Europe, Africa, throughout Asia, in Australasia and the Americas.

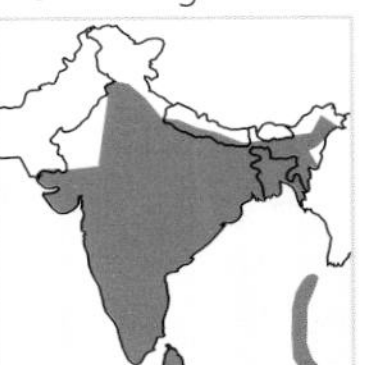

Otto Pfister

Kamal Sahai

CATTLE EGRET

Bubulcus ibis 50 cm F: Ardeidae

Otto Pfister

RK Gaur

Description: Medium-sized, stocky, usually snow white, heron with pronounced jowl and short yellow bill. Very stocky compared to other egrets. In breeding plumage, suffused with orange, bill becomes more orange and head has more bushy look. Sexes alike. Habit of feeding with, and from the backs of, large ungulates is characteristic.

Voice: Croaks at nest.

Habits: Commonest heron. Inhabits wetland, grassland, cultivation (follows plough), garbage dumps, open woodland, and urban parks. Very active feeder, mainly on large insects, such as those disturbed by ungulates. Also eats refuse and carrion. Sociable and confiding. Breeds colonially.

Distribution: Abundant breeding resident throughout region.
Also occurs in Europe, Africa, throughout Asia and in the Americas.

INDIAN POND HERON (Pond Heron, Paddybird)

Ardeola grayii 45 cm F: Ardeidae

RK Gaur

Otto Pfister

Description: Small, dumpy, brown heron with contrasting white wings and tail. Usually earth-brown with heavily dark-streaked buff head, neck and breast. In breeding plumage, back becomes very dark purplish-brown and head, neck and breast warm yellow-buff. Wings and tail always white, strikingly contrasting in flight. Short, thick legs and dark-tipped, yellow bill. Sexes alike.

Voice: A deep croak.

Habits: Occurs wherever there is water, including temporary puddles. Solitary but several birds may feed close together. Roosts communally. Inconspicuous until flies. Sits hunched waiting for aquatic and insect prey. Nests in tree colonies.

Distribution: Common breeding resident throughout region.
Also occurs in Iran and Myanmar.

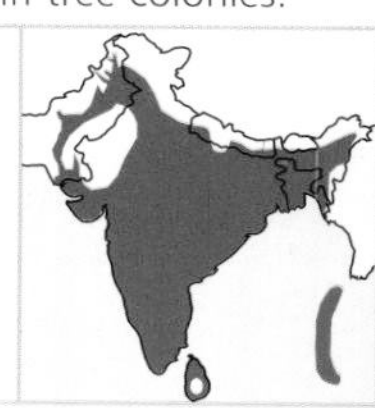

LITTLE HERON (Striated, Little Green Heron)

Butorides striatus 45 cm F: Ardeidae

Description: A small, dark heron with black crown, cheek lines and bill. Mainly dark grey, sometimes with a purplish flush. Upperparts glossy, dark green with buff feather edgings. Appears black in shadows. Dark pink legs. Sexes alike. Immature streaked.

Voice: Occasional sharp **Jackdaw**-like *kyow* when flushed.

Habits: Solitary, crepuscular and extremely secretive so probably overlooked. Inhabits well-vegetated or rocky pools, streams and mangroves. Normally hidden in vegetation, often on a low bare branch from which it spears passing fish, amphibians and insects. Nests singly low in waterside shrubbery. Flies low over water for short distance if flushed.

Toby Sinclair

Distribution: Rather scarce breeding resident throughout but rare in the Northwest.
Also occurs in Africa, W, N, E and SE Asia, Australasia, Indian and Pacific Ocean Islands.
Closely related to the Green Heron *B. virescens* in America.

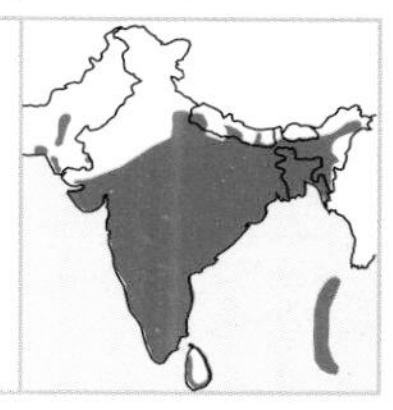

BLACK-CROWNED NIGHT HERON (Night Heron)

Nycticorax nycticorax 60 cm F: Ardeidae

Description: Small, thick-set heron with striking black, grey and cream plumage. Adults have black crown (with two long white plumes) and back, grey wings and tail and cream underparts. Large ruby irises. Short, thick legs and neck. Sexes alike. Juvenile (also shown) buff-spotted and streaked brown. Second year birds all grey above.

Voice: Abrupt crow-like *kwark*, particularly when leaving roost.

Habits: Inhabits all wetland types, including paddy and mangroves. Mainly nocturnal. Small flocks hide in day in thick foliaged trees near water. Leaves roost at dusk to feed on fish and amphibians. Often sits in water up to belly, waiting to strike. Nests low down in mixed colonies.

Otto Pfister

Distribution: Locally common breeding resident throughout lowlands.
Also occurs in Europe, Africa, Central, E and SE Asia and the Americas.

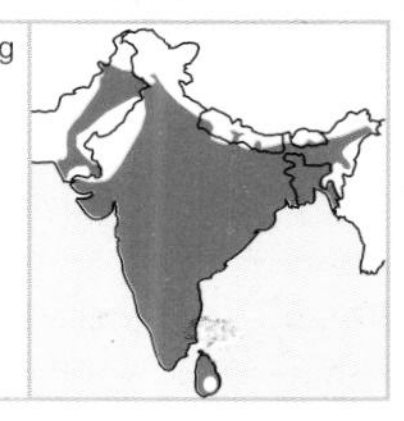

LITTLE BITTERN

○ *Ixobrychus minutus* 35 cm F: Ardeidae

RK Gaur

Description: A very small, black and buff heron with large, yellow patches in black wings. Crown, back and tail also black in male and underparts warm buff. Wing pattern striking in flight. Female and immature streaked and browner.

Voice: In flight a nasal *krak*, sometimes extended. Male's, mainly nocturnal, breeding call is a repeated, far-carrying *hogh*.

Habits: Solitary and crepuscular inhabitant of reedbeds near rivers and jheels, sometimes quite small ones. Usually only seen flying with fast, short wing beats between feeding sites. Feeds on fish, amphibians and insects. Nests on ground in reeds.

Distribution: Very local breeding resident mainly in Kashmir and Assam. Scattered records from elsewhere, mainly in north.
Also occurs in Europe, Africa, W and Central Asia and Australasia.

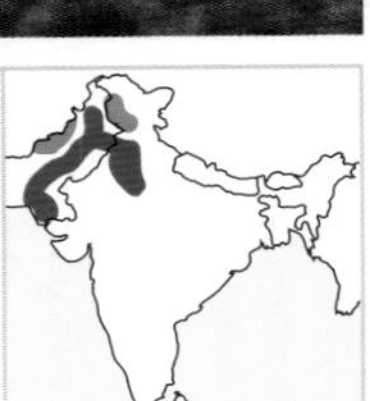

YELLOW BITTERN

⊙ *Ixobrychus sinensis* 38 cm F: Ardeidae

Otto Pfister

imm

Otto Pfister

Description: A very small, black and dark buff heron. Very similar to **Little** but buff-brown back. Black flight feathers, tail and crown contrast with buffish body. Rear neck darker, and foreneck whiter, with faint streaking. Female and immature, more rufous streaked.

Voice: Rarely a sharp *kark*. Male's breeding call is a quiet *uu uu*.

Habits: Inhabits reedbeds and wet scrub near rivers and jheels and wet paddy. Solitary but breeding densities can be quite high. Usually seen in feeding flights, particularly when feeding young in reedbed nests. Feeds on fish, amphibians and insects.

Distribution: Rather local breeding resident mainly in the Northeast, Bangladesh, Sri Lanka and Indo-Gangetic Plains. Mainly monsoon breeding visitor to the Northwest.
Also occurs in N, E and SE Asia.

CINNAMON BITTERN (Chestnut Bittern)

Ixobrychus cinnamomeus 38 cm F: Ardeidae

Description: A very small, cinnamon-rufous heron. Male brighter and rather paler, especially on throat. Female has well-streaked neck and immature is even darker and more streaked. Appears larger, plainer and more rounded-winged than **Yellow** in flight.

Voice: Male's breeding call is a loud *kok kok*.

Habits: Habitat and behaviour as **Yellow** and often seen in same places, although usually less numerous. May fly quite high to feeding sites.

Kamal Sahai

Distribution: Rather local breeding resident mainly to eastern coast, the Northeast, parts of the Indo-Gangetic Plains, Bangladesh and Sri Lanka. Appears to be mainly a monsoon breeding visitor to the Northwest. Also occurs in China and SE Asia.

BLACK BITTERN

Dupetor flavicollis 58 cm F: Ardeidae

Description: A medium-sized, very dark heron, with a bold, yellow neck stripe. Upperparts dull black, browner on female. Below whitish-streaked rufous and black. Belly grey. Immature more streaked and upperpart feathers have buff fringes.

Voice: Occasional harsh *ker*. Male's breeding call is a deep boom like a distant foghorn.

Habits: Inhabits reedbeds with bushes and bushy margins of canals, rivers and jheels. Crepuscular, solitary and extremely secretive. May fly quite high to feeding sites. Often hunts from bushes, sitting motionless for hours waiting for fish or amphibian prey. Usually nests low in a waterside bush.

Sujan Chatterejee

Distribution: Scarce breeding resident throughout. Most frequent in the Northeast, Sri Lanka and extreme southwest. Appears to be mainly a monsoon breeding visitor to the Northwest. Also occurs in China and SE Asia.

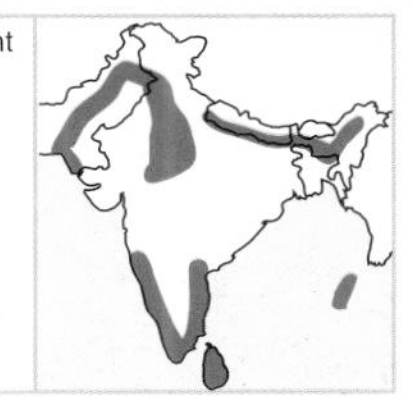

GREAT BITTERN (Bittern, Eurasian Bittern)

Botaurus stellaris 75 cm F: Ardeidae

Tim Loseby

Description: Large, bulky, buff-brown, thick-necked heron with black crown and moustache. Cheeks plain brown, throat whitish. Finely chevroned with black above and black-striped down foreneck. Greenish legs and powerful, yellowish bill. Heavy owl-like flight with neck hidden. Larger and paler above than juvenile **Black**, larger, darker and buffer than juvenile **Night Heron** which lacks black cap. Juvenile **Purple Heron** is more uniform sandy and more slender.

Voice: A deep croak in flight. Far-carrying booming when breeding.

Habits: Inhabits large wet reed-beds. Hunts stealthily on water's edge for fish and aquatic animals. Very secretive and keeps to cover, unless flying from site to site.

Distribution: Rare but widespread winter visitor throughout lowlands of Pakistan (where may breed in Baluchistan), India and Nepal. Vagrant in Sri Lanka and Bangladesh. Probably much overlooked.
Also occurs in Europe, Africa and W, Central, N and E Asia.

GREATER FLAMINGO (Flamingo)

Phoenicopterus ruber 140 cm F: Phoenicopteridae

Otto Pfister

Description: Huge, very long-necked and long-legged, pink waterbird. Long, thin, curved neck with a rather large head and thick, black-tipped, pale, banana-shaped bill. Pinkish-white body and contrasting red wing coverts and black flight feathers. Long pink legs. Sexes alike. Immatures dingy grey-white and often much smaller than adults. Flies with neck and legs fully outstretched.

Voice: Feeding flocks may make a subdued goose-like honking.

Habits: Breeds on saline or brackish lagoons but will feed in any shallow water including estuaries, lakes, rivers and flooded fields. Feeds on insects, seeds, molluscs and crustaceans. Very sociable. Sometimes quite confiding if undisturbed.

Distribution: Breeds in Kutch and wanders widely throughout region particularly to east coasts and the Northwest. Sporadic in appearance.
Also occurs in Europe, Africa, Central Asia and Central and South America.

LESSER FLAMINGO

Phoenicopterus minor 100 cm F: Phoenicopteridae

Description: A large, long-legged and long-necked, pink waterbird. Small and thicker-necked than **Greater** with, usually, darker pink plumage and with black-tipped, dark red bill, red eyes and facial skin. Smaller bill more kinked. Immatures darker than **Greater**. Beware small young **Greaters** in adult flock, which can be the same size.

Voice: Feeding flocks may make a goose-like honking sound, higher-pitched than **Greater**.

Habits: More restricted to salt and brackish water than **Greater** feeding on algae and diatoms. Fresh-water sightings should be checked carefully. Feeds with bill upside down and swept from side to side. Nests colonially on ground. Very sociable.

Distribution: Breeds in Kutch but some wander particularly to east coasts and the Northwest.
Also occurs in Africa.

Otto Pfister

GLOSSY IBIS

Plegadis falcinellus 60 cm F: Threskiornithidae

Description: A fairly large, rather spindly, all dark waterbird with a long, thin, decurved brown bill. Dark purplish-brown with chestnut hue when breeding. Glossed purple and green, particularly on wings. White area round eyes and, in non-breeding plumage, whitish head streaking. Laboured flight with outstretched neck and legs. Sexes alike.

Voice: Usually silent.

Habits: Inhabits all types of shallow freshwater, particularly with floating vegetation. Often feeds on water hyacinth beds. Feeds mainly on invertebrates, especially grasshoppers. May form large flocks at favoured feeding sites. Nests in mixed colonies in trees. Mixes with other water-birds but rather shy.

Otto Pfister

Distribution: Fairly common but local breeding resident subject to wandering. Most frequent in north and west.
Also occurs in Europe, Africa, throughout Asia, in Australasia and N America.

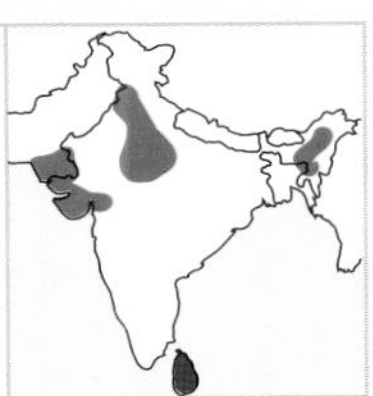

BLACK-HEADED IBIS (White, Oriental White Ibis)

Threskiornis melanocephalus 75 cm F: Threskiornithidae

Kamal Sahai

Description: A large, rather stocky, white waterbird with a bare, black head and upper neck and a thick, black, decurved bill. All white plumage contrasts with black legs, head and bill. Extent of bare neck varies. Has long, drooping, back plumes, grey in wings and variable yellow wash in breeding plumage. Sexes alike. Immature has black wing tips and black only on head. Flies heavily with neck and legs outstretched.

Voice: Usually silent.

Habits: Inhabits all types of shallow water including the coast. Also flooded and quite dry fields near water. Feeds on aquatic animals. Nests colonially in trees with other waterbirds. Very sociable but shy.

Distribution: Local but fairly common breeding resident throughout except for east and the Northeast.
Also occurs in E and SE Asia..

BLACK IBIS (Red-naped Ibis)

Pseudibis papillosa 70 cm F: Threskiornithidae

Otto Pfister

Description: A large, stocky, black bird with red nape and white wing patches. Bill thick, dark and decurved. Rather short red legs. Green and purple gloss. Sexes alike. Immatures lack red nape. From **Glossy** by shape, red legs and nape and white in wings. Broad-winged and heavy in flight with bill and neck outstretched.

Voice: Rather noisy. Haunting gull-like *kyow kyow,* most often in flight or during night.

Habits: Inhabits cultivation, jheel and marsh edges and garbage dumps. Often seen sitting by village ponds but not tied to water. Singly or in small groups. Feeds on reptiles, fish, insects and amphibians. Nests alone, high in isolated tree or on pylon. Often perches high.

Distribution: Widespread but scarce breeding resident throughout lowlands of India and Nepal. Rare or unknown elsewhere.
Also occurs in SE Asia.

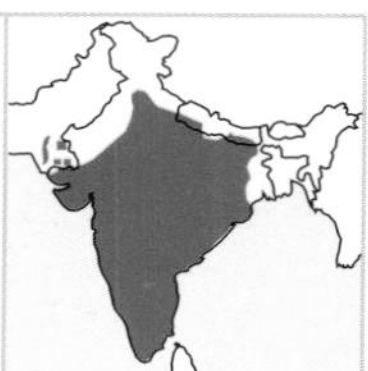

EURASIAN SPOONBILL (Spoonbill, White Spoonbill)

Platalea leucorodia 85 cm F: Threskiornithidae

Description: Large, white waterbird with long, uniquely spatulate bill. All white with black legs and black bill, usually with yellowish spoon. In breeding plumage, warm yellow wash to breast and thick crest. Sexes alike. Immature has black wing tips and paler bill. Flies with much gliding, legs and neck outstretched. Latter held low and bird looks as though straining to keep bill up.

Voice: Clatters bill and grunts at nest.

Habits: Inhabits all types of shallow water. Also the coast. Spends much of day sleeping on one leg. Feeds on aquatic animals by sweeping bill from side to side through water. Usually in small flocks. Nests colonially in trees with other waterbirds.

Distribution: Locally common breeding resident and winter visitor mainly to the Northwest, west, south and Sri Lanka. Also occurs in Europe, N Africa and W, Central and E Asia.

Otto Pfister

Bikram Grewal

GREAT WHITE PELICAN (White, Rosy, Eastern White)

Pelecanus onocrotalus 175 cm F: Pelecanidae

Description: A huge, white waterbird with black flight feathers and long, pouched bill. Large body, small head and huge bill. All white body usually with pinkish tinge. Bare yellow patch round eye and yellow pouch. Short legs usually pink but can be grey. Small white crest in breeding plumage. Black secondaries diagnostic in flight.

Voice: Excited grunts when feeding.

Habits: Inhabits large jheels, dams and rivers where there are plenty of fish. Sociable, often feeding in co-ordinated groups that drive the fish into an enclosed circle. Looks huge when sitting upright on ground. Strong flyer with neck held in. Regularly soars high on thermals in flocks.

Distribution: Breeds in Kutch. Otherwise widespread but very local winter visitor mainly to the Northwest. Also occurs in Europe, Africa and W, Central and E Asia.

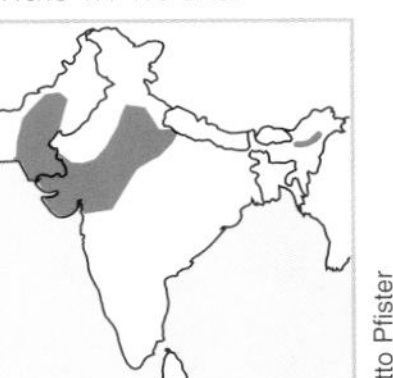

Otto Pfister

Otto Pfister

DALMATIAN PELICAN

Pelecanus crispus 180 cm F: Pelecanidae

X

Otto Pfister

Description: A huge, white waterbird with dusky flight feathers and long, pouched bill. As **Great White** but slightly grey-tinged and with more restricted, yellow eye patches and, usually, rather scruffy, curled crest. Generally looks more "plumey". Short legs, grey. In breeding season pouch turns red and eye patches purple. In flight dusky white underwings with darker trailing edges. Primaries dark grey above.

Voice: Silent when not at nesting colonies.

Habits: Inhabits large water bodies with fish. Behaviour as **Great White**, with which small numbers often mix. Large flocks now rare, but odd individuals wander so it could turn up anywhere.

Distribution: Globally threatened and now very local winter visitor mainly to Pakistan and northwest India.
Also occurs in Europe, Central Asia and China.

SPOT-BILLED PELICAN (Grey Pelican)

Pelecanus philippensis 150 cm F: Pelecanidae

Otto Pfister

Description: A huge, greyish-white waterbird with long pouched bill, spotted along the upper mandible. Smaller than, but similar to, the other two pelicans. Grey flush to white plumage with pinkish hue to back and flanks. Short brownish crest. Tends to look rather scruffy. Purple and yellow bare skin round eyes and dark pink pouch in breeding plumage. Paler when not breeding and pink hues lost. In flight, dark primaries and dusky brown underwing flight feathers.

Voice: Silent when not at nesting colonies.

Habits: Inhabits large freshwater bodies with fish. Behaviour as other pelicans. Nests in traditional colonies in trees, often in villages.

Distribution: Globally threatened and very local breeding resident in parts of south India, Sri Lanka and the North-east. Wanders widely in non-breeding season so could turn up anywhere.
Also occurs in Iran and SE Asia, but now very rare outside India.

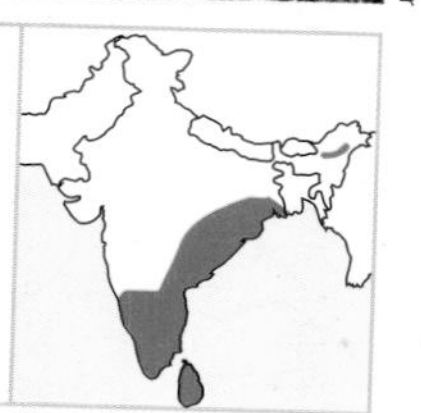

PAINTED STORK

Mycteria leucocephala 95 cm F: Ciconiidae

Description: A huge, erect waterbird with white and black plumage. Largely white with black barring on wings and breast and black flight feathers. Pinkish wash on lower back. Long, slightly decurved, yellow bill, bare reddish face and pink legs. Sexes alike. Immature dusky grey. Flies with neck and legs extended.

Voice: Voiceless. Clatters bill at nest.

Habits: Inhabits larger wetlands, mudflats, salt pans and estuaries. Gregarious, nesting high in tree colonies with other waterbirds. Feeds on fish and other aquatic animals in shallow water. Stately, slow walk. Often perches on bare trees. Frequently soars high on thermals.

Distribution: Locally common breeding resident throughout lowlands including the coast. Scarce or rare in Pakistan, Bangladesh and the Northeast.
Also occurs in China and SE Asia.

Otto Pfister

Otto Pfister

ASIAN OPENBILL (Openbill, Asian Openbilled Stork)

Anastomus oscitans 80 cm F: Ciconiidae

Description: A large, erect, black and white waterbird with a long, permanently open bill. Rather stocky. Smoky-grey wash in non-breeding plumage. Glossy black flight feathers and black tail. Bill pinkish-grey and both mandibles curved to leave a distinctive "nutcracker" gap. Legs reddish. Sexes alike. Immature dark brownish-grey. Flies with neck and legs extended.

Voice: Voiceless. Clatters bill at nest.

Habits: Inhabits larger wetlands. Rare on coast. Gregarious and nests in tree colonies often with other waterbirds. Feeds mainly on molluscs which it cracks open with its specially adapted bill. Frequently soars high on thermals.

Distribution: Locally common breeding resident throughout lowlands but rare in Pakistan.
Also occurs in SE Asia.

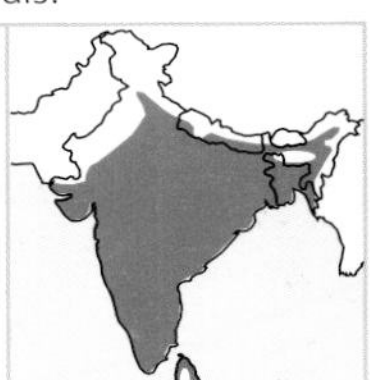

Otto Pfister

BLACK STORK

Ciconia nigra 100 cm F: Ciconiidae

Otto Pfister

Otto Pfister

Description: A huge, erect, black and white waterbird with long red bill and legs. All black with green and purple gloss and white lower breast and belly. White triangles on underwing obvious in overhead flight. Sexes alike. Immature browner with grey-green bill and legs. Flies with neck and legs extended. Tail sometimes has some white in it.

Voice: Voiceless.

Habits: Solitary or in small groups. Very wary. Inhabits marshes, jheels and particularly lakes in wooded areas. Feeds on amphibians and insects in shallow water or waterside grass. Stately, measured walk. Often soars high on thermals.

Distribution: Scarce winter visitor mainly to northern areas.
Also occurs in Europe, Africa and W, Central, N and E Asia.

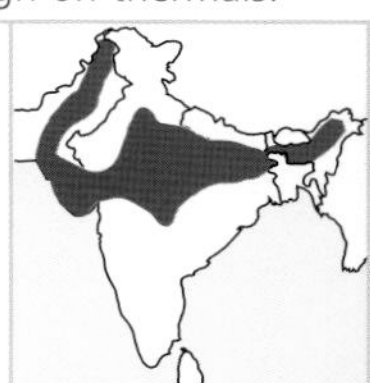

WOOLLY-NECKED STORK (White-necked Stork)

Ciconia episcopus 85 cm F: Ciconiidae

Otto Pfister

Description: A large, black and white stork with a fluffy white neck. Crown and upperparts black-glossed purple and green. May appear brownish in distance. Black tail, long white uppertail coverts, white belly and vent. Broad wings entirely dark except for some white in inner webs. Legs red, bill black with some red. Sexes alike. Immature browner. Flies with legs and neck extended.

Voice: Voiceless. Clatters mandibles at nest.

Habits: Inhabits marshes, wet grassland, cultivation and rivers, usually near trees. Not shy. Usually solitary, in pairs or small parties. Feeds on reptiles, amphibians and insects. Nests high in tree. Frequently soars high on thermals.

Distribution: Locally common breeding resident throughout lowlands but very rare in Pakistan and Bangladesh.
Also occurs in Africa and SE Asia.

WHITE STORK

Ciconia ciconia 105 cm F: Ciconiidae

Otto Pfister

Description: A huge, erect, black and white stork with long red bill and legs. Body, including tail white. Flight feathers black. Dark irises and eye shadow. Long thick breast feathers. Sexes alike. Immature is dingier and has brown bill and legs. Flies with neck and legs extended. Very rare, globally threatened **Oriental Stork** *C. boyciana* has only been recorded in the Northeast. It is larger, has white webs to flight feathers, black bill, white irises and red eye shadows.

Voice: Voiceless. Clappers bill.

Habits: Inhabits cultivation and open grassy marshes, feeding on amphibians, reptiles and insects, particularly grasshoppers. Sociable and not shy. Often soars high on thermals in flocks.

Distribution: Scarce winter visitor mainly to south India. Passage records widespread especially from the Northwest.
Also occurs in Europe, Africa, W and Central Asia.

BLACK-NECKED STORK

Ephippiorhynchus asiaticus 140 cm F: Ciconiidae

Description: A huge, erect, black and white stork with red legs and powerful, upturned black bill. White with green-glossed black head, neck and tail. Broad black bar across white wings creates unique pattern. Flies with long neck and legs outstretched. Sexes alike but male has dark, while female has yellow, irises. Immature brownish.

Voice: Voiceless but clatters mandibles.

Habits: Inhabits large wetlands including rivers. Usually in pairs or solitary. Eats fish, reptiles, amphibians and waterbirds. Slow, stately walk. Wades deeply in water. Stands motionless, striking at any prey that comes close. Soars high on thermals. Builds huge stick nest alone in large tree.

Distribution: Increasingly rare breeding resident throughout Indian lowlands. Most frequent in north. Still breeds in Sri Lanka.
Also occurs in SE Asia and Australasia.

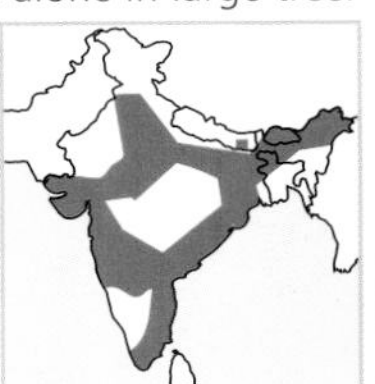

Otto Pfister

Kamal Sahai

LESSER ADJUTANT (Lesser Adjutant Stork)

Leptoptilos javanicus 115 cm F: Ciconiidae

Otto Pfister

Description: A huge, black and white stork with bare pinkish-orange head and neck and powerful, pink bill. Wings, neck ruff and mantle blackish with coppery spots on wings in breeding plumage. White underparts and armpits. Slight scruffy black crest on rear crown and black down on rear neck. No neck pouch. Characteristic hunched posture with neck withdrawn as it is in flight, like a heron.

Voice: Silent. Clatters bill at nest.

Habits: Inhabits marshes, jheels and mangroves. Usually solitary and very shy. Feeds on aquatic animals and grasshoppers by walking slowly on ground or through shallow water to locate prey. Small colonies or individual pairs nest high in forest trees.

Distribution: Globally threatened and now rare but widespread breeding resident throughout lowlands except Pakistan. Most frequent in Nepal and the Northeast. Does wander outside breeding season.
Also occurs rarely in China and SE Asia.

GREATER ADJUTANT (Adjutant, Adjutant Stork)

Leptoptilos dubius 150 cm F: Ciconiidae

James Hancock

Description: Huge, grey and white stork with bare pink head, neck and pendulous neck pouch and massive pale pink bill. Thick white neck ruff and white underparts. Dark grey flight feathers and light grey upperwing panel and undertail, differing from **Lesser**. Also head and neck barer and bill even more powerful. Posture and flight as **Lesser**.

Voice: Silent apart from croaks and bill-clattering.

Habits: Inhabits marshes, jheels, cultivation and urban garbage dumps. Formerly renowned scavenger in towns. Also eats aquatic animals and birds. Usually in small groups, standing motionless or sedately walking over ground or through water. Nests high in trees alone or in colonies.

Distribution: Globally threatened and now very rare except in Assam where locally common. Could still occur anywhere as it wanders after breeding but numbers very small.
Also occurs very rarely in SE Asia.

HOODED PITTA (Green-breasted Pitta)

Pitta sordida 19 cm F: Pittidae

Description: A stocky, bright green and black bird with a very short tail. Large head and bill. Head black with chocolate brown crown and nape. Emerald green above with blue shoulders and rump. Paler green below with black and red belly and vent. Large white patches in black flight feathers. Sexes alike. Juvenile duller with mottled head.

Voice: Explosive, double whistle *whee whee* and a harsh *skyew*.

Habits: Inhabits the undergrowth of forests and thick scrub. Favours dark, damp, leaf-strewn areas. Feeds on ground invertebrates with bounding hops. Calls and roosts in trees. Secretive. Nests low down.

Morten Strange

Distribution: Rare and very local breeding summer visitor to northern foothills from Himachal Pradesh to the Myanmar border including Nepal and Bangladesh. Appears to winter in SE Asia. Probably overlooked. Also occurs in SE Asia, New Guinea and western Pacific Islands.

INDIAN PITTA

Pitta brachyura 19 cm F: Pittidae

Description: Multi-coloured, short tailed bird with long, pale legs. Green above with blue shoulders and uppertail coverts with white wing spots. Below creamy-buff with red vent. Broad, black stripe through eyes to nape with white above and below. Black stripe on buff crown. Sexes alike.

Voice: Noisy. Penetrating whistle *wheeet peu* and *weee*.

Habits: Inhabits all types of forest and thick scrub, favouring shady, damp areas. Eats invertebrates on ground, making long hops and searching leaf litter. Flies quietly to tree branch if disturbed, but prefers hopping away or freezing if seen. Calls from low tree perch. Solitary or in pairs. Small flocks on migration. Domed nest low down in thick cover.

Kamal Sahai

Distribution: Fairly common endemic local migrant and breeding resident. Breeds in northern India and Nepal south to Goa. Winters in south India and Sri Lanka. Nocturnal migrant often attracted to lights so could appear anywhere en route.

SILVER-BREASTED BROADBILL (Hodgson's Broadbill)

Serilophus lunatus 19 cm F: Eurylaimidae

Morten Strange

Description: A large-headed, mainly grey and chestnut, arboreal bird. Yellow eye rings and black stripe back from eyes. Grey head and underparts. Greyish mantle becoming rich chestnut lower down. Black wings with blue and white wing patches and black tail. Female has faint, white band across breast. Thick bluish bill.

Voice: Mostly a squeaky *pee ou*.

Habits: Inhabits lowland and foothill forests, keeping to lower canopies and tree edges. Sluggish and crepuscular. Sits very upright. Feeds on insects on foliage or by fly-catching in air. Usually in pairs or small groups. Often quite tame. Nest an untidy pouch with side entrance and characteristic "tail" of grass.

Distribution: Scarce breeding resident now restricted to extreme northeastern India along Myanmar border.
Also occurs in China and SE Asia.

ASIAN FAIRY BLUEBIRD (Fairy Bluebird)

Irena puella 28 cm F: Irenidae

Asad Rahamani

Description: A large, strikingly blue and black arboreal bird. Male unmistakable; deep velvet black with glistening blue rear neck, mantle, wing and tail coverts. Female greenish-blue with black lores and wings. Black bill and red irises.

Voice: Noisy. A clear liquid *wit weet,* sometimes extended. Also *wee chu*.

Habits: Inhabits evergreen forests and shola forests. Arboreal and active, feeding on insects, nectar and berries. Sometimes descends to bushes, especially for berries. Forms small parties after breeding and joins mixed hunting groups. Nests high in tree fork.

Distribution: Locally fairly common breeding resident subject to local movements. Occurs in the Northeast, Western Ghats, a few peninsula hills, Sri Lanka and the Andaman and Nicobar Islands.
Also occurs in SE Asia.

BLUE-WINGED LEAFBIRD (Gold-mantled, Jerdon's)

Chloropsis cochinchinensis 18 cm F: Irenidae

Description: A grass, green arboreal bird. Male all green with yellow border to face and black throat with deep blue malar streaks. Blue in wings and tail sides (restricted to shoulder patch in southern *C. c. jerdoni,* which may be separate species). Female has blue throat. Slightly decurved black bill.

Voice: Noisy. Various chuckles and churrs mixed with mimicry of **Drongos** and **Bulbuls**.

Habits: Inhabits well-wooded country, including gardens and villages. Feeds high in trees on nectar, fruit and insects. Drives other birds from flowering trees. Solitary or in pairs but also frequently in mixed hunting groups. Often difficult to spot in foliage. Nests high in a tree.

Morten Strange

♀

Morten Strange

Distribution: Locally common breeding resident throughout peninsula and northeastern India, and Sri Lanka. Commoner in hills.
Also occurs in SE Asia.

GOLDEN-FRONTED LEAFBIRD (Chloropsis)

Chloropsis aurifrons 19 cm F: Irenidae

Description: A grass green, arboreal bird with a bright orange forehead. All green with a small blue shoulder patch. Yellowish-edged black throat with blue moustachial stripe, and chin in northern race. Sexes similar. Slightly decurved black bill.

Voice: Noisy. As **Blue-winged** but more regularly, a **Black Drongo**-like *che chew*.

Habits: Inhabits deciduous and evergreen forests. Arboreal and active but inconspicuous among foliage. Feeds on nectar and insects. Behaviour and nesting as **Blue-winged** but much more of a forest bird.

Nikhil Devasar

Distribution: Fairly common breeding resident subject to local movements. Found in Himalayan foothills east from Uttaranchal to the Northeast, the Western Ghats and less commonly the Eastern Ghats and associated hill ranges. Also in Bangladesh and Sri Lanka.
Also occurs in SE Asia.

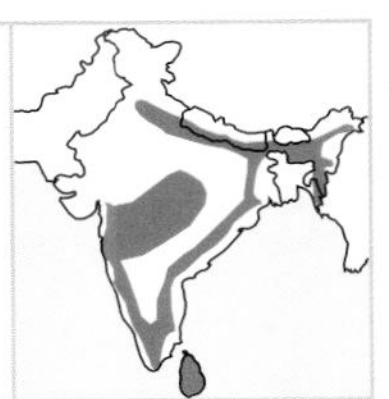

ORANGE-BELLIED LEAFBIRD (Orange-bellied Chloropsis)

Chloropsis hardwickii 19 cm F: Irenidae

Morten Strange

Description: A grass green, arboreal bird with blue and orange underparts. Very colourful. Green with yellowish crown and nape and extensive bluish-black throat and upper breast with contrasting, blue moustachial stripes. Purplish-blue wing and tail edgings, blue wing shoulders. Lower breast to vent warm orange. Female lacks throat and breast patch and is much less orange below. Slightly decurved, black bill.

Voice: Noisy. As other leafbirds.

Habits: Inhabits mainly evergreen forest but also deciduous. Habits, food and nesting as other leafbirds.

Distribution: Locally common breeding resident subject to altitudinal movements, in Himalayan foothills east from Himachal Pradesh to the North-east and including Nepal. Generally higher than other leafbirds.
Also occurs in China and SE Asia.

RUFOUS-TAILED SHRIKE (Isabelline Shrike)

Lanius isabellinus 17 cm F: Laniidae

Mohit Aggarwal

imm

Mohit Aggarwal

Description: Small, pale brown shrike with rusty tail. Variable depending on race, age and sex. Always contrasting tail and small bill. Sandy-brown above and buff below. Thin, white supercilia. Adult male of *L. i. phoenicuroides* is darker with small white patch at base of darker primaries, usually black eye mask and lores. Rufous wash to crown. *L. i. isabellinus* is paler with pale lores. Females and juveniles often without white in wing.

Voice: Can be noisy. A grating *kerishk kerishk*.

Habits: Inhabits dry thorn-scrub and edges of fields, often near water. Perches prominently and catches insects, small reptiles, mammals and birds from perch. Sometimes impales prey on thorns.

Distribution: Breeds in western Pakistan. Common winter visitor to Pakistan and northwest India, vagrant further south and east.
Also occurs in Africa, W and Central Asia and China.

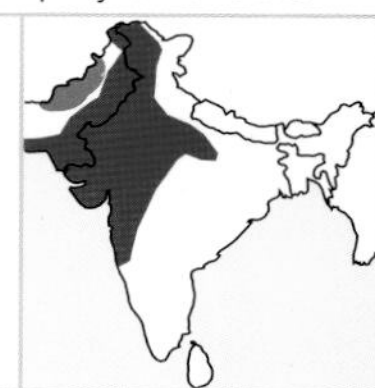

BROWN SHRIKE

Lanius cristatus 19 cm F: Laniidae

Sunit Sen

Description: Small, but stocky, warm brown shrike with powerful bill. Told from **Rufous-tailed** by large head and neck, larger bill (pale in female), concolorous, darker brown upperparts and more graduated tail. Warm rufous-brown from crown to tail. Black eye mask (in male only), deep buff below with whiter cheeks and throat. More prominent white supercilia.

Voice: Noisy. Grating call similar to **Rufous-tailed** and chattering song.

Habits: Inhabits forest edges, thick scrub and open woodland. Similar to **Rufous-tailed** in habits but also hunts within bushes. Can be very territorial and quite tame. Sometimes feeds after dark.

Distribution: A common winter visitor to eastern and southern lowlands including Bangladesh and Sri Lanka. Rare and irregular in the Northwest where there may be confusion with Rufous-tailed. Unknown in Pakistan. Also occurs in N, E and SE Asia.

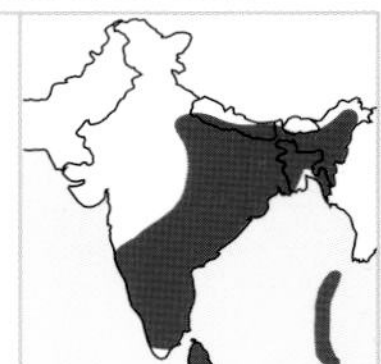

Sunit Sen

PHILIPPINE SHRIKE (race of Brown Shrike)

Lanius (cristatus) lucionensis 19 cm F: Laniidae

Description: A small, brown shrike with grey crown and nape. Still considered a sub-species of **Brown** to which it is similar in structure but bill even more powerful and head looks larger. Distinctly greyish crown, nape and supercilia contrast with brown mantle. There is also a reddish cast to rump and tail but not as markedly contrasting as in **Rufous-tailed**. Immature more heavily barred below.

Voice: As **Brown**.

Habits: As **Brown** but possibly favours more open, drier thorn scrub. Little studied.

Morten Strange

Distribution: Local winter visitor to extreme east, Sri Lanka, Andaman and Nicobar Islands. Can occur in same areas as Brown Shrike. Also occurs in China and SE Asia.

BAY-BACKED SHRIKE

Lanius vittatus 18 cm F: Laniidae

Sumit Sen

Description: A small, chestnut-backed shrike with black forehead and eye mask. Strikingly coloured, delicately built and fine-billed. Pale grey crown and nape contrasting with maroon-chestnut back, black wings and tail. White primary patch, rump, tail tip and outer tail feathers. Pale buff below with warm golden flush on flanks. Sexes alike. Juvenile, barred greyish brown.

Voice: Harsh churring. Extended, rather quiet warbling song with mimicry.

Habits: Inhabits dry scrub, open woodland and the edges of cultivation. Perches on low bushes and wires looking for, mainly insect, prey to pounce on. Flies low between perches. Territorial and not shy. Nests in thorn trees.

Distribution: Fairly common resident throughout lowlands subject to local movements. Rare vagrant in the Northeast, Bangladesh and Sri Lanka. Also occurs in Iran and Afghanistan.

LONG-TAILED SHRIKE (Black-headed, Rufous-backed)

Lanius schach 25 cm F: Laniidae

Kamal Sahai

Sumit Sen

Description: Large grey and rufous shrike with very long tail. Races vary but most have grey crown and back with varying amounts of pale rufous on lower back, vent and flanks. Warm buff underparts. Tail black with white borders. Black eye mask but whole crown and nape black in eastern race *L. s. tricolor* (also shown). Black wings with small white primary patch. Sexes alike. Juvenile scaly.

Voice: Noisy. A grating *shrek*. Song is quiet, pleasing warbling including mimicry.

Habits: The commonest shrike. Inhabits open country with bushes and trees. Frequently perches on road-side wires to locate small bird, mammal and insect prey. Territorial and often confiding.

Distribution: Common and widespread breeding resident throughout most of lowlands subject to local movements. Mainly winter visitor to east and south. Also occurs in Central, N and SE Asia and New Guinea.

GREY-BACKED SHRIKE (Tibetan Shrike)

Lanius tephronotus 25 cm F: Laniidae

Description: A large, stocky, dark grey shrike with rufous uppertail coverts and belly. Superficially similar to **Long-tailed** but more proportionate tail, darker grey, no rufous on upperparts and forehead grey not black. Head and bill larger. Usually lacks white primary patch on blackish wings.

Voice: A grating *shrek*. Mimics other birds.

Habits: Inhabits open areas, usually with trees, but also in open high altitude steppe. Sometimes winters in well wooded areas such as parks and gardens. Habits as **Long-tailed** and very territorial, even in winter quarters.

Nikhil Devasar

Distribution: Locally common breeding summer visitor to Himalayas from Ladakh eastwards to the northeastern hills. Winters lower down to plains including Bangladesh.
Also occurs in China and SE Asia.

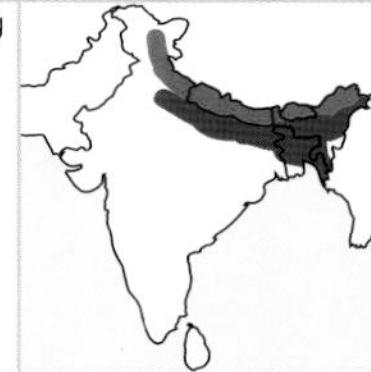

SOUTHERN GREY SHRIKE (Great Grey Shrike)

Lanius meridionalis 25 cm F: Laniidae

Description: Large grey, black and white shrike. Grey above with white scapulars and large, white patches in black wings. White-edged black tail and white below. Black forehead and mask, grey crown and white throat. Winter vagrant *L. m. pallidirostris,* paler grey with white forehead and supercilia and restricted black ear coverts. May be separate species.

Voice: Noisy. An extended *priiit* call. Subdued squeaky song.

Habits: Inhabits dry, open country with thorn-scrub and edges of cultivation. Frequently perches on wires as well as bushes and low trees. Preys on small birds, mammals, reptiles and insects. Larders surplus prey on thorns. Territorial and often mobbed by small birds. Nests in thorn trees.

Sumit Sen

Distribution: Locally common breeding resident and winter visitor in northern and western lowlands of India and Pakistan. Rare further east and south. Subject to local movements and perhaps decreasing as a breeding bird. Also occurs in S Europe, N Africa and W and Central Asia.

EURASIAN JAY (Jay)

Garrulus glandarius 33 cm F: Corvidae

Otto Pfister

Description: A pink crow with blue wing panels. All pinkish-brown (redder in eastern race), paler below with black moustache, wings and tail. Black-barred blue panels in wings. Striking white rump contrasts with black tail in flight. Bill black. Sexes alike.

Voice: Noisy. A screeching *schaaach*, chuckles and whistles. Mimics other birds.

Habits: Inhabits thick broad-leafed forest, particularly oaks as it is partial to acorns. Also eats fruits, nuts and insects and will rob other birds' nests of eggs and young. Feeds in trees and on ground, usually in noisy parties. Inquisitive but shy, though will enter gardens. Rather laboured, uneven flight. Nests low in tree.

Distribution: Locally common breeding resident of northern mountains from north Pakistan to the Burmese border. Moves lower down in winter.
Also occurs in Europe and throughout Asia except Indonesia.

BLACK-HEADED JAY (Black-throated, Lanceolated Jay)

Garrulus lanceolatus 33 cm F: Corvidae

Otto Pfister

Otto Pfister

Description: A pink crow with a crested black head. Body pinkish-buff, greyer on rump and mantle. Wings a complicated mixture of black, blue and white. Tail black-barred blue. Throat white-streaked contrasting with black head. Bill yellow. Sexes alike.

Voice: Usually a single harsh *kraaa*.

Habits: Inhabits mixed temperate forests, particularly oaks but often in more open areas than **Eurasian**. Less shy and often around human habitation. More frequently feeds on ground. Food, flocking and nesting as **Eurasian**, with which it often mixes.

Distribution: Locally common endemic breeding resident in northern mountains from west Pakistan east to Nepal. Some move lower in winter depending on harshness of weather.

YELLOW-BILLED BLUE MAGPIE (Gold-billed Magpie)

Urocissa flavirostris 66 cm F: Corvidae

Description: Very long-tailed, blue crow with powerful yellow bill. Blue above with black and white edges to white-tipped, graduated tail and on wing tips. Head and neck black with small, white nape crescent. White below. Bill golden yellow and legs red. Eastern nominate race greyer above and yellower below. Sexes alike.

Voice: Noisy and mimics other birds. A complex series of squeals, whistles and squawks.

Habits: Inhabits broadleafed and coniferous forests, open woodland and gardens. Feeds in trees and on ground; tail raised as it hops with bouncy gait. Eats small birds, eggs, reptiles, small mammals, fruit and nuts. Usually in pairs or small parties. Flies low in "follow my leader" fashion.

Otto Pfister

Distribution: Locally common breeding resident of northern mountains from north Pakistan to Burmese border. Usually higher than Red-billed. Moves lower in winter.
Also occurs in SE Asia.

RED-BILLED BLUE MAGPIE (Red-billed Magpie)

Urocissa erythrorhyncha 70 cm F: Corvidae

Description: A very long-tailed, blue crow with a powerful, red bill. Larger than **Yellow-billed** and bluer above (purplish in eastern race). White nape patch larger. Black sub-terminal band to white-tipped tail. Red legs. Sexes alike. Juvenile has greyish bill, brownish head and back and is very similar to **Yellow-billed**.

Voice: A loud *pit* and a sharp *kwerer pig pig*. Other calls similar to **Yellow-billed**.

Habits: Inhabits mixed forest, cultivation and gardens at a lower altitude than **Yellow-billed**. Food, nesting and behaviour as **Yellow-billed** but less shy.

Otto Pfister

Distribution: Locally common breeding resident in foothills of northern mountains from Himachal Pradesh east to Burmese border.
Also occurs in China and SE Asia.

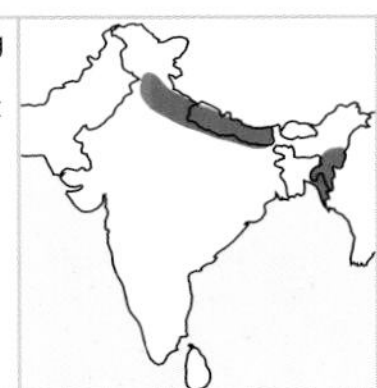

COMMON GREEN MAGPIE

Cissa chinensis 38 cm F: Corvidae

Morten Strange

Description: A large, green crow with a red bill. Grass green (which fades to pale blue in captivity) with a black stripe from bill to short crest on rear crown, Red legs and irises. Wings black and white-tipped chestnut. Long, graduated tail, edged black and white. Sexes alike.

Voice: Very variable. Most commonly a loud *kik wee* note. Complex mixture of whistles, squeals and wails. Also mimics.

Habits: Inhabits dense broad-leafed forests, particularly ravine sides. Feeds low down in foliage or ground in pairs or small groups. Joins mixed hunting grous. Shy, keeping in dense cover. Eats large insects, reptiles, amphibians and small birds. Nests low in tree or bamboo.

Distribution: Locally common breeding resident in northern mountains from Uttranchal to the Myanmar border, and also Bangladesh.
Also occurs in SE Asia.

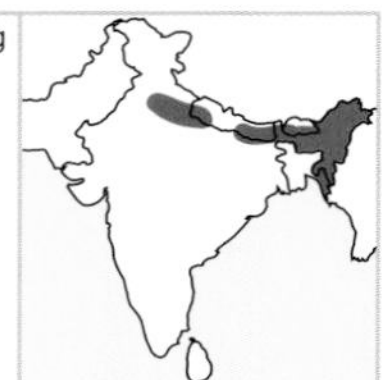

RUFOUS TREEPIE (Indian Treepie)

Dendrocitta vagabunda 50 cm F: Corvidae

Otto Pfister

Description: A long-tailed, grey and rufous crow. Body cinnamon-rufous, paler below with dark grey head and black and white wings. Pale grey (black below) graduated tail, edged and broadly tipped with black and with white sub-terminal area. Sexes alike.

Voice: Noisy and variable. Often a loud, fluty *goo ge lay*.

Habits: The commonest treepie. Inhabits anywhere with trees, including open forest and urban parks and gardens. Mainly arboreal but will feed on ground. Eats insects, reptiles, small birds and their eggs, fruit, nectar and garbage. In pairs or small groups. Also in mixed hunting groups. Bold and inquisitive although often hidden in foliage. Nests high in tree.

Distribution: Common breeding resident throughout lowlands and foothills. Scarcer in the Northeast and unknown in Sri Lanka.
Also occurs in SE Asia.

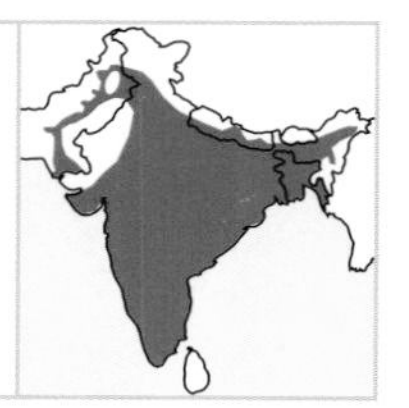

GREY TREEPIE (Himalayan Treepie)

Dendrocitta formosae 40 cm F: Corvidae

Description: A long-tailed, grey and brown crow. Dusky grey face and grey head, neck and underparts. Rufous vent. Graduated grey tail black-edged and tipped. Brownish back. Black wings with small white patch. Smaller and duller than **Rufous**. Sexes alike.

Voice: Noisy and variable. Most commonly a loud, clanking *klok ti klok ti ti*.

Habits: Inhabits forests and secondary growth. Less often near human habitation than **Rufous**. Behaviour and food as **Rufous** but shyer. Arboreal, often in groups or mixed hunting groups. Seems to seek the company of drongos.

Otto Pfister

Distribution: Locally common breeding resident in northern hills from extreme northeastern Pakistan to the Myanmar border. Also the Eastern Ghats. Also occurs in China and SE Asia.

WHITE-BELLIED TREEPIE (Southern Treepie)

Dendrocitta leucogastra 50 cm F: Corvidae

Description: A white, grey and chestnut crow with a very long, black-tipped and pale grey-edged tail. Black face contrasts with white head and underparts. Chestnut mantle and black wings with white patches. Sexes alike.

Voice: Noisy. Various loud, harsh and metallic notes, including *tikituk tikituk* and *kreah kreah*.

Habits: Inhabits evergreen hill forest, thick scrub and plantations. Sociable, often with **Greater Racket-tailed Drongos,** which they imitate freely as well as other forest species. Feeds mainly on invertebrates at lower and middle levels. Nests in shrub.

Clement M Francis

Distribution: Local breeding endemic resident in southern Western Ghats.

ANDAMAN TREEPIE

Dendrocitta bayleyi 36 cm F: Corvidae

Balacandran

Description: A long-tailed, dark crow with startlingly yellow irises. Black face, wings and tail. Dark grey head and neck. Deep brown back and chestnut underparts. Contrasting white wing patches. Smaller and slighter than other treepies. Sexes alike.

Voice: Noisy and variable. Most commonly a fluty *ke chew* and a rasping call.

Habits: Inhabits evergreen and coastal forests. Behaviour as other treepies and seems to seek company of **Andaman Drongos**.

Distribution: Fairly common breeding endemic resident of the Andaman Islands.

BLACK-BILLED MAGPIE (Eurasian Magpie, Magpie)

Pica pica 52 cm F: Corvidae

Otto Pfister

Otto Pfister

Description: A medium-sized, long-tailed, black and white crow. Black all over with green and purple glossed tail, blue-glossed wings, white rump, belly and scapulars. Outer wing shows much white in webs in flight. Short rounded wings produce fluttering flight. Also makes long glides. Hops jauntily with tail raised and flicked.

Voice: A hoarse, staccato *tchrak rak rak*. Also various grating and whining sounds.

Habits: Inhabits open, wooded country, cultivation and near human habitation. Alert, bold and inquisitive. Usually in loose groups, feeding on anything animal or vegetable. Robs nests of other birds. Builds domed, stick nest high in trees.

Distribution: Locally common breeding resident in northern mountains from western Pakistan to Himachal Pradesh and in Bhutan and perhaps Sikkim. Some local altitudinal movements in winter.
Also occurs in Europe, N Africa, throughout mainland Asia and in N America.

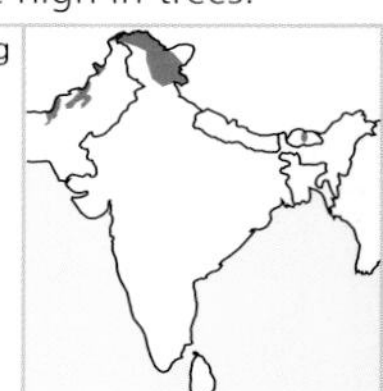

HUME'S GROUNDPECKER (Ground-chough)

Pseudopodoces humilis 20 cm F: Corvidae

Description: A very small crow with an upright stance. Sandy, brown with whitish underparts. Tail pale with dark brown central feathers. White tips to wing shoulder and black lores. Black decurved bill. Long, dark legs and upright stance resembles a large wheatear. Bouncing hops. Weak, fluttering flight on short, rounded wings. Sexes alike.

Voice: A weak repeated *cheep*.

Habits: Inhabits high altitude steppe with low scattered bushes. Often around human habitation and quite confiding. Active feeder on ground searching for insects by probing earth and exploring rock crevices. Hops away, rather than flying. Bobs while perching. Nests at end of self-excavated tunnel in bank.

Otto Pfister

Distribution: Locally common breeding resident in northern Nepal and Sikkim. Rare in Ladakh. Usually above 4000 m. Also occurs in Central Asia and China.

RED-BILLED CHOUGH (Chough)

Pyrrhocorax pyrrhocorax 45 cm F: Corvidae

Description: A medium-sized, glossy, black crow with long, red, decurved bill. Legs pinkish. Short tail has square ends. Longer wings and deeper fingers than **Yellow-billed**. Underwing coverts contrast with greyer-looking flight feathers. Sexes alike. Juvenile less glossy with straighter, shorter yellowish bill similar to **Yellow-billed**.

Voice: Noisy. A high-pitched ringing *cheeo cheeo* Also a hard *jack* and *kew kew*.

Habits: Inhabits high mountain cultivation and pastures, often near human habitation. Gregarious and tame. Probes ground for invertebrates and eats grain and berries. Spectacular acrobatics, rising on thermals and calling. Nests on cliffs.

Distribution: Common breeding resident of northern mountains from western Pakistan east to Arunachal Pradesh. Moves lower in winter. Also occurs in Europe, Africa, W, Central, N and E Asia.

Otto Pfister

YELLOW-BILLED CHOUGH (Alpine Chough)

Pyrrhocorax graculus 38 cm F: Corvidae

Otto Pfister

Description: Medium-sized, glossy black crow with short, yellow bill. Apart from bill, similar to **Red-billed** and best told at distance on shape and voice. Shorter wings with bulging trailing edges and longer, narrow-based tail with rounded ends. Distinguished from **Eurasian Jackdaw** by contrasting black under-wing coverts and greyer flight feathers.

Voice: Less noisy than **Red-billed**. A high rolling *krerrr*, high *kee u* and deep *kruu*.

Habits: As **Red-billed,** with which it sometimes mixes, but usually at higher altitudes. Often very tame. Feeds as **Red-billed** but takes more household scraps. Large acrobatic flocks often perform very close to ground and cliff faces. Nests on cliffs.

Distribution: Common breeding resident of northern mountains from western Pakistan east to Arunachal Pradesh. Many move lower in winter. Also occurs in Europe, Africa, W and Central Asia.

EURASIAN JACKDAW (Jackdaw, Common Jackdaw)

Corvus monedula 33 cm F: Corvidae

Otto Pfister

Description: A small, black crow with grey nape and white irises. All black including crown and face. Grey rear neck can look quite silvery. No contrast under wings, unlike the **Choughs**. Short black bill and black legs. In flight, appears short-headed and stocky. Sexes alike.

Voice: Noisy. An abrupt *jack* and repeated *kyaa* are the commonest calls.

Habits: Inhabits mountain cultivation and pastures usually near cliffs. Also cultivation in winter in plains. Very gregarious; large flocks often wheel around cliffs. Jaunty, rolling walk and often runs on ground. Probes for invertebrates. Also eats other birds' eggs and young, seeds and household scraps. Nests in hole.

Distribution: Locally common breeding summer visitor to northern Pakistan and Kashmir. Winters in Pakistan plains in variable numbers, individuals sometimes occurring as far south and east as Uttar Pradesh.
Also occurs in Europe, Africa, N, W and Central Asia.

HOUSE CROW (Indian House Crow)

Corvus splendens 43 cm F: Corvidae

Description: Medium-sized, slim grey and black crow with long, black bill. Black face and body contrast with grey neck and breast but less so in southern races. Long, powerful bill readily distinguishes from **Eurasian Jackdaw**. Sexes alike.

Voice: Very noisy. Most commonly *kaw kaw*. Also a rolling *kurrrr*.

Habits: One of the region's most familiar, if least-loved, birds. Bold, street-wise and inquisitive. Mostly inhabits the environs of human habitation including urban centres and cultivation. Very gregarious, forming huge roosts and feeding flocks. Feeds on almost anything, including garbage and carrion. Inveterate robber of birds' nests and people's shops. Nests in trees.

Otto Pfister

Distribution: Abundant breeding resident throughout lowlands and foothills. A most adventitious species which by hitching lifts on ships is colonising as far as northern Europe and Africa.
Also occurs in China and SE Asia. "Artificially" in Europe, Africa and W Asia.

ROOK

Corvus frugilegus 48 cm F: Corvidae

Description: A large, glossy black crow with bare whitish skin round bill base. May show violet gloss. Long, slender, unfeathered bill also has whitish base. Often has small throat pouch. High crown peak and deep, shaggy belly and thighs give different profile to similar-sized crows. Fingered, straight-edged, narrow-based wings and slightly wedge-shaped tail give different flight shape. Sexes alike. Juvenile has black bill.

Voice: A hoarse *karrh karrh*.

Habits: Inhabits cultivation and pastures. Very gregarious feeding and roosting in flocks, often with **Eurasian Jackdaws**. Walks rather cockily about fields feeding on invertebrates, for which it probes deeply. Also eats young shoots and seeds.

Otto Pfister

Distribution: Locally common winter visitor to western and northern Pakistan and Kashmir. No recent substantiated records from elsewhere in India.
Also occurs in Europe, Africa and elsewhere in mainland Asia.

CARRION CROW (includes Hooded Crow)

Corvus corone 48 cm F: Corvidae

Otto Pfister

Description: A large, glossy black crow with flat head and rather slim, black bill. These features distinguish it from **Large-billed Crow**. Bill lacks arch and has basal feathering around nostrils. **Hooded Crow** *C. c. cornix* (which may be a separate species) grey with black head and breast patch, wings and tail. Broad, even-edged wings and square-ended tail. Sexes alike. Lazy flight.

Voice: A deep harsh *karr karr*.

Habits: Inhabits open country and cultivation, often near human habitation. Shy and wary. Singly or in pairs but roosts communally. Feeds with other species, especially on carrion. Eats almost anything. Inquisitive feeder, searching methodically.

Distribution: A scarce breeding resident in northern Pakistan and Kashmir. Hooded Crow is scarce winter visitor to same area but common in Ladakh.
Also occurs in Europe, N Africa and W, Central and E Asia.

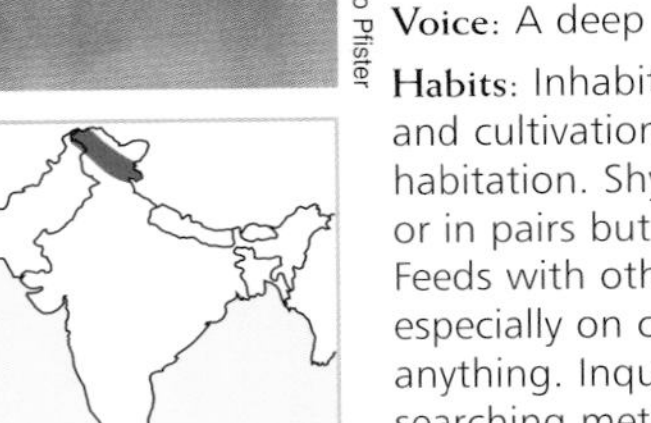

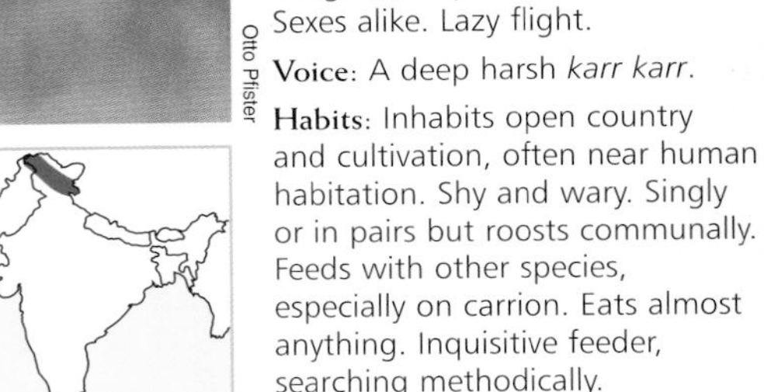

LARGE-BILLED CROW (Jungle, Black Crow)

Corvus macrorhynchos 48 cm F: Corvidae

Otto Pfister

Otto Pfister

Description: A large, glossy black crow with steep forehead and large, curved bill. Himalayan races larger with more powerful bill so separation from **Raven** needs care. Bill and head shape are best features. Smaller-billed lowland race often called **Jungle Crow** (lower photo) and perhaps a separate species, readily separated from **House Crow** on size and all-black plumage.

Voice: A deep *khaa khaa*. Deeper than **House**.

Habits: Inhabits well-wooded country, including forest edges and urban parks. Bold and inquisitive but less so than **House**. Singly or in pairs, but roosts communally, often with other species. Eats anything, including carrion. Nests high in tree.

Distribution: Common breeding resident throughout hills and lowlands. In Pakistan only in northern mountains. Also occurs in N, E and SE Asia.

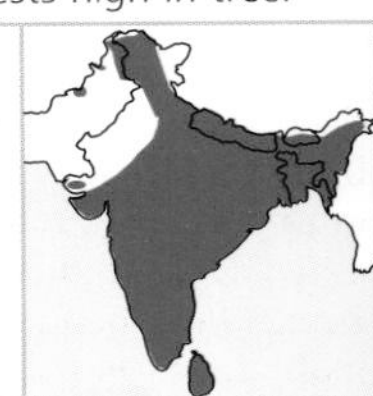

COMMON RAVEN (Raven)

Corvus corax 69 cm F: Corvidae

Otto Pfister

Description: Huge, glossy black crow with deep, black bill. Shaggy throat. In flight long, angled, rather narrow, well-fingered wings, protruding neck and long distinctly wedge-shaped tail. Soars high with wings flat or lowered. High powerful flight and impressive aerobatics by breeding pairs includes flying on back.

Voice: A very deep and far-carrying *prurp prurp*.

Habits: Inhabits desert and semi-desert, high mountain deserts and pastures. Himalayan race often near human habitation. Usually singly or in pairs but wandering groups in winter and at carcasses. Eats mainly carrion. Waddling walk. Life-long pairs have same nesting territory every year. Nests on lone tree or cliff.

Distribution: Locally common breeding resident in Pakistan and northern mountains east to Arunachal Pradesh. Some move to northern plains in winter. Also occurs in Europe, N Africa, W, N, Central and E Asia and N and Central America.

ASHY WOODSWALLOW (Ashy Swallow-shrike)

Artamus fuscus 19 cm F: Corvidae

Morten Strange

Description: A small, grey, large-headed bird with thick, bluish bill. Grey head with blackish face. Pinkish-grey body, paler below. Short, square tail darker with white tip. White vent and rump crescent. Long wings reach tail-end and are triangular in flap and glide flight. At rest resembles top-heavy martin. Sexes alike.

Voice: A harsh *chek chek* mostly in flight.

Habits: Inhabits open country with trees, particularly palms. Usually in small parties and rather crepuscular. Spends much time on overhead wires or bare branches, launching into air after insects and returning to same perch. Often sit huddled together. Pumps tail when perched. Nests high, mainly in palm tree.

Distribution: Locally common breeding resident in southern and eastern peninsular and northeast India, Sri Lanka and Bangladesh. Wanders elsewhere.
Also occurs in China and SE Asia.

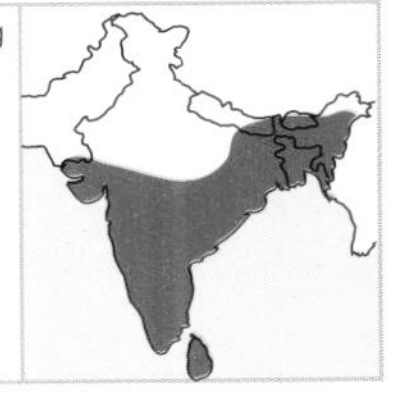

WHITE-BREASTED WOODSWALLOW (White-rumped)

Artamus leucorynchus 18 cm F: Corvidae

Morten Strange

Description: A small, pale, short-tailed and large-headed bird with a stout, black-tipped, bluish bill. Similar to **Ashy** but with darker head, white underparts and rump. Long wings, short tail and sweeping martin-like flight. Appears top-heavy, particularly when perched. Sexes alike.

Voice: As **Ashy**.

Habits: As **Ashy** but mainly in forest clearings.

Distribution: Common breeding resident restricted to the Andaman Islands.
Also occurs in Indonesia, the Philippines and Borneo, Australasia and various Pacific islands.

EURASIAN GOLDEN ORIOLE (Golden Oriole)

Oriolus oriolus 25 cm F: Corvidae

Kamal Sahai

♀

Kamal Sahai

Description: A medium-sized, bright yellow and black arboreal bird with pink bill. Male golden-yellow with black wings and tail and black ring round eyes. Tail has yellow apical tips and wings have yellow wing patches and tips to primaries. Female and juvenile greener with streaked underparts, getting yellower with age (as shown).

Voice: Male has fluting *weela we oo* and single *weeoo* and *weee*. Grating, cat-like *kerach* by both.

Habits: Inhabits open forest, groves, wooded parks and gardens. Feeds high in canopy, mainly on caterpillars and other insects, also fruit and nectar. Bathes frequently by plunge-diving. Solitary or in pairs. Difficult to spot in foliage. Nests high in tree.

Distribution: Common breeding resident throughout most of peninsula. Winter visitor to extreme south and Sri Lanka. Mainly monsoon breeding visitor to northern plains and Pakistan.
Also occurs in Europe, Africa, W and Central Asia.

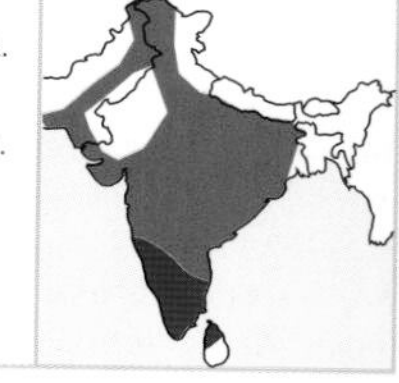

BLACK-NAPED ORIOLE

Oriolus chinensis 25 cm F: Corvidae

Description: A medium-sized, black and yellow arboreal bird with a stout pink bill. Bill size and broad black stripes through eyes round to nape separate from very similar, northeastern **Slender-billed Oriole** *O. tenuirostris* whose head stripes are narrow. All yellow with black in flight feathers and black tail with yellow apical patches. Female greener yellow. Juvenile streaked below with diffuse eyestripe.

Voice: Calls similar to **Eurasian** but less musical. **Slender-billed** also has a diagnostic hard *kick* note.

Habits: As **Eurasian**. Often in rubber plantations.

Otto Pfister

Distribution: Breeds on Andaman and Nicobar Islands. Scarce winter visitor to Kerala and southeast Bangladesh with scattered records elsewhere in peninsula.
Also occurs in E and SE Asia.

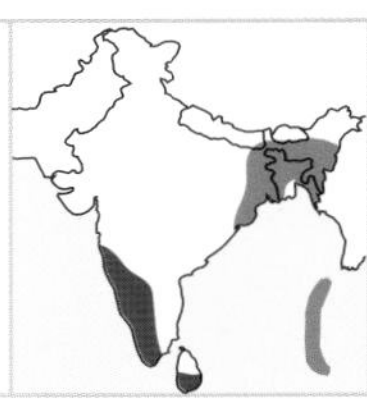

BLACK-HOODED ORIOLE (Black-headed Oriole)

Oriolus xanthornus 25 cm F: Corvidae

Description: A medium-sized, bright yellow and black, arboreal bird with a black head and neck. Golden-yellow body. Flight feathers black with yellow edgings and primary patches. Tail yellow with black central feathers. Bill deep pink. Female greener above. Juvenile has white-speckled throat and greyer hood.

Voice: Fluty *tu yow yow* and *we hoo*. Also harsh *kerwark*.

Habits: Inhabits well-wooded country including urban parks and gardens. Unusually confiding and obvious for an oriole. Active and lively, often seen chasing each other through trees. Perches on high bare branches, calling and sun-bathing. Food and nesting as **Eurasian**. Frequently joins mixed hunting groups.

Nikhil Devasar

Distribution: Common breeding resident in plains and hills but restricted to Himalayan foothills in the Northwest. Makes local movements. Unknown in Pakistan.
Also occurs in SE Asia.

MAROON ORIOLE

Oriolus traillii 28 cm F: Corvidae

Nikhil Devasar

Description: A large, dark oriole with chestnut-red tail. Powerful, decurved, grey bill and whitish irises. Male has maroon body and black head and wings. Female blackish above with dull maroon back and heavily brownish-streaked, white underparts.

Voice: Most commonly a cat-like squawking *meow*. Also a fluty *pi lio ilo*.

Habits: Inhabits thick broad-leafed forest, usually keeping well-hidden in canopy, where it feeds on invertebrates, nectar and fruit. Comes into open to feed on Silk Cotton and other nectar (as shown) when it can be confiding. Usually in pairs or singly, often in mixed hunting groups. Nests high in tree.

Distribution: Locally common breeding resident in northern hills from Himachal Pradesh east to the Myanmar border. Moves lower down to foothills and nearby plains,including Bangladesh, in winter.
Also occurs in S and E China and SE Asia.

LARGE CUCKOOSHRIKE

Coracina macei 28 cm F: Corvidae

Otto Pfister

♀

Kamal Sahai

Description: Fairly large, grey bird with dark mask. Heavy-looking with stout black bill. Bluish-grey with darker wings and paler belly fading to white undertail coverts. Male of northern race *C. m. nipalensis* (shown) has no barring below. Female is barred from breast to vent. Peninsula race male is variably barred.

Voice: Noisy. A loud ringing whistle *te treeee*, the second note inflected and drawn out.

Habits: Inhabits well-wooded country including gardens. Conspicuous, calling from topmost branches. Undulating flap and glide, **Shikra**-like flight. Flicks wings alternately on alighting. Usually in pairs or small parties. Feeds on large insects.

Distribution: Fairly common endemic breeding resident in plains and hills but restricted to Himalayan foothills in the Northwest. Makes local movements. Now very rare in Pakistan.

BLACK-WINGED CUCKOOSHRIKE (Smaller Grey)

Coracina melaschistos 22 cm F: Corvidae

Description: A medium-sized, dark cuckooshrike with unbarred, grey underparts. Male is dark grey above with contrasting black wings and tail. Underside of tail shows broad white feather tips. Female paler with very faint barring on underparts.

Voice: Loud descending *twit twit to we* and *peeo peeo peeo*.

Habits: Inhabits open forest, forest edge and groves. Feeds on invertebrates in canopy and undergrowth, singly or in pairs. Joins mixed hunting groups. Undulating flight. Nests in tree.

Jan Willem de Besten

Distribution: Local breeding resident in northern mountains from north Pakistan east to the Myanmar border. Winters lower down in foothills and plains mainly in Bangladesh and Northeast India south to Orissa. Scattered winter records throughout peninsular India.
Also occurs in China and SE Asia.

BLACK-HEADED CUCKOOSHRIKE

Coracina melanoptera 20 cm F: Corvidae

Description: A medium-sized, black and grey bird. Male has black hood and white-edged black wings and tail. Grey mantle and breast. White underparts. Female browner grey with black barring from throat to belly and, usually, whitish supercilia.

Voice: Not noisy. A clear whistling and a hard *pit pit pit*.

Habits: Inhabits well-wooded country including gardens. Rather shy. Often in arboreal mixed hunting groups, where it feeds on insects, fruit and nectar deep among the foliage. Nests high on tree branch.

Balachandran

Balachandran

Distribution: Local endemic breeding resident from Himalayan foothills (where it is summer visitor only) south through peninsula. Rare in the Northeast and only a scarce winter visitor to the Northwest. Unknown in Pakistan.

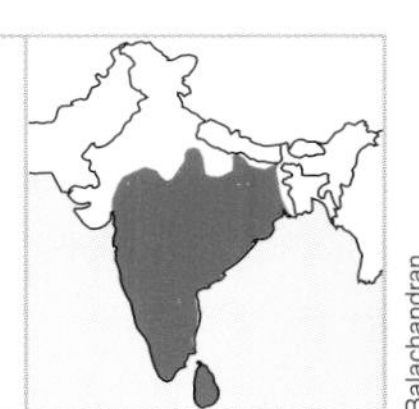

ROSY MINIVET

Pericrocotus roseus 18 cm F: Corvidae

Jan Willem den Besten

Description: A slim, pale minivet with muted colours. Male brownish-grey above with black wings and tail. Red wing patch and tail edgings as **Long-tailed.** Pale grey rear neck and crown, whitish face and pale pink-flushed underparts. Female is very pale yellow below and has greenish-yellow rump; otherwise like **Small** but larger.

Voice: Not well known. A squeaky, trilling note recorded.

Habits: A poorly known species. Inhabits forest, woodland and gardens. Behaves as other minivets feeding in canopy down to bush level on invertebrates. Usually in small parties and apparently more sluggish than other minivets. Nests in tree.

Distribution: Scarce and very local breeding resident and summer visitor in northern mountains from northern Pakistan east to the Myanmar border. Some at least winter in Bangladesh and India south to Kerala but records very sporadic. Most frequently recorded in the Northeast and Eastern Ghats. Also occurs in China and SE Asia.

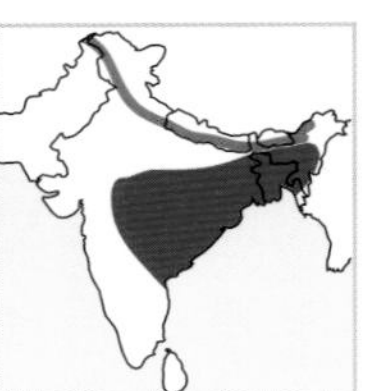

SMALL MINIVET (Little Minivet)

Pericrocotus cinnamomeus 15 cm F: Corvidae

Nikhil Devasar

Otto Pfister

Description: Small, black, grey and orange arboreal bird with long tail. Male has grey head and mantle (dark in southern race), black face and wings and tail. Orange rump and wing patch. Orange breast fading to yellow on belly but in northern race *P .c. pallidus* (shown) this very limited and most of underparts white. Female yellowish-white below and grey above with yellow in wings. Southern race female, orange-yellow below.

Voice: A high, thin *swee swee*.

Habits: The commonest minivet inhabiting open wooded country including groves and gardens. Flits rapidly through canopies in small parties feeding on insects in foliage. Nests in tree.

Distribution: Fairly common breeding resident throughout most of lowlands. Also occurs in SE Asia.

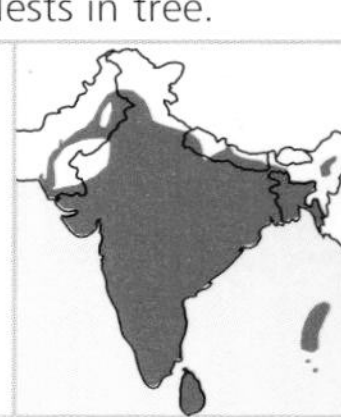

⊙ WHITE-BELLIED MINIVET (Jerdon's Minivet)

Pericrocotus erythropygius 15 cm F: Corvidae

Otto Pfister

Description: A small, black, white and orange bird. Male strikingly similar to a male **Stonechat** with a long tail. Blue-glossed black with broad white wing patch, outer tail feathers and breast to vent. Bright orange rump and breast patch. Female brown above, white below with white wing patch, pale orange rump and white-bordered black tail.

Voice: Calls *tseep tseep* in flight.

Habits: Inhabits open dry woodland and thorn-scrub. Feeds low in trees and bushes on insects, much lower than other minivets. Usually in pairs. Feeds actively, moving rapidly through habitat. Nests low in thorn bush.

Distribution: Rare breeding resident of north-central India lowlands and Gujarat. Wanders outside breeding range in winter but within lowlands. Also occurs in Myanmar.

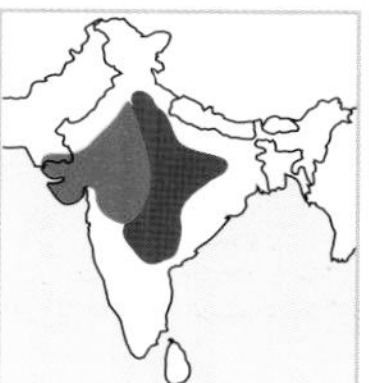

⊙ GREY-CHINNED MINIVET (Yellow-throated Minivet)

Pericrocotus solaris 17 cm F: Corvidae

Description: A small, dark grey and scarlet, arboreal bird with grey cheeks. Male greyer than **Scarlet** above with obvious grey cheeks and chin and orange-yellow throat. Similar shaped red patch in wings. Female similar to female **Long-tailed** but lacks yellow on supercilia and forehead.

Voice: A thin *tsee sip*.

Habits: Inhabits hill forests. Feeds actively, mainly in tree canopy in small, calling parties. Sometimes briefly descends to eye level. Eats insects. Nests high in trees.

Morten Strange

Morten Strange

Distribution: A local breeding resident in mountains east from central Nepal to Arunachal Pradesh and the Myanmar border hills.
Also occurs in China and SE Asia.

LONG-TAILED MINIVET

Pericrocotus ethologus 18 cm F: Corvidae

Otto Pfister

♀

Nikhil Devasar

Description: Small, black and red arboreal bird. Similar to **Scarlet** but smaller body and longer tail. Male deeper red and has red line on tertials creating U-shaped patch on black wings. Head and upperparts black, tail with red outer tail feathers. Underparts red. Male **Short-billed** *P. brevirostris* lacks red on tertials. Female similar to others but paler yellow below, limited yellow on forehead and supercilia and grey on cheeks.

Voice: Distinctive descending whistle *pee ru*.

Habits: Inhabits broadleafed and coniferous hill forests when breeding. Winters in open woodland and groves. Nomadic in winter. Feeds in parties moving rapidly through canopy. Eats insects.

Distribution: Common breeding resident and summer visitor to northern hills from western Pakistan east to the Myanmar border and Bangladesh. Winters widely in plains south to Maharashtra.
Also occurs in Afghanistan, China and SE Asia.

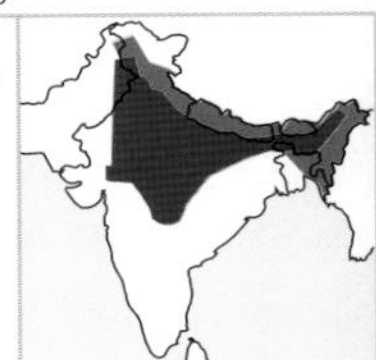

SCARLET MINIVET (Orange Minivet)

Pericrocotus flammeus 20 cm F: Corvidae

Otto Pfister

♀

Otto Pfister

Description: A small, black and red arboreal bird. The largest and stockiest minivet, the male best told from others by one or two isolated red spots on secondaries and more orange-red underparts. Black head and underparts with red wing patch and outer tail feathers. In female, red is replaced with bright yellow with extensive yellow on forecrown.

Voice: A loud, penetrating whistle *twee twee*.

Habits: Inhabits all types of forest feeding in canopies, often in large parties in winter, in typical minivet fashion. Nests high in trees.

Distribution: Common breeding resident in Himalayan foothills from Kashmir west to the Myanmar border and locally in peninsular hills, Bangladesh and Sri Lanka. Some disperse to plains in winter.
Also occurs in SE Asia.

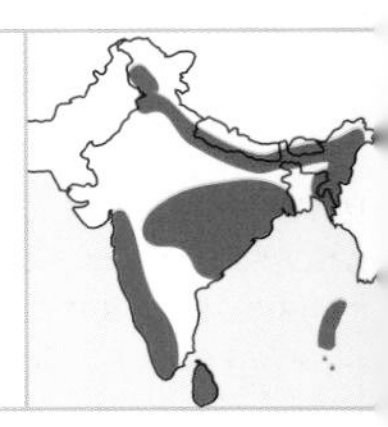

BAR-WINGED FLYCATCHER-SHRIKE

Hemipus picatus 14 cm F: Corvidae

Description: A small, black and white arboreal bird. Black crown and rear neck with mantle brown in Himalayan race (shown). Black in southern races. Long white patches on black wings show as an obvious V in rear views. Tail black with white borders. Throat white, contrasting with rest of underparts which have a pinky-grey wash. Female browner.

Voice: Noisy. Persistent *tsit it it tsit it it*, *si si si si* and a short *chip*.

Habits: Inhabits open forest and forest edge including adjoining scrub. Feeds on insects in noisy, active flocks, often in mixed hunting groups in canopies and lower down. Often fly-catches. Nests high in tree.

Nikhil Devasar

Distribution: Fairly common breeding resident in foothills from Himachal Pradesh east to the Myanmar border, the Eastern and Western Ghats and Sri Lanka. Makes altitudinal and other local movements.
Also occurs in SE Asia.

YELLOW-BELLIED FANTAIL (Yellow-bellied Flycatcher)

Rhipidura hypoxantha 13 cm F: Corvidae

Description: A small, olive and yellow bird with a broad blackish, white-tipped tail. Olive above with thin, white wing bar. Head yellow with olive rear crown and dark mask and eye stripes. Bright buttercup yellow below. Flirts large tail displaying broad white tips. Sexes alike.

Voice: A persistent *sip sip sip*. Also a musical *wee too wee too wee*.

Habits: Inhabits mainly broad-leafed forest, often near water. Extremely active, seeking insects by flicking through foliage and frequently fly-catching, returning to its perch like a typical fly-catcher. Usually seen singly or in pairs, sometimes in mixed hunting groups. Nests at middle level in tree.

Otto Pfister

Distribution: Common breeding resident of foothills from extreme northeastern Pakistan to the Myanmar border. Many move lower down in winter when may occur in northern plains.
Also occurs in China and Myanmar.

WHITE-THROATED FANTAIL (Fantail Flycatcher)

Rhipidura albicollis 17 cm F: Corvidae

Otto Pfister

Description: Small, brown and white bird with broad, white-tipped tail. Northern race (shown) all dark brown with contrasting white throat and supercilia and white tips to all but two central tail feathers. Peninsular race *R. a. albogularis* (which may be a separate species) has white spotted brown breast with rest of the underparts white. Constantly flirts broad tail. Sexes alike.

Voice: Noisy. Rather weak descending *tri ri ri ri ri* and a harsh *chukrr*.

Habits: Inhabits shady forest, scrub and gardens, favouring well-wooded ravines. Forages for insects at middle and lower levels around tree trunks. Usually in pairs or singly. Often confiding. Active.

Distribution: Fairly common breeding resident of foothills from extreme northeastern Pakistan to the Myanmar border, throughout the Northeast and most of the peninsula. Absent from Sri Lanka. Makes altitudinal and local movements.
Also occurs in China and SE Asia.

WHITE-BROWED FANTAIL (Fantail Flycatcher)

Rhipidura aureola 17 cm F: Corvidae

Kamal Sahai

Description: A small, brown and white bird with a broadly white-tipped, brown tail. Dark brown above and on head with fine white wing spots and broad white supercilia and moustachial stripes. Black throat is often lightly spotted. Constantly flirts broad tail. Sexes alike.

Voice: Noisy. A high, undulating *chee chee chee wee chee chi*. A grating *chuk chuk chuker*.

Habits: Inhabits open woodland, groves, gardens including trees in villages. Often in trees near water, but generally in drier habitats than **White-throated**. Feeds on insects fairly low down in foliage and often on ground. Usually solitary or in pairs. Very active and often confiding. Nests low down in tree fork.

Distribution: Common breeding resident throughout lowlands except the Northeast and Bangladesh. Makes local movements and appears rather sporadically in some areas.
Also occurs in Myanmar.

BLACK DRONGO (King Crow)

Dicrurus macrocercus 31 cm F: Corvidae

Description: Glossy black bird with long, forked tail. Blue gloss on head and breast and diagnostic white spot at base of bill. Eyes dark red. Tail is deeply forked except when moulting. Sexes alike. Immature whitish on belly and tail may show little or no fork.

Voice: Noisy. A ringing *tiu tiu* very like **Shikra**. Harsh *cheece cheece*. Squeaks and chatters.

Habits: Commonest drongo. Inhabits open, wooded country, urban parks and cultivation. Perches on ground, wires and bare branches. Usually solitary, sometimes in small groups. Pounces on aerial and ground insects from perch; follows livestock and ploughs. Aggressive to larger birds. Nests in tree.

Otto Pfister

Mohit Aggarwal

Distribution: Abundant breeding resident throughout lowlands. Only in north of Sri Lanka.
Also occurs in Iran, Afghanistan, China and SE Asia.

ASHY DRONGO (Grey Drongo)

Dicrurus leucophaeus 30 cm F: Corvidae

Description: A dull black bird with a long, forked tail. Duller black than **Black** with greyer underparts. No spot at bill-base and eye light red. Tail fork usually deeper and more splayed. Juvenile greyer.

Voice: Noisy and varied. Often *cha ke wip*, *kit whew* and *cheece cheece chichuk tilililili*.

Habits: Inhabits forests, and more open woodland in winter. Partly crepuscular. Hunts mainly aerial insects from bare branches in tree-tops. Usually solitary, in pairs or small parties. Noisy and bold, attacking passing raptors and corvids in particular. As with all drongos, other small birds often nest near them for protection. Nests high in tree.

Nikhil Devasar

Distribution: Fairly common breeding summer visitor to northern foothills from Pakistan east to the Myanmar border. Winters throughout plains and peninsula including Sri Lanka.
Also occurs in Iran, Afghanistan, China and SE Asia.

WHITE-BELLIED DRONGO (White-vented Drongo)

Dicrurus caerulescens 24 cm F: Corvidae

Bikram Grewal

Description: A dark grey bird, usually with white belly and vent. One of the smallest drongos with shorter, less forked tail. Variably dark grey often with bluish sheen on breast. Most have belly and vent white but a Sri Lankan race only has white vent. Care should be taken not to confuse with juvenile **Black**; a common error. Sexes alike.

Voice: Varied whistling and grating notes.

Habits: Inhabits open woodland, plantations and gardens. Habits as other drongos and equally crepuscular. Often joins mixed hunting groups. Feeds on insects and flower nectar. Frequently fly-catches. Nests high in tree.

Distribution: Uncommon endemic breeding resident in plains and hills throughout peninsula and northern plains. Also Sri Lanka. Not known in Pakistan, Bangladesh or the Northeast.

BRONZED DRONGO (Little Bronzed Drongo)

Dicrurus aeneus 24 cm F: Corvidae

Sumit Sen

Description: A rather small, strongly glossed, black bird with a shallow tail fork. Small size and distinct blue-green spangling, especially from crown to mantle and on breast. Tail relatively short with shallow splayed fork. Heavy black bill. Sexes alike.

Voice: Noisy. Varied whistles and grating calls. Also *cheet chi chew*.

Habits: Inhabits shady parts of thick forest and plantations. Very active. Usually in pairs or singly. Often joins mixed hunting groups. Feeds mainly on insects. Nests high in tree fork.

Distribution: Fairly common breeding resident in foothills from Punjab east to the Northeast including Bangladesh then south through the Eastern Ghats and throughout the length of the Western Ghats. Makes some local movements.
Also occurs in China and SE Asia.

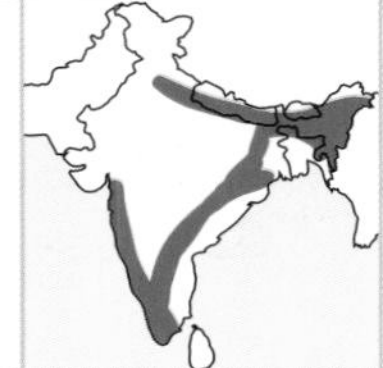

LESSER RACKET-TAILED DRONGO (Small)

Dicrurus remifer 28 cm (plus 20 cm streamers) F: Corvidae

Description: A medium-sized, glossy black drongo with long streamers. Best distinctions from larger **Greater** are flat, oval shaped rackets extending on both sides of steamer shafts and square-ended, not forked, tail. Rackets appear like two large insects following bird in flight. Also has much smaller crown tuft, making crown appear flat. Sexes alike.

Voice: Noisy like **Greater**. Musical whistles and screeches. Mimics other birds.

Habits: Inhabits shady under-storeys of forests often near streams and in mixed hunting groups. Usually in pairs, feeding on invertebrates. Sits quietly but often fly-catches. Nests in tree.

Morten Strange

Distribution: Locally common breeding resident in northern foothills from Uttaranchal east to the Myanmar border. Winters lower down and in Bangladesh.
Also occurs in China and SE Asia.

SPANGLED DRONGO (Hair-crested Drongo)

Dicrurus hottentottus 32 cm F: Corvidae

Description: A large, glossy drongo with incurved outer tail feathers. Bright spangling on head and neck and a wispy crest. Tail long and broad with markedly incurved outer tail feathers. When these are missing tail looks square ended. Powerful, decurved, crow-like but pointed bill. Sexes alike.

Voice: Noisy. A clanking *klaa klok* and a rising *chit we*. Also mimicry and various whistles.

Habits: Inhabits forest and forest edge, particularly in flowering trees, such as Silk Cotton. Feeds mainly on nectar but also inver-tebrates. Quite confiding. Very sociable often in mixed hunting groups. However individuals will zealously defend a favoured flowering tree against all-comers. Nests high in tree.

Nikhil Devasar

Distribution: Common breeding resident in northern foothills through to the Northeast then south in hills and nearby plains from Bangladesh down eastern India to south Tamil Nadu and north through Kerala to north Maharastra.
Also occurs in China and SE Asia.

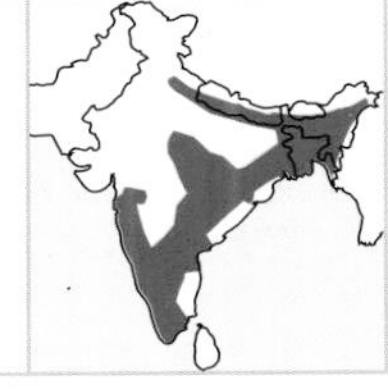

ANDAMAN DRONGO

Dicrurus andamanensis 35 cm F: Corvidae

Otto Pfister

Description: A brownish-black bird with recurved tail forks and a rather long, heavy, decurved black bill. Similar to **Black** but a brownish tinge to plumage and distinctly recurved forks on longer tail. Bill size is also useful. Slight hair like crest on forehead. Sexes alike.

Voice: Noisy. Various hard, metallic calls and a *chyu* note.

Habits: Inhabits forest, often in quite large groups and sometimes in mixed hunting parties. Feeds on insects, often by fly-catching. Nests high in tree fork.

Distribution: Common endemic breeding resident restricted to the Andaman Islands where it is the only common drongo. Also occurs in the Coco Islands of Myanmar.

GREATER RACKET-TAILED DRONGO

Dicrurus paradiseus 32 cm (plus 30 cm streamers) F: Corvidae

Otto Pfister

Description: A glossy black bird, usually with very long wire-like outer tail feathers ending in twisted rackets. Has rather scruffy, prominent crest on forecrown (reduced in southern races). All except the Sri Lankan wet zone race (which has a twisted tail fork) have the streamers. The racket is teardrop-shaped, on one side of the shaft only and twisted. Sexes alike.

Voice: Very noisy. Very varied with whistling and screeching. Much mimicry.

Habits: Inhabits forests and plantations. Bold and obvious. Behaves as other drongos, often in small groups and mixed hunting groups. Often low-down. Partly crepuscular.

Distribution: Locally common breeding resident throughout lowlands and hills of most of peninsula, the Northeast, Nepal, Bangladesh and Sri Lanka. Also occurs in China and SE Asia.

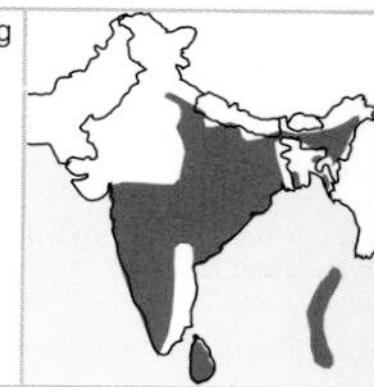

BLACK-NAPED MONARCH (Azure Flycatcher)

Hypothymis azurea 16 cm F: Corvidae

Description: A small, bright blue flycatcher-like bird. Sub-specific variation in depth and extent of blue. Male usually bright blue with black rear crown patch, upper bill base and narrow gorget (absent in Sri Lankan race). Belly and vent greyish-white. Female is duller with greyish mantle and lacks black.

Voice: Noisy. Grating, interrogative *chwich chweech*?. And an extended *weee weee weee*.

Habits: Inhabits forest, secondary growth, plantations and particularly bamboos, usually feeding in canopy in mixed hunting groups. Descends to lower levels. Active; fly-catching, hovering and frequently flirting tail. Usually singly or in pairs. Nests in foliage at middle levels.

Otto Pfister

Distribution: Fairly common resident in plains and hills east and south of Uttar Pradesh and Gujarat including Bangladesh and Sri Lanka. Makes erratic local movements and may turn up rarely the Northwest of usual range. Also occurs in China and SE Asia.

ASIAN PARADISE-FLYCATCHER (Paradise Flycatcher)

Terpsiphone paradisi 20 cm (male's tail 30 cm more) F: Corvidae

Description: Small, black-headed, rufous or white bird with extremely long central tail feathers. Black-crested head, broad bluish bill and blue-rimmed eyes. Rufous phase is all rufous with white under-parts. White phase is all white, flecked black with black wings. Female and juvenile are as rufous phase but with short tail and crest.

Voice: A hard, grating *chwae* and *wee por weli wee por weli*. Melodious slow, warbling song.

Habits: Inhabits shady areas in woodland, gardens and plantations, often near water. Feeds by fly-catching at all levels. Usually singly or in pairs but joins mixed hunting groups. Upright stance and undulating flight. Shy. Nests high in tree fork.

Kamal Sahai

Otto Pfister

Distribution: Fairly common throughout. Mainly breeding summer visitor to northern and peninsular hills, wintering in plains. Mostly resident in extreme south, Sri Lanka and Bangladesh.
Also occurs in Afghanistan, China, Korea and SE Asia.

COMMON IORA (Iora)

Aegithina tiphia 14 cm F: Corvidae

Sumit Sen

Description: A small, yellow and green arboreal bird with white-marked black wings. Northern race, non-breeding males and females have similar plumage although males have black tails. Pale green on crown and mantle with white shoulder patch and white-edged black wings. Southern breeding males have black crowns and either black or black and yellow mantles. All have yellow below.

Voice: Loud musical whistles; *weeeeee chu* and *peee ou*.

Habits: Inhabits forest edges, groves and wooded country. Usually singly or in pairs actively hunting the canopy for flower nectar, insects and caterpillars, and calling frequently. Joins mixed hunting groups. Nests in tree.

Distribution: Common breeding resident in most of region except Pakistan and northwest India. Also occurs in China and SE Asia.

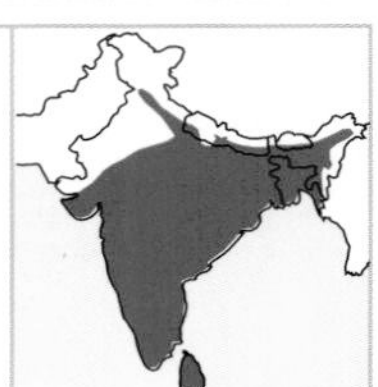

MARSHALL'S IORA (White-tailed Iora)

Aegithina nigrolutea 14 cm F: Corvidae

Otto Pfister

Description: A small, yellow and green arboreal bird with white and black tail. Tail, which has variable amounts of white, is best distinction from **Common**. Otherwise female and non-breeding male almost identical. Breeding male (shown) has black crown, yellow collar and black mantle variably marked with yellowish-green. Wings as **Common**. Note some authorities only recognise this as a sub-species of **Common**.

Voice: Similar to **Common** but also a diagnostic nasal *tzee cheert*.

Habits: Inhabits open woodland and thorn scrub, often drier habitats than **Common**. Food, behaviour and nesting as **Common**.

Distribution: Now a rare and apparently declining endemic breeding resident confined to northwest India. Earlier records from further south and east.

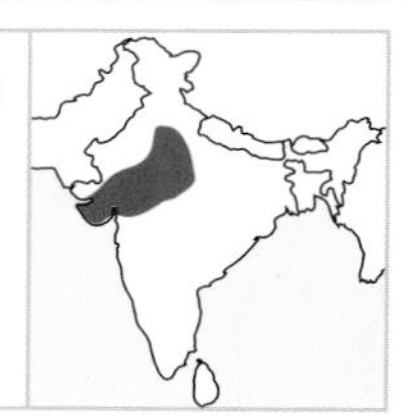

COMMON WOODSHRIKE (Lesser Woodshrike)

Tephrodornis pondicerianus 16 cm F: Corvidae

Description: Small, brown and white bird with white outer tail feathers. Grey wash to crown and nape, broad white supercilia and dark brown face mask. Upperparts unmarked pale brown except for white rump band and in outer tail feathers. Underparts whitish. Sexes alike. Larger **Large Woodshrike** *T. gularis* of northern and peninsular hills lacks pale supercilia and white in tail.

Voice: Noisy. Mainly a short *wheet wheet* and a rising *we we we wee?*. Also a short trill.

Habits: Inhabits open woodland, groves, parks and gardens. Feeds on insects and nectar at all levels in trees, often on trunks. Joins mixed hunting groups. Pairs or small parties move rapidly through trees. Nests in tree fork.

Otto Pfister

Distribution: Common breeding resident throughout lowlands. Also occurs in SE Asia.

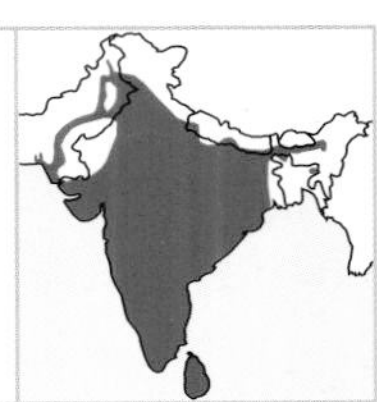

WHITE-THROATED DIPPER (White-breasted Dipper)

Cinclus cinclus 20 cm F: Cinclidae

Description: A dumpy, brown and white waterbird with short tail. Brown head, mantle and belly, rest of upperparts, blackish-brown Gleaming white below. Rare brown morph *sordidus* (also shown) darker than **Brown**. White eye membranes. Short tail often cocked. Long legs. Sexes alike.

Voice: A loud piercing *dzittz* given mainly in flight and audible above sound of running water.

Habits: Inhabits higher altitude fast-flowing streams and waterfalls and glacial lakes. Very active and territorial. Feeds on aquatic insects by walking on stream bottom, floating with wings outstretched on rising. Flies fast and low. Bobs and sways on boulders. Nests low in rock face.

Distribution: Locally common breeding resident of northern mountains from north Pakistan to the Myanmar border. Scarce in southern Himalayan ranges. Moves lower down in winter.
Also occurs in Europe, N Africa, W and Central Asia and China.

Otto Pfister

Otto Pfister

BROWN DIPPER

Cinclus pallasii 20 cm F: Cinclidae

Otto Pfister

Description: A dumpy, all-brown waterbird with a short tail. Completely warm chocolate brown but beware *sordidus* form of **White-throated**. Otherwise identical to that species in shape.

Voice: A short *dzit dzit*, less piercing than **White-throated**, and given mainly in flight. Audible above the sound of running water.

Habits: Inhabits fast-flowing rocky streams and waterfalls at lower altitudes than **White-throated** but there is a zone of overlap. Behaviour, food and nesting as **White-throated**.

Distribution: Common breeding resident in northern mountains from north Pakistan to the Myanmar border. Moves lower down in winter.
Also occurs in Japan, China and SE Asia.

RUFOUS-TAILED ROCK THRUSH (Rock Thrush)

Monticola saxatilis 19 cm F: Muscicapidae

Joanna Van Gruisen

Otto Pfister

Description: A stocky, short-tailed ground thrush in which the breeding male has bluish grey head, breast and mantle, white back, blackish wings and chestnut tail with brown central feathers and chestnut underparts. Female and winter male dark, scaly brown with pale chestnut underparts.

Voice: A squeaky *whit* and a hard *chak chak*. Song is a quiet extended melody, often given in song flight.

Habits: Inhabits open, dry, rocky hillsides feeding on ground singly or in pairs. Shy, but frequently perches on boulders wagging tail. Feeds on insects and seeds. Nests among boulders.

Distribution: Scarce breeder in Baluchistan. Winters elsewhere in Pakistan east to Ladakh and occasionally Nepal.
Also occurs in Europe, Africa, W and Central Asia and China.

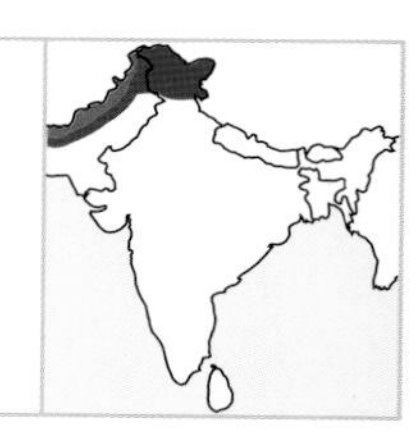

BLUE-CAPPED ROCK THRUSH (Blue-headed)

Monticola cinclorhynchus 17 cm F: Muscicapidae

Description: A small, bright blue and chestnut arboreal thrush with large white wing patches. Male has bright blue crown, throat and wing shoulders. Darker cheek stripes, mantle and wings and chestnut rump and underparts. Female pale brown above and finely barred below with no head markings.

Voice: A short *gwink gwink* and *pree pree*. Song is short and fluty.

Habits: Inhabits all types of open woodland including pine forest. Arboreal usually feeding singly or in pairs on insects on branches and trunks. Occasionally feeds on ground but nests there among rocks or roots. Perches erect and often conspicuously, usually wagging tail.

Bikram Grewal

Distribution: Common breeding summer visitor to northern foothills from north Pakistan east to Assam. Winters mainly in Western Ghats and the North-east but scattered records throughout peninsula and north.
Also occurs in Myanmar.

CHESTNUT-BELLIED ROCK THRUSH

Monticola rufiventris 24 cm F: Muscicapidae

Description: Large, chestnut and dark blue thrush. Male has deep blue upperparts, darker on wings and throat and deep chestnut underparts. Non-breeding male has buff fringes to upperparts. Female greyish brown with buff cheek patches on dark face and heavily dark-barred underparts. Care needed to distinguish female from *Zoothera* thrushes. Immature scaly buff and brown.

Voice: Harsh *schrrk* and *tikk* calls. Song a rather quiet, undulating and fluty melody.

Habits: Inhabits open forest, including pines, on rocky slopes. Usually perches prominently on tree-tops, slowly wagging tail. Feeds singly or in pairs on insects on ground. Nests close to ground.

Distribution: Fairly common breeding resident in northern mountains from Pakistan east to the Myanmar border. Winters lower down.
Also occurs in China and SE Asia.

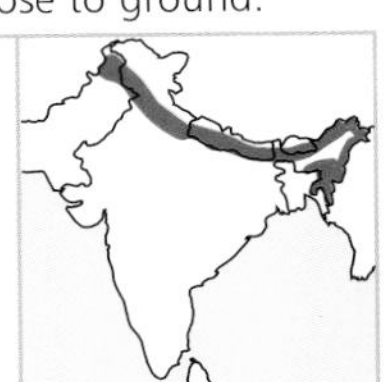

Nikhil Devasar

♀

Otto Pfister

BLUE ROCK THRUSH

Monticola solitarius 23 cm F: Muscicapidae

Otto Pfister

Description: Large, blue-grey ground thrush with a long, dark bill. Male all deep blue-grey (can look black at distance) with darker wings and tail. Female dark brown above, finely barred below and flushed blue. Some birds in the Northeast may show chestnut undertail.

Voice: Rather quiet. A disyllabic, piping *ve eet* and a deep *chuk*. Song has loud, spasmodic, tremulous phrases.

Habits: Breeds on dry, rocky slopes. Winters on rocky hills, quarries, rocky seashores, ruins and buildings. Nest hidden in rocks. Feeds singly or in pairs on insects on ground. Shy but perches prominently on buildings and rocks. Often has nuthatch-like stance on side of rock.

Distribution: Common breeder in northern mountains from Baluchistan east to central Nepal. Winters locally in foothills and throughout peninsula but rare in northern plains and Sri Lanka. Also occurs in Europe, Africa, W, N, Central, E and SE Asia.

MALABAR WHISTLING THRUSH

Myophonus horsfieldii 25 cm F: Muscicapidae

Clement m Francis

Description: Large, plump, blackish ground thrush with shining blue forehead and shoulders. Bill black with tuft of black feathers at base of upper mandible. Sexes alike. Smaller endemic **Sri Lanka Whistling Thrush** *M. blighi* is browner with blue supercilia. Female brown with blue on shoulders only.

Voice: High-pitched extended *kreeee* and rambling, mellow whistling song during monsoon.

Habits: Inhabits damp evergreen forest and plantations mainly along streams and near culverts and waterfalls. Rather shy but active. Often raises tail and stretches legs when perched. Feeds singly or in pairs on ground on insects, crustaceans and amphibians. Nests low in rocks.

Distribution: Fairly common endemic breeding resident throughout Western Ghats south from Mt Abu and also the Satpura range. Moves lower down in winter.

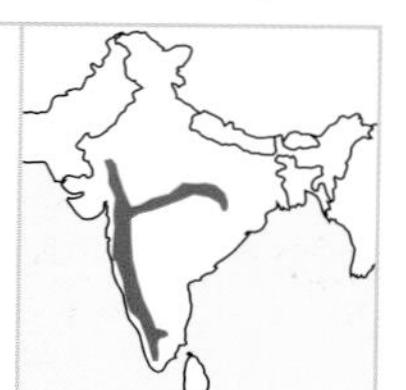

BLUE WHISTLING THRUSH (Whistling Thrush)

Myophonus caeruleus 33 cm F: Muscicapidae

Description: Large, bulky, purple-blue ground thrush with a mainly yellow bill. Whole body spangled with lighter, brighter blue. Forehead, shoulders, wing and tail edges bright blue. Sexes alike. Juvenile duller with dusky bill.

Voice: Loud, piercing *tzeet* and *zee zeee.* Rambling, whistling song, sounding almost human.

Habits: Inhabits damp forests and other wooded areas, usually near water and often in gorges or by road culverts. Perches on stream boulders and low branches. Noisy, bold, usually approachable. Often close to habitation, even entering buildings. Usually singly or in pairs feeding on insects, crustaceans and amphibians on ground. Nests in rocks, streamside roots or buildings.

Otto Pfister

Distribution: Common breeding resident in northern mountains from Pakistan east to the Myanmar border. Moves lower down in winter.
Also occurs in Central and SE Asia and China.

ORANGE-HEADED THRUSH (Ground Thrush)

Zoothera citrina citrina 21 cm F: Muscicapidae

Description: Small, orange and grey ground thrush. Deep orange head and underparts. White vent and shoulder patch. Bluish-grey upperparts. Female browner. Andaman and Nicobar races lack white on shoulder.

Voice: Shrill *kreee* call and loud rich song, often with mimicry of other birds.

Habits: Inhabits forests, plantations and wooded patches, often near water in shady spots. Feeds, usually in pairs or singly, unobtrusively on ground, usually under cover. Remains motionless if disturbed, either on ground or after flying into tree. Feeds on invertebrates in leaf litter. Crepuscular. Nests low in tree.

Sumit Sen

Distribution: Nominate race is a local breeding summer visitor to northern foothills from north Pakistan east to Arunachal Pradesh and winters in plains and eastern peninsula, south to Sri Lanka. Also resident in the Northeast including Bangladesh. Andaman race is resident.
Also occurs in China and SE Asia.

WHITE-THROATED THRUSH (Orange-headed Ground)

Zoothera (citrina) cyanotus 21 cm F: Muscicapidae

Otto Pfister

Description: A compact, very beautiful grey and orange thrush. Very like **Orange-headed** (with which it is currently considered conspecific) but has white face and throat and two blackish-brown vertical lines through cheeks. These can appear as horizontal lines in flight. Has white bars on wing shoulders. Female browner above.

Voice: As **Orange-headed**.

Habits: As **Orange-headed** but particularly favours dark understoreys of plantations. May be seen in same localities in winter but any differences in behaviour not noted. Nests in low tree.

Distribution: This form is a locally common endemic breeding resident in peninsular foothills north to Gujarat, Madhya Pradesh and Orissa. Most common along west coast. Not recorded in Sri Lanka.
Other related forms occur in China and SE Asia.

PLAIN-BACKED THRUSH (Plain-backed Mountain)

Zoothera mollissima 27 cm F: Muscicapidae

Tim Loseby

Description: Large, rich brown thrush with densely black-barred, white underparts. Barring is crescent-shaped. Upperparts plain brown with no, or indistinct, wingbars. Sexes alike. Broad white bands on underwings show in flight.

Voice: Usually silent. Alarm call is a rattling *churr*. Song has short mellow, descending phrases.

Habits: Inhabits open woodland, grassy, fallow and bushy country feeding in pairs or small groups on invertebrates on ground. Very shy and unobtrusive, flying up into tree if disturbed and remaining motionless. Sings high, but hidden, in tree. Nests on or near ground.

Distribution: Local breeding resident to northern mountains from north Pakistan to northeast India. Winters lower down.
Also occurs in Tibet, China and SE Asia.

LONG-TAILED THRUSH (Long-tailed Mountain Thrush)

Zoothera dixoni 27 cm F: Muscicapidae

Description: A large, dark thrush with distinct buff wing bars. Most similar to **Plain-backed** but obviously longer tail and less barred below. Dark greyish-brown above and mainly white below with black barring and spotting on face, breast and flanks. White band across underwings. Sexes similar.

Voice: A rambling song with scratchy and fluty notes.

Habits: Inhabits forest undergrowth and shrubbery feeding on invertebrates on ground. Often near streams. Shy and inconspicuous. Usually singly or in small groups. Nests in tree.

Jan Willem den Besten

Distribution: Scarce breeding resident in northern mountains from Himachal Pradesh east to the Myanmar border. Moves lower down in winter.
Also occurs in Tibet, S China and SE Asia.

SCALY THRUSH (White's, Speckled, Golden Mountain)

Zoothera dauma 26 cm F: Muscicapidae

Description: Large, black-scaled, olive brown ground thrush. Nominate race most spangled with dark ear crescent. Western Ghats race darker, less spangled with longer bill. Sri Lankan race smaller and darker still, with short tail and longer bill. May all be separate species. Black underwings have two white bars. Sexes alike.

Voice: Rather quiet. Call a slurred *tzee*. Song haunting series of spaced phrases from high perch.

Habits: Inhabits forest and woodland often in shady areas near water. Feeds singly or in pairs on invertebrates on ground. Very shy and retiring, sweeping up into tree when disturbed and remaining motionless. Flight bounding. Gait often crouching.

Goren Ekstrom

Distribution: Fairly common breeding summer visitor to northern mountains from north Pakistan east to northeast India. Winters locally in foothills and plains of northern peninsula. Also local resident in southern Western Ghats. Rare resident in Sri Lankan hills.
Also occurs in N, E and SE Asia.

TICKELL'S THRUSH (Indian Grey Thrush)

Turdus unicolor 21 cm F: Muscicapidae

Otto Pfister

♀

Otto Pfister

Description: Rather small, plain thrush, usually with yellow bill and pale legs. Male grey above with a bluish tinge and greyish-white below. Female brown with some dark throat streaking. Whitest on belly and vent. Rufous underwing coverts show in flight.

Voice: Soft *chuk chuk*. Repetitive song.

Habits: Inhabits mainly open forest and groves as well as orchards and cultivation with trees. Feeds singly, in pairs or scattered parties, on ground invertebrates. Hops openly, stopping with cocked head to listen for prey. Also eats tree fruits and berries. Usually not shy but flies up into trees if disturbed. Nests in trees and bushes.

Distribution: Fairly common endemic breeding resident in northern hills from northern Pakistan east to Bhutan and wintering in foothills and plains of northeast and northern India south to Andhra Pradesh and Goa.

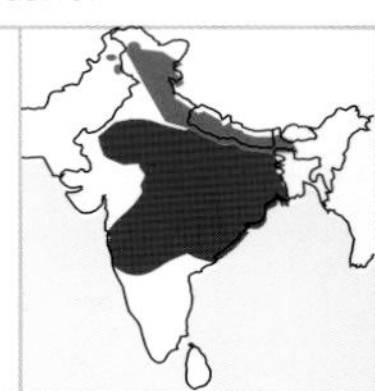

GREY-WINGED BLACKBIRD

Turdus boulboul 28 cm F: Muscicapidae

Otto Pfister

Description: Large, long-tailed, dark thrush with pale grey wing panels. Male sooty black with a bright yellow bill. Female brown. Both have grey panel on wings and yellow eye-rings to dark irises. Similar **White-collared Blackbird** *T. albocinctus* is all black in male and brown in female with broad white shawl and no grey.

Voice: A deep *chuk chuk* and a very melodious song.

Habits: Inhabits open woodland and scrub usually singly, in pairs or small parties, sometimes with other thrushes. Feeds on invertebrates on ground or fruit and berries on trees. Wary, flying up into trees if disturbed. Often sings from high tree perch. Nests low in trees or in banks.

Distribution: Fairly common breeding resident from Pakistan to the Myanmar border. Moves lower down in winter. Also occurs in China and SE Asia.

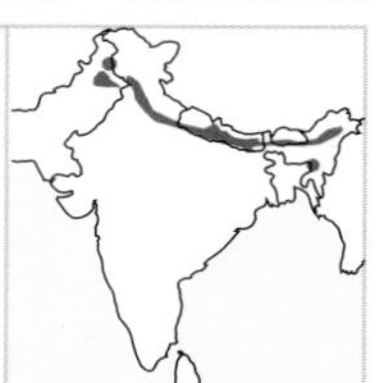

EURASIAN BLACKBIRD (Blackbird, Common Blackbird)

Turdus merula 26 cm F: Muscicapidae

Description: Large, long-tailed, dark thrush with orange bill and yellow eye rims. Males of most races black or dark brown, females brown with throat streaking. Race *nigropileus* (shown below and may be distinct species) purplish-brown with grey wings and tail and, in males, black head.

Voice: Noisy. Hard *chek chek*, rushed when going to roost. Also quiet *tsih*. Fluting song.

Habits: Open forest, woodland, plantations and fields. Wary, flying into cover if disturbed. Feeds on invertebrates on ground singly or in groups. Also eats berries and fruit in trees. Often sings from high perch. Nests low in trees or banks.

Morten Strange

Balachandran

Distribution: Fairly common breeding resident in northwestern hills from north Pakistan to Uttaranchal and in parts of the Northeast. Southern races are locally common, breeding in Western and Eastern Ghats and Sri Lanka. Winter sparsely in nearby foothills and plains.
Also occurs in Europe, N Africa, W and Central Asia and China.

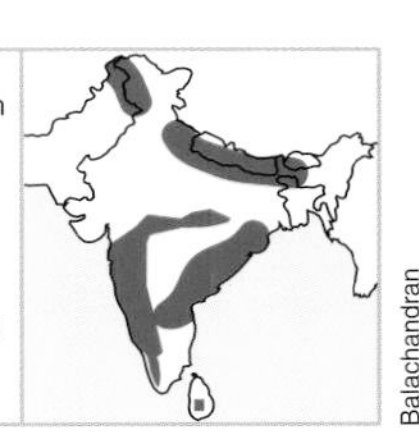

EYE-BROWED THRUSH (Dark Thrush)

Turdus obscurus 23 cm F: Muscicapidae

Description: A small, orange and grey thrush with distinct white supercilia and black lores. Male has small white patches below eyes, grey head and neck and brownish-grey upperparts. Female and immature less grey with blackish-bordered white throat so similar to **Tickell's**. All have peachy orange flanks and breast which is diagnostic. Small bill with black upper mandible and tip. Rare north-eastern **Grey-sided Thrush** *T. fea* is plainer and browner and grey below.

Voice: A thin *teseep teseep*. Also chatters.

Habits: Inhabits open forest, feeding on ground on insects, often with other thrushes. Hops. Wary, flushing rapidly into tree canopy.

Morten Strange

Distribution: Locally common winter visitor to northeastern hills and east Nepal. Rare further south in peninsula India, Bangladesh and Sri Lanka.
Also occurs in N, E and SE Asia.

DARK-THROATED THRUSH (Red/ Black-throated)

Turdus ruficollis 25 cm F: Muscicapidae

Toby Sinclair

♀

Nikhil Devasar

Description: A large, plump, brown thrush with a dark throat. Two subspecies occur. Male of black-throated form has black throat and upper breast, often mottled. Female has blackish mottling there. Male of red-throated form has brick-red throat and upper breast and red in tail. Female (lower photo) is mottled chestnut on throat.

Voice: A drawn out *schweee* and a chuckling *wheech which*.

Habits: Inhabits open woodland, forest edges, scrub and cultivation feeding on invertebrates on ground. Hops boldly with upright stance, flying into trees if disturbed. Gregarious, often mixing with other thrushes.

Distribution: Fairly common winter visitor to most of Pakistan, N India and Nepal but in variable numbers. Records south to Madhya Pradesh and Orissa. Red-throated form (which may be separate species) mainly in NE India. Also occurs in N and Central Asia and China.

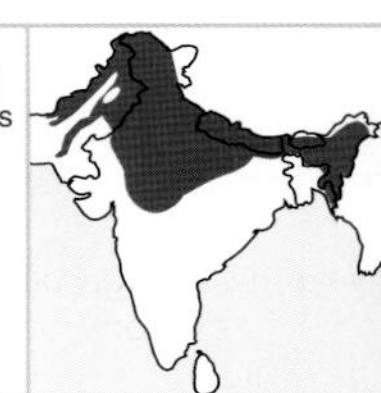

FIELDFARE

Turdus pilaris 25 cm F: Muscicapidae

Otto Pfister

Description: A large, chestnut and grey thrush. Pale blue-grey crown, nape and rump contrast with black tail and chestnut mantle. Heavily black-streaked, deep yellow throat and upper breast. White on rest of underparts, heavily streaked on flanks. Orange in bill. Sexes similar.

Voice: A hard *shack shack* mainly in flight. Also a squeaky *tee*.

Habits: Inhabits open cultivation, pastures and orchards. Feeds on invertebrates in open, hopping boldly. Also feeds on berries. Rather shy and flies into trees if alarmed. Most likely to be seen with other thrushes in our region.

Distribution: A very rare vagrant from northern Asia. Only one record, from Uttar Pradesh, but should be looked for in winter in northern foothills. Also occurs in Europe, N and Central Asia.

MISTLE THRUSH (Stormcock)

Turdus viscivorus 27 cm F: Muscicapidae

Description: Large, pale grey-brown thrush with bold black spotting on white underparts. Rump browner and often yellowish on lower flanks. Dark patch on ear coverts and breast sides. White apical tips to long tail and white underwings. Sexes alike. Juvenile heavily mottled whitish above. Confusion possible with **Scaly** and **Long-tailed**.

Voice: A harsh rattle, usually given in flight. Song is mellow and haunting.

Habits: Inhabits open mountain forest, grasslands and scrub. Feeds on invertebrates on ground and berries, particularly mistletoe, in trees. Usually singly or in pairs. Wary, flying high between food sites. Sings from exposed perch. Nests high in tree.

Otto Pfister

Distribution: Fairly common breeding resident in northern mountains from western Pakistan to western Nepal. Moves lower down in winter.
Also occurs in Europe, W, N and Central Asia.

GOULD'S SHORTWING

Brachypteryx stellata 13 cm F: Muscicapidae

Description: A small, very dark, dumpy, terrestrial thrush with long legs and a short tail. Dark chestnut above and grey-black from face to vent with distinct white spotting on flanks and belly. Sexes alike. Juvenile more streaked.

Voice: Rather quiet. Calls *tik tik* and *zee zee*. A rising high-pitched song in which the notes run together as it gets louder.

Habits: Inhabits dense undergrowth of evergreen forest feeding on invertebrates on ground. Favours damp ravines with rhododendrons. Usually singly or in pairs. Generally unobtrusive but will come into open and perch on bushes more than other **Shortwings**. Nest unknown.

Peter Morris

Distribution: Rare presumed breeding resident of Himalayan forests from Uttaranchal east to Arunachal Pradesh including Nepal and Bhutan. Winters lower down in foothills.
Also occurs in Tibet, Myanmar and Vietnam.

WHITE-BELLIED SHORTWING (Rufous-bellied)

Brachypteryx major 15 cm F: Muscicapidae

Tim Loseby

Description: Small, dark blue, dumpy, terrestrial thrush with long legs and short tail. Slaty-blue with rufous washed white lower breast to vent in nominate race (shown). *Albiventris* race white below with blue flanks and vent. Red irises. Sexes alike. Similar northeastern **Lesser Shortwing** *B. leucophrys* paler slaty-blue with whitish underparts. Female brown.

Voice: Call piercing whistle *wheep*. Song a varied series of loud whistles.

Habits: Inhabits shady evergreen forest, particularly sholas, and ravines. Usually singly or in pairs, feeding on ground invertebrates. Unobtrusive. Comes onto paths in evening. Nests low down.

Clement A. Francis

Distribution: Globally threatened rare endemic breeding resident of extreme southwestern hills. Nominate race in southern Karnataka and Nilgiris, *Albiventris* race (which may be a separate species) in southern Kerala and Palni Hills.

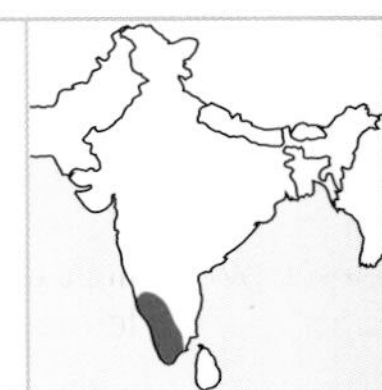

WHITE-BROWED SHORTWING

Brachypteryx montana 13 cm F: Muscicapidae

Morten Strange

Description: A small, dark blue thrush with long, white supercilia. This is adult male plumage. Female brown with orange on face. Immature plumage (in which males sometimes breed) as female but with white supercilia. Dark legs.

Voice: A hard *tak* and *tt tt tt* Surprisingly loud song.

Habits: Inhabits damp forest undergrowth often near streams or in ravines. Very shy, hopping quietly but rapidly away if disturbed. Usually singly. Feeds on ground invertebrates. Nests low down.

Distribution: Locally common breeding resident in northern mountains from central Nepal east to the Myanmar border. A few records further west in India. Winters lower down.
Also occurs in China and SE Asia.

SPOTTED FLYCATCHER

Muscicapa striata 15 cm F: Muscicapidae

Description: A rather large, long-winged and long-tailed flycatcher with a large, rounded head. Pale grey-brown, whiter below. Crown, throat and breast lightly streaked. Whitish edges to flight feathers. Fairly long, black bill and short, black legs. Sexes alike. Juvenile, spotted buff-white above and scaly below.

Voice: A high *zee* and a stuttered *zee tic*.

Habits: Inhabits open forest and groves. Usually solitary. Perches upright and usually fairly high, launching into air after insects and returning to same perch. Sometimes forages in foliage. Flicks wings and tail when calling. Nests low against tree trunk, in crevice or on ledge.

Otto Pfister

Distribution: Local breeding summer visitor to Pakistan mountains. Scarce autumn passage migrant (en route to Eastern Africa) in rest of Pakistan and northwest India. Returns west of Indus. Also occurs in Europe, Africa, W and N Asia.

DARK-SIDED FLYCATCHER (Sooty, Asian Sooty)

Muscicapa sibirica 13 cm F: Muscicapidae

Description: A medium-sized, slim dark flycatcher with long primary projections. Dark grey-brown above with thin buff wing stripe. Below all dusky brown with white central belly, neck stripes and throat. Obvious dark moustachial stripes. White eye ring and lores. Bill short and broad, usually all-black or with small yellow base to lower mandible. Black legs. Sexes alike. Juvenile dark brown-speckled buff with a more prominent orange wing bar.

Voice: Usually silent but has a quiet sibilant song.

Habits: Inhabits clearings in and edges of hill forests. Usually solitary. Feeds as **Spotted** but often crepuscular. Unobtrusive but quite confiding. Nests on tree branch or in hole.

Otto Pfister

Distribution: Fairly common breeding summer visitor to northern mountains from north Pakistan east to Arunachal Pradesh. Moves lower in autumn but winter quarters uncertain. Possibly mainly northeast India. Also occurs in N, E and SE Asia.

ASIAN BROWN FLYCATCHER (Brown Flycatcher)

Muscicapa dauurica 14 cm F: Muscicapidae

Otto Pfister

Description: Medium-sized, stocky flycatcher with large head and eye, and short tail. Paler and browner than **Dark-sided** with whiter underparts and only diffuse brown wash on flanks and breast. Faint, dark moustachial stripes. Thin, white wing stripes and obvious whitish eye rings, lores and throat. Shorter primary projection and larger bill with prominent yellow base to lower mandible. Black legs. Sexes alike. Juvenile paler, especially below, and with whiter spotting.

Voice: Usually silent but call *tzet*. Song loud and trilling.

Habits: Inhabits open deciduous forests, plantations, groves and gardens. Feeds as **Spotted** but less aerial and more crepuscular. Usually singly. Perches rather low.

Distribution: Locally common breeding summer visitor to northern hills from extreme northern Pakistan east to Bhutan. Also breeding resident in some peninsular hills, notably Western Ghats. Winters in central and southern peninsula and Sri Lanka.
Also occurs in N, E and SE Asia.

RUSTY-TAILED FLYCATCHER (Rufous-tailed Flycatcher)

Muscicapa ruficauda 14 cm F: Muscicapidae

Jan Willem den Besten

Description: A small, brown flycatcher with a rufous tail. Warm brown upperparts and rather dark brownish throat to breast with only belly to vent whitish. Ill-defined supercilia and eyerings. Rufous rump and notched tail recalling a redstart. Distinct, pale orange lower mandible. Rear crown often raised. Short dark legs. Sexes alike.

Voice: A plaintive *peu peu* call. Also churrs.

Habits: Inhabits open forest, clearings and edges. Feeds on invertebrates among foliage, rarely fly-catching from perch. Usually singly and easily overlooked but confiding. Bobs and flicks wings. Nests on ground and in tree.

Distribution: Fairly common endemic breeding summer visitor to northern mountains from north Pakistan east to eastern Nepal. Commoner in west. Winters mainly in southwest India but widespread passage records from elsewhere in India.

BROWN-BREASTED FLYCATCHER (Layard's)

Muscicapa muttui 13 cm F: Muscicapidae

Description: Upper bird in photo. Medium-sized, stocky flycatcher with large head and eye, all yellow lower mandible and yellowish legs. Warmer brown above than **Asian Brown** (lower bird) and darker on breast and flanks contrasting with white throat. Whitish eye rings and lores. Sexes alike.

Voice: Usually silent but call *zit*. A weak, warbling song.

Habits: Inhabits dense cover of evergreen forests, often along rivers. Usually singly and low down, feeding on insects in foliage or by short aerial sallies. Crepuscular, unobtrusive and territorial. Nests low in bush or on bank.

Balachandran

Distribution: Uncommon endemic breeding summer visitor to north-eastern hills. Winters in Western Ghats and Sri Lanka.

FERRUGINOUS FLYCATCHER

Muscicapa ferruginea 13 cm F: Muscicapidae

Description: A medium-sized, stocky, rusty-brown flycatcher with large head and eye. Warm rufous-brown above, more orange on rump and tail and on prominent wing bars. Head contrasting grey. White throat and belly contrast with pale orange breast, flanks and vent. Prominent white eye rings. Pale legs. Sexes alike. Juvenile spotted rufous-orange.

Voice: Usually quiet. High-pitched trilling song.

Habits: Inhabits dense cover of hill forests. Usually singly and low down, feeding on insects both in foliage and by aerial sorties. Crepuscular and unobtrusive. Nests in moss-covered trees.

Morten Strange

Distribution: A scarce and local breeding summer visitor to northeast Indian hills and Himalayas east from central Nepal. Probably winters in SE Asia.
Also occurs in SE Asia.

RUFOUS-GORGETED FLYCATCHER (Orange-gorgeted)

Ficedula strophiata 14 cm F: Muscicapidae

Bikram Grewal

Description: A medium-sized, dark flycatcher with a deep orange breast patch. Long wings and tail. Dark olive-brown above with black rump and tail and extensive white tail patches, very obvious in flight. Male has black throat and white forehead stripe and blue-grey lower breast. In female face and throat greyer, breast patch paler and white stripe is indistinct.

Voice: Calls a harsh *trrrt* and a metallic *pink pink*. Quiet song.

Habits: Inhabits dense forest undergrowth and scrub. Feeds on insects in foliage or from aerial sorties. Flicks and fans tail. Usually singly. Unobtrusive except in flight. Nests on ground or in tree hole.

Distribution: Locally common breeding summer visitor to northern hill forests from Himachal Pradesh east to the Myanmar border. Moves lower down into foothills in winter.
Also occurs in China and SE Asia.

RED-THROATED FLYCATCHER (Red-breasted)

Ficedula parva 12 cm F: Muscicapidae

Otto Pfister

Sumit Sen

Description: Small, brown flycatcher with white basal patches in black tail. Most individuals are creamy or greyish white below but older males have orange throat patch. Male also has greyer head and neck. White eye rings. Small black, pale-based bill. Eastern race *F. p. abicilla* (possibly a separate species) has all-black bill and rump and, in older males, grey round larger throat patch.

Voice: Noisy. A trilled *trrr* and hard *tik tik*.

Habits: Inhabits open wooded country, groves, parks and gardens. Feeds on insects at middle levels, on ground and in air. Perches with high cocked tail and drooped wings. Active but often rather shy. Best located by calls.

Distribution: Common winter visitor throughout most of India, Nepal and Bangladesh. Passage migrant only in Pakistan. Rare in Sri Lanka and extreme south India.
Also occurs in Europe, N, Central and E Asia and Myanmar.

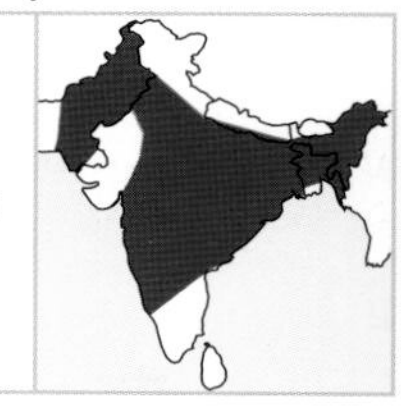

WHITE-GORGETED FLYCATCHER

Ficedula monileger 11 cm F: Muscicapidae

Description: A tiny, brown flycatcher with a black-bordered, white throat. Paler brown below. Western race has whitish supercilia, eastern (shown) warm buff. Pink legs. Large head and bill and short tail. Sexes alike.

Voice: A short *tik*, a rattle and a thin whistle. Wheezy short song.

Habits: Inhabits dense undercover of mountain forests. Feeds on invertebrates on or close to ground. Fly-catches and flirts tail open but usually shy and difficult to observe. Nests near ground.

Sujan Chatterjee

Distribution: Rare breeding resident from western Nepal east to Arunachal Pradesh. Eastern race rather more frequent in northeastern hills to the Myanmar border.
Also occurs in SE Asia.

SNOWY-BROWED FLYCATCHER (Rufous-breasted Blue)

Ficedula hyperythra 11 cm F: Muscicapidae

Description: A tiny, blue and orange flycatcher with short white supercilia. Male resembles larger **Indian Blue Robin** with its blue head and upperparts and deep orange from chin to breast, paler below. White patches at base of tail. Female warm dark brown above with rich rusty supercilia, face and breast. Both have pink legs.

Voice: A thin, repeated *sip* and *tsit*. Also *sit si sii* song.

Habits: Inhabits the bushy undergrowth of forest, particularly favouring bamboo and ravines. Feeds on insects unobtrusively on or near ground. Nests in low hole.

Sujan Chatterjee

Distribution: Locally common breeding resident in northern mountains from Uttaranchal east to the Myanmar border. Moves lower down in winter.
Also occurs in SE Asia.

LITTLE PIED FLYCATCHER (Westermann's Flycatcher)

Ficedula westermanni 10 cm F: Muscicapidae

Otto Pfister

Description: A small, stocky, black and white flycatcher with a large head. Small black bill. Male black above with broad white stripes on head, wings and tail sides and white underparts. Female and juvenile warm brown above with thin buff wing bar, rusty rump, white throat and buffish under-parts. As such, very similar to female **Slaty-blue** which lacks wing bar.

Voice: Rather silent. A mellow *tweet* and a thin, high song.

Habits: Inhabits tree canopies in forests and open wooded areas. Feeds very actively on insects in leaves and on bark. Singly or in pairs and often with mixed hunting groups. Nests on ground, usually on banks.

Distribution: Locally common breeding resident and altitudinal migrant in Himalayas from Himachal Pradesh to the Myanmar border. Winters in foothills and, more rarely, eastern plains south to southern peninsula. Also occurs in China and SE Asia.

ULTRAMARINE FLYCATCHER (White-browed Blue)

Ficedula superciliaris 10 cm F: Muscicapidae

♀

Otto Pfister

Otto Pfister

Description: Small, stocky, arboreal flycatcher. Male rich blue above with large blue breast side patches. Western race (shown) has prominent white eyebrows and small white basal patches to tail. This white usually lacking in eastern race. White below. Female slaty-brown above and on breast side patches, buffish below. Sometimes with bluish wash on rump and tail.

Voice: Noisy. Calls soft *tik tik* and rattling *chrrr*. Song soft, trilling, repeated *te che pur*.

Habits: Inhabits tree canopies in forests and open wooded areas, including gardens. Feeds on insects in foliage and air, usually singly or in pairs. Also in mixed hunting groups. Nests in bank or tree hole.

Distribution: Locally common breeding resident and summer visitor to northern mountains from north Pakistan to the Myanmar border. Winters in foothills and sparsely in plains, south to southern peninsula.
Also occurs in Tibet and China.

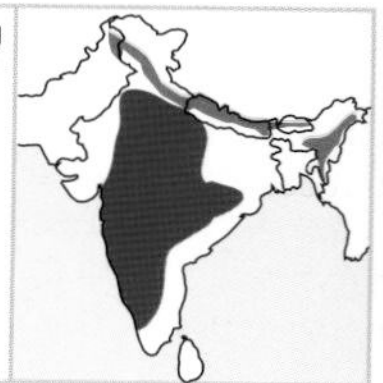

SLATY-BLUE FLYCATCHER

Ficedula tricolor 10 cm F: Muscicapidae

Description: Small, slim, terrestrial flycatcher with long tail. Male dark blue above, brighter on white-based crown. Western race (shown) has white throat and greyish-white underparts. Eastern race has pale orange throat and orange washed underparts. Western female warm brown above with rufous rump, buff underparts and contrasting white throat. Eastern all rich buff below.

Voice: Call *ee tik-tik*. Trilling whistles.

Habits: Inhabits the dense understoreys of forests and scrub but also, in winter, long grass and sugarcane. Feeds singly or in pairs on insects, on or close to ground. Often cocks tail and can look very chat-like. Nests in low hole or crevice.

Tim Loseby

♀

RK Gaur

Distribution: Common breeding resident and altitudinal migrant in Himalayas from north Pakistan east to the Myanmar border. Winters in foothills and, rarely, nearby plains. Also occurs in Tibet, China and SE Asia.

BLACK AND ORANGE FLYCATCHER

Ficedula nigrorufa 13 cm F: Muscicapidae

Description: A dumpy, large headed, strikingly coloured flycatcher. Male is deep tomato soup orange with black hood and wings. Female much duller with buffish face.

Voice: A soft but penetrating *peee* and a hard *zit zit*. High pitched, metallic song.

Habits: Inhabits dense undergrowth, including bamboo, in evergreen forests and plantations particularly in ravines and on slopes of sholas. Feeds on insects in low foliage and on ground. Makes short, low aerial sorties. Usually singly or in pairs but sometimes joins mixed hunting groups. Confiding and approachable. Nests low down in vegetation.

S Elamon

Distribution: Locally common breeding endemic resident restricted to southern Western Ghats and adjacent hills.

VERDITER FLYCATCHER

Eumyias thalassina 15 cm F: Muscicapidae

Otto Pfister

Description: A medium-sized, slim, turquoise flycatcher with a black mask. Wings darker and vent white-barred. Black bill and legs. Female greyer. **Nilgiri** is darker and bluer, with white tail base patches. Sri Lankan **Pale Blue** *E. sordida* is paler ashy-blue.

Voice: Usually silent. A faint *chwee* call and a hurried, descending musical song.

Habits: Inhabits open wooded country, including gardens and mangroves. Feeds on insects mainly by aerial sorties from exposed branches or wires. Active, confiding and conspicuous. Usually singly or in pairs. Nests low in a hole.

Distribution: A common breeding summer visitor to northern hills from extreme north Pakistan east to the Myanmar border. Winters in foothills, Bangladesh and Western Ghats, less commonly throughout peninsula and northwest India.
Also occurs in SE Asia.

NILGIRI FLYCATCHER

Eumyias albicaudata 15 cm F: Muscicapidae

Ron Saldino

Description: A medium-sized, dark blue flycatcher with small, white basal spots on tail. Male deep greenish-blue, greyer on belly and with white scaling on vent. Bright blue fore-crown and supercilia. Female dark greenish-grey. Both much darker than similar species.

Voice: Calls *tsik tsik*, *chip chip*. Sings slow hesitant warble of up to ten notes.

Habits: Inhabits evergreen forest and forest edges often near streams. Also well-grown plantations. Feeds, usually in pairs, on insects which it catches in aerial sorties, mostly from canopy. Nests low down.

Distribution: Globally near-threatened but fairly common endemic breeding resident restricted to the Western Ghats from south Karnataka southwards.

LARGE NILTAVA

Niltava grandis 21 cm F: Muscicapidae

Description: A large, dark blue flycatcher. Male all dark blue, much brighter on crown, neck sides and shoulder. Throat and breast blackish. Female dull rufous-brown with more rufous tail and wings, blue neck patches and buff underparts, including throat.

Voice: Three or four ascending whistling notes *doo ree me (me)*.

Habits: Inhabits lower storeys of moist forest, sitting upright and inactive for long periods. Feeds on invertebrates on ground or in air. Usually in pairs or singly. Also eats berries. Song attracts attention. Nests near ground.

Morten Strange

Distribution: Locally common breeding resident in northern hills from central Nepal east to the Myanmar border. Moves lower down in winter. Also occurs in China and SE Asia.

SMALL NILTAVA

Niltava macgrigoriae 11 cm F: Muscicapidae

Description: A very small, dark blue flycatcher. Male identical to, but much smaller than, **Large Niltava** except lower breast and belly are paler greyish blue, palest in western populations. Female also similar to **Large** but paler buff on belly. Blue neck patches often difficult to see.

Voice: A high-pitched undulating *twee twee it wee*. Also churrs.

Habits: Inhabits shady shrubberies and understoreys of damp forests often along streams or in ravines. Also grass jungle in winter. Active but often skulking and difficult to see well. Feeds on insects in air and on ground. Also berries. Usually singly or in pairs. Nests near ground.

Otto Pfister

Otto Pfister

Distribution: Fairly common endemic breeding resident of northern mountains from Uttaranchal east to the Myanmar border. Moves lower down in winter.

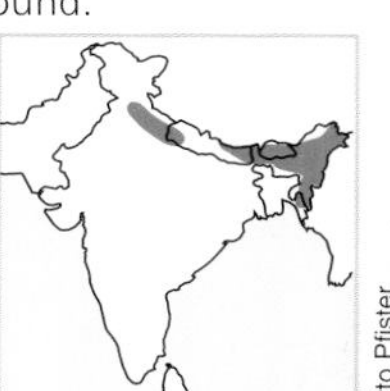

RUFOUS-BELLIED NILTAVA (Beautiful Niltava)

Niltava sundara 15 cm F: Muscicapidae

Otto Pfister

Description: Medium-sized, blue and orange flycatcher. Male deep blue, much brighter on crown, neck sides, shoulders and rump. Darker than other blue and orange flycatchers, with bright blue patches diagnostic. Female dull rufous-brown, more rufous on tail and with blue neck patches. Contrasting white, oval, lower throat patch.

Voice: A harsh, squeaky song. Churring notes and a sharp *psi psi* call.

Habits: Inhabits understorey of forest. Rather inactive and very unobtrusive, though occasionally flicks wings and bobs body. Usually singly or in pairs. Feeds on invertebrates in air or on ground. Nests near ground.

Distribution: Fairly common breeding resident in northern hills from north Pakistan east to the Myanmar border. Moves lower down in winter. Also occurs in China and SE Asia.

BLUE-THROATED FLYCATCHER

Cyornis rubeculoides 14 cm F: Muscicapidae

Balachandran

Description: A small, blue and orange flycatcher with a blue throat. Male bright blue, brightest on forehead with orange upper breast (sometimes extending onto flanks) and contrasting white belly to vent. Blue throat distinguishes from **Tickell's**. Female is brown above, more rufous on rump and tail with pale orange throat and upper breast. Lores, belly and vent white. Female **Tickell's** is bluish above.

Voice: Rapidly delivered warbling song. Calls a harsh *chrr* and a hard *tak tak*.

Habits: Inhabits damp forest, particularly overgrown ravines, feeding on invertebrates at lower levels. Usually singly or in pairs. Makes aerial sorties and flies to ground for prey. Nests low down.

Distribution: Locally breeding summer visitor to northern hills from north Pakistan east to the Myanmar border. Winters in foothills, southwestern India north to Goa and Sri Lanka with scattered records elsewhere. Also occurs in China and SE Asia.

HILL BLUE FLYCATCHER (Large-billed Blue)

Cyornis banyumas 15 cm F: Muscicapidae

Description: A large-billed, blue and orange flycatcher. Male very similar to **Tickell's** but larger and deeper blue above with more black on cheeks. Orange on underparts variable but often extends down flanks as a wash. Female dark brown above with rusty tail and warm buff below. Both have rather long black bill and pinkish-brown (not black) legs.

Voice: Hard *tak* and short rattle. Warbling song more complex than **Tickell's.**

Habits: Inhabits thick forest feeding unobtrusively on invertebrates at lower levels in canopy. Fly-catches. Nests low down.

Morten Strange

Distribution: Rare breeding resident in northern mountains from central Nepal east to the Myanmar border. Winters lower down.
Also occurs in SE Asia.

TICKELL'S BLUE FLYCATCHER (Orange-breasted Blue)

Cyornis tickelliae 14 cm F: Muscicapidae

Description: Small, blue and orange flycatcher with orange throat. Male as **Blue-throated** but less bright blue and orange throat and upper breast. White belly clearly demarcated. Female dull bluish above with faint orange throat to breast. Whitish lores. Sri Lankan race *C. t. jerdoni* darker blue, more orange below.

Voice: An undulating metallic trill usually of six, but sometimes up to ten, notes. Calls harsh *chrr* and hard *tak tak*.

Habits: Inhabits dry, open forest and woodland, groves and gardens, often near water. Feeds on aerial insects from low perches. Active and fairly confiding. Usually singly or in pairs but joins mixed hunting groups. Nests low down.

Otto Pfister

Distribution: Locally common breeding resident of low hills south of Gujarat and Uttaranchal, including Sri Lanka. Winters lower down in plains. Rare in the Northwest, the Northeast and Bangladesh.
Also occurs in China and SE Asia.

GREY-HEADED CANARY FLYCATCHER

Culicicapa ceylonensis 9 cm F: Muscicapidae

Otto Pfister

Description: A tiny, green, grey and yellow flycatcher. Distinctive upright stance. Rather large, slightly crested, grey head and grey breast contrast with green upperparts and yellow underparts. Sexes alike. Stance and behaviour distinguish from similarly coloured warblers.

Voice: Noisy. A much repeated, loud and questioning *whi che chee?*. Also calls *chik chik*.

Habits: Inhabits open forest, plantations, groves and gardens. Feeds at all levels on insects, catching them by aerial sorties or by foraging. Very active, inquisitive and confiding. Frequently perches in open. Often in mixed hunting groups but also in pairs or singly. Nests against tree trunk or rock.

Distribution: Common breeding summer visitor to northern hills from north Pakistan to the Myanmar border and most other hills of India, Bangladesh and Sri Lanka. Winters lower down and also throughout plains, where widespread but local.
Also occurs in China and SE Asia.

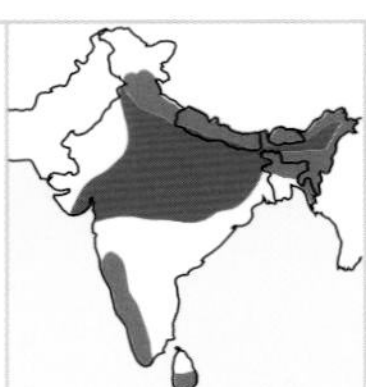

SIBERIAN RUBYTHROAT (Eurasian Rubythroat)

Luscinia calliope 15 cm F: Muscicapidae

Otto Pfister

Description: A perky, plain olive brown chat with white supercilia. Male has black lores and black-bordered white cheek stripe, deep red throat with variable back border. Breast and flanks washed olive buff. Belly white. Female has face buff and throat white. Unmarked short tail. Pale legs.

Voice: A hard *chak*, a rising *ee lu* and a low churr. Song a conversational, harsh but varied warbling.

Habits: Inhabits deep, damp, shady cover in open woodland, scrub and thick grasses, often near water. Very skulking but given time, appears hopping or, briefly, running with tail often cocked high and wings held below body. Feeds, usually singly, on invertebrates on ground.

Distribution: Local winter visitor from Siberia throughout foothills and plains from Punjab and Nepal south to Andhra Pradesh and east to Bangladesh and northeast India. Not recorded in Pakistan or Sri Lanka.
Also occurs in N, E and SE Asia.

WHITE-TAILED RUBYTHROAT (Himalayan Rubythroat)

Luscinia pectoralis 15 cm F: Muscicapidae

Description: Perky, dark grey chat with white-edged, blackish tail. Male has striking white supercilia with deep red throat, broadly bordered black over whole breast. Only Tibetan race, also shown, which winters in the Northeast, has white cheek stripes. Female dark grey above with shorter supercilia and white throat contrasting with greyish breast band. Legs black.

Voice: A harsh *ker* and *chruk*. Warbling song.

Habits: Inhabits forest edge, thick scrub and long grass. Less skulking than **Siberian** (but still wary), often perching openly and given to blatant territorial displays, even when wintering. Feeds, in confident long-legged hops, on insects on ground.

Tim Loseby

Otto Pfister

Distribution: Fairly common breeding resident in northern mountains from north Pakistan east to Arunachal Pradesh. Winters lower down in foothills.
Also occurs in Russia, Central Asia and China.

BLUETHROAT

Luscinia svecica 15 cm F: Muscicapidae

Description: Perky, mud-brown chat with red-based, dark tail and distinct white supercilia. Adults have variable orange and black breast band and blue throat with red, white or blue centre. Juvenile has blackish necklace. Greyish-white on rest of underparts. Looks very dark in flight. Long thin black legs. Yellow base to lower mandible.

Voice: A harsh *kerruck* and a short *chak chak*.

Habits: Inhabits damp scrub, reed beds and cultivation near water. Feeds on invertebrates on ground, usually singly but often several close together. Hops on paths jauntily with tail raised. Sometimes perches on reeds or low bushes. Usually rather secretive but can be quite confiding.

Otto Pfister

Otto Pfister

Distribution: Locally common breeding summer visitor to northern mountains from Pakistan east to Himachal Pradesh. Common and widespread winter visitor throughout plains, becoming much scarcer in south India and a vagrant to Sri Lanka.
Also occurs in Europe, N Africa, W, N and SE Asia and China.

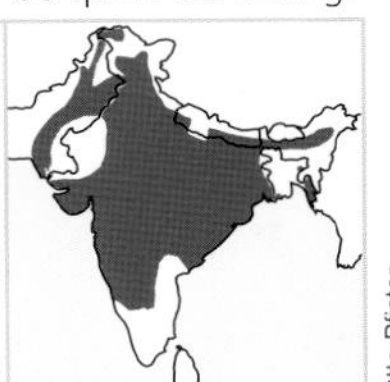

INDIAN BLUE ROBIN (Blue Chat)

Luscinia brunnea 15 cm F: Muscicapidae

Otto Pfister

Otto Pfister

Description: A skulking, blue and orange chat with white supercilia. Male blue-grey above. Black cheeks and deep chestnut-orange underparts, white vent and belly centre. Female brown above with white throat and belly and rufous-buff breast and flanks. Rather short tail and long pale legs.

Voice: A piercing *tsrurr* and a hard *tak tak*. Song is musical trill preceded by whistles.

Habits: A secretive and difficult to observe ground dweller in thick forest, scrub and plantations. Usually solitary. Flicks wings and conclusion. Horizontal stance. Feeds on ground invertebrates. Nests low down, usually in rhododendron and bamboo.

Distribution: Locally common breeding summer visitor to northern hills from Pakistan east to Burmese border. Winters mainly in foothills of Sri Lanka, southwest and northeast India. Also occurs in Myanmar.

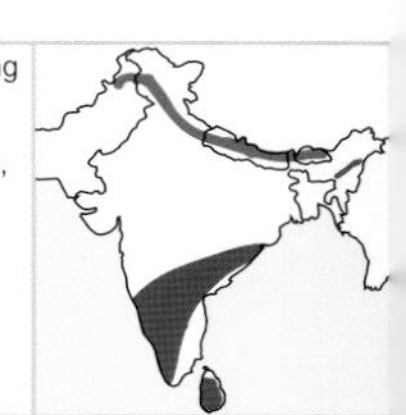

ORANGE-FLANKED BUSH ROBIN (Red-flanked)

Tarsiger cyanurus 15 cm F: Muscicapidae

Otto Pfister

♀

Otto Pfister

Description: A blue-backed chat with orange flanks and white throat. Male deep blue above, including tail, with variable white supercilia. Underparts greyish-white with obvious white throat and orange flanks. Autumn male, female and juvenile have blue rump and tail and orange flanks. Otherwise brown above and dingier below. Long dark legs.

Voice: A throaty *tok tok* and a whistled *whet*.

Habits: Inhabits forest undergrowth, feeding on invertebrates, on ground and in air. Sometimes perchs quite high in trees, particularly singing males. Flicks wings and tail. Solitary and rather shy. Nests close to ground.

Distribution: Locally common breeding resident in northern hills from Pakistan to Arunachal Pradesh. Winters in foothills. Also occurs in N, E and SE Asia.

GOLDEN BUSH ROBIN

Tarsiger chrysaeus 15 cm F: Muscicapidae

Description: A uniquely coloured, golden-orange and black chat. Male has blackish tail, wings and cheeks with brownish mantle and crown. Contrasting golden-orange rump, basal tail patches, scapulars and supercilia. Underparts all golden-orange. Female much browner above. Legs long and pink. Long tail.

Voice: Calls include a throaty *trrrr* and a hard *tchek tchek*.

Habits: Inhabits open scrub and forest undergrowth, feeding mainly on invertebrates on ground or in air. Very skulking and usually solitary. Nests in bank or on ground near or above treeline.

Jan Willem den Besten

Distribution: Locally common breeding resident in northern hills from Pakistan (where very rare) east to the Myanmar border. Winters lower down but usually above 1500 m.
Also occurs in China and SE Asia.

WHITE-BROWED BUSH ROBIN

Tarsiger indicus 15 cm F: Muscicapidae

Description: A slaty-blue and orange chat with long, white supercilia. Male similar to **Indian Blue** but longer tailed, with longer supercilia. Much darker above and all orange below. Cheeks black. Female dull brown above and orange-buff below. Has fine white supercilia. Dark legs. Rare **Rufous-breasted Bush Robin** *T. hyperythrus* has bright blue supercilia and shoulders and white vent and belly. Female has bluish tail and no supercilia.

Voice: A churring *trrr*.

Habits: Inhabits the damp undergrowth of dense forest feeding on invertebrates on ground or in air. Solitary and usually secretive. Nests in bank.

Otto Pfister

Distribution: Scarce breeding resident of northern hills from Uttaranchal east to the Myanmar border. Winters lower down in foothills.
Also occurs in China and SE Asia.

ORIENTAL MAGPIE ROBIN (Dayal, Asian Magpie Robin)

Copsychus saularis 20 cm F: Muscicapidae

Otto Pfister

♀

Otto Pfister

Description: A long-tailed, black and white chat. Male glossy blue-black above with white wing patches and outer tail feathers. Lower breast to vent white. Female much greyer. Cocks and fans long tail frequently, usually with wings drooped.

Voice: Noisy. A slurred *chrrr* in alarm. Contact call is a plaintive *sweee*. Rich piping song.

Habits: One of the region's most familiar birds and the national bird of Bangladesh. Inhabits all types of wooded country including parks and gardens. Frequently perches high, particularly when singing, but feeds mainly on ground on invertebrates. Usually in pairs. Aggressively territorial and quite confiding. Nests in hole in tree, bank or building.

Distribution: Abundant breeding resident of plains and foothills throughout region except Pakistan where it is very local. Also occurs in China and SE Asia.

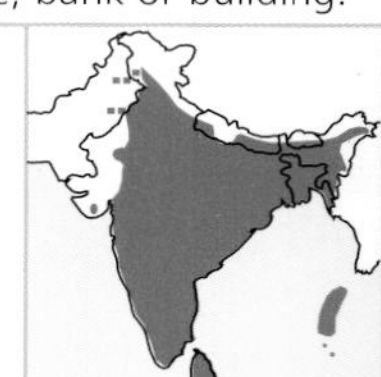

WHITE-RUMPED SHAMA (Shama)

Copsychus malabaricus 25 cm F: Muscicapidae

Otto Pfister

♀

Otto Pfister

Description: A very long-tailed, black and orange chat with a large, white back patch. Black tail graduated and edged with white. Male deep orange from lower breast to vent. Female paler and greyer above. Frequently cocks tail.

Voice: A harsh *chrrr* call. Very rich song often with mimicry.

Habits: Inhabits the dense undergrowth of forests, favouring bamboo. Usually close to ground on which it feeds on invertebrates. Rather secretive except when in territorial display. Best located by frequent crepuscular singing. Nests low down.

Distribution: Locally common breeding resident in northern foothills from Uttaranchal east to the Myanmar border, peninsular hills and Sri Lanka. Also occurs in China and SE Asia.

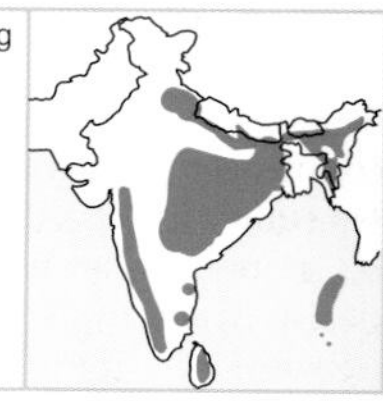

INDIAN ROBIN (Black Robin, Indian Chat)

Saxicoloides fulicata 16 cm F: Muscicapidae

Otto Pfister

Nikhil Devasar

Description: A perky, dark chat with rufous belly and undertail coverts. Northern males are dark brown from crown to mantle and glossy black below and on wings and tail. Southern males are all black above. Both have small white wing patch. Female brownish-grey with the same rufous under tail. Flicks tail high over back on alighting.

Voice: An extended *sweech* and a harsh *churrr* call. Song a simple, rather quiet, melody.

Habits: Inhabits dry, often rocky, wooded areas, scrub, cultivation and gardens. Feeds on invertebrates on ground. Hops with head and tail held high. Reasonably confiding but can be secretive. Solitary or in pairs. Nests low down.

Distribution: Common breeding endemic resident throughout plains except in the Northeast and Bangladesh. Most common in north of range.

RUFOUS-BACKED REDSTART (Eversmann's Redstart)

Phoenicurus erythronota 16 cm F: Muscicapidae

Jan Willem den Besten

Description: A medium-sized, grey, black and orange chat. Larger than most redstarts. Breeding male has grey crown and broad black face mask. Wings black with white bars and patches. Body rather mottled orange, much paler on belly and vent. Tail brown with orange outer feathers. Becomes much paler and more mottled in winter. Female brown with orange rump and outer tail and two buff wing bars.

Voice: A soft croaking *trrr* and *few eet* calls.

Habits: Inhabits dry, rocky, hilly country, scrub and the edges of hill cultivation. Feeds on insects on ground, often flying from low perch. Upright posture. Flicks tail upwards but does not shiver it.

Distribution: Locally common but erratic winter visitor to Pakistan and northern Indian hills east to west Nepal. Also occurs in Iran and Central Asia.

BLUE-CAPPED REDSTART (Blue-headed Redstart)

Phoenicurus coeruleocephalus 15 cm F: Muscicapidae

Jan Willem den Besten

Description: A medium-sized, black and white chat with a black tail. Male all black above and on breast, with contrasting blue-grey crown and white in wings. Underparts white. The only red-start without a reddish tail. Female dark grey with two buff wing bars, rufous rump and black tail.

Voice: A soft *tik tik tik* and a warbling song.

Habits: Inhabits open forest and scrub on rocky slopes in summer. Winters in open forest and secondary cover. Often near water. Feeds on invertebrates on low branches and the ground, often dropping from a low perch. Shivers and slowly wags tail. Often confiding. Nests on ground.

Distribution: Locally common breeding resident in northern hills from north Pakistan to Bhutan. Commoner in west. Winters lower down occasionally moving into northern plains as far south as Delhi.
Also occurs in Central Asia.

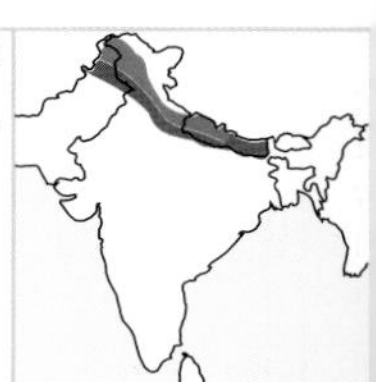

BLACK REDSTART

Phoenicurus ochruros 15 cm F: Muscicapidae

Otto Pfister

♀
Kamal Sahai

Description: Grey, black and rusty-red chat with dark-centred, rufous tail and rufous rump. Males are either grey above with black face and throat (Western race) or all black above with a greyish forecrown (Eastern race, which is the most common in winter). Deep rusty-red below. Females have same tail pattern but are dull rufous-brown. Sits erect with characteristically shivering tail.

Voice: A thin *tsip* often followed by *tic tic tic*.

Habits: Inhabits open cultivation and woodland. Perches on low branches, rocks and wires, flying to ground to pick up invertebrate prey, then returning to perch. Usually solitary and rather shy. Territorial even in winter. Nests among rocks.

Distribution: Common breeding summer visitor to northern mountains from Pakistan east to Sikkim. Winters in foothills and throughout most of region except the extreme south and Sri Lanka. Commonest in north.
Also occurs in Europe, N Africa, W and Central Asia, China and Myanmar.

HODGSON'S REDSTART

Phoenicurus hodgsoni 15 cm F: Muscicapidae

Description: A grey, black and orange chat with a dark-centred, rufous tail and rufous rump. Male all grey above apart from tail and rump, with a small white wing patch and white forehead. Black throat and upper breast, orange underparts. Female has same rufous tail and vent but is overall greyish. Very rare **Common Redstart** *P. phoenicurus* has no white in wing and more restricted black area on throat.

Otto Pfister

Voice: A rippling *prtt* and a harsh *churrr*.

Habits: Inhabits stony river beds, scrub, open woodland, grassland and cultivation. Feeds on insects in air and on ground. Shivers tail less than **Black** but otherwise similar in behaviour.

Distribution: Locally common winter visitor to northern hills from Uttaranchal east to the Myanmar border.
Also occurs in China and Myanmar.

WHITE-WINGED REDSTART (Guldenstadt's Redstart)

Phoenicurus erythrogaster 18 cm F: Muscicapidae

Description: A large, black, white and orange chat with a plain orange tail. Male black above and on throat, with large, white wing patches and crown. Orange underparts, rump and tail. Female brown above and warm buff below with all orange rump and tail.

Otto Pfister

Voice: A soft *lik* and a hard *teek teek*.

Habits: Inhabits stony hillsides, pastures and river beds. Feeds on invertebrates on ground and through aerial sallies. Perches low. Nests among rocks.

Distribution: A locally common breeding resident in northern mountains from Pakistan east to Arunachal Pradesh. Winters lower down in foothills.
Also occurs in W and Central Asia and China.

BLUE-FRONTED REDSTART

Phoenicurus frontalis 15 cm F: Muscicapidae

Otto Pfister

♀

Otto Pfister

Description: A blue and orange chat with black-centred and black-tipped orange tail. Male has blue head and upperparts and orange underparts. Female grey-brown above, buff below with rufous belly and vent. Flicks tail up and down; does not shiver it.

Voice: A hard *tik* and a rapid *trrrr*.

Habits: Inhabits mountain scrub, bushes near cultivation and open woodland. Perches upright and low down, feeding on invertebrates on ground. Also berries. Usually singly or in loose parties. Nests low down.

Distribution: Common breeding resident in northern mountains from Pakistan east to the Myanmar border. Winters lower down in foothills.
Also occurs in Tibet, China and SE Asia.

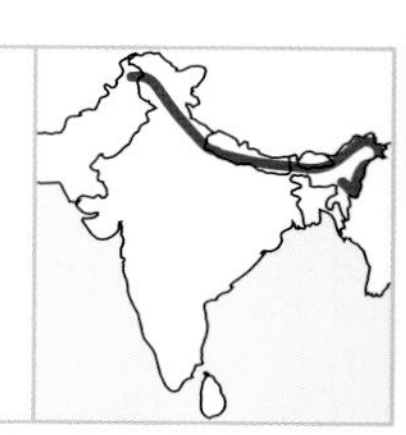

WHITE-CAPPED WATER REDSTART (River Chat)

Chaimarrornis leucocephalus 19 cm F: Muscicapidae

Otto Pfister

Description: Large black and rufous river chat with rufous tail, broadly-tipped black. White crown, black back, wings and breast. Rufous underparts. Perches with tail raised and wings drooped. Sexes alike.

Voice: A plaintive *tseee* and a softer *psst psst*. Whistling song.

Habits: Inhabits rocky streams and rivers, waterworks with fast- flowing water and alpine pastures. Singly or in pairs, actively feeding on invertebrates, in air or from stones. Also eats berries. Often confiding. Perches on stones mid-stream, bobbing body and flicking and fanning wings and tail. Flies low over water. Very territorial but often seen close to **Plumbeous**. Nests in hole in bank or in rocks.

Distribution: A common breeding resident to northern hills from Pakistan east to the Myanmar border. Winters lower down in foothills sometimes extending into nearby plains along rivers or canals.
Also occurs in Central and SE Asia and China.

PLUMBEOUS WATER REDSTART (Plumbeous Redstart)

Rhyacornis fulginosus 12 cm F: Muscicapidae

Otto Pfister

Description: A small, dumpy grey chat. Male lead grey all over with rufous rump, tail and vent. Female grey above with white rump and white-edged black tail and two narrow wing bars. Below white-spotted grey. Juvenile as female but browner and more speckled.

Voice: A sharp *kreee* and a snapping *tzit tzit*. Jingling song.

Habits: Inhabits rocky streams, rivers and waterworks with fast-flowing water. Often with **White-capped**. Energetically feeds on invertebrates on stones or in air. Also eats berries. Usually in pairs and very confiding. Fans and wags tail. Territorial and crepuscular. Nests in hole in bank or bridge.

♀

Otto Pfister

Distribution: Common breeding resident of northern hills from Pakistan east to the Myanmar border. Winters lower down in foothills and rarely extending into nearby northern plains along rivers and canals. Also occurs in Tibet, China and SE Asia.

WHITE-TAILED ROBIN (White-tailed Blue Robin)

Myiomela leucura 18 cm F: Muscicapidae

Description: A large blue-black chat with broad white edges to long, broad, black tail. Male inky-blue with brighter blue forehead and shoulders. Female brown with white lower throat patch and brown breast band. Lowers and spreads tail. The rare but similar, northeastern **Blue-fronted Robin** *Cinclidium frontale* has entirely black, graduated tail.

Sujan Chatterjee

Voice: Calls quiet *tuk* and a thin whistle. Undulating, rapid song.

Habits: Inhabits damp undergrowth and bamboo clumps in evergreen forests, often near streams or in ravines. Shy, keeping close to ground where it feeds on invertebrates. Flies up into trees if disturbed. Singly or in pairs. Nests low down.

Distribution: Generally scarce breeding resident of northern hills from Uttaranchal east to the Myanmar border. Also much lower in northeast Bangladesh. Winters lower down in foothills and rarely in nearby plains. Also occurs in SE Asia.

GRANDALA (Hodgson's Grandala)

Grandala coelicolor 23 cm F: Muscicapidae

Peter Morris

Description: A large, purplish-blue, thrush-like chat of the high mountains. Male all shining purplish-blue with blackish wings and tail. Much brighter than similar sized **Blue Rock Thrush**. Female and immature white-streaked brownish-grey with white patches on wings and bluish wash on rump.

Voice: A sharp *jeeu jeeu*. Song an extension of this.

Habits: Inhabits rocky hillsides and pastures above tree-line. Usually in (often large) flocks, catching invertebrates on wing or on ground. Also eats berries. Flocks often feed wheeling high in the sky like **Starlings**. Perches upright on rocks and hops on ground. Rarely descends below treeline. Nests on rock ledges.

Distribution: Locally common breeding resident of high mountains from Kashmir east to Arunachal Pradesh. Descends a little lower in winter but always remains in mountain zone.
Also occurs in Tibet, China and Myanmar.

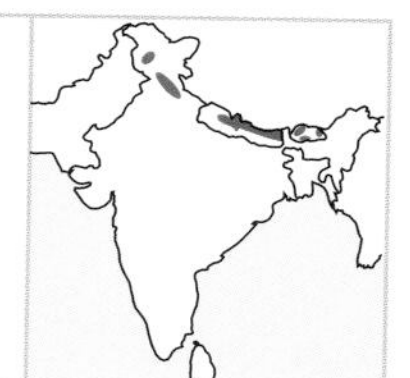

LITTLE FORKTAIL

Enicurus scouleri 12 cm F: Muscicapidae

Otto Pfister

Description: A very small, black and white river chat with a short forked tail and pink legs. Black with high white forehead and white belly. Long white wingbar and white tail edges. Sexes alike.

Voice: Rather quiet. An infrequent *tzittzit*.

Habits: Inhabits rocky streams and rivers particularly near waterfalls. In winter will occupy slower rivers and lakesides. Feeds on invertebrates on stream rocks, in air and even under water. Wags and fans tail. In spite of striking plumage, easy to overlook. Singly or in pairs and rather shy. Nests in hole in bank, often behind waterfall.

Distribution: Fairly common breeding resident in northern mountains from Pakistan east to the Myanmar border. Moves lower down to foothills in winter. Also occurs in Central Asia and China.

BLACK-BACKED FORKTAIL

Enicurus immaculatus 25 cm F: Muscicapidae

Description: A large, black and white river chat with a long forked tail and pink legs. All black above with narrow white face mask, broad white wing bars and white-tipped and barred tail. White below. Sexes alike.

Voice: A whistled *dew*, sometimes followed by a higher *seee*.

Habits: Inhabits small, shaded rocky streams in forest. Feeds, singly or in pairs, on invertebrates on stones, in air and under water. Moves restlessly with slowly swaying tail, which is usually held slightly raised. Shy, flying low upstream with repeated calls if disturbed. Nests in hole or on ledge by water.

Toby Sinclair

Distribution: Locally common breeding resident in northern foothills from Uttaranchal east to the Myanmar border. Rare in west of range. Winters lower down including Bangladesh. Also occurs in SE Asia.

SLATY-BACKED FORKTAIL

Enicurus schistaceus 25 cm F: Muscicapidae

Description: A large, grey, black and white river chat with long forked tail and pink legs. Very similar to **Black-backed,** but with grey crown and mantle. Contrasting black throat and usually rather less white on forehead. Bill longer. Sexes alike.

Voice: A high-pitched *tsee*, *teenk* and *cheet*.

Habits: As **Black-backed** inhabiting the same sort of shaded rocky streams in forest. Also wooded lakes and ravines. Nests low down.

Morten Strange

Distribution: Scarce breeding resident in northern mountains from Uttaranchal east to the Myanmar border. Winters lower down including nearby plains. Odd records from Bangladesh. Also occurs in SE Asia.

SPOTTED FORKTAIL

Enicurus maculatus 25 cm F: Muscicapidae

Otto Pfister

Description: Large, black and white river chat with long forked tail and pink legs. Similar to **Black-backed,** but fore-crown as well as forehead white and fine white spots on black mantle and white collar. Sexes alike. Rare, north eastern **White-crowned Forktail** *E. leschenaulti* has whole crown white and an unspotted black mantle.

Voice: A shrill *jreee*, mainly in flight.

Habits: The commonest forktail. As **Black-backed**, inhabiting the same sort of shaded water courses in forest.

Distribution: Fairly common breeding resident of northern mountains from Pakistan east to the Myanmar border. Moves lower down in foothills in winter. Also occurs in Tibet, China and SE Asia.

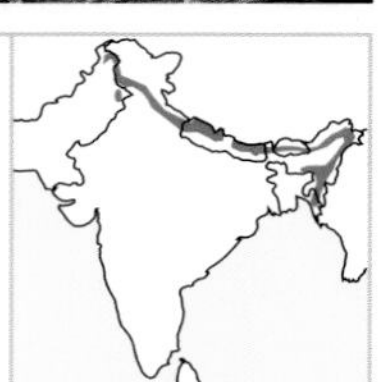

STOLICZKA'S BUSHCHAT (White-browed)

Saxicola macrorhyncha 15 cm F: Muscicapidae

Nikhil Devasar

♀

Nikhil Devasar

Description: Small, slim, long-legged and long-billed chat with white in outer tail feathers. Wheatear-like. Male has blackish ear coverts, contrasting whitish supercilia and white wing patches. White throat contrasts with peachy breast. Dark blackish-brown upperparts, paler and more streaked in winter. Female streaked brown with marked supercilia and less white in tail.

Voice: Very quiet. A soft *prupp prupp* call.

Habits: Inhabits dry semi-desert with grass clumps and scattered bushes. Perches low and flies to ground or into air for insects, returning to regular perches. Territorial in small area. Unique "puff and roll" display on ground. Often confiding.

Distribution: Globally threatened. Very local. Now known mainly in winter in western Rajasthan deserts though recently rediscovered in eastern Haryana in spring. Breeding areas and nest unknown. Formerly more widespread in Pakistan and northwest India.
Historical records from Afghanistan, otherwise a regional endemic.

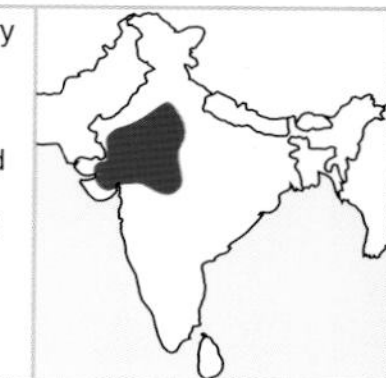

COMMON STONECHAT (Siberian, Collared Stonechat)

Saxicola torquata 13 cm F: Muscicapidae

Nikhil Devasar

Nikhil Devasar

Description: A small, dumpy, variable chat with a short black tail. In breeding plumage male has black head with white half-collar, white rump and wing patches. Dark brown above and orange on breast. Browner and drabber in non-breeding plumage. Several races occur. Female streaked brown with rufous rump and short, pale supercilia. Deep buff below with paler throat.

Voice: A double, repeated *wheel tak*.

Habits: Inhabits dry, open country with scrub, including fallow land. Also reedbeds. Usually in pairs perching conspicuously and flying to ground or into air for invertebrate prey. Flicks wings and flicks and fans tail. Confiding. Nests on ground.

Distribution: Common breeding summer visitor to northern hills from Baluchistan east to Nepal. Winters commonly throughout most of plains except extreme south India and Sri Lanka. Winter numbers augmented by north Asian races. Also occurs in Europe, Africa and throughout mainland Asia.

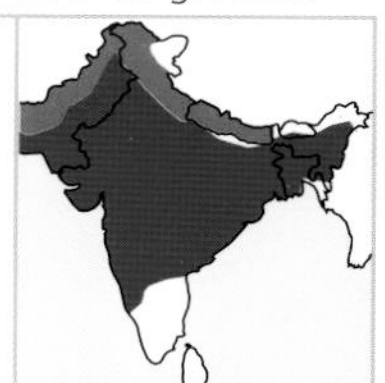

WHITE-TAILED STONECHAT (White-tailed Bushchat)

Saxicola leucura 13 cm F: Muscicapidae

Nikhil Devasar

Nikhil Devasar

Description: Small, dumpy chat with extensive white in tail. Male very similar to **Common** except for white in tail and darker, smaller and more clearly defined orange breast patch, surrounded by white. Female very similar to **Common** but greyer, paler below and with greyish in tail. Note white in tail usually only obvious in flight.

Voice: A short *pseep* and a harder *kek kek kek*.

Habits: Inhabits reedbeds and tall grass usually near rivers. Sometimes in adjoining scrub (often tamarisk) and cultivation. Frequently shares habitat with **Common**. Behaviour similar but more frequently feeds among reeds and thus more readily overlooked. Nests in base of reed clump or bush.

Distribution: Local breeding, near-endemic resident in Indus and Upper Ganges river systems and the terai east to Assam and patchily south to north Orissa.
Also occurs in Myanmar.

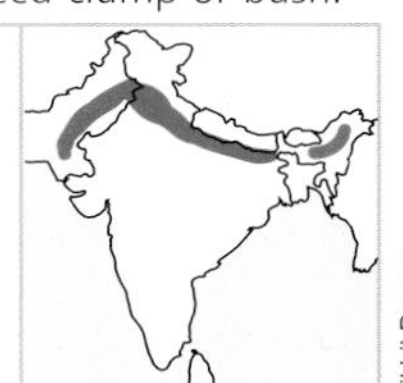

PIED BUSHCHAT (Pied Stonechat)

Saxicola caprata 13 cm F: Muscicapidae

Otto Pfister

♀

Otto Pfister

Description: A small, dumpy, black and white chat. Male all black with white rump and vent and, often hidden, small wing patches. Northern races have white on lower breast and belly. Female dark brown with rusty rump and flanks, pale vent and throat and black tail. Similar to **Indian Robin**.

Voice: A plaintive *cheep cheep, tree*. Short song.

Habits: Inhabits open country with bushes including cultivation, the edges of reedbeds and semi-desert. Often near water and villages. Garden bird in hills. Strongly territorial and usually in pairs. Perches prominently, including on wires, flying to ground or in air for invertebrates. Very confiding. Nests on ground or in hole in bank or structure.

Distribution: Common breeding resident throughout most of plains and foothills but rare in the Northeast. Restricted to hills in extreme south India and Sri Lanka.
Also occurs in W and SE Asia and New Guinea.

GREY BUSHCHAT (Dark-grey Bushchat)

Saxicola ferrea 15 cm F: Muscicapidae

Nikhil Devasar

♀

Otto Pfister

Description: Small, slim, grey chat. Male dark-streaked grey above, darkest on wings and tail which has white borders. Obvious white supercilia and throat and black cheeks. Pale grey below. Female superficially resembles **Common Wood-shrike** in pattern. Brown above with broad, buff supercilia and darker cheeks. Rump and tail edges rufous. Warm buff below.

Voice: Vocal. A plaintive *praee* and an abrupt *zee chunk*. Also *tak tak*. Loud, short song.

Habits: Inhabits bushy country and forest edges, including hill cultivation and gardens. Territorial and usually in pairs. Feeds on insects in air or ground. Perches prominently, often on wires.

Distribution: Common breeding resident in northern hills from Pakistan east to the Myanmar border. Winters lower down in foothills and occasionally in plains south to Mysore.
Also occurs in China and SE Asia.

NORTHERN WHEATEAR (Wheatear)

Oenanthe oenanthe 15 cm F: Muscicapidae

Description: A small, terrestrial chat with a black and white tail. Male grey above with black cheeks, wing and tail. Tail base and rump and narrow supercilia white. Throat and upper breast peachy when fresh. Female and juvenile much duller, without face markings but same tail pattern.

Voice: A hard *chak*, often preceded by *wheet*.

Habits: Inhabits open, usually stony, country and cultivation. Feeds on invertebrates on ground, running and stopping to pick up prey. Erect carriage when perched on low eminences. Bobs body and flicks and fans tail. Flies off low if approached too close.

Otto Pfister

Distribution: Scarce passage migrant (may have bred) in western and northern Pakistan en route to eastern Africa. Rare vagrant to northern India, the Maldives and Nepal.
Also occurs in Europe, Africa, N and W Asia and N America.

VARIABLE WHEATEAR (Eastern Pied Wheatear)

Oenanthe picata 17 cm F: Muscicapidae

Description: A perky, small, black and white terrestrial chat. Male black with white vent (*opistholeuca*), white breast to vent (*pictata;* shown) or white crown and breast to vent (*capistrata*). Female also varies but usually grey above and whitish below. Both sexes have extensive white rump and black central tail bar and black tip.

Voice: A hard *chek chek* and a low whistle. Song full of mimicry.

Habits: Inhabits dry, open country. Often perches on low bushes or walls from which it flies to feed on invertebrates on ground or in air. Territorial but rather shy. Usually solitary. Nests in hole in ground.

Distribution: Common breeding summer visitor to northern and western Pakistan. Winters in plains there and less commonly in northwest India. Decidedly rare further east and south.
Also occurs in Iran and Central Asia.

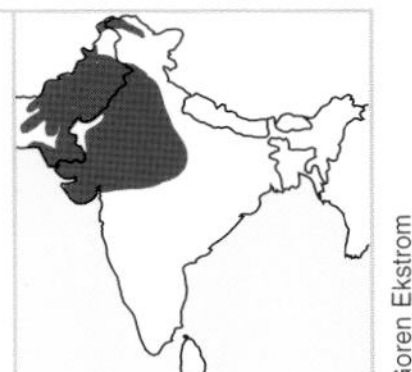

Otto Pfister

Goren Ekstrom

PIED WHEATEAR (Pleschanka's Pied Chat)

Oenanthe pleschanka 15 cm F: Muscicapidae

Otto Pfister

Description: A small, perky, black and white terrestrial chat with a buff-white crown. Similar to **Variable** of *capistrata* race but smaller. More white in outer tail feathers, often reaching tail tip in places. Thin, black edge to outer tail. Black throat more limited. Female and juvenile browner with mottled blackish throat and same tail pattern.

Voice: A throaty *trrtt* and a hard *tak tak*.

Habits: Inhabits dry, open, rocky country, perching prominently on rocks and buildings from which it preys on aerial and terrestrial invertebrates. Usually solitary. Territorial. Nests among stones.

Distribution: Locally common breeding summer visitor to northern hills from Pakistan east to Himachal Pradesh. High altitudes than Variable. Scarce passage migrant elsewhere in Pakistan and, very rarely, northern India, the Maldives, Sri Lanka and Nepal, en route to eastern Africa.
Also occurs in Europe, Africa, W, N and Central Asia and China.

DESERT WHEATEAR

Oenathe deserti 15 cm F: Muscicapidae

Otto Pfister

Otto Pfister

Description: A small, sandy and black terrestrial chat. Tail wholly black contrasting with white vent and rump. Male sandy-brown with black throat and wings, mottled whitish in non-breeding plumage. Female has mottled black cheeks and whitish edges to wings. Both duller in winter (female shown).

Voice: Calls an occasional *ch chett*.

Habits: Inhabits open dry, rocky, or sandy country including dry fallow. Feeds on invertebrate prey; running after it or by fly-catching in air. Lively and fairly confiding but well camouflaged. Flies away low when disturbed, often then crouching behind a stone or low plant. Usually solitary. Nests in hole.

Distribution: Locally common breeding summer visitor to dry northern hills and plateaux from north Pakistan east to Nepal. Winters in plains of Pakistan and northwest India with occasional records from further south and east as far as Sri Lanka and Bangladesh.
Also occurs in Europe, Africa, W and Central Asia and China.

ISABELLINE WHEATEAR

Oenanthe isabellina 17 cm F: Muscicapidae

Description: A slim, upright, pale terrestrial chat with a black and white tail. Sexes similar. Pale sandy-brown, slightly darker on wings, and with black lores and alulas. White basal patches on black tail. Looks long-legged because of very upright stance. Paler than female **Northern** and **Desert**. Similar coloured **Rufous-tailed** *O. xanthoprymna* is plumper and has darker wings and rufous rump and tail sides.

Voice: Calls *chak chak* and *tew*.

Habits: Inhabits mainly sandy semi-desert and overgrazed pasture, breeding in rocky gulleys and plateaux. More terrestrial than other wheatears, mainly catching invertebrate prey on ground. Solitary and territorial. Shy. Nests in deep rodent holes.

Otto Pfister

Distribution: Locally common breeding summer visitor to Baluchistan. Winters fairly commonly in Pakistan and northwest India south to Gujarat with scattered records south to the Maldives and Sri Lanka and east to Sikkim. Also occurs in Europe, Africa, W and Central Asia and China.

BROWN ROCK-CHAT (Indian Chat)

Cercomela fusca 17 cm F: Muscicapidae

Description: A fairly large, upright, dark brown chat. All dark brown with rufous tinge to underparts and blackish tail. Sexes alike. Darker than female **Indian Robin** which has different posture and rufous under tail.

Voice: Calls harsh *chaeck* and a whistling *cheee*. A short, melodious song with mimicry.

Habits: Inhabits buildings in towns and villages, ruins, quarries, rocky hills and cliffs. Perches high on vantage point, often sitting for long periods. Usually in pairs and very territorial. Especially confiding, even entering occupied houses. Feeds on invertebrates by flying down from perch. Nests on ledges and in crevices, frequently inside buildings.

Otto Pfister

Distribution: Locally common endemic breeding resident in plains and foothills of northwest India from Punjab south to Maharastra and east to Bihar. Rare in north Pakistan and southern Nepal. Also occurs in Europe, Africa, W and Central Asia and China.

ASIAN GLOSSY STARLING (Stare, Philippine Glossy)

Aplonis panayensis 20 cm F: Sturnidae

Morten Strange

Description: A small, glossy black starling with red irises. Green gloss on adults. Immature heavily streaked below. Dark bill. Slim and rather short-tailed. Sexes alike.

Voice: A loud whistle and a hard *ink* call.

Habits: Inhabits open forest, clearings, secondary growth, plantations and gardens. Feeds mainly in canopy on fruit, nectar and insects and often with other starlings. Gregarious and approachable. Nests colonially in hole in tree or building.

Morten Strange

Distribution: Common breeding resident in Andaman and Nicobar Islands. Local breeding summer visitor in parts of Assam. Scarce migrant elsewhere in the Northeast, including Bangladesh.
Also occurs in SE Asia..

CHESTNUT-TAILED STARLING (Grey-headed Mynah)

Sturnus malabaricus 21 cm F: Sturnidae

Otto Pfister

Description: A pale, grey and chestnut, arboreal starling with white irises. Head and neck pale grey (white in southern race), dark grey upperparts, chestnut, brown-centred, tail and chestnut underparts. Chestnut sometimes rather pale. Usually with whitish head and neck hackles. Yellow bill with blue base. Sexes similar.

Voice: Noisy. Metallic, whistling calls and warbling song.

Habits: Inhabits open woodland, cultivation with trees and village groves. Usually in noisy flocks, feeding restlessly on flower nectar, berries and insects in treetops, although will come to ground occasionally. Often in mixed hunting groups. Nests in hole in tree.

Distribution: Locally common breeding resident in the Northeast and south-west. Some move into central and western India in winter with vagrant records north to Punjab and south to Sri Lanka.
Also occurs in China and SE Asia.

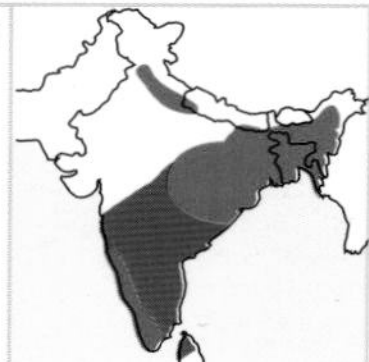

BRAHMINY STARLING (Black-headed, Brahminy Mynah)

Sturnus pagodarum 20 cm F: Sturnidae

Description: Small, stocky, grey and orange starling with glossy black, crested crown. Upperparts grey with broad, white-tipped and white-edged, black tail. Under-parts warm orange with white vent. Irises pale, yellow bare patch behind eyes, bill yellow with blue base and legs yellow. Sexes alike. Juvenile has no crest and buffer. Broad blackish wings.

Voice: Noisy. Various chirping, warbling and whistling notes, often including mimicry.

Habits: Inhabits open woodland, dry scrub, gardens and groves. Usually in pairs and small parties. Feeds on fruit and flower nectar in treetops, often with other species. Also on invertebrates on ground, where it struts purposefully. Nests in tree hole.

Kamal Sahai

Distribution: Common breeding resident throughout much of plains and foothills of region. Rare or local in Pakistan, Sri Lanka, Nepal, Assam and Bangladesh. Wanders widely. Also occurs in Afghanistan.

ROSY STARLING (Rosy Pastor, Rose-coloured Starling)

Sturnus roseus 22 cm F: Sturnidae

Description: A pink and black starling with shaggy black crest. Pink body and glossy black head, neck, wings, vent and tail. Black much duller in non-breeding plumage when pink appears dirty white. Sexes alike. Bill yellow, legs pink. Juvenile buff with brown wings and yellow bill.

Voice: Noisy. Chatters, screams and whistles when feeding and in flight.

Habits: Inhabits open woodland and thorn scrub, grassland and cultivation. Usually in active flocks feeding on berries or flowers in trees and bushes or on ground. Also in air on insects, especially locusts, the swarms of which it follows. Often rather shy and difficult to track as flocks wheel off suddenly. Roosts communally.

Otto Pfister

Distribution: Locally common winter visitor to peninsula, Gujarat and Sri Lanka. Passage migrant in Pakistan and the Northwest. Rare in the Northeast including Bangladesh. Often leaves as late as May and returns as early as late June depending on breeding conditions in central Asia. Also occurs in E Europe, W and Central Asia.

COMMON STARLING (Starling, Eurasian Starling)

Sturnus vulgaris 20 cm F: Sturnidae

Otto Pfister

Description: Small, iridescent black, mainly terrestrial starling with long, pointed bill. Breeding plumage black with purple, blue or green gloss (depending on race) and fine white spots on mantle and vent. Bill then yellow. In non-breeding plumage, black is duller and white spotting more extensive. Bill then black and face greyish. Juvenile streaked greyish-brown. Short tail and pointed wings in rushing flight.

Voice: Noisy. Rich medley of squeaking, clicking and whistling.

Habits: Inhabits damp grassland and cultivation, roosting communally in reed beds or groves. Feeds on invertebrates, mainly on ground in hurried, running flocks. Shy and inclined to take off in wheeling flocks.

Distribution: Local breeding resident in Indus Valley and breeding summer visitor to Kashmir. Local winter visitor to plains of Pakistan and northern India becoming much scarcer in peninsula and in north east. Vagrant to Bangladesh. Unknown in Sri Lanka. Also occurs in Europe, N Africa, W, N and Central Asia. Introduced in N America and elsewhere.

ASIAN PIED STARLING (Pied Mynah)

Sturnus contra 23 cm F: Sturnidae

Otto Pfister

Description: A black and white, mainly terrestrial, starling with long, pointed, red-based yellow bill and orange patches round eyes. Black above, including head, neck and upper breast, with white cheeks, wing stripes and rump and greyish-white underparts. Yellow legs. Sexes alike.

Voice: Noisy. Whistling, chuckling and screaming notes.

Habits: Inhabits towns, villages, grassland and arable. Feeds on invertebrates on ground in pairs or scattered flocks, often using animals' backs as perches. Also eats fruit and grain. Roosts communally in reedbeds and groves. Usually confiding. Pair-bond strong. Makes untidy nest in open branches.

Distribution: Common breeding resident of northern plains from eastern Pakistan east to northeast India and Bangladesh and south to northern Andhra Pradesh. Rare straggler further west and south.
Also occurs in SE Asia.

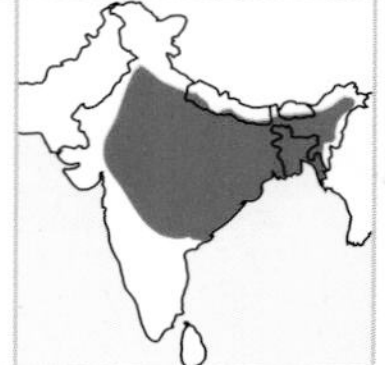

COMMON MYNAH (Mynah)

Acridotheres tristis 23 cm F: Sturnidae

Description: A stocky, purplish-brown starling with large, white wing patches. Black head, flight feathers and white-tipped tail. Vent white. Yellow bill and yellow patches behind eyes. Sexes alike.

Voice: Noisy. A harsh *chake chake* and various chattering, whistling and gurgling notes, often made by pairs together.

Habits: Inhabits towns and villages and associated cultivation, parks and gardens. Bold, aggressive and inquisitive, even entering houses. Feeds on invertebrates, fruit, nectar, grain and human garbage, strutting haughtily on ground. Usually in pairs. Roosts communally in reedbeds or groves. Very territorial, pairs often fighting each other. Nests in hole.

Otto Pfister

Distribution: Common breeding resident throughout plains and lower hills.
Also occurs in Afghanistan and SE Asia. Introduced to S Africa.

BANK MYNAH

Acridotheres ginginianus 21 cm F: Sturnidae

Description: A stocky, bluish-grey starling with deep orange bill and eye patches. Noticeably smaller and greyer than **Common**. Crown, wings and tail black. Tail tip and wing patches ginger. Sexes alike. Juvenile paler and browner.

Voice: Noisy. Similar to **Common** but more conversational and softer.

Habits: Inhabits towns, villages and cultivation. Often round tea-stalls and markets. Gregarious, nesting in colonies in holes in banks and cuttings and roosting in reedbeds and groves. Extremely bold and confiding. Feeds on insects, vegetable matter and human garbage. Waddles and runs on ground and frequently rides on animals' backs.

Distribution: Locally common endemic breeding resident in most of northern plains south to Maharastra and Orissa. Somewhat patchy in distribution and commonest in major river valleys.

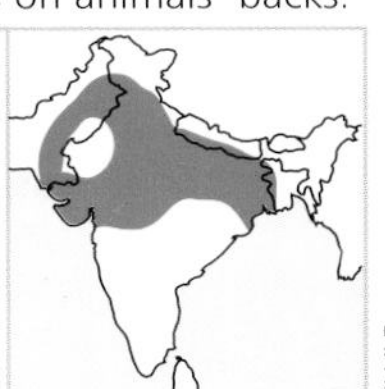

Otto Pfister

Nikhil Devasar

JUNGLE MYNAH

Acridotheres fuscus 23 cm F: Sturnidae

Otto Pfister

Description: A stocky, grey starling with no eye patches and a distinct forehead tuft. Northern race greyer, southern race browner. Black crown and cheeks, black wings with white patches and white-tipped black tail. Irises and bill yellow, latter with blue base. Sexes alike.

Voice: Noisy. Similar to **Common** but higher with more whistling.

Habits: Inhabits cultivation, plantations, forest edge, scrub and the outskirts of towns and villages. Less associated with man than **Common** and **Bank** and less confiding. Also less communal except when nesting in colonies in holes in trees, banks or walls. Feeds on invertebrates, fruit, grain but much less so on human garbage.

Distribution: Locally common breeding resident in northern hills from Pakistan east to Arunachal Pradesh and south through Bangladesh to Orissa and again down western seaboard from Gujarat to Kerala and western Tamil Nadu.
Also occurs in SE Asia.

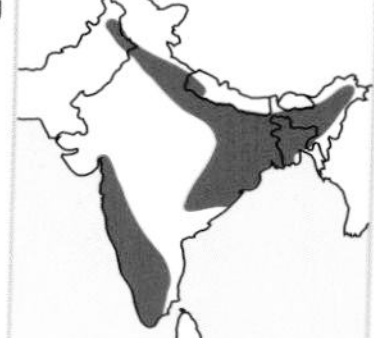

WHITE-VENTED MYNAH (Great, Orange-billed Jungle)

Acridotheres cinereus 23 cm F: Sturnidae

Morten Strange

Description: A glossy, dark grey starling with white vent, wing patches and tail tips. Similar to **Jungle** but darker, with more of a frontal crest. Orange bill, legs and irises. Sexes alike. **Collared Mynah** *A. albocinctus* has large whitish neck patches and white-barred vent.

Voice: Noisy. Similar to **Common** but harsher.

Habits: Inhabits open wooded country, cultivation, villages and parks. Frequently feeds and roosts with other mynahs but usually in smaller numbers. Feeds on fruits and nectar in canopy and on invertebrates on ground, often accompanying cattle or other herbivores. Nests in tree or bank hole.

Distribution: Locally common breeding resident in north-eastern India. Rare in Bangladesh.
Also occurs in SE.Asia.

HILL MYNAH (Grackle, Talking Mynah)

Gracula religiosa 29 cm F: Sturnidae

Description: A large, glossy black, arboreal starling with prominent white wing patches. Yellow head wattles and powerful orange bill. Short tail and broad wings. Northern race (shown) larger with more powerful bill. Wattles differently shaped in southern race (which may be separate species, *G. indica*). Latter occurs with endemic **Sri Lankan Mynah** *G. ptilogenys* which has wattle restricted to nape and darker bill.

Voice: Famously vocal and an exceptional mimic. Screeches, whistles, gurgles and croaks.

Habits: Inhabits tropical forest and plantations. Feeds on fruit in canopy, often with other species. Usually in pairs or small parties. Perches high on bare branches. Nests in high tree hole.

RK Gaur

Distribution: Northern race locally common breeding resident in northern hills from Uttaranchal east to the Myanmar border, Bangladesh and Eastern Ghats. Southern race locally common breeding resident in Western Ghats and Sri Lanka.
Also occurs in SE Asia.

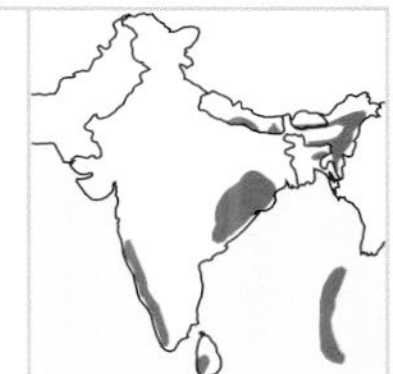

CHESTNUT-BELLIED NUTHATCH

Sitta castanea 12 cm F: Sittidae

Description: Small, dumpy, chestnut and grey woodpecker-like bird with a very short tail. Chestnut below, darkest in peninsular races. White cheeks and black eye lines. Dark blue-grey above. Short, sharp bill. Northern races have larger bill and white scalloping on vent. Females paler.

Voice: Noisy. Hard *chit chit* loud *tzsib*, trilling *wee wee wee*.

Habits: Inhabits mainly deciduous forests and groves. Feeds singly, in pairs or in mixed hunting groups, high in trees on insects, seeds and nuts. Often secures hard food to hammer it open. Clings to bark and moves both upwards and downwards in jerky hops. Nests in tree hole, plastering edges with mud to required size.

Otto Pfister

Distribution: Locally common breeding resident to northern foothills and eastern peninsular hills from Uttaranchal east to the Myanmar border and south to Andhra Pradesh. Also in Western Ghats and extreme northern Pakistan.
Also occurs in SE Asia.

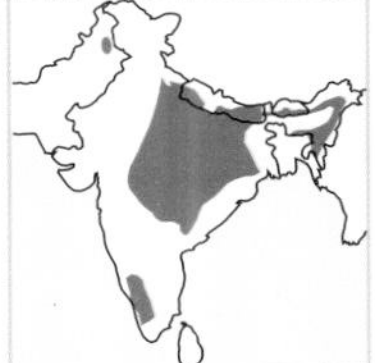

WHITE-TAILED NUTHATCH

Sitta himalayensis 12 cm F: Sittidae

Chandan Choudhuri

Description: Small, dumpy, grey and orange woodpecker-like bird with white in tail. Paler than **Chestnut-bellied** with pale orange from chin to vent, no white scalloping on vent and a smaller bill. White in tail often difficult to see. Sexes alike. Similar but larger northwestern **Kashmir Nuthatch** *S. cashmirensis* has no white in tail. Northeastern **Chestnut-vented** *S. nagaensis* is darker grey above, white below with white-spotted, chestnut vent and chestnut flanks.

Voice: A hard *chit* and whistles.

Habits: Inhabits broadleafed and coniferous forests. Behaviour, food and nesting as **Chestnut-bellied** but particularly favours moss-covered branches.

Distribution: Common breeding resident in northern hills from Himachal Pradesh to the Myanmar border. Winters lower down in foothills. Also occurs in SE Asia.

VELVET-FRONTED NUTHATCH (Velvet-fronted Blue)

Sitta frontalis 10 cm F: Sittidae

Morten Strange

Description: A tiny, bright blue and grey bird with a short, pointed, red beak. Violet-blue above with a black forehead and eye stripes. Blackish wings. Pale pinkish-grey below with white throat. Irises yellow. Female similar but lacks eye stripes.

Voice: Loud rapid repeated *chwit chwit* and *sit sit sit sit*.

Habits: Inhabits open forest and plantations. Feeds on insects high in trees, often in mixed hunting groups but also explores fallen wood. Favours moss-covered trunks and branches. Very active, moving rapidly from tree to tree. Nests in tree hole.

Distribution: Common breeding resident in northern foothills from Uttaranchal east to the Myanmar border. Also in several peninsular hill ranges, including the Western and Eastern Ghats, and Sri Lanka. Also occurs in SE Asia.

WALLCREEPER

Tichodroma muraria 17 cm F: Sittidae

Description: Long-billed, grey bird with rounded, crimson, black and white wings. Body grey with throat black in breeding season and white in non-breeding season. Tail white-tipped black. White scalloping on vent. Bill long, dark and decurved. Jerky **Hoopoe**-like flight. Sexes similar.

Voice: Usually silent. Occasionally cheeps.

Habits: Inhabits cliffs and gorges in breeding season. In winter also ruins, cuttings and rocky river beds. Singly or in pairs. Moves with jerky hops on rock face or among stones on ground, constantly flicking wings as it searches for insects. Shy and inconspicuous until it flies. Flies vertically up or high along gorge or river course.

Otto Pfister

Otto Pfister

Distribution: Locally common breeding resident in northern mountains from Pakistan east to Arunchal Pradesh. Winters lower down in river valleys and foothills and sometimes adjoining plains south to Rajasthan.
Also occurs in Europe, W and Central Asia and China.

EURASIAN TREECREEPER (Common, Northern)

Certhia familiaris 12 cm F: Certhiidae

Description: A small, slim, brown and white bird with a slender, decurved bill. Intricately patterned buff and brown above. Unbarred tail and usually rufous rump. White supercilia and in cheeks. White underparts, dusky on vent. Sexes alike.

Voice: A high *tzee tzee* note and trilling song.

Habits: Inhabits mountain coniferous and mixed forest. Hunts invertebrates in bark by spiralling jerkily upwards and along all sides of branches, using stiff tail as prop. Singly, in pairs or in mixed hunting groups. Active, flying to base of new tree with undulating flight as soon as one tree is finished. Fairly confiding but inconspicuous. Nests behind loose bark or in tree crevice.

Goren Ekstrom

Distribution: Locally common breeding resident in northern mountains from Pakistan east to Arunachal Pradesh. Moves lower down in winter.
Also occurs in Europe and W, N, Central and E Asia.

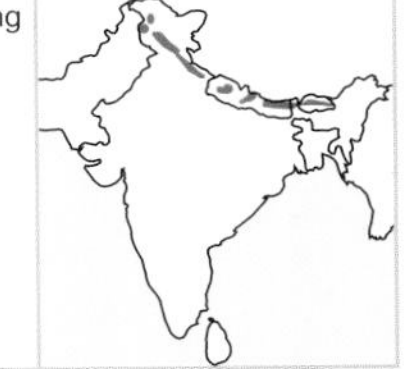

BAR-TAILED TREECREEPER (Himalayan Treecreeper)

Certhia himalayana 12 cm F: Certhiidae

Otto Pfister

Description: A small, slim, brown and white bird with stiff, barred brown tail. Very similar to **Eurasian** apart from tail. Bill slightly longer and rump usually duller. Sexes alike.

Voice: A thin *tsiu* call. Trilling song.

Habits: Inhabits mainly coniferous, rhododendron and birch forest. Winters in all types of trees including groves, gardens and orchards. Behaviour, food and nesting as **Eurasian**.

Distribution: Common breeding resident in northern mountains from Pakistan east to western Nepal. Also Arunchal Pradesh. Winters lower down in foothills.
Also occurs in W and Central Asia. China and Myanmar.

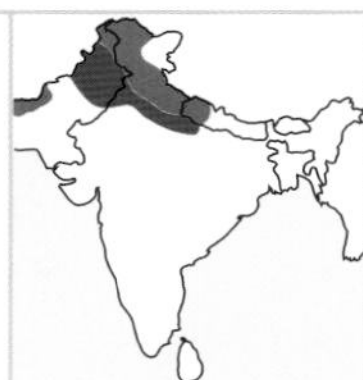

RUSTY-FLANKED TREECREEPER (Nepal Treecreeper)

Certhia nipalensis 12 cm F: Certhiidae

Otto Pfister

Description: Very similar to **Eurasian** but with extensive orange-rufous on flanks, dark ear coverts and a shorter bill. Sexes alike. Similar **Brown-throated Treecreeper** *C. discolor* has longer bill and a pale brown or cinnamon wash on throat and breast.

Voice: A thin *sit sit* and an accelerating trilling song.

Habits: Inhabits oak and mixed forests. Behaviour, food and nesting as **Eurasian**.

Distribution: Locally common breeding resident from Uttaranchal east to Arunachal Pradesh. Winters lower down in foothills.
Also occurs in Tibet and Myanmar.

SPOTTED CREEPER (Grey Creeper)

Salpornis spilonotus 13 cm F: Certhiidae

Description: A plump, brown and buff bird with broad, white supercilia. Dark wings and tail broadly barred white. Rest of upperparts spotted buff. Buff underparts strongly barred and mottled with brown. Stocky, short-tailed shape recalls **nuthatches** but plumage closer to **treecreepers**. Bill thin and decurved. Sexes alike.

Voice: A thin *see ee* and a deep *kek kek kek*. Whistling song.

Habits: Inhabits open woodland and groves, favouring trees with deep-fissured bark such as mangoes and babul. Feeds, usually singly, on invertebrates in bark at all levels in tree. Climbs rapidly up and under trunks and branches, often fluttering down to change position. Does not spiral. Nests close to trunk.

Goren Ekstorm

Distribution: Strangely rare and very local breeding resident in northern peninsula from Haryana, Uttar Pradesh and Bihar south to Maharastra and Andhra Pradesh.
Also occurs in Africa, where also very local.

WINTER WREN (Wren, Northern Wren)

Troglodytes troglodytes 9 cm F: Certhiidae

Description: A tiny, plump, dark brown bird with short tail, usually held sticking up. Wings held low. Well-barred on wings, tail and underparts. Pale supercilia. Rather long, pointed bill and long, pale legs. Eastern race very dark. Sexes alike.

Voice: Noisy. An angry *zirrr* and *tzit*. Loud, rapid song.

Habits: Inhabits rocky areas with scrub and low growth in open forests, often near water and habitation. Feeds secretively and very actively, hunting insects in low tangles in a rather mouse-like way. Usually singly but may roost communally in a hole. Often sings from low but prominent perch. Bobs when alarmed. Nests in low hole or crevice.

Otto Pfister

Distribution: Locally common breeding resident in northern mountains from western Pakistan east to Arunachal Pradesh. Moves lower in winter.
Also occurs in Europe, N Africa, W, N, Central and E Asia, Myanmar and N America.

FIRE-CAPPED TIT

Cephalopyrus flammiceps 9 cm F: Paridae

Joanna Van Gruisen

Description: A tiny, green and yellow tit, the male with a red crown and throat in breeding plumage. Green above with two thin, white wing bars. Deep yellow below in male, paler in female and almost white in juvenile. Beady black eyes and very short pointed bill. **Yellow-browed Tit** *Sylviparus modestus* is similar but duller, with stubby bill and faint whitish wingbars. No red.

Voice: Soft *tsit* and *tsee tsee tsee*. Song a quiet *we two we two we two*.

Habits: Inhabits forest feeding on invertebrates, nectar and pollen, usually high in canopy but will descend to low growth. Gregarious and joins mixed hunting groups. Easily overlooked.

Distribution: A local breeding summer visitor to northern mountains from Pakistan east to Arunachal Pradesh. Winters in foothills and rarely in plains south to Maharastra and Madhya Pradesh.
Also occurs in Tibet, China and Myanmar.

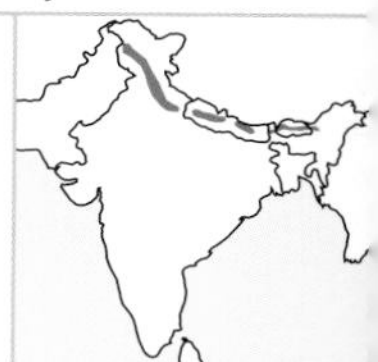

RUFOUS-NAPED TIT (Simla Black Tit, Dark-grey Tit)

Parus rufonuchalis 13 cm F: Paridae

Otto Pfister

Description: A plump, dark tit with extensive black bib from throat to breast. Black, pointed crown, dark grey upperparts and belly. Rufous on nape, breast sides and vent. White cheeks. Sexes alike. Larger than similar species.

Voice: Noisy. Calls *tsee tsee peeou*, *peep* and *seep*.

Habits: Inhabits mainly open coniferous forest, feeding on invertebrates in canopy, on trunks and branches and on the ground. Gregarious, often with other tit species or mixed hunting groups. Tends to feed lower and be less restless. Nests in hole in bank or hidden among stones.

Distribution: Fairly common breeding resident in northern mountains from Baluchistan east to western Nepal. Winters lower down in foothills.
Also occurs in W and Central Asia.

RUFOUS-VENTED TIT (Rufous-bellied Crested, Black Tit)

Parus rubidiventris 10 cm F: Paridae

Description: A very small, plump, dark tit with black throat. Very similar to larger **Rufous-naped,** but overlapping western race has pinkish-rufous belly (grey in eastern race). Bib much less extensive. Nape all white. Vent always rufous. Sexes alike.

Voice: Noisy. A thin *seet* and *psst*. Rattling song.

Habits: Inhabits open coniferous and broad-leafed forests, particularly rhododendron. Behaviour, nest and food as **Rufous-naped**, with whom it often mixes in west where both occur together. More active when feeding. Will also nest in hole in building or tree.

Otto Pfister

Distribution: Common breeding resident in northern mountains from Himachal Pradesh east to Nagaland. Winters lower down in foothills. Also occurs in Tibet, China and Myanmar.

SPOT-WINGED TIT (Crested Black, Black-crested Tit)

Parus melanolophus 11 cm F: Paridae

Description: A small, dark tit with two distinct white-spotted wing bars. Apart from these spots, very similar to grey-bellied eastern race of **Rufous-vented,** although has rufous flanks. Sexes alike.

Voice: Noisy. Calls *te tui* and *tzee tzee tzee*.

Habits: Inhabits mainly open coniferous forests. Behaviour and food as **Rufous-naped** but more active, tending to stay higher in trees. Nests in tree hole.

Otto Pfister

Distribution: Common breeding resident in northern mountains from northwest Pakistan east to central Nepal. Winters lower down in foothills. Also occurs in Afghanistan.

GREY-CRESTED TIT (Brown Crested, Crested Brown Tit)

Parus dichrous 12 cm F: Paridae

Jan Willem den Besten

Description: A small, grey and orange tit with dark-edged, whitish moustache extending round neck as a collar. Mottled grey cheeks and throat and dark recurved eyestripes. Underparts pale orange buff. Sexes alike.

Voice: Thin *zai* and *ti ti ti ti* calls and *chea chea* alarm. Song *wee wee tz tz tz.*

Habits: Inhabits hill forests. Feeds from middle storeys to ground on invertebrates, often in mixed hunting groups. Rather quiet and inconspicuous. Nests in tree hole.

Distribution: Locally common breeding resident in northern mountains from Kashmir east to Bhutan. Commonest in east. Winters lower down.
Also occurs in China and Myanmar.

GREAT TIT (Grey Tit)

Parus major 13 cm F: Paridae

Kamal Sahai

Description: Large, grey, black and white tit with large, white cheek patches. Black crown and surrounds to cheeks and black stripe from throat to vent. Grey above with darker, white-edged tail and darker wings with white wing bars. Whitish below. Sexes similar. Rare northwestern **White-naped Tit** *P. nuchalis* is black above with large, white wing patches.

Voice: Noisy. Repeated *wee chi chee* is commonest of many calls.

Habits: Inhabits open forest, groves and gardens. Feeds usually low down on invertebrates and fruit. Holds latter in feed to peck at it. Usually in pairs or singly but will join mixed hunting groups. Confiding and very active. Nests in a hole.

Distribution: Common breeding resident in northern foothills from Baluchistan east to the Northeast and throughout much of central and western peninsula and Sri Lanka, particularly in hills. Some move into plains in winter.
Also occurs in Europe, Africa and throughout Asia.

GREEN-BACKED TIT

Parus monticolus 13 cm F: Paridae

)escription: A large, green, /ellow and black tit with white :heeks. Very similar to **Great** but orighter with green mantle and /ellow underparts. Shows two ving bars. Sexes similar.

Voice: Noisy. Calls similar to **Great**. Include a loud *teacher eacher* note and a repeated *vhitee whittee*.

Habits: Inhabits moist forest, eeding mainly near the ground on invertebrates and fruit.)ften in mixed hunting groups. 3ehaviour and nesting as **Great**.

Otto Pfister

Distribution: Locally common breeding resident to northern mountains from north Pakistan east to the Myanmar border. May move lower down in foothills in winter.
Also occurs in China and SE Asia.

BLACK-LORED TIT (Yellow-cheeked Tit)

Parus xanthogenys 14 cm F: Paridae

)escription: A large, black and ellow tit with high-peaked lack crown. Yellow cheeks and nderparts. Black line through ye, neck stripe and stripe from hin to belly. Greenish mantle nd white-edged, blue-grey wings nd tail. Two wing bars. Sexes imilar. Northeastern **Yellow-cheeked Tit** has yellow lores nd black striping on mantle.

Voice: Noisy. Commonly a epeated *towit towit churr* and *vicheewe wicheewe*.

Habits: Inhabits open forest and voodland, forest edge, groves, ardens and plantations. Feeds on invertebrates and fruit, usually n upper canopy and often in nixed hunting groups. Active. Jests in tree hole.

Otto Pfister

Distribution: Common breeding endemic resident in northern mountains from extreme northern Pakistan east to West Bengal. Also in northern peninsular hills and all down west coast from Gujarat south to Kerala.

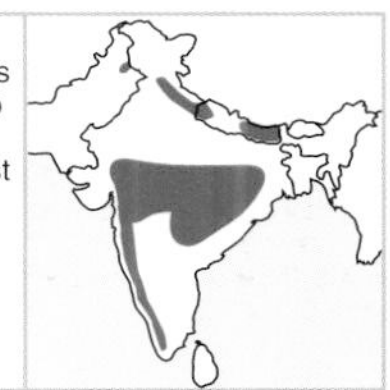

YELLOW-CHEEKED TIT (Black-spotted Yellow Tit)

Parus spilonotus 14 cm F: Paridae

Morten Strange

Description: A crested, black and yellow tit. Very similar to more widespread **Black-lored**. Differs in having yellow face with black restricted to short lines back from eyes. Greenish mantle, more heavily streaked blackish and rump grey. Wing bars white not yellow. Sexes similar.

Voice: **Great Tit**-like song *chee chee pui*. Calls *sit*, *si si si* and *chrrrr*.

Habits: Inhabits open forest feeding on invertebrates and seeds at lower levels. Often in mixed hunting groups and a typical tit in behaviour. Nests in tree hole.

Distribution: Locally common breeding resident in northern mountains from eastern Nepal east to the Myanmar border. Moves lower down in winter. Also occurs in China and SE Asia.

SULTAN TIT

Melanochlora sultanea 20 cm F: Paridae

Simon Harrap

Description: A very large, black and yellow tit with a large feathery yellow crest. Apart from this, all black with yellow lower breast to vent. Rather heavy, finch-like bill and long tail. Female duller black with olive cast to throat and upper breast.

Voice: Noisy. Loud call *cheerie cheerie* most frequent. Also *churrs*.

Habits: Inhabits mainly evergreen, broadleafed, mountain forests. Feeds on invertebrates and fruit, mainly in canopy and often in mixed hunting groups. Rather slower moving than other tits but still acrobatic. Nests in tree hole or crevice.

Distribution: Scarce resident in northern foothills from central Nepal east to the Myanmar border. Also occurs in China and SE Asia.

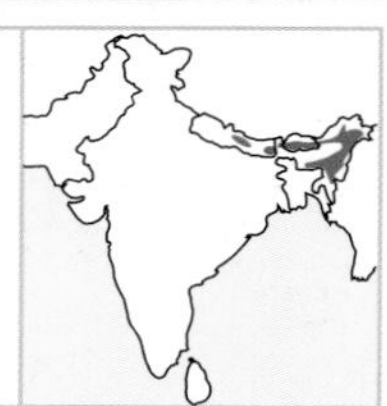

BLACK-THROATED TIT (Red-headed Tit)

Aegithalos concinnus 10 cm F: Aegithalidae

Description: A tiny, chestnut, grey and black tit with a white-bordered black throat. Chestnut crown, upper breast and flanks, paler on belly. Broad, black stripes through eyes. White irises. Grey upperparts. Tail has white tip and outer feathers. Head pattern distinguishes from four similar species of Aegithalidae in region. Sexes alike. Juvenile lacks black throat.

Voice: Soft but insistent *trrr trrr* and *chek chek*.

Habits: Inhabits open forest, gardens and forest edges. Very active, confiding and inquisitive. Small parties, sometimes with mixed hunting groups, feed on invertebrates and fruit at all levels. Nests in low bush.

Otto Pfister

Distribution: Common breeding resident in northern mountains from northern Pakistan east to the Myanmar border. Moves lower down in foothills in winter.
Also occurs in China and SE Asia.

SAND MARTIN (Collared Sand Martin)

Riparia riparia 13 cm F: Hirundinidae

Description: A small, brown swallow with a distinct breast band. Muddy-brown above with slightly darker wings. White below with white throat clearly demarcated by brown breast band. Clearly forked tail and dark underwings. Gliding swallow-like flight. Northwestern **Pale Martin** *R. diluta* similar in flight but much paler with indistinct breast band, more diffuse brown on cheeks and shallower tail fork.

Voice: A dry, quiet *trssh* among flocks.

Habits: Inhabits the vicinity of water where it feeds on aerial insects, usually close to the surface and often with other swallows. Regularly perches on wires and tall reeds. Sociable. Nests in tunnels in sand banks.

Otto Pfister

Distribution: Scarce breeding resident in Assam. Local winter visitor to Bangladesh. Scarce and irregular passage migrant and perhaps winter visitor in Northwest.
Also occurs in Europe, Africa, W, N, Central and SE Asia, China and the Americas.

PLAIN MARTIN (Brown-throated Sand Martin)

Riparia paludicola 12 cm F: Hirundinidae

Otto Pfister

Description: A small, brown swallow with pale brownish throat and upper breast. Intensity of brown varies and may be slightly darker on breast sides but never shows a breast band and throat never white. Broader-based wings than **Sand** produce a characteristic fluttering, bat-like flight. Shallow tail fork.

Voice: A distinctive abrupt, rather high *brret* call in flight.

Habits: Inhabits vicinity of water, catching aerial insects often quite high in the sky. Gregarious, often with other swallows. Regularly perches on wires, tall reeds and even bare sand or mud. Nests in tunnels in vertical sand or mud banks.

Distribution: Common breeding resident in plains and foothills throughout northern parts of region north of Maharastra. Scattered records further south.
Also occurs in Africa and SE Asia.

EURASIAN CRAG MARTIN (Northern Crag Martin)

Hirundo rupestris 15 cm F: Hirundinidae

Otto Pfister

Otto Pfister

Description: A medium-sized, dark brown swallow with greyish-white underparts. Dark vent with pale scalloping. White spots across spread, barely forked tail. Contrasting dark underwing coverts. Stocky and bull-necked with broad-based, pointed wings. More widespread and often urban **Dusky Crag Martin** *H. concolor* is smaller and almost uniform very dark brown including throat.

Voice: Short *plee*, musical *weeeh* and a harsh *tshrr*.

Habits: Inhabits rocky gorges and cliffs and old hill-forts. Feeds on aerial insects usually close to the rock face. Sociable but not often with other swallows. Nests on rock face under overhang.

Distribution: Fairly common but local breeding resident and summer visitor in western Pakistan and northern mountains from north Pakistan to Bhutan. Winter visitor to Western Ghats.
Also occurs in Europe, N Africa, W and Central Asia.

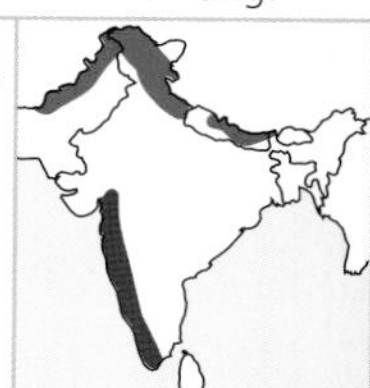

BARN SWALLOW (Common Swallow, Swallow)

Hirundo rustica 18 cm (including tail streamers) F: Hirundinidae

Description: A large, blue-black swallow with deeply forked tail. Tail streamers vary in length, being longest in males and absent in juveniles. White band across undertail. Underparts white. Rufous in race *tytleri*. Forehead and throat red. Complete blue breast band. Long pointed wings.

Voice: A hard *vit* in flight and a twittering rambling song. Calls vary with object of alarm.

Habits: Inhabits cultivation, towns and villages usually near or over water. Feeds on aerial insects in, usually low, fast, powerful flight. Gregarious, often with other swallows perched on wires (as shown). Roosts in reeds communally. Nests against wall or beam in building or under a bridge.

RK Gaur

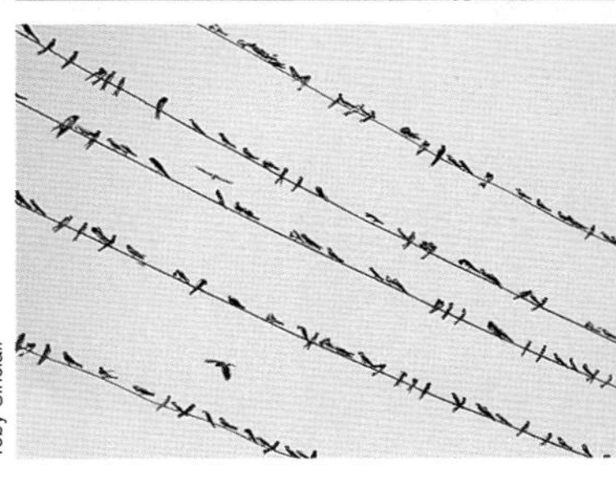

Toby Sinclair

Distribution: Common breeding resident to northern mountains from western Pakistan east to the Myanmar border. Common winter visitor throughout rest of region.
Also occurs in Europe, Africa, throughout Asia, in New Guinea and the Americas.

PACIFIC SWALLOW (House Swallow, Hill Swallow)

Hirundo tahitica 13 cm F: Hirundinidae

Description: A small, blue-black swallow with a short, forked tail. Rufous forehead to breast. No breast band. Dusky greyish-white belly and blackish vent with white scalloping. Grey underwing coverts.

Voice: A hard *qwik qwik*.

Habits: Inhabits open country in hills, often around houses and plantations. Also along coast and rivers in Andamans. Feeds sociably on aerial insects. Flight less sweeping than **Barn**. Often perches on wires and bare branches. Nests against wall in building or against bank or culvert.

Morten Strange

Distribution: Scarce breeding resident in southwestern hills. Commoner in Andamans and Sri Lankan hills.
Also occurs in SE Asia, Australasia and Pacific Islands.

WIRE-TAILED SWALLOW

Hirundo smithii 14 cm (including tail streamers) F: Hirundinidae

Otto Pfister

Description: A small, blue-black swallow with very long and fine tail streamers in adults. Streamers difficult to see at distance. Red crown and black band through eye to nape. Throat to vent pure white with small black shoulder patches. Underwing coverts and most of undertail white. Juvenile and moulting adult lack streamers.

Voice: *chirik wik* and *chichip*. Twittering song.

Habits: Inhabits cultivation and villages near water and all types of wetland. Feeds aerially on insects in low graceful flight. Usually in pairs or small parties. Roosts with other swallows in reed beds and bushes. Often perches on wires. Nests on wall in building, under eaves, bridge or rock overhang.

Distribution: Locally common breeding resident to most of region but absent from much of the Northeast, south, Sri Lanka and Bangladesh. Summer visitor to much of Pakistan and northwest India.
Also occurs in Africa, W and SE Asia.

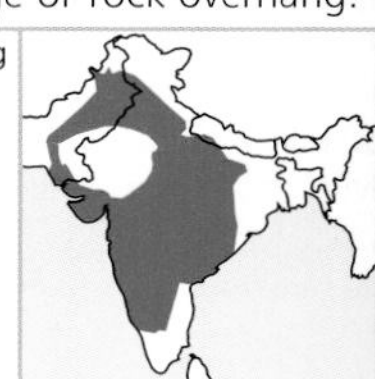

RED-RUMPED SWALLOW (Striated Swallow)

Hirundo daurica 18 cm F: Hirundinidae

Otto Pfister

Otto Pfister

Description: A large, rufous and blue-black swallow with deeply forked tail. Rufous cheeks and collar, pale rufous rump (sometimes looks whitish) and black vent. Streaking on white underparts varies from slight in west to heavy in east. Sri Lankan race has underparts unstreaked deep rufous. Larger, northeastern **Striated Swallow** *H. striolata* heavier streaked with incomplete collar.

Voice: A distinct *treep* call and a rather harsh twittering song.

Habits: Inhabits open cultivation, pastures, clearings and rocky areas, often far from water. Feeds on aerial insects high in sky. Gregarious mixing with other swallows. Roosts communally.

Distribution: Locally common resident and partial migrant in most of India, Nepal and Sri Lanka. Northern mountain breeding birds move into plains in winter. Rare in the Northeast and scarce winter visitor to Bangladesh and much of Pakistan.
Also occurs in Europe, Africa, Central and E Asia and Myanmar.

◉ STREAK-THROATED SWALLOW (Indian Cliff Swallow)

Hirundo fluvicola 12 cm F: Hirundinidae

Description: A small, dusky swallow with notched, long, broad tail. Blackish-blue mantle, brown wings, rump and tail and chestnut crown. Heavily streaked blackish on face, throat and upper breast. Brown under tail. Often looks very dark. Note right-hand bird in photo is a **Plain Martin** with which the browner juvenile might be confused. Weak, martin-like flight.

Voice: Calls *chrrp* and *trr trr*.

Habits: Inhabits open country and cultivation near water. Gregarious often with other swallows, particularly **Plain Martins**. Feeds on aerial insects, usually low over water. Frequently perches on wires. Nests under bridges, sometimes in towns.

Otto Pfister

Distribution: Locally common breeding resident in Indo-Gangetic Plains from Pakistan east to Uttar Pradesh and south through western peninsula to Karnataka. Mainly summer visitor to northern areas. Scattered records elsewhere in India but vagrant to the Northeast and extreme south, Sri Lanka and Bangladesh. Also occurs in Afghanistan.

⊙ NORTHERN HOUSE MARTIN (Common House Martin)

Delichon urbica 14 cm F: Hirundinidae

Description: Dumpy, black and white swallow with white, wrap-around, rump. Dark blue-black above. White below. Legs white feathered. Deep tail fork. **Asian House Martin** *D. dasypus* has less tail fork, blackish underwing coverts, dingy white rump and vent. **Nepal House Martin** *D. nipalensis* has square tail, dark underwings, black throat and vent.

Voice: Noisy. Soft twittering chirps and a short *prrit prrit*.

Habits: Inhabits mountain valleys with cultivation, cliffs and gorges. Feeds gregariously on aerial insects, often high in the sky. Mixes with swifts and other swallows. Nests colonially against rock faces, under ledges and on outside walls of buildings.

Otto Pfister

Distribution: Local breeding summer visitor to northern mountains from north Pakistan east to Himachal Pradesh. Scarce passage migrant throughout India but perhaps overlooked because it flies so high. Some winter in Western Ghats. Also occurs in Africa, Europe and mainland Asia.

GOLDCREST (Golden-crested Wren, Kinglet)

Regulus regulus 8 cm F: Regulidae

Goren Ekstrom

Description: A tiny, plump, greenish warbler-like bird with a black-bordered orange crown. Female's crown yellow. Pale face, large dark eye with white eyering. Short, black moustache. Below off-white. Yellowish white wing bars and wing feather edges. Tiny, black bill and pinkish-grey legs. Juvenile has no markings on greyish head.

Voice: Noisy. A persistent, very thin *tsi tsi* call which is extended into a undulating song which rushes to a trilling finish.

Habits: Inhabits mountain coniferous forest. Feeds very actively in canopy on insects. Acrobatic and hovers frequently. Often in mixed hunting groups, particularly with warblers, tits and treecreepers. Nests in conifer.

Distribution: Locally common breeding resident in northern mountains from Pakistan east to Arunachal Pradesh. Winters lower down but above foothills. Also occurs in Europe, W, N, Central and E Asia.

STRIATED BULBUL (Striated Green Bulbul)

Pycnonotus striatus 20 cm F: Pycnonotidae

Sujan Chatterjee

Description: A crested, streaked, olive green bulbul. Yellow lores, chin and vent. Heavily streaked whitish on head, mantle and yellowish on underparts. Pointed crest. Sexes alike

Voice: Chatters. Calls *tee wut*, *pik pik* and *chee tu*.

Habits: Inhabits forest where it feeds mainly on fruit in canopy. Also fly-catches. Often in small parties. Strong flight like **Black Bulbul**. Nests in bush.

Distribution: Locally common breeding resident in northern mountains from central Nepal to the Myanmar border. Moves lower down in winter. Also occurs in China and SE Asia.

GREY-HEADED BULBUL

Pycnonotus priocephalus 19 cm F: Pycnonotidae

Description: A medium-sized, crestless, green and grey bulbul with a thick yellowish bill and white irises. Quite different from other bulbuls. Thick-necked, broad-tailed appearance. Often perches with wings drooped and tail raised. Lime green above, yellower below. Forecrown and forehead bright yellow. Rear crown and neck lavender grey. Black eye rings and throat patch. Grey rump with sparse short black barring. Tail grey with darker outer feathers. Sexes alike.

Voice: Calls *chaik*, a buzzing *dzee* and a high *tweep*.

Habits: Inhabits forest with dense understorey and thickets. Feeds in canopy and in dense growth on fruit, often with other bulbuls. Shy and unobtrusive.

Balachandran

Distribution: Locally common endemic breeding resident to forests of southwest India from Goa southwards.

BLACK-HEADED BULBUL

Pycnonotus atriceps 18 cm F: Pycnonotidae

Description: A medium-sized, crestless, green and black bulbul with an obvious, broad, yellow tip to its broad tail. Black head and throat. Dark olive green above with bright yellowish-green wing patches and yellow belly. Some birds greyer. Tail green with black sub-terminal bar before yellow tip. White irises, Sexes alike.

Voice: A plaintive whistle *whiwhi tyee* and metallic *chirp*.

Habits: Inhabits open forest, scrub and gardens, feeding on fruit and insects at all levels. Usually in pairs or small parties. Nests low down in thick cover.

Morten Strange

Distribution: Locally common breeding resident in Bangladesh. Status uncertain in neighbouring parts of northeast India and in Andamans but possibly similar.
Also occurs in SE Asia.

BLACK-CRESTED BULBUL (Black-headed Yellow Bulbul)

Pycnonotus (melanicterus) flaviventris 18 cm F: Pycnonotidae

Toby Sinclair

Description: A medium-sized, black, yellow and green bulbul with a long crest. Olive green above and yellow below. Tail plain olive brown and irises white. Black, crested head distinguishes from **Black-headed**. Throat black.

Voice: Song is *weet tre trippy weet*, the last three notes repeated.

Habits: Inhabits forest with thick undergrowth, scrub and orchards. Feeds on fruit and insects, usually high in canopy and often in pairs or small groups. Rather shy but inquisitive. Nests in thick cover.

Distribution: Locally common breeding resident in Himalayan foothills from Himachal Pradesh east to the Myanmar border, Bangladesh and eastern Ghats and associated hills.
Other species occur in SE Asia.

RUBY-THROATED BULBUL (Black-crested Bulbul)

Pycnonotus (melanicterus) gularis 18 cm F: Pycnonotidae

Balachandran

Description: A crestless, black, yellow and green bulbul with a red throat. Similar to **Black-crested** (with which it is considered a sub-species) in all respects of plumage except for head pattern. Closely related (and considered conspecific) **Sri Lankan Bulbul** *P. (m) melanicterus* also has no crest, but throat is yellow and irises brown.

Voice: Song is a loud, musical ascending *whe we de we*. Frequently calls a churring *prrp prrp*.

Habits: Inhabits forest and thickets, feeding usually high in canopy on fruit and insects. Sometimes descends into thick undergrowth. Usually in pairs or small groups. Secretive but fairly confiding. Nests in thick cover.

Distribution: Locally common breeding resident in southwest Indian plains and hills from Goa and northern Karnataka south through Kerala and western Tamil Nadu. The State bird of Goa.

RED-WHISKERED BULBUL

Pycnonotus jocosus 20 cm F: Pycnonotidae

Description: A slim, jaunty bulbul with prominent, black crest. Dark brown above with white tail tips and underparts. Brown side breast patches. Red vent. Black crown and nape and moustachial lines. White cheeks and red patches behind eyes. Sexes alike.

Voice: Noisy. A loud, liquid *peter who?* often preceded by *tik, whicher tee* and *pik ter welu*.

Habits: Inhabits open forest, scrub, groves, gardens and parks including in urban areas. Feeds on fruit, nectar and insects at all levels including the ground. Usually in small groups or pairs. Often indulges in slow wing raising display when meeting, accompanied by much calling. Confiding and inquisitive, perching openly. Nests in low cover.

Otto Pfister

Distribution: Common breeding resident of northern Indian plains and foothills from southern Punjab east to the Myanmar border. Also in much of eastern and western peninsula but absent from much of northern peninsula.
Also occurs in China and SE Asia.

WHITE-EARED BULBUL (White-cheeked Bulbul)

Pycnonotus leucotis 20 cm F: Pycnonotidae

Description: A black-headed bulbul with white cheeks and a yellow vent. Brown above with white-tipped tail. Pale buff below. Black head slightly domed. Chin black. Sexes alike.

Voice: Usually a loud *whichyu whichyu,* sometimes extended to more phrases. Calls a liquid *pip pip*.

Habits: Inhabits dry open woodland and scrub, often on rocky ground. Also mangroves, urban gardens and parks. Feeds usually at all levels on fruit, nectar and insects. Sociable, confiding, inquisitive and lively, often perching openly. Mixes with other bulbuls. Powerful undulating flight. Nests in low cover.

Otto Pfister

Distribution: Locally common breeding resident in plains of Pakistan and northwest India south to Gujarat and east to Delhi and western Uttar Pradesh.
Also occurs in W Asia.

HIMALAYAN BULBUL (White-cheeked Bulbul)

Pycnonotus leucogenys 20 cm F: Pycnonotidae

Otto Pfister

Description: A black-headed bulbul with long, recurved crest. Similar to closely related **White-eared** but white cheek patches have curved, irregular shape and black crescents in the corners. Crest, crown and nape brownish, not black. Black round eyes and on chin and throat. Sexes similar. Note hybrids with **Red-vented** known from Pakistan.

Voice: Noisy. Song variable, loud, musical *we wid we de up* and *wu wik wik ker*. Calls *plee plee* and *wik wik wiker*.

Habits: Inhabits open dry scrub and gardens. Often in towns and villages. Obvious and confiding. Perches openly including on wires. Usually in pairs or small groups. Feeds on insects, nectar and fruit. Nests in low cover.

Distribution: Common breeding endemic resident in northern foothills from Pakistan east to Arunachal Pradesh. Some move lower down in winter.

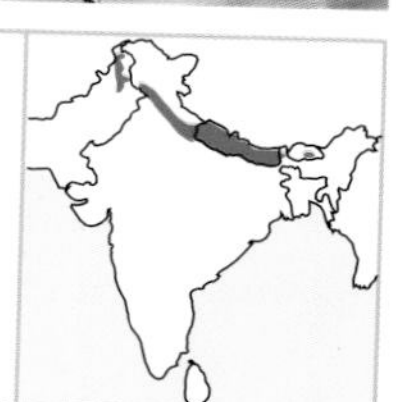

RED-VENTED BULBUL

Pycnonotus cafer 20 cm F: Pycnonotidae

Otto Pfister

Otto Pfister

Description: A dark, sooty-brown bulbul with red vent. All dark brown with paler feather edgings, giving scaly appearance to back and breast. Underparts paler but head black with peaked crown. Rump whitish. Plain brown tail has black sub-terminal band and white tips. Two races shown. Eastern race much darker. Sexes alike.

Voice: Noisy. Commonest call is loud, insistent *we we wool*. Also *peep peep* and *jit jew*.

Habits: Inhabits open wooded country including urban gardens and parks. The most familiar bulbul. Bold and inquisitive. Usually in pairs or small parties. Pairs indulge in chattering and wing raising. Feeds on insects, nectar and fruit.

Distribution: Common breeding resident throughout the plains and foothills of the region including Sri Lanka.
Also occurs in Myanmar and China.

YELLOW-THROATED BULBUL

Pycnonotus xantholaemus 20 cm F: Pycnonotidae

Description: A crestless, greenish-grey bulbul with bright yellow throat, vent and tail tip. Head and wing edgings yellowish-green and tail quite dark greenish-grey, especially towards tip. Breast grey, belly and flanks whitish. Sexes alike.

Voice: Lively *whichit woo ichit wee wee*.

Habits: Inhabits thorn scrub with scattered trees on rocky hillsides. Feeds on fruit and insects usually in pairs or small groups. Shy but sometimes perches in open. Nest not well known but those few found always close to or on ground.

R Vijaykumar Thondaman

Distribution: Extremely local endemic breeding resident restricted to a few hills in southern Andhra Pradesh, eastern Karnataka and Tamil Nadu.

YELLOW-EARED BULBUL

Pycnonotus penicillatus 20 cm F: Pycnonotidae

Description: An olive green and yellow bulbul with a complex head pattern. Black crown, eye and moustachial stripes. White tufts in front of eyes, yellow below eyes and yellow tufts on ear coverts. Throat white. Sexes alike.

Voice: A loud, mellow *wheet wit wit* and a *crr crr* call.

Habits: Inhabits lower and middle levels of wet forest and nearby gardens. Feeds mainly on fruit, for which it enters canopy. Usually in pairs or small parties but larger flocks congregate on fruiting trees. Rather shy. Nests in tree.

Tim Loseby

Distribution: A common endemic breeding resident restricted to the wet zone of Sri Lanka.

FLAVESCENT BULBUL (Blyth's Bulbul)

Pycnonotus flavescens 22 cm F: Pycnonotidae

Morten StrangeS

Description: A large, uncrested, dull olive bulbul with yellow vent. Stubby, black bill and black lores contrasting with short, white supercilia. Yellowish wing patches. Sexes alike. Similar **Olive Bulbul** *Iole virescens* is smaller, longer and paler billed, lacks yellow vent and supercilia and has a reddish-brown tail.

Voice: Variations on a lively *chi witty witty wit tu.* Harsh repeated alarm *drr dreet.*

Habits: Inhabits thick cover of forests and scrub. Feeds on fruit and insects at all levels. Rather shy and difficult to see. Nests low down in bush.

Distribution: Rarely recorded presumed breeding resident in north-eastern India and perhaps adjoining parts of Bangladesh. Also occurs in SE Asia.

WHITE-BROWED BULBUL

Pycnonotus luteolus 20 cm F: Pycnonotidae

Clement M Francis

Description: A rather plain bulbul with white eyebrows and eye rings. Dark eye stripes and moustache but otherwise a dull bird. Plain olive brown above and dusky white below with yellow vent and yellowish on chin sides. Sexes alike.

Voice: Noisy. Short explosive bubbling song. A harsh *churr.*

Habits: Inhabits thick scrub, gardens and forest edge. Shy, keeping to cover in pairs or small groups. In early morning sometimes sings from higher exposed perch. Feeds on invertebrates, nectar and fruit. Nests in bush.

Distribution: Locally common endemic breeding resident in peninsular India south from Gujarat, Madhya Pradesh and Orissa. Also Sri Lanka.

YELLOW-BROWED BULBUL

Iole indica 20 cm F: Pycnonotidae

Description: A striking yellow and green bulbul with black bill and eyes. Olive green above including crown and buttercup yellow below including face. Sexes alike. Juvenile duller with rufous in wings.

Voice: Noisy. A fluty *whit wee* which is extended into song. Also harsh *churrs*.

Habits: Inhabits forest and secondary cover. Feeds on berries and invertebrates, often in mixed hunting groups, and usually at middle or lower levels. Lively and not shy. Nests in bush.

Clement M Francis

Distribution: Common endemic breeding resident in Western Ghats and associated hills south from Maharastra.

ASHY BULBUL (Brown-eared Bulbul)

Hemixos flavala 20 cm F: Pycnonotidae

Description: A crested, grey bulbul with large, lime green wing patches. Darker face with brown cheeks and white throat. Outer wings and tail dark grey. Sexes alike.

Voice: Noisy. A liquid, descending *tew de de do it* and variants.

Habits: Inhabits forest and plantations. Feeds on fruit, nectar and insects in canopy and middle levels, often in flocks. Fly-catches. Nests low down.

Morten Strange

Distribution: Common breeding resident in northern foothills from Uttaranchal east to the Myanmar border. Also Bangladesh. Winters lower down, sometimes in plains. Also occurs in China and SE Asia.

MOUNTAIN BULBUL (Rufous-bellied Bulbul)

Hypsipetes mcclellandii 23 cm F: Pycnonotidae

Morten Strange

Description: A large, olive and rufous bulbul with a shaggy brown crown. Cheeks and breast pale rufous-streaked with buff. Throat streaked white and often puffed out. Belly and vent buff. Upperparts olive green. Long powerful bill. Sexes alike.

Voice: Noisy with a variety of calls. Most commonly a metallic *tsyi tsyi* and *cheep har lee*.

Habits: Inhabits forest and secondary growth with scattered trees, feeding mainly on fruit, mostly in canopy. Usually in pairs or small groups. Nests high in tree.

Distribution: Fairly common breeding resident in northern mountains from Uttaranchal east to the Myanmar border and perhaps Bangladesh. Moves lower down to foothills and even adjacent plains in winter. Also occurs in SE Asia.

BLACK BULBUL (Grey Bulbul)

Hypsipetes leucocephalus 23 cm F: Pycnonotidae

Otto Pfister

Description: Large, crested, blackish bulbul with red bill and legs. Colour ranges from slaty grey with black crown in north to black in Sri Lanka and Western Ghats. Whitish vent. Bill long and slightly decurved. Tail notched in square end. Sexes alike.

Voice: Very noisy. Wide variety of loud whistles and screeches, including mewing *weenh* and repeated *we weu*, *wer kiyu*.

Habits: Arboreal, inhabiting canopies of forest, shade and garden trees. Usually in noisy flocks. Feeds on fruit, nectar and insects, moving restlessly from tree to tree with direct flight. Often perches on, and fly-catches from, high exposed branches. Bold but sometimes difficult to approach. Nests in high tree.

Distribution: Common breeding resident in northern mountains from Pakistan to the Myanmar border, Western Ghats and associated hills and Sri Lanka. Winters lower down in foothills and adjacent plains. Also occurs in Indian Ocean Islands, China and SE Asia.

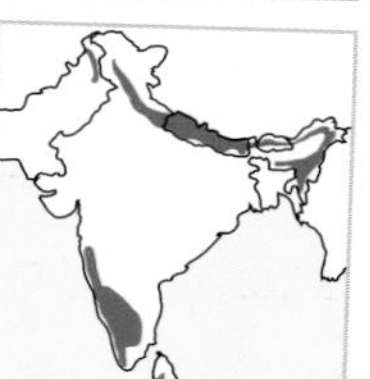

ZITTING CISTICOLA (Streaked Fantail Warbler)

Cisticola juncidis 10 cm F: Cisticolidae

Description: Tiny, black-streaked buff warbler with a white-tipped tail. Tail very short in breeding plumage. Head and neck streaked. Underparts and face plain buff with long whitish supercilia. Pale legs. Sexes alike.

Voice: Noisy. Diagnostic song is insect-like, oft-repeated *zit zit zit*, usually delivered in bouncing circular song flight high in sky, when can be difficult to spot.

Habits: Inhabits all types of damp grassland, paddy and other irrigated cereals and reed-beds. Usually in pairs or family parties. Feeds on insects, creeping through grass like a mouse. Makes short, low, jerky flights on whirring wings with spread tail. Sometimes perches high on grass stem or overhead wire.

Otto Pfister

Distribution: Common breeding resident throughout plains and hills but rare in parts of Pakistan southeast and parts of northeast India.
Also occurs in Europe, Africa, E and SE Asia and Australasia.

BRIGHT-HEADED CISTICOLA (Bright-capped Cisticola)

Cisticola exilis 10 cm F: Cisticolidae

Description: Very similar to **Zitting**, particularly in non-breeding plumage, but always has unstreaked, chestnut hind neck, supercilia and rump and indistinct brownish tips to tail. In breeding plumage, male has unstreaked, creamy crown in the Northeast, chestnut in Western Ghats. Sexes alike.

Voice: Sings from perch or in rapid display flight, *bzzeee joo ee*. Calls *nyaae*.

Habits: Inhabits tall, dry grassland and scrub, usually on hillsides. Drier habitat than most **Zitting**. Behaviour and food as **Zitting** but display flight less bouncy and ends with a steep dive. Low flight more direct. Nests in grass as does **Zitting**.

Morten Strange

Distribution: Scarce and local breeding resident in northern terai from Uttaranchal east to Arunachal Pradesh, northeastern plains and hills including parts of Bangladesh and some south-western hills.
Also occurs in China, SE Asia and Australasia.

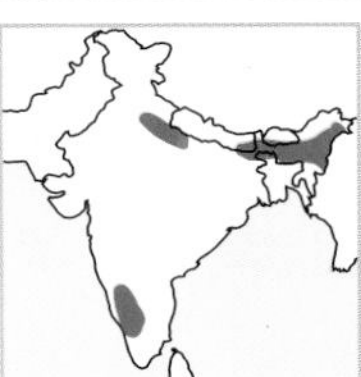

STRIATED PRINIA (Brown Hill Warbler/Prinia)

Prinia criniger 16 cm F: Cisticolidae

Otto Pfister

Description: A fairly large, hill warbler with a very long, graduated tail. In non-breeding plumage, rufous-brown with light streaking on head, breast and upperparts. Greyer brown in breeding plumage and streaking more obvious. Indistinct dark tips to tail, dark strong bill and flesh legs. Sexes alike. Similar **Rufous-vented Prinia** *P. burnesii* paler, usually has rufous vent and is very local in northern wetlands.

Voice: A grating *chitzweet, chitzweet* and a hard *chak*.

Habits: Inhabits grassy and scrub-covered hillsides and terraced cultivation. Usually solitary or in pairs, feeding low down on insects and quite skulking. Males sing openly from bush tops and wires. Nests low down in cover.

Distribution: Common breeding resident in northern hills from Pakistan east to the Myanmar border. Moves lower down to foothills in winter. Also occurs in Myanmar and China.

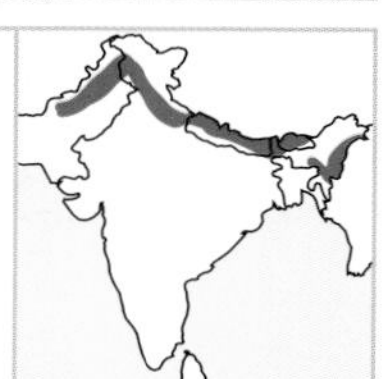

HILL PRINIA (Black-throated, White-browed)

Prinia atrogularis 17 cm F: Cisticolidae

Morten Strange

Description: Large, long-tailed brown warbler. In breeding plumage, western race greyer with black from chin to breast, white-spotted on lower breast. Eastern race (shown) browner, rufous on crown and with black restricted to chin and throat and white-speckled breast band. In non-breeding plumage in both races most of black disappears. Obvious white supercilia and lack of streaking distinguish from **Striated**. Sexes alike.

Voice: Calls a rapid, soft *tptptptptp* and *pri pri pri pri*. Song repeated *tulip tulip*.

Habits: Inhabits scrub and grass on hillsides, forest edges. Feeds low down, singly or in pairs on invertebrates. Skulking but will sing from exposed perches.

Distribution: Locally common breeding resident in northern foothills from east Nepal to the Myanmar border. Also occurs in Tibet, China and SE Asia.

GREY-BREASTED PRINIA (Franklin's Wren-Warbler)

Prinia hodgsonii 11 cm F: Cisticolidae

Description: A small warbler with bright pink legs, fine, dark bill and red irises. The smallest prinia with proportionally the shortest tail. Grey above with rufous patch in wings. Graduated tail has black and white feather tips. In breeding season has greyish breast band which weakens to a greyish wash when not breeding. Sexes similar.

Voice: Noisy. Commonest call is a loud *zee zee zee*.

Habits: Inhabits the scrubby undergrowth of open forest and woodland, forest edges and thick scrub. Often in canopy and therefore higher than most prinias. Feeds on insects, usually in small active groups. Inquisitive and confiding. Nests in low cover.

Nikhil Devasar

Distribution: Common breeding resident throughout most of lowlands but very local in most of Pakistan, Bangladesh and parts of southeast India.
Also occurs in China and SE Asia.

GRACEFUL PRINIA (Streaked Wren-Warbler)

Prinia gracilis 13 cm F: Cisticolidae

Description: A small, pale, long-tailed warbler with lightly streaked crown and mantle. Legs pink, fine bill dark and irises orange. Pale sandy-brown above. Face and underparts, unstreaked pale buff. No obvious supercilia. Graduated tail has black and white tips. Sexes alike.

Voice: Calls a nasal *szeep* and has a burring *trrit trrit trrit* song. Snaps wings in display to make a rapid clicking sound.

Habits: Inhabits long grass and scrub, particularly tamarisks and usually near water. Feeds actively on insects, often on ground. Confiding, frequently perching in open. Usually in pairs or small parties. Nests low down in cover.

Otto Pfister

Distribution: Locally common breeding resident in Pakistan and the Northwest Indian lowlands east to Nepal and parts of the Northeast.
Also occurs in N Africa and W Asia.

JUNGLE PRINIA (Jungle Wren-Warbler, Large Prinia)

Prinia sylvatica 15 cm F: Cisticolidae

Otto Pfister

Description: A fairly large, long-tailed warbler with short supercilia. Bill dark, long and stout. Irises red. In breeding plumage dark grey-brown above and off-white below. Graduated tail has prominent white tips and whitish outer feathers. More rufous-brown in non-breeding plumage. Sexes similar. Uniform wings, larger bill, short supercilia and voice distinguish from similar but smaller **Plain**.

Voice: Loud, rhythmic song *zee tu zee tu*.

Habits: Inhabits dry scrub and grass, often in rocky areas. In pairs or small parties. Feeds on insects low down but sings from high perch. Secretive but will appear in open briefly. Nests low down.

Distribution: Locally common endemic breeding resident in lowlands and hills throughout most of India and Sri Lanka. Rare in northwest India. Restricted to extreme west Nepal and absent from Pakistan and probably Bangladesh.

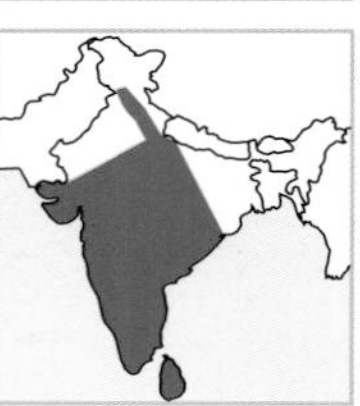

YELLOW-BELLIED PRINIA (Yellow-bellied Wren-Warbler)

Prinia flaviventris 13 cm F: Cisticolidae

Otto Pfister

Description: A small, long-tailed warbler with lemon yellow belly and vent. Crown grey and upperparts olive grey. Usually has distinct white supercilia. White chin to breast. Long graduated tail has no distinctive markings and is often held well-cocked over back. Sexes alike.

Voice: Song a musical *twee dulu lu lee,* often from high reed. Calls *tzee* and *chink chink*. Snaps wings.

Habits: Inhabits reed beds, tall grass and tamarisk scrub near water. Active, inquisitive and confiding, often appearing high on a reed stem to look around and call. Feeds on insects low in cover usually singly or in pairs. Nests low down.

Distribution: Locally common breeding resident in northern river valleys in Indo-Gangetic Plain and the Northeast including Bangladesh coast. Also occurs in China and SE Asia.

ASHY PRINIA (Ashy Wren-Warbler)

Prinia socialis 13 cm F: Cisticolidae

Description: Small, dark, long-tailed warbler with deep orange belly. In breeding plumage, head and mantle dark grey, wings and tail rufous. Latter graduated with black and white tips. Browner above when not breeding and with short, white supercilia. Irises usually deep red. Sexes similar.

Voice: Noisy. Distinctive *jimmy jimmy jimmy* song and nasal *pee pee pee* call. Snaps wings.

Habits: Most catholic of prinias, inhabiting all types of scrub and grassland, including reed beds, gardens and undergrowth in open forest. Feeds singly or in pairs low down on insects. Often on ground. Active, inquisitive and confiding, often raising and waving tail and flicking wings. Jerky flight. Nests low down.

Otto Pfister

Distribution: Common breeding near-endemic resident in lowlands and hills throughout most of India (except parts of the Northwest, Northeast and east), Sri Lanka and Nepal. Local in Pakistan and Bangladesh.
Also occurs in Myanmar.

PLAIN PRINIA (Plain, White-browed Wren-Warbler)

Prinia inornata 13 cm F: Cisticolidae

Description: Small, long-tailed warbler with prominent white supercilia. Pale brown above with rufous wing edgings. More rufous in non-breeding plumage. Whitish below with buff flanks. Graduated tail has whitish outer feathers and buff tips. Sexes alike. Similar **Rufous-fronted** *P. buchanani* has rufous crown and broad white tips to tail and is restricted to thorn-scrub.

Voice: Noisy. Plaintive *tee tee tee* and *krrk krrk*. Song is wheezy, insect-like *tlik tlik tlik*.

Habits: Inhabits all types of damp grassland, crops and scrub. Singly or in pairs, feeding secretively on insects low down. Active and inquisitive. Low, jerky flight. Rotates tail, making it appear barely affixed. Nests low down.

Sumit Sen

Distribution: Common breeding near-endemic resident in plains and lower hills throughout most of region.
Also occurs in Myanmar.

ORIENTAL WHITE-EYE (White-eye, Small White-eye)

Zosterops palpebrosus 10 cm F: Zosteropidae

Otto Pfister

Description: A small, plump, greenish-yellow bird with broad, white rings round black eyes. Greenish-yellow above, more yellow on throat and upper breast and rump. Greyish-white on rest of underparts. Black lores, bill and legs. Sexes alike. Endemic **Sri Lankan White-eye** *Z. ceylonensis* is larger, darker above with a longer bill and broken eye rings.

Voice: Noisy. Mournful *tseer* contact notes. Tinkling song.

Habits: Inhabits wooded areas including forests, parks and gardens. Often in quite large parties; sometimes in mixed hunting groups. Lively, moving restlessly through foliage at all levels feeding on insects, nectar and fruit. Confiding. Nests in trees.

Distribution: Common breeding resident throughout in plains and hills. More local in Pakistan.
Also occurs in China and SE Asia.

CHESTNUT-HEADED TESIA (Ground Warbler)

Tesia castaneocoronata 8 cm F: Sylviidae

Otto Pfister

Description: A tiny, apparently tailless, ground-dwelling warbler with a bright chestnut head. Olive green above and yellow below from chin to vent with greyish flanks. Chestnut crown and cheeks contrast with white spots behind black eyes. Legs pink. Sexes alike.

Voice: A surprisingly loud *tziet* or *tsit* and variations. Song *tsee tsu wit*.

Habits: Inhabits the ground and low growth, beneath thick forest undergrowth particularly ferns and bamboos near streams or in ravines. Feeds actively, usually singly or in pairs, on invertebrates hopping boldly about and bobbing when alarmed. Secretive but inquisitive. Rarely flies. Nests low down in cover.

Distribution: Locally common breeding resident in northern mountains from Himachal Pradesh east to Arunachal Pradesh and in parts of the northeastern hills. Moves lower down into foothills in winter.
Also occurs in China and SE Asia.

BROWNISH-FLANKED BUSH WARBLER

Cettia fortipes 11 cm F: Sylviidae

Description: A small, dark warbler with obvious buff supercilia. Dark reddish-brown above and greyish-white below with brown flanks and legs. Sexes alike. **Pale-footed** *C. pallidipes* is smaller, brighter above and whiter below and with pale legs. **Aberrant Bush Warbler** *C. flavolivacea* also has pale legs but is dark olive above and yellow below with distinct yellow supercilia.

Voice: A loud *chuk* and *wheeeeel cheet u*.

Habits: Inhabits thick undergrowth, under or on the edge of forest and plantation trees. Feeds unobtrusively and usually singly, on invertebrates close to the ground. Normally located by loud calls. Nests low down in thick cover.

Goren Ekstrom

Distribution: Locally common breeding resident in northern mountains from Pakistan east to Arunachal Pradesh and the northeastern hills. Winters lower down in foothills.
Also occurs in Myanmar and China.

GREY-SIDED BUSH WARBLER (Rufous-capped)

Cettia brunnifrons 11 cm F: Sylviidae

Description: A small, brown warbler with a chestnut crown. Rich russet-brown above, brighter still on crown. White supercilia and chin to belly. Extensive grey on flanks and brown vent. Legs brown. Sexes alike. **Chestnut-crowned Bush Warbler** *C. major* is much larger and paler with a duller olive brown back, showing contrast with chestnut crown.

Voice: A thin *tzik,* extended to *wichichichit whooeeu whooweeou* in song.

Habits: Inhabits grassy scrub. Feeds low down on invertebrates, usually singly. Can be confiding but usually secretive, except when singing. Nests low down in thick cover.

Goren Ekstrom

Distribution: Locally common breeding resident in northern mountains from Pakistan east to Arunachal Pradesh. Winters lower down in foothills.
Also occurs in Myanmar.

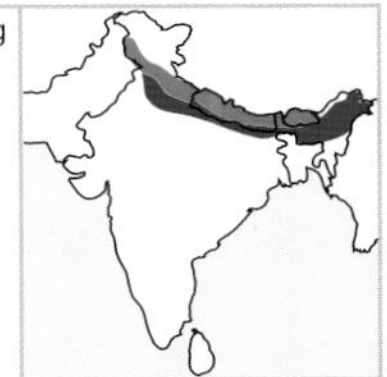

LANCEOLATED WARBLER (Streaked Grasshopper)

Locustella lanceolata 12 cm F: Sylviidae

Tim Loseby

Description: A small, heavily streaked warbler. Brown above with heavy blackish streaking. Whitish below with warm buff flanks and vent. Distinct and extensive black streaking on breast and flanks. Vent may appear unstreaked. Blackish tertials, clearly edged buff. This, the heavier streaking, shorter tail and bull neck are best distinctions from similar **Grasshopper**. Immature duller, less streaked.

Voice: Hard *clik* and *chrrr*.

Habits: Inhabits dense grass, stubble and thick bush. Feeds low down on invertebrates. Extremely secretive, scuttling rapidly along ground like a small rodent. Able to conceal itself in very short grass. Rarely flies, so very difficult to observe well.

Distribution: Rare winter visitor mainly to lowlands of Pakistan and northern India. Much overlooked so status uncertain.
Also occurs in N, E and SE Asia.

GRASSHOPPER WARBLER

Locustella naevia 13 cm F: Sylviidae

Gertrud Denzau

Description: A very secretive, streaked warbler with long streaked, undertail coverts. Olive brown streaked darker above with a rather plain head. Below usually unstreaked whitish or yellowish. Long graduated and rounded tail. Pink legs. Sexes alike. **Rusty-rumped** *L. certhiola* has dark crown, marked creamy supercilia and a very rusty rump.

Voice: A sharp *pstt*. A churring rhythmic insect-like song, unlikely to be heard in our region.

Habits: Inhabits damp grassland, reedbeds and paddy. Creeps low down feeding on insects. Difficult to observe. Crepuscular.

Distribution: Widespread and much overlooked winter visitor throughout. Most regularly seen in the southwest.
Also occurs in Europe, Africa, W, N and Central Asia.

MOUSTACHED WARBLER (Moustached Sedge Warbler)

Acrocephalus melanopogon 13 cm F: Sylviidae

Description: Small, dark, short-winged warbler with bold, square-ended, creamy supercilia. Black-bordered streaked crown. Above streaked olive brown with unstreaked rusty rump. Cheeks greyish, throat white and below warm buff. Sexes alike. Often cocks and flicks tail. **Black-browed Reed Warbler** *A. bistrigiceps* has similar head pattern but paler and unstreaked.

Voice: A deep *truk* and a short *trik* which may be extended into a rattling trill.

Habits: Inhabits reedbeds, particularly typha. Also in water-side bushes. Feeds on invertebrates low down in reed litter and on floating vegetation. Secretive and unobtrusive but will climb to top of reeds to call.

Balachandran

Distribution: Mainly a scarce but probably overlooked winter visitor to Pakistan and northwest India south to Madhya Pradesh and east to Uttar Pradesh. Commonest in Pakistan where may still breed in Baluchistan. Also occurs in southern Europe, N Africa, W and Central Asia.

PADDYFIELD WARBLER

Acrocephalus agricola 13 cm F: Sylviidae

Description: A long-tailed, brown, wetland warbler with marked supercilia. Rufous-brown above, particularly on rump, with short wings and long, rounded tail. Buffish below. Whitish supercilia extend well beyond eyes and bordered above with black. Bill dark above and pale below with dark tip. Sexes alike. Similar northeastern **Blunt-winged** *A. concinens* is even shorter-winged and has unbordered, indistinct supercilia.

Voice: Calls *chek* and *cherr*.

Habits: Inhabits reedbeds, paddy, sugar-cane and mangroves, rarely in adjoining scrub. Feeds low down on invertebrates, usually singly, although several may be in vocal contact. Not easy to observe but can be confiding.

Nikhil Devasar

Distribution: Locally common but overlooked winter visitor to plains of southern Pakistan (has bred in Baluchistan), northern, western and northeastern India and southern Nepal. Common elsewhere as a passage migrant.
Also occurs in Iran and Central Asia.

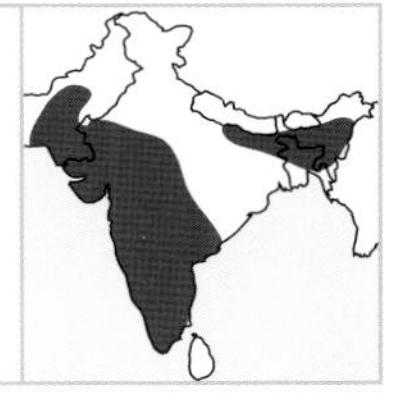

BLYTH'S REED WARBLER

Acrocephalus dumetorum 13 cm F: Sylviidae

Balachandran

Description: A plain warbler with low, sloping forehead. Plain greyish, olive brown above with no contrast. Long-looking, rounded tail (often half cocked) and long undertail coverts. Short, bulging, white supercilia to eyes and thin, white eye rings. Whitish below with greyish flanks. Grey legs. Long-looking bill, dark above and pale below. Sexes alike. Juvenile slightly rustier above.

Voice: Distinctive, frequently uttered *thik* or *zuk* contact call. Also harsher *tersh* alarm.

Habits: Inhabits well-wooded, often quite dry, country feeding on invertebrates at all levels, including canopies, bushes and ground. Rarely in reedbeds. Flicks wings and fans tail.Usually singly or in loose groups.

Distribution: Common winter visitor to the plains of southern peninsular India, Bangladesh and the Northeast. Very common passage migrant in Pakistan and northern India.
Also occurs in Europe, W, N and Central Asia.

CLAMOROUS REED WARBLER (Indian Great Reed)

Acrocephalus stentoreus 19 cm F: Sylviidae

Otto Pfister

Description: A very large, wetland warbler with long, thin bill. Brown above and buffish below with white throat and short, white supercilia. Long tail with rounded tip. Short wings. Similar, local **Great Reed** *A. arundinaceus* and **Oriental Reed** *A. orientalis* have thicker bills and prominent supercilia. Former long-winged, latter throat streaking. Sexes alike.

Voice: Noisy. Loud *chak* and *kurrr* calls. Deep, staccato song *tek tek karra karra kareet kareet kareet* and variants.

Habits: Inhabits reedbeds and adjoining scrub, and mangroves. Sometimes in quite dry areas. Feeds ponderously on insects usually low down but often in bushes and waterside trees. Not shy. Nests in reeds.

Distribution: Locally common breeding resident in parts of Pakistan, N and S India and Sri Lanka. More widespread winter visitor.
Also occurs in W and SE Asia and Australasia.

THICK-BILLED WARBLER

Acrocephalus aedon 19 cm F: Sylviidae

Balachandran

Description: A very large, plain warbler with obvious, beady, black eyes. Short, thick bill with distinct curve to upper mandible and pale lower mandible. Plain face with no supercilia but pale lores. Greyish-brown above but more rusty on rump. Tail long and very rounded. Wings very short. Sexes alike.

Voice: Call a loud *tschack* and *chack*.

Habits: Inhabits all types of low scrub and grass often, but not exclusively, in wetlands. Feeds on invertebrates low down in vegetation, hopping through in a crouched, rather babbler-like fashion with tail held low. Secretive and usually singly.

Distribution: Scarce and local, but probably overlooked, winter visitor mainly to plains of south Nepal, northeast and south India and Bangladesh.
Also occurs in N, Central, and SE Asia.

BOOTED WARBLER (Booted Tree Warbler)

Hippolais caligata 12 cm F: Sylviidae

Tim Loseby

Description: Small, plain warbler with dark-edged supercilia. Warm brown above with white-edged, square-ended tail and short undertail coverts. White supercilia usually have dusky borders. Buff underparts and white throat. Bill dark above, pale below with tiny dark tip. Legs pinkish-brown, feet often dark grey. Resembles **Chiffchaff** and closely related **Sykes's Warbler**. **Paddyfield** has longer, rounded tail and undertail coverts, shorter wings, more obvious supercilia and is usually rustier brown.

Voice: A clicked *zek* call.

Habits: Inhabits low scrub, tamarisk, thorn trees and crops. Actively feeds on invertebrates, often in loose parties and at all levels.

Distribution: Winters in Pakistan, western, southern and central India and Sri Lanka. Widespread on passage elsewhere. Lack of differentiation from Sykes's in past means that the true winter range not certain.
Also occurs in Central and N Asia.

SYKES'S WARBLER (Booted Warbler)

Hippolais rama 12 cm F: Sylviidae

Tim Loseby

Description: Small, plain warbler with whitish outer tail feathers to square-ended tail. Very similar to, and recently separated from, **Booted**. Differs in having shorter wings and longer tail, flatter crown, colder grey above, whiter below and and shows less dark crown edges. Rarely shows "booted" effect of darker toes. Similar to **Chiffchaff** but broader-based bill has pinkish-yellow lower mandible and legs pale pinkish-grey. Sexes similar.

Voice: Clicking *chuk* call, softer than **Booted**.

Habits: Inhabits dry scrub. Breeds in waterside tamarisks and reeds. Feeds very actively on invertebrates and nectar, often in loose parties and at all levels. Flicks tail upwards.

Distribution: Common breeding summer visitor to Baluchistan and Sind in Pakistan and perhaps the Punjab in India. Winter visitor and passage migrant to lowlands of Pakistan, India and Sri Lanka, most common in peninsular India.
Also occurs in Central Asia and Iran.

COMMON TAILORBIRD (Tailorbird)

Orthotomus sutorius 13 cm F: Sylviidae

Otto Pfister

Description: A small, long-tailed, greenish warbler with rusty fore-crown. Bright green above, whitish below. Long, thin, pale bill and pink legs. Tail, longest in breeding male, usually cocked and waved. Dark feather bases on throat sides when calling may confuse with **Dark-necked**.

Voice: Very noisy. Main call loud, abrupt *wit wit wit wit*. Song repeated *pitchit pitchit*.

Habits: Inhabits all types of scrub including forest edges, gardens and mangroves. Usually singly or in pairs feeding unobtrusively, mainly low down on invertebrates and nectar. Confiding and inquisitive. Flies weakly with tail raised. Stitches two leaves into famous pouched nest, low in shrub or pot-plant.

Distribution: Common breeding resident throughout most of plains and hills of region.
Also occurs in China and SE Asia.

DARK-NECKED TAILORBIRD (Black-necked Tailorbird)

Orthotomus atrogularis 13 cm F: Sylviidae

Description: A small, long-tailed, greenish warbler with a large, blackish throat patch. Very similar to **Common** but more extensive chestnut crown and lores and yellow vent and bend of wings. Male has black throat patch and longer tail. Beware **Common** revealing dark feather bases when, and soon after, singing but these are usually restricted to neck sides.

Voice: A loud, nasal *krrri krrri krrri* and *tew*.

Habits: Inhabits dense scrub, forest undergrowth and forest edges. Behaves and nests as **Common** but much less confiding. Call attracts attention.

Morten Strange

Distribution: Scarce and local breeding resident in plains and lower hills of northeastern India and Bangladesh.
Also occurs in China and SE Asia.

WHITE-BROWED TIT WARBLER (Stoliczka's Tit Warbler)

Leptopoecile sophiae 10 cm F: Sylviidae

Description: A tiny, dark warbler with obvious, whitish supercilia. Rufous crown and grey-brown mantle and wings. Rump and flanks bluish. Dark tail with white outer tail feathers. Underparts dark, variably washed purple-pink. Female paler overall. Eastern race darker still in both sexes.

Voice: A soft *teet* and a hard *tzrit*. Loud chirping song.

Habits: Inhabits montane scrub and bushes by streams. Feeds on invertebrates in small parties or mixed hunting groups. Very active and acrobatic but rather shy. Cocks tail. Nests low down in bushes.

Tim Loseby

Distribution: Locally common breeding resident in high mountains from northern Pakistan east to western Nepal. Moves lower down in winter.
Also occurs in Central Asia and China.

COMMON CHIFFCHAFF (Brown Leaf Warbler)

Phylloscopus collybita 11 cm F: Sylviidae

Otto Pfister

Otto Pfister

Description: Small, plain warbler with black bill and legs. Rather dull greyish-brown above, often with olive wash. Sometimes shows very slight wing bars and, more often, yellow at shoulder. Dark eye-stripes and whitish supercilia. Pale greyish-white below with no yellow. Sexes alike.

Voice: Rather quiet. Occasionally a plaintive inflected *wheep*, *zit* or *peeu*.

Habits: Inhabits all types of wooded country, particularly near water. Also reedbeds and crops. Feeds actively at all levels but usually low down, even on ground. Usually in small parties. Fly-catches, often over water, and hovers to pick from leaves. Flicks wings and tail frequently.

Distribution: Common winter visitor throughout northern plains including Pakistan, Nepal, Bangladesh and India north of Karnataka and Orissa.
Also occurs in Europe, N Africa and W, N and Central Asia.

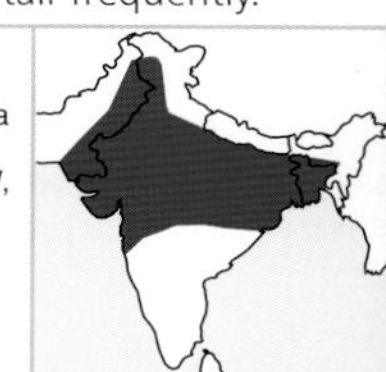

MOUNTAIN CHIFFCHAFF

Phylloscopus sindianus 11 cm F: Sylviidae

Otto Pfister

Description: A small, plain warbler, very similar to **Common Chiffchaff**. Usually slightly darker brown above with darker flanks. Never shows any olive above or yellow on shoulder but often barely distinguishable except by call. Sexes alike. Similar, but smaller, northwestern **Plain Warbler** *P. neglectus* is browner with shorter tail and wings and tiny bill.

Voice: Noisier than **Common** with a distinctive, plaintive, disyllabic *tiss yip* or *swe eet* call.

Habits: Inhabits montane scrub and open woodland to breed, often near water. Usually winters close to water, often feeding low down close to surface. Behaviour as **Common**. Nests close to ground.

Distribution: Locally common breeding summer visitor to northern mountains from northern Pakistan east to Himachal Pradesh. Winters mainly in foothills and also plains of Pakistan and extreme northern India. Records as far south as Delhi and probably much overlooked.
Also occurs in Central Asia.

SMOKY WARBLER (Smoky Willow, Leaf Warbler)

Phylloscopus fuliginventer 10 cm F: Sylviidae

RK Gaur

Description: A tiny, very dark warbler with, usually, yellowish supercilia. Dark brown above, dusky yellowish-buff below. Dark eyestripes and short yellowish or whitish (in eastern race) supercilia. Dark legs. Pale base to lower mandible. Sexes alike. Similar **Dusky Warbler** *P. fuscus* is paler brown above and whiter below, with prominent supercilia and eye stripes and pale legs.

Voice: A low *tsrrk* and *chup*.

Habits: Inhabits montane scrub and boulder fields in summer Feeds on ground. Winters in low waterside herbage. In winter usually feeds, on invertebrates, close to water in half-submerged tangles. Sometimes fly-catches over water. Secretive and rather furtive. Nesting habits not known.

Distribution: Scarce breeding summer visitor to mountains of Nepal, Bhutan and Sikkim. Winters sparsely in plains as far west as eastern Rajasthan and Delhi and in the northeast.
Also occurs in Tibet and China.

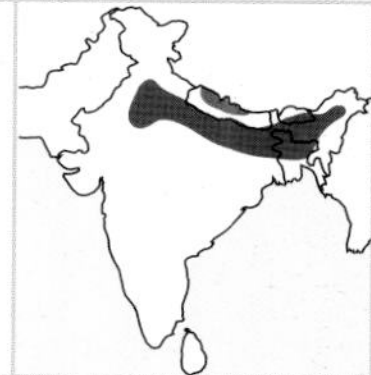

TICKELL'S LEAF WARBLER (Tickell's Warbler)

Phylloscopus affinis 11 cm F: Sylviidae

Otto Pfister

Description: A small, green warbler with very prominent, yellow supercilia. Rather dark olive green above with no wing bars. Deep yellow on throat and breast as supercilia. Paler on belly. Prominent dark eye stripes. Pale legs and lower mandible. Sexes alike. Beware **Green Warbler** *P. trochiloides nitidus* lacking wing bars but legs always dark.

Voice: A hard *chip* or *tsip*.

Habits: Summers in montane scrub. Winters in forest edge, scrub and gardens. Feeds on invertebrates, usually quite close to the ground, flitting though the herbage singly or in small parties. Confiding and conspicuous. Nests close to ground.

Distribution: Common breeding summer visitor to northern mountains from Pakistan east to Bhutan. Winters mainly in Bangladesh and northeastern and southwestern Indian hills but more widespread on passage and could be found anywhere in peninsular India in winter.
Also occurs in China and SE Asia.

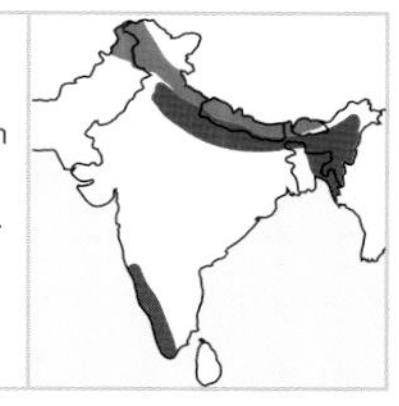

SULPHUR-BELLIED WARBLER (Olivaceous Leaf Warbler)

Phylloscopus griseolus 11 cm F: Sylviidae

Jan Willem den Besten

Description: A small, rather dark warbler with prominent bright yellow supercilia. Dark brownish-grey above with no wing bars. Dull oily-yellow below. Bright supercilia verging on orange in front of eyes. Pale legs. Sexes alike. Not as dark above as **Smoky** or **Dusky Warblers**.

Voice: A liquid *dip* or *pik*.

Habits: Inhabits rocky areas with scrub and scattered trees. Also old ruins and open woodland in winter. Characteristically feeds, on invertebrates, on or close to ground, creeping over rocks, tree trunks and even walls like a nuthatch. Usually singly or in pairs. Confiding but unobtrusive. Nests in low vegetation.

Distribution: Locally common breeding summer visitor from north Pakistan east to Himachal Pradesh. Winters sparsely in north and central Indian plains as far south as Andhra Pradesh and rarely even further south.
Also occurs in Central Asia.

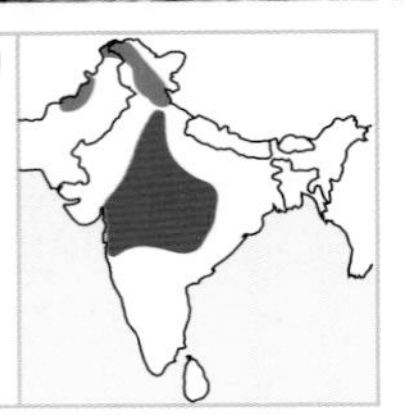

BUFF-BARRED WARBLER (Orange-barred Leaf Warbler)

Phylloscopus pulcher 10 cm F: Sylviidae

Jan Willem den Besten

Description: A small, brightly coloured warbler with orange wing bars. Dark green above with yellow rump and supercilia. Indistinct grey crown stripe. At least one prominent, orange wing bar. Obvious white in outer tail making tail appear white from below. Yellowish underparts, pale legs and bill base. Sexes alike.

Voice: A high *swit*.

Habits: Inhabits high altitude shrubbery and forest in summer moving to broadleafed forests in winter. Feeds actively on invertebrates in canopy, often in mixed hunting groups. Frequently flicks tail open. Rarely hovers. Nests low down.

Distribution: Common breeding resident in northern hills from Himachal Pradesh east to the Burmese border. Moves lower down in winter.
Also occurs in S China and SE Asia.

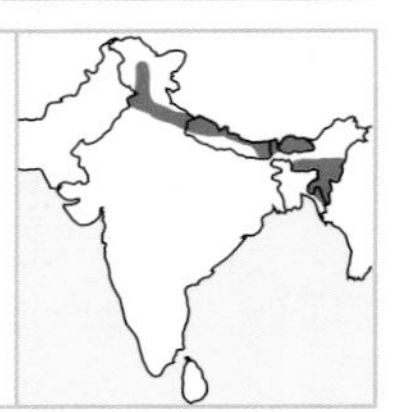

ASHY-THROATED WARBLER (Grey-faced Leaf Warbler)

Phylloscopus maculipennis 9 cm F: Sylviidae

Description: A tiny, bright yellow and green warbler with distinctive grey face and throat. Olive green above with yellow rump and double wing bars. White outer tail feathers obvious when hovering. Dark crown sides and eyestripes and long, greyish supercilia and crown stripe. Similar to **Lemon-rumped** and **Buff-barred** but they lack the grey. Sexes alike.

Voice: A sharp *zit*, often repeated. Song a rhythmic *sweechu sweechu*.

Habits: Inhabits forest and secondary growth. Feeds very actively on invertebrates, singly or in mixed hunting groups at all levels. Often hovers. Nests high in tree.

Sujan Chatterjee

Distribution: Locally common breeding resident in northern mountains from Himachal to Arunachal Pradesh. Also perhaps in northeast hills. Winters lower down.
Also occurs in China and SE Asia.

LEMON-RUMPED WARBLER (Pallas's Leaf, Pale-rumped)

Phylloscopus chloronotus 9 cm F: Sylviidae

Description: A tiny, green and yellow warbler with a distinct, pale yellow rump. Dumpy. Olive green above, whitish below sometimes with pale yellow wash. Distinct yellowish supercilia and central crown stripe, edged dark. Long, recurved, dark eyestripes. Dusky cheeks. Two yellow wing bars. No white in tail. Pale legs. Dark bill. Sexes alike.

Voice: A high *tsip* call. Long trilling songs.

Habits: Inhabits hill forests. Feeds on invertebrates very actively in canopy and sometimes in undergrowth. Frequently hovers, revealing rump. Usually in mixed hunting groups. Can be confiding. Nests in tree. Recently separated from eastern **Pallas's Warbler** *P. proregulus*.

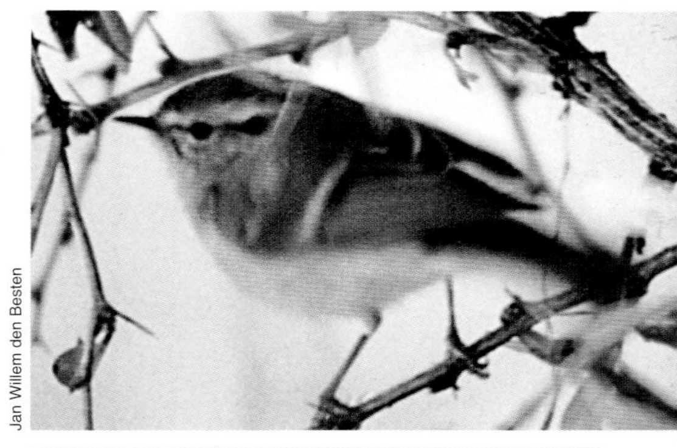

Jan Willem den Besten

Distribution: Common breeding resident in northern hills from north Pakistan east to Arunachal Pradesh. Winters lower down and in northeastern hills. Vagrant in plains south to Delhi and Bangladesh.
Also occurs in Afghanistan and SE Asia.

HUME'S WARBLER (Plain Leaf, Yellow-browed Warbler)

Phylloscopus humei 10 cm F: Sylviidae

Otto Pfister

Description: A dumpy, grey-green, arboreal warbler with wing bars. Rather dull greyish-green above with one or two short whitish wing bars and pale edges to tertials. Murky white below. Prominent whitish supercilia and dark eyestripes. All dark bill and legs. Sexes alike. More eastern **Yellow-browed** *P. inornatus* (once considered conspecific) much brighter with very prominent tertial edgings, pale legs and base of lower mandible.

Voice: Noisy. Penetrating disyllabic *tzz wip*, uttered constantly. **Yellow-browed** calls *tsweet*.

Habits: Inhabits wooded areas, city parks and gardens. Feeds actively on invertebrates in tree canopy, often in mixed hunting groups. Confiding but restless.

Distribution: Common breeding summer visitor to northern mountains from north Pakistan east to Bhutan. Common winter visitor to northern plains of Pakistan, lowlands of Nepal and northern and eastern India. Rare in west and south. The commonest wintering warbler in Delhi.
Also occurs in Central Asia and China.

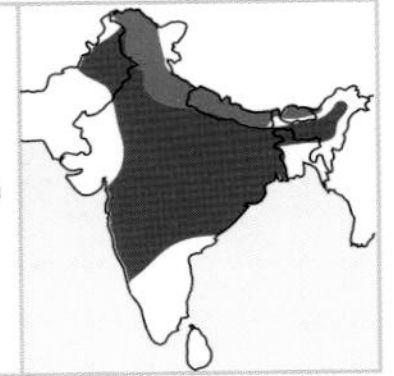

GREENISH WARBLER (Dull Green, Greenish Leaf Warbler)

Phylloscopus trochiloides 11 cm F: Sylviidae

Jan Willem den Besten

Description: A small, green, slim, arboreal warbler with brown legs. Dull greenish above and whitish below. Prominent long, sweeping supercilia. Usually only one wing bar shows. Sexes alike. Possibly conspecific **Green Warbler** is greener above and yellower below, especially on supercilia, throat and breast and shows two yellowish wing bars.

Voice: Noisy. A loud slurred disyllabic *tis lee*.

Habits: Inhabits all types of wooded country, including gardens. Feeds on invertebrates, mainly in the canopy but also lower down. Active and restless. Frequently fly-catches. Usually in scattered parties or mixed hunting groups. Nests low down.

Distribution: Common breeding resident in northern mountains from Pakistan east to the Burmese border. Winters widely in penisular lowlands and Sri Lanka. Common passage migrant elsewhere but occasionally winters.
Also occurs in Europe, W, Central and SE Asia and China.

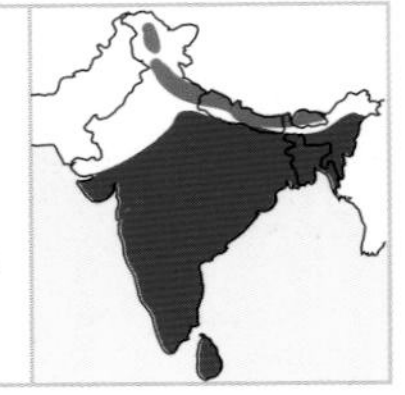

LARGE-BILLED LEAF WARBLER (Large-billed Tree)

Phylloscopus magnirostris 13 cm F: Sylviidae

Description: A large, arboreal leaf warbler with a heavy, dark bill. Olive green above, noticeably darker on crown. One obvious and, sometimes, a second slight wing bar on each wing. Murky white below with yellowish wash and faint breast streaking. Very striking, long yellowish-white supercilia and dark eye stripes. Bill has orange base to lower mandible. Legs dark grey. Sexes alike.

Voice: Noisy. A loud inflected *dir tee* call and sings a loud, descending *see si si si si*.

Habits: Inhabits forest feeding, usually singly, on invertebrates high in the canopy. Often near streams. Rather less active than relatives. Nests close to ground.

Goren Ekstrom

Distribution: Locally common breeding summer visitor to northern mountains from Pakistan east to Bhutan. Winters mainly in southwest Indian and Sri Lankan hills but also in Bangladesh and northeast India. Recorded elsewhere in peninsula mainly on passage.
Also occurs in Tibet and China.

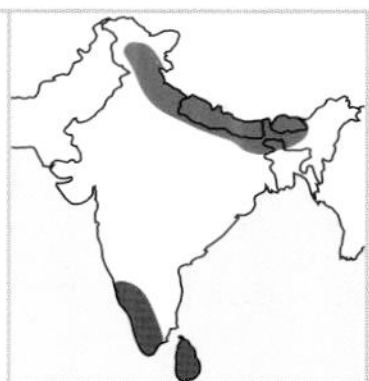

TYTLER'S LEAF WARBLER (Slender-billed Warbler)

Phylloscopus tytleri 10 cm F: Sylviidae

Description: A tiny, plain warbler with a long, black bill. Unmarked olive green above. Whitish below sometimes with pale yellow wash. Long, whitish supercilia and dark eyestripes. Dusky cheeks. Short tail. Dark legs and long thin and mainly dark bill. Sexes alike.

Voice: An extended *sooeeet*. Also a hoarse *huweest* and song a repeated *pi tsi pi*.

Habits: Inhabits forest undergrowth and conifers. Feeds very actively on invertebrates on leaves and stems and sometimes fly-catches. Nests in tree. Winter habits poorly known.

Jan Willem den Besten

Distribution: A scarce endemic breeding summer visitor restricted to north Pakistan and Kashmir. Winters mainly in Western Ghats but passage records elsewhere, particularly from northwest India in autumn. Probably overlooked.

WESTERN CROWNED WARBLER (Large Crowned Leaf)

Phylloscopus occipitalis 11 cm F: Sylviidae

Otto Pfister

Description: A stocky, grey-green arboreal warbler with a pale, central crown stripe. Upperparts grey-green, noticeably greener on wings and tail and with two yellowish wing bars on each wing. Striking yellowish supercilia White below. Dark legs and pale lower mandibles. Sexes alike. **Blyth's Leaf Warbler** *P. reguloides* is greener with more obvious yellow wing bars, white in outer tail and yellowish below.

Voice: Rather quiet. A repeated *chit weei*.

Habits: Inhabits forest, feeding or invertebrates actively in canopy and also in shrub layer. Often in small parties and joins mixed hunting groups. Alternately flicks wings half open. Confiding. Nests low down.

Distribution: Common breeding summer visitor to northern hills from Pakistan east to Uttaranchal. Winters mainly in Eastern and Western Ghats. Scattered records elsewhere and more widespread in India on passage. Also occurs in Central Asia.

GOLDEN-SPECTACLED WARBLER (Flycatcher Warbler)

Seicercus burkii 10 cm F: Sylviidae

Sujan Chatterjee

Description: A small, yellow and green warbler with yellow eye rings. Green above with white in outer tail and slight wing bars. Blackish crown stripes with grey central crown. Yellow on face and below. Pale legs. High, rounded crown and small, pale bill. Sexes alike.

Voice: Noisy. Common call is a quiet, sparrow-like *chip chip*.

Habits: Inhabits forest undergrowth and low canopy, feeding very actively on invertebrates in pairs, small parties or as members of mixed hunting groups. Not shy, but often difficult to pinpoint as so active. Nests on ground.

Distribution: Common breeding resident in northern mountains from north Pakistan east to Burmese border. Winters lower down in foothills, sparsely in northern Indian plains and some peninsular hills and in Bangladesh. Also occurs in Tibet, China and SE Asia.

GREY-HOODED WARBLER (Flycatcher Warbler)

Seicercus xanthoschistos 10 cm F: Sylviidae

Description: A small, green and yellow arboreal warbler with striking black-bordered, white supercilia. Central crown and nape grey, contrasting with supercilia, and green mantle. Chin to vent bright yellow. Obvious white in outer tail. *Phylloscopus* warbler shape. Sexes alike.

Voice: Noisy. A constant *psit psit* or *tjee tjee* call. Song is a musical *tee tsi tee tsi* tee and variants.

Habits: Inhabits the middle storey of forests, forest edge and wooded gardens. Actively feeds on insects in pairs, small parties or mixed hunting groups. Frequently fly-catches. Lively and confiding; often the most obvious small bird in the right habitat. Nests on or near ground.

Sujan Chatterjee

Distribution: Common breeding resident in northern hills from north Pakistan east to the Burmese border. Winters lower down in foothills. Also occurs in Myanmar.

WHITE-SPECTACLED WARBLER (Flycatcher Warbler)

Seicercus affinis 10 cm F: Sylviidae

Description: A small, green and yellow warbler with white eye rings. Very similar to **Golden-spectacled,** but grey extends down nape and cheeks greenish. Sexes alike. Similar **Grey-cheeked Warbler** *S. pologenys* has grey head (including lores) and white chin with only indistinct crown stripes.

Voice: A hard *che weet*.

Habits: Inhabits forest undergrowth. Behaviour, nest and food as **Golden-spectacled**.

Sujan Chatterjee

Distribution: Locally common breeding resident in northeastern hills from West Bengal north and east to Burmese border. Winters lower down in foothills and nearby plains. Also occurs in China and SE Asia.

STRIATED GRASSBIRD (Striated Marsh Warbler)

Megalurus palustris 25 cm F: Syvlidae

Otto Pfister

Description: A huge warbler with heavily streaked upperparts. Warm buff above with blackish streaking from crown to rump and barred, graduated tail. Long whitish supercilia and light breast streaking on pale buff underparts. Long pale bill and pale legs. Sexes alike. Beware confusion with **Striated Babbler**.

Voice: Loud *wee choo* call extended into exhuberant, swooping aerial song. Also a hard *chak*.

Habits: Inhabits tall grass, scrub and reedbeds, usually near water. Usually singly or in pairs. Frequently perches high on grass stem. Bold when breeding but furtive when feeding. Feeds on invertebrates low down in vegetation. Nests close to ground.

Distribution: Locally common resident in northern river valleys from eastern Pakistan east along the Ganges floodplain and up the Brahmaputra valley to Arunachal Pradesh. Also the terai in India and Nepal and Bangladesh. Odd records further south. Also occurs in China and SE Asia.

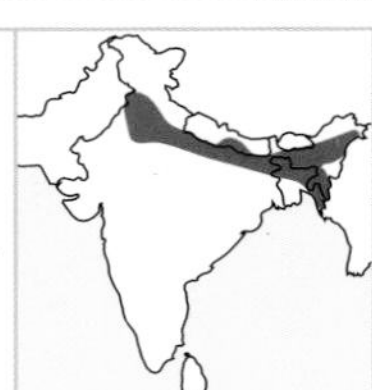

BRISTLED GRASSBIRD (Bristled Grass Warbler)

Chaetornis striatus 20 cm F: Sylviidae

Otto Pfister

Description: Large, streaked warbler with thick, blackish bill. Similar to **Striated** (apart from bill) but smaller and paler streaked. No obvious supercilia. Rump unstreaked. Breast has indistinct necklace of dark, stiffened feathers (presumably origin of name, although also has prominent rictal bristles). Shorter, graduated tail has distinct pale tips. Beware confusion with **Striated Grassbird** and **Striated Babbler**. Sexes alike.

Voice: Distinctive repetitive *trew treuw* song, usually delivered in vertical song flight.

Habits: Inhabits tall, often wet, grassland and reed beds, usually with scattered bushes. Exceptionally furtive, feeding on invertebrates low in grass.

Distribution: Globally threatened endemic breeding resident now restricted to a few sites in the Indus-Ganges and Brahmaputra floodplains in India and in Nepal but very erratically recorded. Unpredictable in appearance. Previously recorded in scattered sites south to Tamil Nadu but some of these records must be questionable.

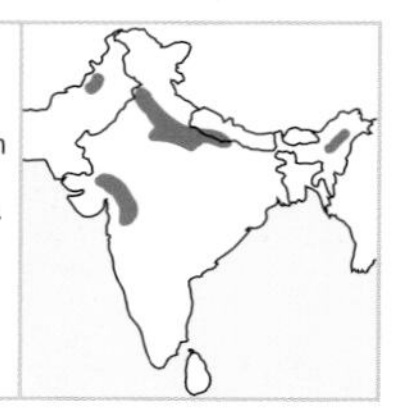

WHITE-THROATED LAUGHINGTHRUSH

Garrulax albogularis 28 cm F: Sylviidae

Description: A large, brown babbler with very obvious white throat and upper breast. Olive grey-brown above with white-tipped, graduated tail. Grey in flight feathers. Pale chestnut vent and belly, greyer lower breast, edged blackish. Rufous fore-head, white irises and extensive black lores. Sexes alike.

Voice: Noisy. A lot of squealing, hissing and chattering and a strident *twitz tzee* alarm call.

Habits: Inhabits the under-growth of forest and scrub, usually in quite large groups, often with other species. Feeds on invertebrates and berries. Confiding and easy to observe, though fast-moving. Nests in shrub or tree.

Tim Loseby

Distribution: Locally common breeding resident in northern hills from north Pakistan east to Assam. Rarer in east of range. Moves lower down in winter. Also occurs in China and SE Asia.

WHITE-CRESTED LAUGHINGTHRUSH

Garrulax leucolophus 28 cm F: Sylviidae

Description: Large, brown babbler with a white head. Dark brown body with chestnut wash on mantle and lower breast. Upper breast and whole head striking white with a prominent crest. Broad black band through eye and grey on nape. Sexes alike.

Voice: Very noisy. Various explosive chattering calls and whistling, often delivered in chorus by flock in a loud crescendo of sound.

Habits: Inhabits thick forest undergrowth, secondary scrub and bamboos. Feeds on inverte-brates and fruit, usually low down in cover and on ground. Will go higher in trees. Always in parties or small flocks which, though noisy, are rather wary. Nests in shrub or tree.

Sanctuary

Distribution: Locally common breeding resident in northern mountains from Himachal Pradesh east to the Burmese border including eastern Bangladesh. Moves lower down in winter. Also occurs in China and SE Asia.

GREATER NECKLACED LAUGHINGTHRUSH

Garrulax pectoralis 29 cm F: Sylviidae

Morten Strange

Description: Large, brown babbler with black breast band. Brown above with darker, white-edged graduated tail and black patches at base of primaries. Black lores, and eye stripes extending round cheeks to form broad black breast band enclosing buff or white throat. Cheeks white, patchy or black. Deep orange below. Pale-based bill and grey legs. **Lesser** *G. monileger* very similar but narrower orange-edged breast band, pale lores, all-dark bill and brown legs. No black in wing.

Voice: Calls *week week week*, whistling and grating notes.

Habits: Inhabits undergrowth of forest where feeds, often with **Lesser**, on invertebrates on ground. Usually in groups. Shy. Nests in bush or low tree.

Distribution: Common breeding resident in northern foothills and nearby plains from west Nepal east to the Burmese border and Bangladesh. Scarce in west.
Also occurs in China and SE Asia.

STRIATED LAUGHINGTHRUSH

Garrulax striatus 28 cm F: Sylviidae

Tim Loseby

Description: Large, dark brown babbler with high-domed crest. Rich chocolate-brown with fine white streaking on head, mantle and underparts. Northeastern race less streaked, particularly on crown and has black eyebrows. Long, plain brown tail. Heavy black bill. Sexes alike. Beware smaller Barwings *Actinodura* in the Northeast which are more rufous with black barring.

Voice: Very noisy. Various cackling, chattering, gurgling notes and squeaky *pseet* call.

Habits: Inhabits forest and scrub, particularly ravines. Feeds on insects and fruit at all levels, often higher than other laughing-thrushes. Usually in pairs, small parties or mixed hunting groups. Fairly confiding. Nests in tree.

Distribution: Locally common breeding resident in northern mountains from Himachal Pradesh to the Burmese border. Moves lower down in winter.
Also occurs in Myanmar.

RUFOUS-CHINNED LAUGHINGTHRUSH

Garrulax rufogularis 22 cm F: Sylviidae

Description: A large, highly patterned babbler with a broad black sub-terminal band on tail. Upperparts uniquely patterned brown, grey and black. Crown black. Chestnut throat and either chestnut or blackish ear coverts. Underparts lightly black-spotted on greyish-buff. Bright chestnut vent. Sexes alike.

Voice: Noisy. Various chatters and squeals and a whistling song *whee whoo weewooweee*.

Habits: Inhabits thick undergrowth and secondary growth. Feeds on invertebrates and fruit on or close to ground. Usually very secretive although will sing in open. In pairs or small parties. Nests in shrub or tree.

Sumit Sen

Distribution: Local and rather scarce breeding resident in northern mountains from Kashmir east to the Burmese border. Moves lower down in winter.
Also occurs in SE Asia.

GREY-SIDED LAUGHINGTHRUSH

Garrulax caerulatus 25 cm F: Sylviidae

Description: A large, grey and rufous babbler with a black face. Upperparts warm rufous-brown. White below with broad grey flanks. White cheek spots, bluish round eyes and scaly rufous crown. Sexes alike. White tail edges in one eastern race.

Voice: Noisy. Musical *klee loo*, *ovik chorrr*, *new jeriko* and *oh dear dear* notes. A *chik chik* alarm.

Habits: Inhabits the dense understorey of forests. Feeds on invertebrates on or near ground in small parties. Nests in bush or small tree.

Sujan Chatterjee

Distribution: Scarce breeding resident from central Nepal east to the Burmese border. More common in east. Moves lower down in winter.
Also occurs in Myanmar and China.

NILGIRI LAUGHINGTHRUSH (Fulvous-breasted)

Garrulax cachinnans 20 cm F: Sylviidae

Tim Loseby

Description: A medium-sized, dark babbler with white eyebrows. Plain dark brown above and rich rufous below. Dark grey crown and black chin, forehead and eyestripes. Sexes alike.

Voice: Very noisy. A squealing "laughter" often in chorus; *pee ko ko* and *kee kee kee*.

Habits: Inhabits forest undergrowth, secondary growth and gardens. Small parties feed mainly on invertebrates on ground under cover. Also takes fruit. Very shy. Nests in tree.

Distribution: Globally threatened. Very local but, in places, common endemic breeding resident in Nilgiri Hills of western Tamil Nadu and nearby parts of Kerala.

GREY-BREASTED LAUGHINGTHRUSH

Garrulax jerdoni 20 cm F: Sylviidae

Joanna Van Gruisen

Description: A medium-sized, brown and rufous babbler with a grey or whitish breast. Paler than **Nilgiri** but similar pattern. Breast and neck sides grey or grey-streaked white, depending on race. Some races lack black chin and eyestripe but all have distinct white eyebrows. Sexes alike.

Voice: Very noisy. A variety of chattering, squeaking and whistling, including *puwee puwee pokee* and *pee koko*.

Habits: Inhabits undergrowth of forest edge, plantations and gardens, particularly wild raspberry along stream sides. Feeds on invertebrates and fruit, mainly on or near ground in small parties or in mixed hunting groups. Very shy.

Distribution: Locally common breeding endemic resident in Western Ghats and Palni Hills in west Tamil Nadu, Kerala and southern Karnataka.

STREAKED LAUGHINGTHRUSH

Garrulax lineatus 20 cm F: Sylviidae

Description: A medium-sized, streaked brown babbler with rufous ear coverts. Mostly lightly white-streaked, greyish-brown but wings and tail more chestnut and tail has grey tip. Sexes alike.

Voice: Noisy. Constant whistling and squeaking notes including *pitt wee err* whistle.

Habits: Inhabits low scrub, forest edge, gardens and roadside grass patches. Often around habitation. Feeds on invertebrates, fruit and human refuse, shuffling along ground, usually in pairs or small parties. Bobs and dips, flicks wings and jerks tail when perched. Short, weak flight, diving for cover. Not shy but rather furtive. Nests in low cover.

Otto Pfister

Distribution: Common breeding resident in northern mountains from western Pakistan east to Arunachal Pradesh. Moves down to foothills in winter.
Also occurs in Central Asia.

VARIEGATED LAUGHINGTHRUSH

Garrulax variegatus 25 cm F: Sylviidae

Description: A large, pale greyish-brown babbler with white irises. Races vary. Brownish-grey above with grey wings with black and white markings and grey, rufous or yellowish-green edgings. Tail dark grey, with paler broad band towards tip and with white tips. Rufous forehead. Black chin and round eyes, white throat. Underparts greyish-buff, more rufous on vent. Sexes alike.

Voice: Noisy. Various whistling and squeaking notes including *weet a weer* and *peet weer*.

Habits: Inhabits thick forest-edge undergrowth and secondary growth, feeding on invertebrates and fruit usually on or close to ground but will go higher up trees. Usually in pairs or small parties.

Distribution: Locally common endemic breeding resident in northern mountains from north Pakistan east to eastern Nepal. Moves lower down in winter.

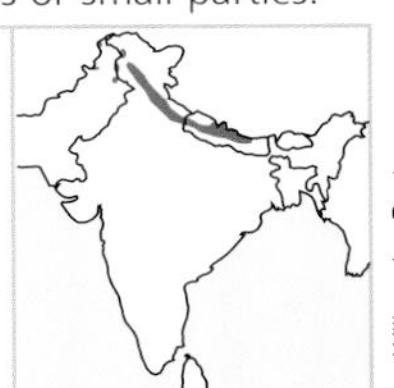

Jan Willem den Besten

BLACK-FACED LAUGHINGTHRUSH

Garrulax affinis 25 cm F: Sylviidae

Otto Pfister

Description: A large, dark babbler with a white-marked, blackish face. Finely grey-scalloped, rufous-brown body with yellowish green wings, edged grey. Tail yellowish-green with broad grey tip. Blackish head with white rear cheeks and malar patches and partial eye rings. Sexes alike.

Voice: Fairly noisy with various chuckles and churrs. Alarm call is *whrrr* and song *wheeit wheeoo woo* and variants.

Habits: Inhabits forest undergrowth and montane scrub, feeding on invertebrates and fruit, usually low down in pairs or small parties. More confiding than most laughingthrushes. Runs, rather than flies, from disturbance. Nests in low growth.

Distribution: Common breeding resident in northern mountains from western Nepal east to Arunachal Pradesh. Winters lower down in foothills.
Also occurs in China and SE Asia.

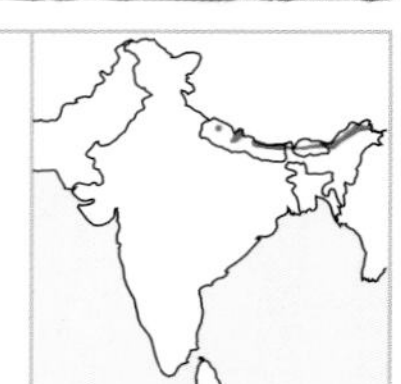

CHESTNUT-CROWNED LAUGHINGTHRUSH

Garrulax erythrocephalus 28 cm F: Sylviidae

Sujan Chatterjee

Description: A large, rufous-brown babbler with bright yellowish-green wings. Rather variable. Crown chestnut and cheeks spotted black. Black throat. Dark scalloping on mantle, neck and breast. Tail has broad, yellowish-green sides. Eastern race is darker and has forecrown and rear cheeks grey. Sexes alike.

Voice: Noisy. Various chuckles and whistles including *wee yu wee wip* and *yu wheeo*.

Habits: Inhabits forest undergrowth and secondary growth, generally keeping low down in pairs or small parties. Feeds on invertebrates and fruit. Very shy. Nests in low cover.

Distribution: Fairly common breeding resident in northern mountains from Kashmir east to the Burmese border. Winters lower down.
Also occurs in China and SE Asia.

RED-FACED LIOCICHLA (Crimson-winged Laughing)

Liocichla phoenicea 23 cm F: Sylviidae

Description: A medium-sized, brown and red babbler with orange-tipped, black tail. Warm brown all over with extensive, red cheek and wing patches and black crown edges and in wings. Orange vent. Sexes alike.

Voice: Calls *tu reew reew* often extended and varied. Also *ji uuu* and a harsh *chrrt chrrt* alarm.

Habits: Inhabits dense undergrowth of forest, often in ravines or along streams. Feeds low down on invertebrates in pairs or small parties, sometimes joining mixed hunting groups. Very secretive and difficult to see well. Nests in bush.

Sujan Chatterjee

Distribution: Locally common breeding resident from extreme east Nepal east to the Burmese border. May no longer occur in Nepal. Moves lower down in winter.
Also occurs in China and SE Asia.

PUFF-THROATED BABBLER (Spotted Babbler)

Pellorneum ruficeps 15 cm F: Sylviidae

Description: Small, brown babbler with strongly spotted underparts. Dull-brown above with chestnut crown, pale brown cheeks and buff supercilia. White throat and rufous-washed flanks. Obvious blackish spots in lines down breast and flanks. Slightly decurved two-toned bill. Sexes alike.

Voice: Loud call *tee teu* often preceded by quieter *te*. Extended loud, mellow song.

Habits: Inhabits thick forest undergrowth and bamboo thickets, particularly in ravines and along streams. Very secretive but sings from mid-level perch. Feeds singly or in pairs on invertebrates, mainly on ground among leaf litter where turning leaves and hopping gait may be the only clue to its presence.

Balachandran

Distribution: Common breeding resident in northern foothills from Himachal Pradesh east to the Burmese border, Bangladesh and peninsular hills including both Eastern and Western Ghats.
Also occurs in China and SE Asia.

RUSTY-CHEEKED SCIMITAR BABBLER

Pomatorhinus erythrogenys 25 cm F: Sylviidae

Nikhil Devasar

Description: A large, russet-brown babbler with powerful, decurved yellow bill. Brown above, more russet on cheeks, neck and flanks. White below with greyish throat edged with black. Sexes alike. Darker brown **White-browed** *P. schisticeps* has long white eyebrows, black eye bands and black or rufous-edged white throat and breast.

Voice: Noisy. A fluty *cue pe cue pe* from male answered by *quip* from female.

Habits: Inhabits forest undergrowth, bamboo thickets and secondary growth, particularly in ravines. Feeds in pairs or small parties on or close to ground on invertebrates and fruit. Extremely secretive, moving rapidly and invisibly through low growth.

Distribution: Common breeding resident from north Pakistan east to Bhutan.
Also occurs in China and SE Asia.

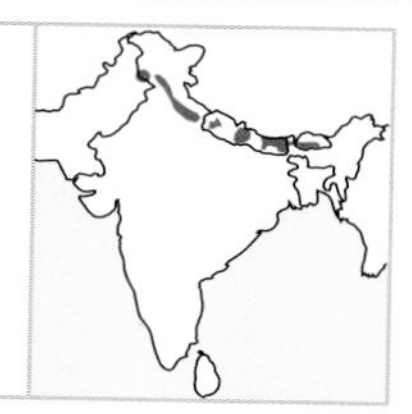

INDIAN SCIMITAR BABBLER

Pomatorhinus horsfieldii 23 cm F: Sylviidae

Ron Saldino

Ron Saldino

Description: Large, dark brown and white babbler with powerful, yellow, decurved bill. Dark brown with brown or grey crown, white supercilia and throat to vent. Dark grey-edged breast and flanks (brighter chestnut in Sri Lanka races). Black cheeks. Sexes alike. Similar northern **White-browed** has larger bill, dark grey crown, rufous or black flanks and breast sides.

Voice: Male calls musical *hup pu pu* answered by female's quiet *kru kru*.

Habits: Inhabits forest and plantation undergrowth and thick scrub, including gardens. Feeds on invertebrates low down in groups, often with other babblers. Extremely secretive. Nests low down.

Distribution: Locally common endemic breeding resident in peninsular hills and Sri Lanka. Occurs as far north as Rajasthan and Orissa.

STREAK-BREASTED SCIMITAR BABBLER

Pomatorhinus ruficollis 19 cm F: Sylviidae

Description: A small, brown scimitar babbler with a yellow, decurved bill. Smaller and paler than **Indian** with white underparts restricted to throat and breast and heavy brown streaking on latter. Olive brown crown. Rufous neck and black cheeks contrast with long white supercilia. Sexes alike.

Voice: Male has musical "off and on" call replied to by female's *kwee kwee*. Song *pooki worki* and variants.

Habits: Inhabits dense forest undergrowth and scrub usually on hillsides. Feeds low down on invertebrates usually in pairs but sometimes with other babblers. Very secretive and difficult to see, unless flying rapidly between cover. Nests low down.

Sujan Chatterjee

Distribution: Locally common breeding resident in northern mountains from Uttaranchal east to the Burmese border. Moves lower down in winter. Also occurs in China and SE Asia.

PYGMY WREN BABBLER (Brown Wren Babbler)

Pnoepyga pusilla 9 cm F: Sylviidae

Description: A tiny, brown babbler with no visible tail. Dark brown above with some buff spots. White or deep buff below with dark scaling and unmarked throat. Long legs. Sexes alike. There are several other **Wren Babblers** and care needs to be taken in identification, given the difficulty of obtaining clear views.

Voice: Song is a repeated drawn-out *tsee tsu*. Calls *tzook*.

Habits: Inhabits damp, ferny undergrowth of forests often among mossy boulders and fallen logs near streams. Feeds very actively on invertebrates on ground, usually singly or in pairs. Very shy and if disturbed tends to hop away rather than fly. Flicks wings. Nests close to ground.

Morten Strange

Distribution: Fairly common breeding resident in northern mountains from western Nepal east to the Burmese border and perhaps Bangladesh. Moves lower down in winter. Also occurs in China and SE Asia.

RUFOUS-CAPPED BABBLER (Red-headed Babbler)

Stachyris ruficeps 12 cm F: Sylviidae

Sujan Chatterjee

Distribution: Scarce breeding resident in northern foothills from extreme eastern Nepal east to the Burmese border. Also in the northern Eastern Ghats in Orissa. Rufous-fronted also found in Bangladesh and south to Andhra Pradesh in the Eastern Ghats. Also occurs in China and SE Asia.

Description: A small, brown babbler with an extensive rufous crown. Yellowish-brown above with rufous on crown extending well onto nape. Warm yellowish below and on face with paler, unstreaked throat. Pale legs. Sexes alike. Very similar **Rufous-fronted** *S. rufifrons* has rufous cap restricted to crown, darker cheeks, more obvious supercilia and white throat and belly. Throat often appears streaked.

Voice: *Wee wee wee wee* call and high-pitched *pi pi pi pi pi pi pi* song.

Habits: Inhabits undergrowth and bamboo clumps of forests. Feeds on invertebrates and berries up to middle levels, in pairs and mixed hunting groups. Active and acrobatic. Nests low down.

BLACK-CHINNED BABBLER (Red-billed Babbler)

Stachyris pyrrhops 10 cm F: Sylviidae

Jan Willem denBesten

Distribution: Fairly common endemic breeding resident in northern mountains from north Pakistan east to eastern Nepal. Moves lower down in winter.

Description: A tiny, short-tailed, brown babbler with black lores and chin. Underparts orange-buff. Sexes alike. Similar larger southern **Dark-fronted Babbler** *Rhopocichla atriceps* has black head or mask (depending on race) and white throat to belly.

Voice: Mellow *wit wit wit wit* and a soft variable *chirrr*.

Habits: Inhabits forest edge and secondary undergrowth, often near streams. Favours bamboo. Feeds in pairs or small parties on invertebrates, often in mixed hunting groups. Active and not shy. Nests low down.

GREY-THROATED BABBLER (Black-throated Babbler)

Stachyris nicriceps 12 cm F: Sylviidae

Description: A small, brown babbler with black and grey head markings. Brown above and buff or orange below. Streaked grey crown with broad black borders and grey eye stripes. Throat grey with variable amounts of black depending on race. Sexes alike. Similar **Black-chinned** has very restricted, black lores and chin and no grey on head.

Voice: A distinctive but variable jingling trill.

Habits: Inhabits the undergrowth of forests, particularly with bamboos and near water. Feeds low down on invertebrates, often in mixed hunting groups. Very secretive. Nests low down.

Morten Strange

Distribution: Locally common breeding resident from central Nepal east to the Burmese border. Also occurs in China and SE Asia.

TAWNY-BELLIED BABBLER (Rufous-bellied)

Dumetia hyperythra 13 cm F: Sylviidae

Description: A small, brown babbler with yellow irises. Chestnut crown, orange-buff underparts and, in most races, white throat and belly. Eastern peninsula race all orange-buff below. Sexes alike. Similar **Rufous-fronted Babbler** has grey-streaked white throat, grey supercilia and buff underparts.

Voice: Rather soft *cheep cheep* call. Whistling song.

Habits: Inhabits secondary scrub with patches of grass. Feeds low down on invertebrates. Usually in small, furtive but noisy parties. Often very difficult to see well. Nests in low cover.

Balachandran

Distribution: Locally common endemic breeding resident in northern foothills from Himachal Pradesh east to West Bengal and south through peninsula including Sri Lanka. Not known in the Northwest or Northeast.

STRIPED TIT BABBLER (Yellow-breasted Babbler)

Macronous gularis 11 cm F: Sylviidae

Morten Strange

Description: A tiny, brown babbler with streaked yellowish underparts. Brown above, more olive on mantle and with chestnut crown. Yellowish supercilia and underparts, the latter finely streaked blackish. Pale irises. Sexes alike.

Voice: Very noisy. A loud, regular and repeated barbet-like *chunk chunk chunk*. Also *chrrr*.

Habits: Inhabits forest undergrowth and bamboo, often feeding fairly high but also on ground. Feeds on invertebrates in pairs or small parties, often in mixed hunting groups. Active but furtive. Nests low down.

Distribution: Common breeding resident in northern foothills from Uttaranchal east through Nepal to Arunachal Pradesh, the whole of the Northeast and the Eastern Ghats. Also occurs in China and SE Asia.

CHESTNUT-CAPPED BABBLER (Red-capped Babbler)

Timalia pileata 17 cm F: Sylviidae

Sujan Chatterjee

Description: A medium-sized, bull-necked, long-tailed babbler with a bright chestnut cap. Olive brown above and buff below. Short, thick bill. Black round eyes and on lores. Contrasting white supercilia, cheeks and throat. Lightly streaked grey on neck and upper breast. Sexes alike.

Voice: Calls *tit tit* and *kerchuk*. Song a high-pitched descending trill.

Habits: Inhabits damp grasslands, reedbeds and adjacent scrub. Feeds on invertebrates low down, usually in small parties. Skulking but climbs grass stems and sometimes appears out in open, especially if grass is wet. Nests low down.

Distribution: Locally common breeding resident in northern floodplains from Uttaranchal east to the Burmese border. Also in Bangladesh. Also occurs in China and SE Asia.

YELLOW-EYED BABBLER

Chrysomma sinense 18 cm F: Sylviidae

Description: A medium-sized, brown, long-tailed babbler with red eye rings and yellow irises. Warm brown above. White below including throat, lores and short supercilia. Belly and flanks buffer. Short, thick, black bill. Sexes alike. Similar, but globally threatened, grassland **Jerdon's Babbler** *C. altirostre* has grey supercilia and lores and greyish breast. Also no eye rings and a paler bill.

Voice: Noisy. A mournful *cheep cheep cheep* call. Trilling song.

Habits: Inhabits undergrowth in open woodland and secondary growth with plenty of long grass. Feeds low down on invertebrates in pairs or small parties. Furtive but inquistive, clambering up grass stems like a warbler. Jerky flight. Nests low down in cover.

Nikhil Devasar

Distribution: Locally common breeding resident throughout region but scarce in Pakistan, Bangladesh and Sri Lanka. Also occurs in China and SE Asia.

COMMON BABBLER

Turdoides caudatus 22 cm F: Sylviidae

Description: Small, slim, long-tailed, streaked babbler with dark irises. Streaked brown above and on breast sides. Dark graduated tail faintly cross-barred. Throat white. Rest of underparts pale buff. Sexes similar. Smaller size and white throat distinguish it from **Striated**.

Voice: Noisy. Fluty, rippling *pieuu u u pie u u* and *wich wich wee wee* and variants.

Habits: Inhabits dry scrub and low secondary growth with tall grass, including urban parks and gardens. Feeds on invertebrates and fruit. Usually in small parties. Flies rather weakly and low, one after the other with gliding between flaps. Hops with loose-looking tail cocked. Wary. Nests in low bush.

Otto Pfister

Distribution: Fairly common breeding resident throughout much of lowlands but absent from Sri Lanka, much of Nepal, eastern and northeastern India and Bangladesh.
Also occurs in Afghanistan and W Asia.

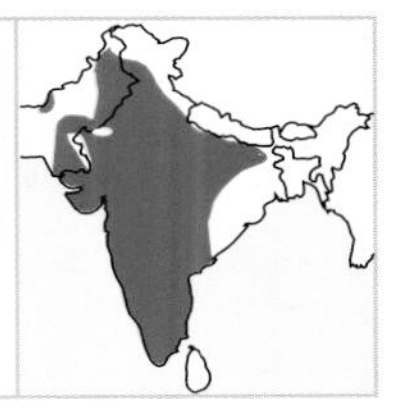

STRIATED BABBLER

Turdoides earlei 25 cm F: Sylviidae

Otto Pfister

Otto Pfister

Description: A fairly large, streaked greyish-brown babbler with yellow irises. Similar to **Common** but rather greyer and much darker buffy-brown below (including throat) with breast streaking. Usually shows faint grey moustache. Sexes alike.

Voice: Noisy. Commonest note is a far-carrying *pee pee pee* usually uttered from high on reed stem. Sometimes slurred to *cheer cheer cheer* and *chip chip*.

Habits: Inhabits wet grass and reedbeds and adjacent scrub. Feeds low down on invertebrates in noisy active parties, threading their way through the cover. Flies in lines over reeds with much gliding. Confiding and inquisitive. Nests low down.

Distribution: Locally common breeding resident in the Indo-Gangetic and Brahmaputra river plains and related wetlands.
Also occurs in Myanmar.

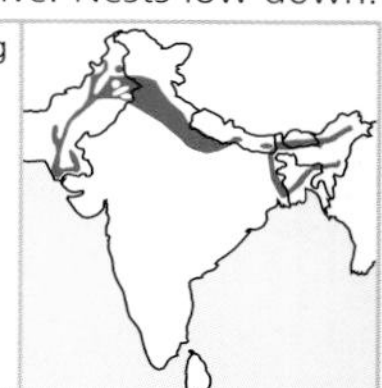

LARGE GREY BABBLER

Turdoides malcolmi 28 cm F: Sylviidae

Sumit Sen

Description: A large, greyish babbler with yellowish-white irises and broad, whitish edges to its very long, dark, graduated tail. Rather pale brownish-grey with some mottling on back. Buffish-grey below. Blackish lores contrast with pale irises and dark-tipped yellowish bill. Sexes alike.

Voice: Very noisy. Constant mournful *kaa kaa kaa* calls and squeaking chattering in chorus.

Habits: Inhabits dry open woodland and scrub often near habitation and cultivation. Moves around in noisy, sweeping groups feeding on invertebrates from ground to treetops. Frequently one individual will perch very high, perhaps as a look-out. Brazen and inquisitive. Nests in low cover or in a tree.

Distribution: Locally common endemic breeding resident in lowlands from Punjab south to Tamil Nadu and east to Bihar and Orissa borders. Rare in Pakistan and Nepal.

JUNGLE BABBLER (Seven Sisters)

Turdoides striatus 25 cm F: Sylviidae

Otto Pfister

Description: Stocky, grey-brown babbler with whitish irises and bright yellow bill. Brown-streaked paler above with greyer head and darker tail. Paler below with pale breast streaking. Short, dark eyebrow gives fierce impression. Western race *somervillei* may be separate species, with chestnut tail and dark brown primaries.

Voice: Very noisy. Harsh *ke ke ke* and chorused hysterical chattering and squeaking.

Habits: Inhabits open woodland, scrub, cultivation, gardens and villages. Feeds on insects mainly on ground in noisy, excited groups. Confiding and inquisitive. Groups mutually preen. Fluffs up rump and droops wings. Nests in bush.

Distribution: Common endemic breeding resident throughout lowlands and foothills except parts of the Northeast and Sri Lanka. Tends to be at higher locations than Yellow-billed where they overlap in south.

Otto Pfister

YELLOW-BILLED BABBLER (White-headed, Southern)

Turdoides affinis 23 cm F: Sylviidae

Description: A fairly large, stocky grey-brown babbler with a pale greyish crown and nape. Very similar to **Jungle Babbler** (which is lower bird in photo) but bill paler yellow and lores pale. Paler crown and nape and dark speckled throat to breast (not in Sri Lankan race). Grey rump and basal part of tail contrast with darker rest of tail, particularly in flight. Distinct grey wing panel. Sexes alike.

Balachandran

Voice: Noisy. More high-pitched and musical than **Jungle**.

Habits: Inhabits open scrub, woodland, cultivation and gardens. Often with **Jungle** but commoner at lower altitudes. Behaviour, food and nest similar.

Distribution: Locally common endemic breeding resident in plains from Goa and Andhra Pradesh southwards including Sri Lanka.

SILVER-EARED MESIA

Leiothrix argentauris 15 cm F: Sylviidae

Morten Strange

Description: Small, brightly coloured babbler with a black crown. Grey above with crimson and yellowish-green wings. Red upper and under tail coverts and dark grey tail. Orange or deep yellow forehead, neck, throat and breast, silvery-white cheeks. Yellow legs and bill. Sexes alike.

Voice: Extended, descending song *weet chewit chewit chewee chewee* and an incessant, short *chirrup* call.

Habits: Inhabits forest edge and plantation undergrowth feeding, usually at middle levels, on invertebrates or fruit, although sometimes in canopy. In pairs, small parties or mixed hunting groups. Active and acrobatic. Fly-catches.

Distribution: Locally common breeding resident in northern foothills and mountains from Uttaranchal east to the Burmese border. Commoner in east. Moves lower down in winter. Also occurs in China and SE Asia.

RED-BILLED LEIOTHRIX (Pekin Robin)

LeIothrix lutea 13 cm F: Sylviidae

Otto Pfister

Description: A small orange and grey babbler with a red bill. Yellowish or greenish-grey above with black and orange in wings (also crimson in eastern race) and dark grey, slightly forked tail. Long chestnut upper tail coverts. Throat bright yellow becoming rich orange on breast. Greyer belly. Creamy round eyes. Dark moustache. Female greyer.

Voice: Noisy. A piping *tee tee tee* and various rattling and explosive notes. Fast fluty song.

Habits: Inhabits forest and plantation undergrowth particularly in ravines. Feeds actively low down on invertebrates in pairs or small parties. Fairly confiding. Nests low down in cover.

Distribution: Locally common breeding resident in northern foothills and mountains from Himachal Pradesh east to Burmese border. More common in east. Moves lower down in winter. Also occurs in China and SE Asia.

⊙ WHITE-BROWED SHRIKE BABBLER (Red-winged)

Pteruthius flaviscapis 16 cm F: Sylviidae

Description: A small, grey, black and white arboreal babbler with rufous wing patches. Male has grey upperparts and white-tipped black wings and tail. Crown black with broad white stripes back from eyes. Whitish below. Female greenish above with grey crown and very indistinct eye stripes. Short, thick bill.

Voice: Quite noisy and varied. Loud song *pik kew weew*. Various churring calls.

Habits: Inhabits canopy and large branches of tall forest trees, feeding on invertebrates in pairs or small parties. Often in mixed hunting groups. Nests high in tree.

Morten Strange

Distribution: Fairly common breeding resident in northern mountains from Pakistan border east to Burmese border. Moves lower down in winter. Also occurs in SE Asia.

⊙ BLACK-EARED SHRIKE BABBLER (Chestnut-throated)

Pteruthius melanotis 11 cm F: Sylviidae

Description: A very small, green and yellow arboreal babbler with short black bill. Green above with blue-grey in wings and nape. Male has white-barred black wing patches, black cheek patches and extensive chestnut throat. Female duller with chestnut moustache and black ear spot. Both have white eye rings. Very rare **Chestnut-fronted** *P. aenobarbus* has chestnut forehead and yellow cheeks.

Voice: Quiet. Sometimes calls *too weet* and *si si si*.

Habits: Inhabits middle and upper stories of forests feeding on invertebrates in pairs, small parties or mixed hunting groups. Very difficult to see well. Nests high in tree.

Distribution: Locally common breeding resident in northern mountains from western Nepal east to the Burmese border. Moves lower down in winter. Also occurs in SE Asia.

Sujan Chatterjee

Morten Strange

WHITE-HOODED BABBLER (White-headed Shrike)

Gampsorhynchus rufulus 23 cm F: Sylviidae

Morten Strange

Description: A large, bull-necked, long-tailed babbler with white head and underparts. Upperparts warm brown. Tail tip and vent buff. Noticeably pale irises, bill and legs. Sexes alike. Juvenile has orange crown and nape and buffer underparts.

Voice: A harsh *kaw ka yawk* and a hard rattling.

Habits: Inhabits forest undergrowth and bamboo thickets. Feeds on invertebrates in small parties or mixed hunting groups. Nests in bush.

Distribution: Rather scarce breeding resident in northern foothills from extreme eastern Nepal to the Burmese border.
Also occurs in SE Asia.

RUSTY-FRONTED BARWING (Spectacled Barwing)

Actinodura egertoni 23 cm F: Sylviidae

Sujan Chatterjee

Description: A large, slim, long-tailed, brown babbler with a shaggy grey crown. All rufous brown with fine black barring on wings and tail. Unstreaked grey head with rufous face separate from other barwings. Pale bill, eye rings and irises. Sexes alike.

Voice: Calls *cheep, de de de tew* and *ti ti ta*.

Habits: Inhabits forest undergrowth and secondary growth. Feeds in pairs or small parties on invertebrates usually low down but also in canopy. Nests in tree or bush.

Distribution: Locally common breeding resident in northern foothills from central Nepal to the Burmese border. Moves into adjoining plains in winter. Also occurs in Myanmar.

BLUE-WINGED MINLA (Blue-winged Siva)

Minla cyanouroptera 15 cm F: Sylviidae

Description: A small, pale blue and brown babbler with prominent black eyes. Pale brown above with greyish-blue edged wings and on square-ended tail. Black-edged and streaked bluish crown. Pale supercilia and very plain face emphasising dark eyes. Underparts pale greyish-buff and bill pale yellow. Sexes alike.

Voice: A whistling *cree cree* and a hard *cheep*.

Habits: Inhabits forest edge and undergrowth, feeding fairly low down on invertebrates. Usually in mixed hunting groups. Active and rather shy. Nests in low cover.

Sujan Chatterjee

Distribution: Local breeding resident in northern mountains from Uttaranchal east to the Burmese border. Moves lower down in winter.
Also occurs in China and SE Asia.

CHESTNUT-TAILED MINLA (Bar-throated Siva/Minla)

Minla strigula 14 cm F: Sylviidae

Description: A small, brightly coloured babbler with a high-domed chestnut crown. Olive brown above with black, white and chestnut patterned wings and a chestnut-centred, black tail with yellow edgings. Dusky cheeks and finely black-barred, white throat. Chin and underparts yellow. Sexes alike. **Red-tailed Minla** has black head with broad, white supercilia.

Voice: Whistling *tsee tsi tsay tsse* and *pseep*.

Habits: Inhabits lower canopy and high undergrowth of forests, actively feeding on invertebrates and fruit in small parties or in mixed hunting groups. Nests low down in cover.

Morten Strange

Distribution: Locally common breeding resident in northern mountains from Himachal Pradesh east to the Burmese border. Moves lower down in winter.
Also occurs in China and SE Asia.

RED-TAILED MINLA

Minla ignotincta 14 cm F: Sylviidae

Ron Saldino

Description: A distinctive, brown and yellow babbler with a black head. Broad white supercilia and white throat. Underparts yellow and mantle rufous in male. Red tail edges and wing panels. White-tipped, black wing coverts and scapulars. Female duller, olive brown on mantle and paler below.

Voice: Noisy and variable. Commonly *wi wi wi*, a repeated *chik* and *chitchitchit*. Loud song *twiyi twiyuwi*.

Habits: Inhabits forests, feeding on invertebrates in canopy in a tit-like way. Also creeps about mossy trunks and branches. Mainly seen in mixed hunting groups. Nests in tree.

Distribution: Locally common breeding resident in northern mountains from central Nepal east to the Burmese border. Moves lower down in winter. Also occurs in China and SE Asia.

RUFOUS-WINGED FULVETTA (Chestnut-headed Tit)

Alcippe castaneceps 10 cm F: Sylviidae

Sujan Chatterjee

Description: A tiny, plump, short-billed babbler with a chestnut crown. Olive brown above with reddish wing patches and black wing coverts. White supercilia and throat. Greyish cheeks and black eye and moustachial stripes. Buff flanks. Sexes alike.

Voice: *tew twee twee* and *cheep*.

Habits: Inhabits forest undergrowth where it feeds on invertebrates and tree sap. Usually in quite large parties, often in mixed hunting groups. Very active. Creeps on mossy trunks. Nests low down.

Distribution: Locally common breeding resident in northern mountains from central Nepal east to the Burmese border. Moves lower down in winter. Also occurs in SE Asia.

⊙ WHITE-BROWED FULVETTA (White-browed Tit Babbler)

Alcippe vinipectus 11 cm F: Sylviidae

Description: A tiny, plump, short-billed babbler with broad, white supercilia. Races vary. Dull brown above with brighter orange-brown wing patches and white edged black flight feathers. Usually whitish below with rusty flanks and vent. Dark-edged, brown crown and dark cheeks. White irises. Sexes alike.

Voice: Calls *chip chip* and *churr* Songs *tsee tsee ter tee* and *pees psee kew*.

Habits: Inhabits forest undergrowth and mountain scrub. Feeds actively on invertebrates usually in small parties, often in mixed hunting groups. Inquisitive and confiding. Nests low down.

Sujan Chatterjee

Distribution: Common breeding resident in northern mountains from Himachal Pradesh east to the Burmese border. Moves lower in winter.
Also occurs in China and SE Asia.

⊙ NEPAL FULVETTA (Nepal Babbler)

Alcippe nipalensis 12 cm F: Sylviidae

Description: A small, plump, short-billed babbler with white eye rings. Mainly brown with contrasting grey head and whitish throat to belly. Thin, dark borders to crown sides. Pale legs. Sexes alike. Peninsular **Brown-cheeked** *A. poioicephala* is similar but larger, with darker legs and a plain grey head.

Voice: Noisy. Trilling *dzi dzi dzi dzi* and *pi pi pi pi* Also *p p p jet*.

Habits: Inhabits forest undergrowth and middle strata. Feeds inconspicuously on invertebrates, usually in small active parties. Often acrobatic. Nests low down.

Sujan Chatterjee

Distribution: Locally common breeding resident in northern mountains from western Nepal east to the Burmese border and Bangladesh. Moves lower down in winter.
Also occurs in Myanmar.

RUFOUS SIBIA (Black-capped Sibia)

Heterophasia capistrata 21 cm F: Sylviidae

Jan Willem den Besten

Description: A medium-sized, orange and grey, arboreal babbler with a black hood. Mainly rufous-orange, darker above and with black and grey wings. Long tail has black sub-terminal and grey terminal band. Sexes alike.

Voice: Noisy. Ringing *tee dee dee dee dee o lu* and *chre chre* alarm.

Habits: Inhabits forest feeding on invertebrates, nectar and fruit in canopy and on large branches and trunks. Also fly-catches. Usually in pairs, small parties or mixed hunting groups. Active, acrobatic, conspicuous and fairly confiding. Nests in tree.

Distribution: Common breeding resident in northern mountains from north Pakistan east to Arunachal Pradesh. Moves lower down in winter. Also occurs in Myanmar.

LONG-TAILED SIBIA

Heterophasia picaoides 30 cm F: Sylviidae

Morten Strange

Description: A large, grey arboreal babbler with a long, graduated, white-tipped tail. Grey head and breast. Browner on back, wings and tail. White wing flashes. Pale grey below. Red irises. Sexes alike.

Voice: Ringing *tsip tsip tsip*.

Habits: Inhabits upper and middle levels of evergreen forest and forest edges. Feeds on nectar, invertebrates and seeds in pairs or single species flocks. Confiding but very active. Nests high in tree.

Distribution: Locally common breeding resident from eastern Nepal to the Burmese border. Moves lower down in winter.
Also occurs in China and SE Asia.

WHITE-NAPED YUHINA

Yuhina bakeri 13 cm F: Sylviidae

)**escription**: A small, brown-
rested babbler with white nape
atch. High, upswept crest more
hestnut. Upperparts lightly
treaked whitish. Throat whitish
nd cheeks streaked. Underparts
uff, lightly streaked on breast.
exes alike. Similar **Striated**
'uhina *Y. castaniceps* lacks white
ape. Always has chestnut cheeks
ut crested crown either chestnut
r grey, depending on race.

/**oice**: A nasal *chew chew* call
nd chattering.

labits: Inhabits middle levels
f forest, actively feeding on
nvertebrates in pairs or in mixed
unting groups. Nests close
o ground.

Tim Loseby

Distribution: Locally common breeding resident in northern foothills and mountains from eastern Nepal east to the Burmese border. Moves lower down in winter.
Also occurs in Myanmar.

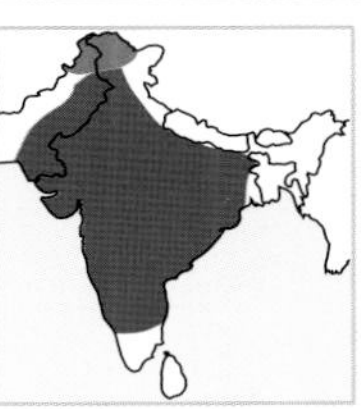

WHISKERED YUHINA (Yellow-naped Yuhina)

Yuhina flavicollis 13 cm F: Sylviidae

)**escription**: A small, grey-brown,
rested babbler with yellow or
ufous (in east) nape. Dark brown
ıpswept crest, paler and greyer
rown. Black moustache. Buffish
vhite underparts with fine white
ɔreast and flank streaking. White
eye rings. Sexes alike.

/**oice**: Noisy. Twittering and
a loud *chee chi choo* and
ɔther notes.

labits: Inhabits middle and
ower levels of forest, sometimes
ascending to canopy. Actively
eeds on invertebrates, fruit and
nectar, usually in mixed hunting
groups. Fly-catches. Confiding.
Nests low down.

Sujan Chatterjee

Distribution: Common breeding resident in northern mountains from Himachal Pradesh east to the Burmese border. Moves lower down in winter.
Also occurs in China and SE Asia.

STRIPE-THROATED YUHINA

Yuhina gularis 14 cm F: Sylviidae

Sujan Chatterjee

Description: A small, brown-crested babbler with a black-streaked throat. Brown above with striking orange wing panels and black flight feathers. Upswept crest all brown. White eye rings. Underparts buffish-orange. Sexes alike.

Voice: A nasal *chay*, *kwee* and *chrr*.

Habits: Inhabits lower and middle levels of forests, particularly rhododendrons. Feeds on invertebrates, fruit and nectar, in pairs or in mixed hunting groups. Nests low down.

Distribution: Locally common breeding resident in northern mountains from Uttaranchal east to the Burmese border. Commoner in east. Moves lower in winter.
Also occurs in China and SE Asia.

RUFOUS-VENTED YUHINA

Yuhina occipitalis 13 cm F: Sylviidae

Sujan Chatterjee

Description: A small, slim, high-crested babbler with a rufous vent. Brown above with contrasting grey-streaked head and very obvious rufous nape and lores. Fine black moustache. Paler, pink-brown below. Sexes alike.

Voice: A rattling *trr trr trr* and *zee zit*.

Habits: Inhabits forest, usually quite high in canopy. Feeds on invertebrates and nectar in small active parties, often in mixed hunting groups. Nests in small tree.

Distribution: Locally common breeding resident in northern mountains from central Nepal to Arunachal Pradesh. Moves lower down in winter.
Also occurs in Tibet, Myanmar and China.

BLACK-CHINNED YUHINA

Yuhina nigrimenta 11 cm F: Sylviidae

Description: A very small, brown and grey, crested babbler with black lores and chin. Short crest blackish on grey head. Whitish throat and warm buff underparts. Red base to bill. Sexes alike.

Voice: Noisy. Thin twittering and buzzing calls including *whee dod dee dee dee*.

Habits: Inhabits undergrowth and middle levels of forests, sometimes in canopy. Feeds actively and acrobatically on invertebrates. In pairs, flocks or mixed hunting groups. Nests low down.

Sanctuary / Nirmal Kaur

Distribution: Local breeding resident in northern mountains from Uttaranchal east to the Burmese border. Moves lower in winter.
Also occurs in China and SE Asia.

BEARDED PARROTBILL (Bearded Tit, Reedling)

Panurus biarmicus 16 cm F: Sylviidae

Description: A long-tailed, orange-brown babbler with a tiny, yellow bill. Black and white in wings and white tips to graduated tail. Male has blue-grey head, white throat and obvious, black moustache. Female has buff head. Young have variable amounts of black streaking on back. Arboreal and rare north-western **White-capped Penduline Tit** *Remiz coronatus* has black head, white crown and much shorter tail.

Voice: A resonant *ping ping*, particularly in flight.

Habits: Inhabits, usually wet, reed and grass beds. Often in flocks feeding actively mainly on seeds. Climbs up and down stems, flicking and opening tail. Low unsteady, whirring flight.

Otto Pfister

Distribution: Rare winter vagrant to northern Pakistan plains.
Also occurs in Europe, Russia, Iran and N Asia.

BROWN PARROTBILL (Brown Suthora)

Paradoxornis unicolor 21 cm F: Sylviidae

Balachandran

Description: A large, long-tailed, brown babbler with a stubby, yellow or grey bill. Greyish wash to slightly streaked brown head and long black lateral crown stripes. Otherwise all rufous brown. Sexes alike. Larger **Great Parrotbill** *Conostoma oemodium* is paler brown, has a much longer, stout, yellow bill, black lores, whitish forehead and no crown stripes.

Voice: Noisy and varied. Commonly a *chirrup* call and a bleating *wheah wheah*.

Habits: Inhabits bamboo stands and forest scrub. Feeds on vegetable matter, particularly bamboo and bracken buds, low down in small skulking groups. Can be acrobatic. Flight weak.

Distribution: Rare breeding resident in northern mountains, mainly from eastern Nepal east to Arunachal Pradesh. Does not move much lower in winter.
Also occurs in Tibet, Myanmar and China.

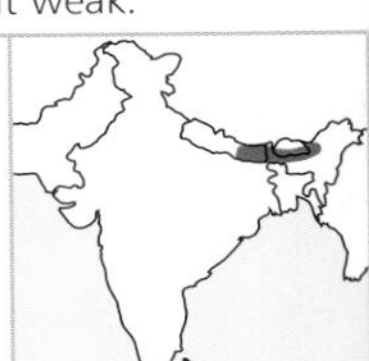

BLACK-THROATED PARROTBILL (Orange Suthora)

Paradoxornis nipalensis 10 cm F: Sylviidae

Sumit Sen

Description: A tiny, dumpy babbler with a stubby, yellow bill. Head pattern (which varies with different races) reminiscent of **Black-throated Tit.** Basically rufous brown with black and white in wings and rather long dark tail. Black-bordered crown either brown or grey. White eye stripes and broad white moustache. Throat black. Cheeks brown or grey and underparts usually grey but brown in one race. Sexes alike.

Voice: Noisy. High *tsi tsi tsi* and twittering.

Habits: Inhabits forest undergrowth, particularly bamboos. Feeds low down on invertebrates in noisy parties, often in mixed hunting groups. Extremely active and acrobatic and not shy.

Distribution: Rare breeding resident in northern mountains from Uttaranchal east to the Burmese border.
Also occurs in China and SE Asia.

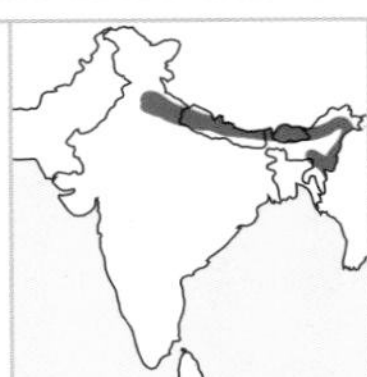

GREATER WHITETHROAT (Common Whitethroat)

Sylvia communis 14 cm F: Sylviidae

Description: A small, brown warbler with a striking white throat contrasting with pinkish-buff underparts. Brown above with rusty wing edgings and white outer tail feathers. Male has high-domed grey head. White eye rings. Vagrant **Garden Warbler** *S. borin* is plain, feature-less brown above with grey neck sides and a thick bill.

Voice: A rasping *charr* and a hard *tak*. Song is short and scratchy often delivered in short song flight.

Habits: Inhabits low scrub and cultivation. Feeds mainly low down on invertebrates and fruit, sometimes in scattered groups. Horizontal carriage. Flight low and laboured. Often perches on bush top with raised crown.

Otto Pfister

Distribution: Has bred in Baluchistan. Fairly common passage migrant through Pakistan and northwest India, presumably en route to and from East Africa.
Also occurs in Europe, Africa, W and N Asia.

LESSER WHITETHROAT (Siberian Lesser Whitethroat)

Sylvia curruca 13 cm F: Sylviidae

Description: A small, greyish warbler with contrasting dark cheek patches. Grey head contrasts with browner mantle and wings. White outer tail feathers to rather short, brown tail. White throat merges into greyer breast and belly. Brownish wash on flanks. Sexes alike. Recently split **Small** or **Desert Lesser Whitethroat** *S. minula* is smaller, finer-billed and much paler sandy-brown with contrasting grey cheeks.

Voice: Noisy. A clicking *tek*. Also churrs. Song is a flourished rattle.

Habits: Inhabits open woodland and scrub, feeding on insects and nectar, usually high in trees but also in bushes. Often in loose groups. Horizontal carriage. Confident direct flight. Rather shy.

Joanna Van Gruisen

Distribution: Common winter visitor to Pakistan and western peninsular and northern India south to north Tamil Nadu and northern Sri Lanka. Absent or rare from most of eastern and northeastern India and Bangladesh.
Also occurs in Europe, N Africa, W and Central Asia and China.

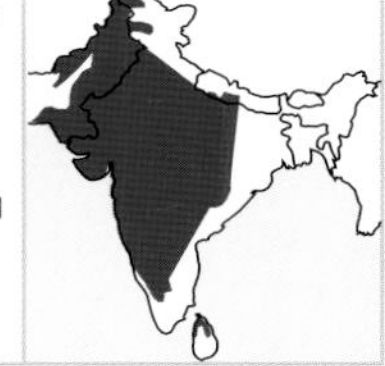

HUME'S LESSER WHITETHROAT (Hume's Whitethroat)

Sylvia althaea 13 cm F: Sylviidae

Otto Pfister

Description: Small, rather dark greyish warbler with even darker cheeks. Very similar to and recently split from **Lesser** but slightly larger and upperparts darker and greyer with minimal contrast between head and mantle. Darker cheeks are less contrasting. Underparts white with grey wash on flanks. Sexes alike. Local western **Menetries's Warbler** *S. mystacea* has blacker head and tail in male, obvious white outer tail feathers on longer, often cocked, tail, pinkish underparts and red eye rings.

Voice: Noisy. As **Lesser** but *tek* call more often followed by low churrs. Song more bubbling.

Habits: Inhabits open woodland and scrub. Behaviour as **Lesser**, with which it sometimes consorts.

Distribution: Uncommon breeding summer visitor to northern hills from western Pakistan east to Kashmir. Winter range appears to mirror that of Lesser but rarely specifically identified when treated as a sub-species so this needs checking.
Also occurs in Iran and Central Asia.

EASTERN DESERT WARBLER (Desert Warbler)

Sylvia nana 11 cm F: Sylviidae

Otto Pfister

Description: A very small, sandy-grey warbler with broad, white edges to rusty tail. Sandy-grey above, brighter on wings and with contrasting rusty rump and central tail. White-edged black alulae. Yellow irises, lower mandible and legs. Sexes alike. Recently split from paler north African race *S. deserti*.

Voice: Chattering *ch errr* call. Jingling song.

Habits: Inhabits sparse scrub in sandy or stony deserts and on saline flats. Also rocky hillsides. Feeds on invertebrates, mainly on ground at base of shrubs where it creeps about like rodent, running across open to new site. Also feeds in shrubs and occasionally perches in open. Constantly flirts tail. Low whirring flight.

Distribution: Locally common winter visitor to dry parts of Pakistan and extreme northwestern India from south Punjab through the Thar to western Gujarat. Vagrant further east.
Also occurs in W and Central Asia.

⊙ EASTERN ORPHEAN WARBLER (Orphean Warbler)

Sylvia crassirostris 15 cm F: Sylviidae

Description: A bulky, grey warbler with striking white irises when adult. Male has black head and grey mantle with slightly browner wings. Female has grey head and darker ear coverts, similar to **Hume's Lesser Whitethroat**. Broad, white sides to long, dark tail. Throat white, underparts with greyish wash and diffuse white scalloping on vent. Strong dark bill. Recently split from western race *S. hortensis*.

Voice: A hard *chak* and a rattling *trurrr*. Loud liquid song.

Habits: Inhabits woodland and scrub, feeding mainly in canopy on invertebrates and nectar. Usually singly or in pairs. Rather ponderous with laboured flight. Cocks tail. Shy and difficult to get close to. Nests low in bush.

Tim Loseby

Distribution: Locally common breeding summer visitor to western Pakistan. Winter visitor mainly to northwestern Deccan of India (Gujarat, Maharastra and Karnataka) but regular on passage further northwest and scattered winter records to east and south. Also occurs in Africa, W and Central Asia.

⊙ SINGING BUSHLARK (Singing Lark)

Mirafra cantillans 14 cm F: Alaudidae

Description: A medium-sized, streaked brown lark with white outer tail feathers. Upperparts heavily streaked and light rufous wing edgings. Slight crest and buff supercilia. Plain cheeks and white throat. Buff below, darker across breast and with spotting limited to sides if present. Stout bill and relatively long tail. Sexes alike.

Voice: High pitched **Skylark**-like song in air or from perch, contains much mimicry.

Habits: Inhabits open dry scrub, fallow and grassland. Feeds singly or in pairs, actively running over ground searching for seeds and invertebrates. Often perches on top of low bush if disturbed or to sing. Nests on ground.

Otto Pfister

Distribution: Very local and patchily distributed breeding resident mainly in north and central peninsular India and parts of Pakistan and Bangladesh. Also occurs in Africa and W Asia.

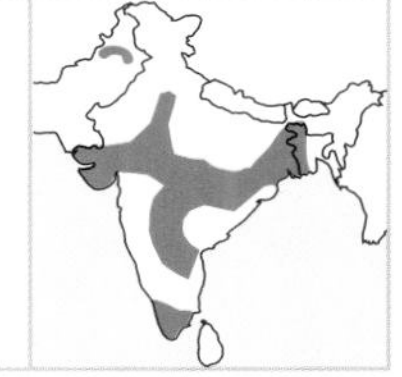

INDIAN BUSHLARK (Red-winged Bushlark, Indian Lark)

Mirafra erythroptera 14 cm F: Alaudidae

Vincent Van Ross

Description: A medium-sized, streaked brown lark with heavily spotted breast. Similar to **Singing** but more rufous in wings and darker cheeks. Outer tail feathers dark buff. Longer-looking stout bill and shorter tail. Sexes alike.

Voice: Calls *sweee* and *trrr weet*. Song whistling *tit tit tit tsweeh tsweeh tsweeh,* started from top of bush and continued in parachuting song flight. Often sings at night.

Habits: Inhabits sparse rocky scrub and barren flats with some bushes and long grass. Feeds on invertebrates and seeds on ground, in pairs or small groups. Furtive but quite confiding, squatting rather than flying when disturbed. Song flight unique. Nests on ground.

Distribution: Locally common endemic breeding resident in plains of northern and central India from Punjab south to Orissa and, in west, northern Tamil Nadu. Also in parts of Pakistan. Absent from southeast, northeast, Nepal, Bangladesh and Sri Lanka.

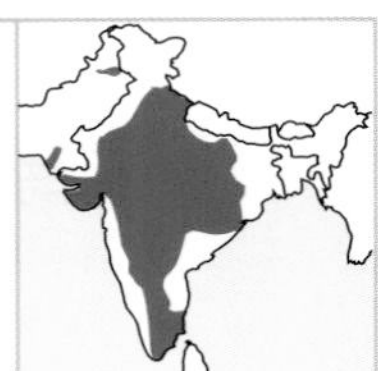

BENGAL BUSHLARK (Assam, Rufous-winged Lark)

Mirafra assamica 15 cm F: Alaudidae

Sumit Sen

Kamal Sahai

Description: A stocky, powerful-billed lark with a short tail. Dark grey-brown above with heavy streaking. Rufous in wings and outer tail feathers. Dark edge to cheeks. Rather dark rufous buff below with heavy breast spotting. Sexes alike. Closely-related and recently split **Jerdon's Bushlark** *M. affinis* is browner and paler on belly.

Voice: A thin high *p zeee* or *tzee tzee* delivered in high, circling song flight. **Jerdon's** has dry rattle delivered from perch.

Habits: Inhabits extensive open grassland and irrigated cultivation, feeding on invertebrates and seeds on ground. In pairs or singly. Nests on ground. **Jerdon's** favours dry scrubby areas.

Distribution: Locally common breeding resident in north Indian plains from Punjab east to the northeast and Bangladesh. Jerdon's is restricted to east (south of Orissa) and southern India and Sri Lanka.
Related races or species occur in SE Asia.

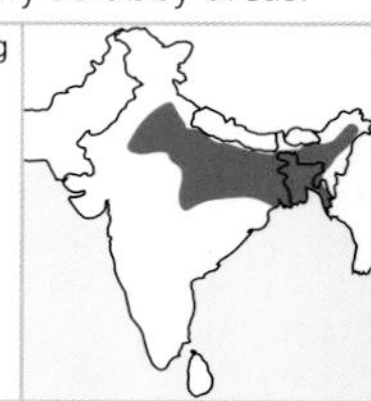

ASHY-CROWNED SPARROW LARK (Finch Lark)

Eremopterix grisea 12 cm F: Alaudidae

Description: A small, stocky, short-tailed lark resembling a **House Sparrow**. Sandy-brown above with darker wing markings. Dark underwing coverts. Pale conical bill. Male has blackish underparts, pale grey crown and nape and black-bordered, whitish cheeks. Female plain, buff on head and below. **Black-crowned** *E. nigriceps* has black crown and nape and grey forehead.

Nikhil Devasar

Voice: Calls *chulp*. Undulating, aerial song *twee deedle deedle*, ends with *deeeeh* as bird plummets to earth.

Habits: Inhabits dry grass plains, mudflats and fallow. Feeds on invertebrates and seeds on ground in roving active groups. Approachable, crouching when disturbed. Nests on ground.

Distribution: Locally common endemic breeding resident throughout plains of region including Bangladesh and Sri Lanka. Absent from Western Ghats and northeast India. Makes local movements particularly into northwest to breed.

RUFOUS-TAILED LARK (Rufous-tailed Finch Lark)

Ammonomanes phoenicurus 16 cm F: Alaudidae

Description: A large, rufous-brown lark with a long, thick bill. Dark brown above with rufous rump and tail which has a broad, blackish terminal band. Underparts more rufous with light breast spotting. Paler supercilia and throat. Sexes alike.

Otto Pfister

Voice: Calls a liquid *qu i qu*. Mellow whistling song delivered in flight, from perch or on ground.

Habits: Inhabits dry, open, often rocky, scrub, fallow and plough. Feeds on invertebrates and seeds on ground in pairs or loose flocks. Can be very inconspicuous but quite confiding. Frequently sits upright and runs erratically. Short, direct, jerky flight with regular glides. Nests on ground.

Distribution: Locally common endemic breeding resident throughout most of peninsular India. Rare or unknown in much of the northern plains and the Northeast, most of Pakistan and Nepal, Bangladesh and Sri Lanka.

BIMACULATED LARK (Eastern Calandra Lark)

Melanocorypha bimaculata 18 cm F: Alaudidae

Tim Loseby

Description: A large, short-tailed, uncrested lark with black upper breast patches. Heavily streaked above with white-tipped, black tail very obvious in flight. Huge yellow bill with dark culmen. Black lores and white supercilia. Wings have no white trailing edges and are greyish below. Rapid **Starling**-like flight. Sexes alike.

Voice: A dry *tripp tripp* and *trr trr treel*. Also a harsh *te tcher*.

Habits: Inhabits dry pastures and stubble, semi-desert and dry mudflats. Feeds on seeds and invertebrates on the ground in restless, noisy flocks often with the two **Short-toed** species. Frequently perches upright when can look huge but usually crouches low when alarmed.

Distribution: Locally common but erratic winter visitor to lowlands of Pakistan and northwest India, south to Maharastra and east to Uttar Pradesh. Also occurs in NE Africa and W and Central Asia.

TIBETAN LARK (Long-billed Calandra Lark)

Melanocorypha maxima 21 cm F: Alaudidae

Otto Pfister

Description: Large, mottled brown lark with long, pale, decurved bill. Long-necked and small-headed. Mottled and streaked dark brown above with dark-centred median coverts and variable blackish patches on breast sides. Greyish-buff below with breast streaking. White trailing edges to secondaries and white tail tip. Sexes alike.

Voice: Calls *tchu lip*. Song is slow and hesitant.

Habits: Inhabits montane grassy marshes, feeding on invertebrates and seeds on ground. Twitches wings when singing. Confiding.

Distribution: Local breeding resident on high plateaux of Ladakh and Sikkim. Also occurs in Tibet and China.

GREATER SHORT-TOED LARK (Short-toed Lark)

Calandrella brachydactyla 15 cm F: Alaudidae

Description: A medium-sized, well-streaked, uncrested lark usually with two black breast patches. Variable and extremely similar to **Hume's** but usually browner and more streaked above with plain brown wing shoulders and obvious dark-centred lesser coverts. Underparts usually look whiter and bill flesh-grey rather than yellow.

Voice: A soft dry rippling *chirip* and *dyu* often combined. Flocks call constantly in flight.

Habits: Inhabits dry pasture, stubble, fallow, semi-desert and dry mudflats. Feeds on seeds and invertebrates on ground in large restless flocks, often mixed with **Hume's.** Inconspicuous until flushed, when flocks wheel off then return to their original site.

Tim Loseby

Distribution: Locally common, often numerous, winter visitor to lowlands of Pakistan, Nepal, Bangladesh and India south to Madhya Pradesh. Much rarer further south and not recorded in Sri Lanka.
Also occurs in Europe, N Africa, W, N and Central Asia, Myanmar and China.

HUME'S SHORT-TOED LARK (Hume's Lark)

Calandrella acutirostris 14 cm F: Alaudidae

Description: Medium-sized, uncrested, greyish-brown lark with conical, black-tipped yellow bill. Lightly streaked greyish-brown above. Whitish below with grey-buff wash on breast and, often, small black breast patches. Cheeks and lores greyish. Sexes alike. Very similar **Greater** is browner with dark-centred median coverts. Bill greyer.

Voice: A rolling *tyrrr*.

Habits: Inhabits montane semi-desert and rocky hills to breed. Winters on open grass flats, river sandbanks, fallow and semi-desert. Feeds on seeds and insects on ground in large restless flocks. Flocks rise suddenly to wheel round then return to their original site.

Otto Pfister

Distribution: Locally common breeding summer visitor to northern hills from Baluchistan east to Sikkim. Particularly in Ladakh. Winters locally in northern India and Nepal south to Madhya Pradesh and east to Bangladesh.
Also occurs in Central Asia.

SAND LARK (Indian Short-toed, Indian Sand Lark)

Calandrella raytal 12 cm F: Alaudidae

Jan Willem den Besten

Description: A small, pale lark with a rather slender bill. Finely streaked sandy-grey above. Whitish below with fine breast streaking. Short crest and tail. Pale cheeks. Sexes alike. **Lesser Short-toed** *C. rufescens* has a thicker bill, white forehead line, projecting primaries and more prominent cheek, mantle and breast streaking.

Voice: A deep *prrr prrr* call. Song, delivered from ground or in song flight, is a rapid, undulating and repeated series of warbles.

Habits: Inhabits sandy and muddy river and lake banks and islands. Also coastal flats and sand dunes. Feeds on invertebrates and seeds, actively on ground in pairs or small flocks. Confiding. Runs erratically.

Distribution: Locally common breeding resident along rivers of Pakistan and north India east to Assam and Bangladesh. Also northwestern coasts. Also occurs in Myanmar.

CRESTED LARK

Galerida cristata 18 cm F: Alaudidae

Otto Pfister

Description: Large, streaked sandy-grey lark with a prominent erectile crest. Heavily streaked above and on breast. Short tail with rufous-buff underwing and outer tail feathers. Powerful, decurved bill. Broad, rounded wings. Sexes alike.

Voice: Plaintive *du ee* or *toolee du deeo*. Varied whistling song includes call notes and mimicry, in flight, from perch or ground.

Habits: Inhabits dry, open country including fallow, jheel and river edges. Feeds on invertebrates and seeds on ground, singly, in pairs or small flocks. Confiding, freezing if approached too closely. Nests on ground.

Distribution: Common breeding resident in plains from Pakistan east to southern Nepal and Bihar and south to Madhya Pradesh and northern Gujarat. Also occurs in Europe, Africa, W, N, Central and E Asia.

MALABAR LARK (Malabar Crested Lark)

Galerida malabarica 15 cm F: Alaudidae

Description: A medium-sized, dark lark with a long crest. Rufous brown with heavy black streaking above and on breast. Rufous-buff below but not as dark as smaller **Sykes's**. Pale rufous outer tail feathers and large bill. Sexes similar.

Voice: Call *chew chew yu*. Song similar to **Crested**.

Habits: Inhabits cultivation, grassy hillsides and open scrub. Feeds on seeds and invertebrates on ground in pairs, singly or in small flocks. Very confiding, usually crouching if alarmed. Sings from perch or in short song flight. Nests on ground.

Clement M Francis

Distribution: Locally common endemic breeding resident in plains and hills of Western Ghats and west coast from Gujarat south to Kerala.
Previously considered a race of mainly African Thekla Lark *G. theklae*.

SYKES'S LARK (Sykes's Crested, Tawny Lark)

Galerida deva 14 cm F: Alaudidae

Description: A rather small, tawny-brown lark with a prominent crest. Tawny below with sparse breast streaking. Richer brown above, well-streaked blackish. Rufous-buff outer feathers to short tail. Short bill. Sexes alike.

Voice: A nasal *dwezee* call. Song similar to **Indian Bushlark**.

Habits: Inhabits dry stony grassland, open scrub and arable cultivation. Often on dark soils. Feeds on invertebrates and seeds singly, in pairs or small parties. Fairly approachable. Nests on ground.

Sanctuary / Ravi Sankaran

Distribution: Local endemic breeding resident in Deccan from Gujarat and south Uttar Pradesh south to northern Karnataka and Andhra Pradesh.

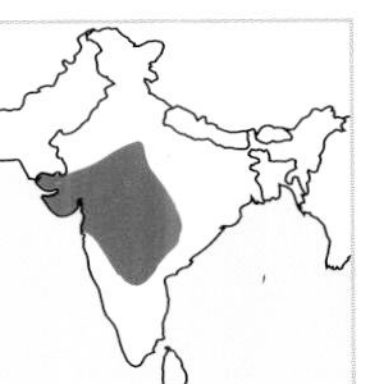

EURASIAN SKYLARK (Skylark, Common Skylark)

Alauda arvensis 18 cm F: Alaudidae

Morten Strange

Morten Strange

Description: A large, streaked brown lark with a short crest. Similar to **Oriental** but obviously larger with clear, white outer tail feathers and trailing edge to wings. No rusty colouring in wings. Looks much whiter below with finer breast and flank streaking. Rather long wings and tail. Sexes similar.

Voice: A loud, dry *chirrup* or a musical *truee*, often repeated and with variants. Famous, continuous song delivered in towering song-flight.

Habits: Inhabits fallow, stubble and young cereal cultivation. Feeds on invertebrates and seeds on ground. Frequently dust-bathes. Often in large flocks. Rather shy and flighty.

Distribution: Locally common but rather erratic winter visitor to lowlands of Pakistan and northern India east to Uttar Pradesh and south to Rajasthan. Also occurs in Europe, N Africa, W, N, Central and E Asia.

ORIENTAL SKYLARK (Eastern, Little Skylark)

Alauda gulgula 16 cm F: Alaudidae

Otto Pfister

Description: Medium-sized, streaked-brown lark with short crest. Dark-streaked, warm brown (more russet in western and Sri Lankan races). Rufous wing edges and buff outer tail feathers. Buff underparts and lightly streaked breast. Fine bill. Sexes alike. Larger **Eurasian Skyark** has longer tail, no rufous in wing, white trailing edges and outer tail feathers and much whiter underparts.

Voice: A short *bazz bazz* flight call. Warbling song usually in high circling song flight.

Habits: Inhabits grasslands and cultivation often near water. Feeds on invertebrates and seeds on ground in pairs or loose flocks. Makes short runs and crouches if disturbed. Nests on ground.

Distribution: Common breeding resident in hiils and lowlands throughout region including Sri Lanka. Also occurs in W, Central and SE Asia and China.

HORNED LARK (Shore Lark)

Eremophila alpestris 19 cm F: Alaudidae

Description: Large, pale lark with unique black markings. Pale rather pinky-brown above with white-edged, black tail and darker wing markings. Whitish below. Black horns on crown, black band from bill through eyes to lower neck and broad black breast band. In female black is duller. Long wings.

Voice: Thin, squeaky *tsee* or *eeh*. Short squeaky song.

Habits: Inhabits montane stony plateaux, slopes and pastures, wintering on fallow and sandy river banks. Feeds actively on seeds and invertebrates on ground in pairs or small flocks. Characteristic, crouching gait but also runs and perches on mounds. Confiding but well-camouflaged. Nests on ground.

Otto Pfister

Distribution: Common breeding resident in high northern mountains from northern Pakistan east to Sikkim. Winters in higher foothills. Exceptional vagrant to plains (e.g., Delhi, September 2001).
Also occurs in Europe, N Africa, W, N, Central and E Asia and the Americas.

THICK-BILLED FLOWERPECKER

Dicaeum agile 9 cm F: Nectariniidae

Description: A tiny, arboreal bird with a stubby, dark bill. Olive grey, often with a green wash. Short white-tipped tail. Indistinct breast streaking and whitish malar stripe. Red irises. Sexes similar.

Voice: Calls *chuk* and *psit* and *ch ch ch ch chrr ch-pss* song.

Habits: Inhabits the canopy of forest, woodland and garden trees. Feeds on fruit, particularly of peepul, banyan and loranthus. Usually singly or in pairs, moving rapidly from site to site on regular beat. Flicks tail from side to side. Best seen by watching favoured tree. Makes suspended purse-like nest.

Morten Strange

Distribution: Local breeding resident in plains and foothills of northern and peninsular India, Nepal and Sri Lanka. Moves into plains in winter. Rare in Pakistan, Bangladesh and much of Northwest, extreme east and south India.
Also occurs in SE Asia.

YELLOW-VENTED FLOWERPECKER

Dicaeum chrysorrheum 9 cm F: Nectariniidae

Morten Strange

Description: A tiny, arboreal bird with heavily streaked underparts. Distinct, black moustache. Rather long, decurved bill, greenish above with dark tail and wings. Bright orange vent and yellowish belly. Red irises. Sexes similar.

Voice: A harsh repeated *tzreep* call.

Habits: Inhabits canopy of forest and orchard trees. Feeds mainly on loranthus in pairs or singly. Flies strongly, covering considerable distances to favourite feeding trees. Nests high in tree.

Distribution: Common breeding resident in hills of extreme northeast India to the Myanmar border. Rare in Nepal and Bangladesh. Also occurs in China and SE Asia.

ORANGE-BELLIED FLOWERPECKER

Dicaeum trigonostigma 9 cm F: Nectariniidae

Morten Strange

Description: A tiny, orange and grey arboreal bird. Male blue-grey on head and wings with darker tail and paler throat. Orange lower mantle to rump and underparts. Female olive with orange tinge on rump and belly.

Voice: A repeated metallic *zip* call. Also a descending *tsi si si si sew* song.

Habits: Inhabits mainly forest edge and mangroves. Behaviour and nesting as other flowerpeckers, feeding mainly in canopy but also close to the ground. Uneven, whirring flight between feeding sites. Very active and confiding.

Distribution: Local breeding resident restricted to coastal Bangladesh and probably neighbouring parts of India. Also occurs in SE Asia.

PALE-BILLED FLOWERPECKER

Dicaeum erythrorynchos 8 cm F: Nectariniidae

Description: Tiny, olive grey flowerpecker with pink bill. Short-tailed and plain. Olive grey above, grey-buff below. Bill short and decurved. Sexes alike. Similar **Plain** has dark bill, pale supercilia and is browner. Similar and widespread **Thick-billed** has thick bill and white-tipped tail.

Voice: Noisy. Surprisingly loud *chik* call often given in flight. Chittering song.

Habits: Inhabits open forest, groves and gardens wherever there is loranthus growing. Feeds very actively, mainly high in canopy on berries and invertebrates. Usually in pairs or small parties, moving rapidly between tree tops. Bounding confident flight. Nests in tree.

Jan Willem den Besten

Distribution: Common breeding resident in plains and low hills of India, Nepal, Sri Lanka and Bangladesh. Extends along Himalayan foothills to Himachal Pradesh but unknown in or rare visitor to most of the Northwest, south to Madhya Pradesh and Gujarat. Also occurs in Myanmar.

PLAIN FLOWERPECKER

Dicaeum concolor 8 cm F: Nectariniidae

Description: A tiny, unmarked, arboreal bird with a short, dark, decurved bill. Note juvenile has paler bill. Olive brown above and grey-buff below so, apart from bill, very similar to **Pale-billed**. Southwestern race darker above and whiter below. Andamans race greener above and yellowish on belly. Sexes similar.

Voice: A sharp repeated *chik* and *pseep* particularly in flight. Extended as a song.

Habits: Inhabits canopy of forest, orchard and garden trees. Also mangroves. Feeds in pairs or small groups mainly on loranthus berries. Flies strongly and noisily from site to site over a large area. Very lively. Nests high in tree.

Otto Pfister

Distribution: Locally common breeding resident in northern foothills from central Nepal east to the Myanmar border, parts of Bangladesh, the Andaman Islands and the Western Ghats.
Also occurs in China and SE Asia.

FIRE-BREASTED FLOWERPECKER (Buff-bellied)

Dicaeum ignipectus 7 cm Nectariniidae

Nigel Redman

Description: Tiny, short-tailed and short-billed arboreal bird. Male dark metallic green above with red breast and buff underparts. Female olive-green above and bright buff below and very similar to **Plain** and **Pale-billed.**

Voice: A penetrating *chip chip* and high-pitched clicking song.

Habits: Inhabits forests, forest edge and orchards, usually feeding actively in canopy on berries, nectar and invertebrates. Singly, in pairs or in mixed hunting parties. All flower-peckers particularly favour loranthus clumps. Moves rapidly from clump to clump, calling excitedly. Nests high in tree.

Distribution: Common breeding resident in northern mountains from Kashmir to the Myanmar border. Moves into foothills in winter.
Also occurs in China and SE Asia.

SCARLET-BACKED FLOWERPECKER

Dicaeum cruentatum 7 cm F: Nectariniidae

Morten Strange

♀

Morten Strange

Description: Tiny, scarlet and black flowerpecker. Male bright red from forehead to rump. Short, black tail and black cheeks, neck and upper breast sides and wings. Underparts greyish-white, darker grey on flanks. Female brownish above and buff below with red rump contrasting with black tail. Juvenile lacks red rump but has red bill.

Voice: Noisy. Penetrating *chip chip* call. Song *tissit tissit* repeated.

Habits: Inhabits open forest, secondary growth and orchards, wherever there is loranthus growing. Feeds on nectar, berries and insects, usually singly, in pairs or small parties. Noisily flies from clump to clump high in canopy. Nests high in tree.

Distribution: Scarce breeding resident in northern hills from eastern Nepal (where rare), northeastern India and Bangladesh.
Also occurs in China and SE Asia.

RUBY-CHEEKED SUNBIRD (Rubycheek)

Anthreptes singalensis 10 cm F: Nectariniidae

Description: A small, short-billed green and yellow sunbird with orange throat and breast. Dark red cheeks of male not easy to see in field. Male is metallic green above and bright yellow below. Female paler and olive green above but also has the orange throat.

Voice: Loud *chirp* and inflected *wee eest* calls.

Habits: Inhabits middle and low levels of forest, forest edge, secondary growth and mangroves. Usually in pairs, small parties or mixed hunting groups. Very active feeding on insects and nectar. Nests low in bush.

Morten Strange

Distribution: Local breeding resident mainly in foothills of northeast India and Bangladesh hills. Occurs as far west as central Nepal with recent records from Uttaranchal and Himachal Pradesh so may be spreading west.
Also occurs in China and SE Asia.

PURPLE-RUMPED SUNBIRD

Nectarinia zeylonica 10 cm F: Nectariniidae

Description: A short-tailed, purple and yellow sunbird. Male has metallic green crown, maroon head and mantle and purple throat, wings and rump contrasting with pale yellow underparts. Female olive-brown above, greyish head and throat and pale yellow below. Similar but much smaller **Crimson-backed** *N. minima* of Western Ghats has crimson breast and mantle. Female has red rump.

Voice: Noisy. Calls *tsiswee tsiswee* and *chip chip*.

Habits: Inhabits forests, secondary growth, parks and gardens. Actively feeds on nectar and invertebrates, usually in pairs and often on low flowers. Nest suspended from tree or bush.

Sumit Sen

Distribution: Common breeding resident in southern half of India, Bangladesh and Sri Lanka south of a line from Mumbai to south Assam. Also occurs in Myanmar.

PURPLE-THROATED SUNBIRD (Van Hasselt's Sunbird)

Nectarinia sperata 10 cm F: Nectariniidae

Morten Strange

Morten Strange

Description: A small, rather short-billed, dark sunbird. Male has purple throat, dark red breast and upper belly and black vent. Metallic green crown and metallic blue-green and black above. Female greenish-olive above and dark buff below often with reddish wash on throat.

Voice: Weak *chip chip* call.

Habits: Inhabits open forest, forest edge, secondary growth and gardens. Feeds actively on nectar and insects usually high in canopy singly or in pairs. Nest supended from tree.

Distribution: Very local breeding resident in Bangladesh north to plains in adjoining Indian states.
Also occurs in SE Asia.

OLIVE-BACKED SUNBIRD

Nectarinia jugularis 11 cm F: Nectariniidae

Morten Strange

Description: A medium-sized, olive green sunbird with a metallic blue-green throat and breast in male. Breast has narrow chestnut band. Rest of underparts bright yellow. Female has throat to vent yellow. Both have dark tail. Eclipse male has dark central throat stripe. Some sub-specific variation on islands.

Voice: Call is *sweet*. Rapid chattering song.

Habits: Inhabits forest, scrub, gardens and mangroves. Feeds, often quite low down, on nectar and insects, singly or in pairs. Confiding. Nest suspended from grass stem or shrub.

Distribution: Common breeding resident restricted to Andaman and Nicobar Islands. One vagrant recorded in southeast Bangladesh. Note this is the commonest sunbird in SE Asia including Myanmar.
Also occurs in China and Australasia.

PURPLE SUNBIRD

Nectarinia asiatica 10 cm F: Nectariniidae

Description: A small, dark sunbird. Male is metallic purple-blue above and on head. Blackish wings and belly and maroon breast band. Two orange and yellow breast tufts which it discloses in display. Eclipse male has dark wings and tail and a long dark purple stripe from chin to belly. Female brownish-grey above, yellowish below. Dark tail.

Voice: Noisy. Loud buzzing *chweet* call extended into hurried, excited song.

Habits: Inhabits all types of open, wooded country, including urban parks and gardens. Feeds singly, in pairs or small groups at all levels, on nectar and invertebrates. Fly-catches. Sings from high perch, often jerking wings. Active and confiding.

Otto Pfister

RK Gaur

Distribution: The most widespread sunbird. Common breeding resident throughout in plains and hills except extreme Northwest and Northeast. Also occurs in W and SE Asia.

LOTEN'S SUNBIRD (Long-billed Sunbird)

Nectarinia lotenia 13 cm F: Nectariniidae

Description: A large, dark sunbird with diagnostic, long, steeply-curved bill. Male metallic blackish-purple on head and upperparts with blacker wings and tail and sooty brown belly. Maroon breast band and yellow breast tufts. In eclipse as female but with blackish stripe from chin to belly. Female dark olive green above and pale yellow below with white tail tips.

Voice: A hard *tchit tchit* call. Measured song *titti titu chewit chewit*.

Habits: Inhabits open wooded country including gardens. Feeds singly or in pairs on nectar (particularly loranthus flowers) and invertebrates. Nest suspended from tree.

Sujan Chatterjee

♀

Balachandran

Distribution: Fairly common endemic breeding resident to southern peninsular India and Sri Lanka, south of Mumbai and southern Andhra Pradesh.

MRS GOULD'S SUNBIRD (Gould's Sunbird)

Aethopyga gouldiae Male 15 cm Female 10 cm F: Nectariniidae

Sujan Chatterjee

Description: A brightly coloured, long-tailed sunbird with a yellow rump. Male has crimson upperparts with purplish crown, cheek patches and throat. Purplish-blue tail with projecting central feathers. Yellow below and on rump. Extreme eastern race has crimson breast. Female short-tailed, olive green above, paler below with white tail tips and yellow rump.

Voice: A sharp repeated *tzit*.

Habits: Inhabits montane forest feeding on nectar and invertebrates at all levels. Favours rhododendrons. Active and not shy. Nests in bush.

Distribution: Locally common breeding resident in northern mountains from Himachal Pradesh east to the Myanmar border. Winters lower down.
Also occurs in Tibet, China and SE Asia.

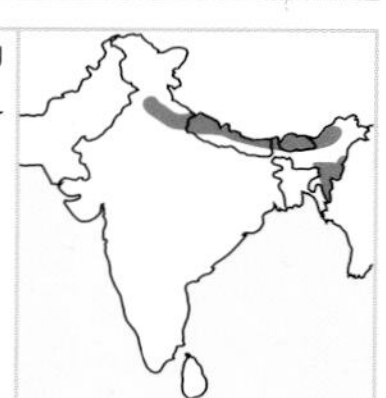

GREEN-TAILED SUNBIRD (Nepal Yellow-backed Sunbird)

Aethopyga nipalensis Male 14 cm Female 10 cm F: Nectariniidae

Otto Pfister

Description: A small, brightly coloured, long-tailed sunbird. Male metallic blue-green crown, throat and tail, yellow rump and underparts and maroon mantle. Female olive with graduated white-edged tail.

Voice: Call *dzit*. Song *tchiss iss iss iss*.

Habits: Inhabits forests, secondary growth and gardens. Feeds at all levels on nectar and invertebrates, often in mixed hunting groups. Nest suspended from tree.

Distribution: Locally common breeding resident in northern mountains from Uttaranchal east to the Myanmar border. Commoner in east. Moves lower down in foothills.
Also occurs in SE Asia.

BLACK-THROATED SUNBIRD (Black-breasted Sunbird)

Aethopyga saturata Male 15 cm Female 10 cm F: Nectariniidae

Description: A very dark, long-tailed sunbird. Male has metallic blue crown and moustache. Blue rump and tail with yellow back patch and maroon mantle. Black face and throat to breast and dull greyish belly to vent. Female drab olive with pale rump.

Voice: A sharp *pzit* call.

Habits: Inhabits mainly low growth in and on edges of forest, secondary growth and gardens. Feeds at rather low levels, as well as in canopy, on nectar and invertebrates. Confiding. Nest suspended rather low in tree.

Morten Strange

Morten Strange

Distribution: Locally common breeding resident in northern mountains from Uttaranchal east to the Myanmar border. Winters lower down in foothills. Also occurs in China and SE Asia.

CRIMSON SUNBIRD (Yellow-backed, Scarlet-breasted)

Aethopyga siparaja Male 15 cm Female 10 cm F: Nectariniidae

Description: Mainly red sunbird with tail length varying with sub-species. Male crimson on mantle and head, scarlet throat and breast with grey belly. Crown, moustache and tail green. Yellow rump patch. Northern races have elongated tail feathers. Western Ghat and Nicobar forms have much shorter tails and some plumage variations. Female dull olive. Eclipse male similar but with yellow rump and reddish from throat to breast.

Voice: *chi chiwee, wepitititi*. Chirping song.

Habits: Inhabits shrubbery in forest, secondary growth, gardens and mangroves. Feeds on nectar and insects, usually singly or in pairs.

Bikram Grewal

Nikhil Devasar

Distribution: Locally common breeding resident in northern hills from Himachal to Arunachal Pradesh and some parts of the Northeast, Bangladesh, Eastern and northern Western Ghats and the Nicobar Islands. Moves lower and into adjacent northern plains in winter, sometimes west to Pakistan. Also occurs in China and SE Asia.

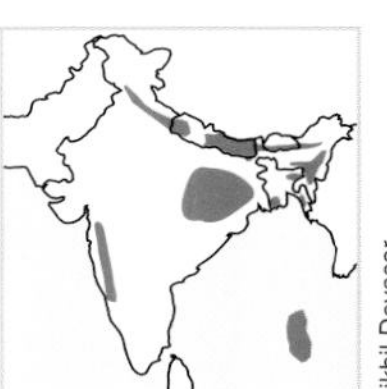

FIRE-TAILED SUNBIRD

Aethopyga ignicauda Male 16 cm Female 11 cm F: Nectariniidae

Balachandran

Otto Pfister

Description: A very long-tailed, red and yellow sunbird. Male scarlet above with yellow rump band and brown wings. Purple crown, face and throat and yellow underparts, more orange on breast. In eclipse has reddish rump and tail sides and collar on otherwise olive plumage. Female dark olive with reddish tail sides and paler rump and belly.

Voice: High pitched *dzidzi dzidzi dzidzi*.

Habits: Inhabits montane shrubbery, moving into forest in winter. Feeds very actively on invertebrates and nectar in pairs or singly. Male waves tail around. Nest suspended from low bush.

Distribution: Locally common high altitude breeding resident in northern mountains from Himachal Pradesh east to the Myanmar border. Winters within upper treeline.
Also occurs in Myanmar and China.

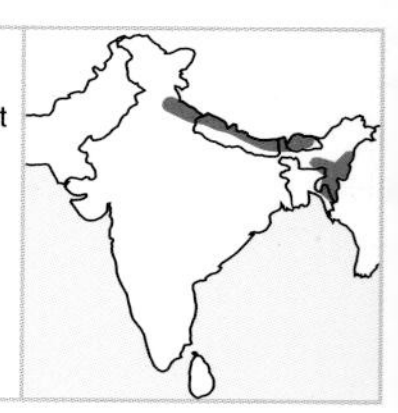

LITTLE SPIDERHUNTER

Arachnothera longirostra 14 cm F: Nectariniidae

Morten Strange

Distribution: Locally common breeding resident in northeastern foothills, Bangladesh and Western Ghats. Also in adjacent plains.
Also occurs in China and SE Asia.

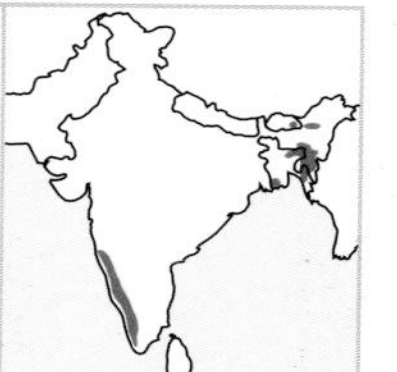

Description: Plump, short-tailed sunbird relative with very long, decurved bill. Olive above with darker, white-tipped tail and more greyish head. White throat and pale yellow underparts. Orange side breast tufts usually hidden. Whitish eye rings. Sexes alike. Confusable with female sunbird.

Voice: Noisy. Repeated *chizt chizt* or *chee chee chee* call. Song *which which which*.

Habits: Inhabits low thick undergrowth in forests and secondary growth often along streams and particularly favouring wild bananas. Feeds singly or in pairs, on nectar, especially of wild bananas. Also invertebrates. Secretive and difficult to see but noisy and active. Nest sewn to underside of large leaf.

STREAKED SPIDERHUNTER

Arachnothera magna 18 cm F: Nectariniidae

Description: A large, heavily streaked sunbird relative with a very long, decurved bill. Bright olive green above and whitish below with bold blackish streaking all over. Pinkish-orange legs.

Voice: Call a hard *chirrick chirrick*. Also chatters.

Habits: Inhabits forest and secondary growth, particularly wild bananas. Feeds very actively on nectar (especially of banana flowers) and invertebrates at all levels including the canopy. Singly, in pairs or in mixed hunting parties. Fast, confident, undulating flight between feeding places. Nest sewn to underside of a large leaf.

Morten Strange

Distribution: Locally common breeding resident in northern foothills from central Nepal east to the Myanmar border. Most frequent in the east. Moves lower down and into plains in winter.
Also occurs in China and SE Asia.

HOUSE SPARROW

Passer domesticus 15 cm F: Passeridae

Description: Brown and grey sparrow with stout bill. Male has grey crown, rump. Black lores and throat to upper breast, forming variably sized and blotched bib. Nape plain chestnut. Cheeks whitish, underparts greyish. Upperparts dark-streaked chestnut-brown. Female buffer with dark streaking above, pale supercilia, unstreaked buff below.

Voice: Noisy especially in display or at roosts. Steady series of *chirrups*, *cheeps*, chattering calls.

Habits: Inhabits urban centres, villages, arable cultivation and reedbeds, where feeds on seeds in large flocks. Also eats human refuse, invertebrates and green shoots. Feeds on ground and on plant heads. Nests in hole.

Otto Pfister

Nikhil Devasar

Distribution: Common breeding resident throughout plains and foothills. Rare in parts of the Northwest and the Northeast.
Also occurs in Europe, Africa, W, N and Central Asia, Myanmar and Indonesia. Introduced to the Americas and Australia.

SPANISH SPARROW

Passer hispaniolensis 16 cm F: Passeridae

Otto Pfister

Description: A chestnut and black sparrow with a stout bill. In breeding plumage male has plain chestnut crown and nape, chestnut and white in wings, heavily black-streaked back and black eye stripe and bib extending as lines of black spots down to belly. Contrasting white cheeks and white supercilia. Chestnut and black fade in non-breeding season and black bill becomes yellow. Female as **House** but more marked supercilia and back braces and lightly streaked underparts.

Voice: Noisy. More metallic *che uit* chirping than **House**.

Habits: Inhabits arable cultivation, semi-desert and reedbeds, feeding on seeds in large flocks, often mixing with **House**.

Distribution: Locally common but erratic winter visitor to plains of Pakistan and northwest India mainly east to Haryana and south to Rajasthan. Vagrant in Uttar Pradesh and Nepal.
Also occurs in Europe, N Africa, W and Central Asia.

SIND SPARROW (Sind Jungle Sparrow)

Passer pyrrhonotus 12 cm F: Passeridae

Nikhil Devasar

♀

Nikhil Devasar

Description: Slim, brown and grey sparrow with neat, rectangular black bib. Male has grey crown and nape, bordered with chestnut stripe that curls partly around pale grey cheeks, concolorous with underparts. Chestnut upperparts including rump, streaked blackish. Female like **House** but with bolder sweeping supercilia, grey cheeks and buff underparts. Pale chestnut lesser wing coverts.

Voice: Noisy. Distinct rocking *cheepa cheepa* call with periodic *tsweep* note.

Habits: Restricted to babuls and other river, canal and lakeside trees and nearby grass and reed beds. Feeds in small parties on ground on seeds and invertebrates in trees. Builds untidy nest in low tree.

Distribution: Local breeding near-endemic restricted range species in Indus valley extending into Indian Punjab and recently colonising Haryana via canal systems.
May also occur in E Iran.

RUSSET SPARROW (Cinnamon Tree Sparrow)

Passer rutilans 15 cm F: Passeridae

Description: A bright chestnut sparrow with, usually, yellowish-buff underparts. Male is chestnut crown to mantle with light streaking. Neat black bib and distinct, white wing bars. Female similar to **House** but brighter with distinct, dusky eye stripes, long, pale, supercilia, chestnut wing shoulders and rump and yellowish-buff underparts.

Voice: High *cheep* and *swee* calls.

Habits: Inhabits open forest, clearings and edges and the neighbourhood of upland settlement and cultivation. Feeds in pairs or small parties on seeds on ground. Also invertebrates. Confiding. Nests in hole in trees, walls and thatch.

Nikhil Devasar

Jan Willen den Besten

Distribution: Locally common breeding resident in northern mountains from north Pakistan east to the Myanmar border. Moves lower in winter. Also occurs in Tibet, E and SE Asia.

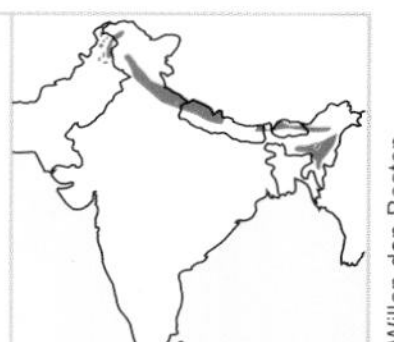

EURASIAN TREE SPARROW (Tree Sparrow)

Passer montanus 14 cm F: Passeridae

Description: A brown and buff sparrow with a rounded, chestnut crown. Both sexes (which unusually for *Passer* sparrows are alike) have discrete, black cheek spots and small, black bib. Cheeks, neck and wing bars white. Underparts buffish-grey. Neater than **House** or **Russet**. Northwestern sub-species, duller and paler.

Voice: A sharp *chip* or *tet* and a nasal *tsuwitt*.

Habits: Inhabits neighbourhood of habitation and cultivation, and is the urban sparrow in the Northeast. Feeds in flocks on seeds often mixed with other sparrows. Confiding. Nests in hole in trees, walls or thatch, sometimes colonially.

Otto Pfister

Distribution: Locally common breeding resident in western Pakistan then from Uttaranchal east to the Myanmar border in foothills and plains. Also occurs in Europe and throughout Asia.

CHESTNUT-SHOULDERED PETRONIA

Petronia xanthocollis 14 cm F: Passeridae

RK Gaur

Otto Pfister

Description: Slim, plain brownish-grey sparrow with rather long tail. Male has prominent chestnut shoulder patches bordered with black and white. Also an indistinct lemon-yellow throat patch. Both markings duller in otherwise identical female. Bill yellow, black in male in breeding season. Rather pale and featureless in some lights. Pipit-like flight.

Voice: Noisy. A rather high *cheep cheep* song. Also calls *cheellup* and *chelp*.

Habits: Inhabits open woodland, forest and thorn-scrub, often near cultivation. Usually arboreal. Feeds on invertebrates, leaves and nectar, in small groups. Difficult to spot in foliage. Also feeds on ground on seeds, mixing with other seed eaters.

Distribution: Locally common resident and local migrant throughout most of Indian plains and hills except extreme Northwest and Northeast. Summer visitor to Pakistan. Vagrant to Sri Lanka and unknown in Bangladesh.
Also occurs in Africa, W and Central Asia.

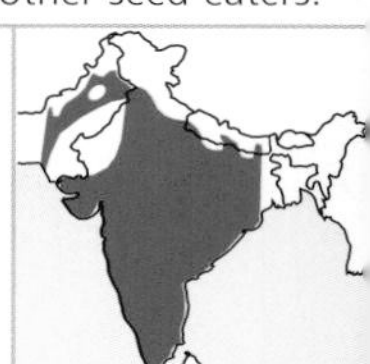

TIBETAN SNOWFINCH (Black-winged Snow Finch)

Montifringilla adamsi 17 cm F: Passeridae

Otto Pfister

Description: A large, pale sparrow with black and white wings and tail. Greyish head and buffish underparts. Small, black bib. Brown mantle. Black wings have broad, white basal patches while black tail has white outer feathers. Both very obvious in flight. Black bill when breeding, otherwise pale. Sexes alike.

Voice: Calls *pink pink*. Staccato chattering song.

Habits: Inhabits barren montane deserts, often near habitation. Feeds on seeds and invertebrates, often at the edge of snow-melt. Frequently perches prominently and can be very confiding. Runs and walks. Flight undulating, low and powerful with gliding. Nests in hole in ground or rock.

Distribution: Locally common breeding resident patchily at high altitudes from Ladakh east to eastern Nepal. Commonest in west of range. Moves lower down in mountains in winter. Also occurs in Tibet and China.

PLAIN-BACKED SNOWFINCH (Blandford's Snowfinch)

Pyrgilauda blanfordi 15 cm F: Passeridae

Description: Pale sparrow with black face markings. Buffish-grey above with cinnamon on neck and mantle. Whitish-buff below. Dark tail with white subterminal patch. No wing bars. Black chin and line through black forehead. Black round eyes extends up to crown as two short lines. Juvenile lacks these face markings. Similar **Rufous-necked** *P. ruficollis* lacks most of black face markings. Chestnut above and on neck.

Voice: Rapid twittering.

Habits: Inhabits montane steppe, often near marshes and lakes. Feeds actively on ground on invertebrates and seeds. Flocks in winter with other sparrows. Confiding. Nests in pika burrow.

Otto Pfister

Otto Pfister

Distribution: Very local breeding high altitude resident in Ladakh and a few sites in Nepal and Sikkim. Moves slightly lower down in winter.
Also occurs in Tibet and China.

FOREST WAGTAIL

Dendronanthus indicus 17 cm F: Passeridae

Description: A pale, woodland wagtail with striking black and white wings. Pale greyish-brown above with white supercilia and a white-edged black tail. Wings black with complex white barring. Underparts white with prominent black double breast bars. Pink legs. Sexes alike.

Voice: A hard *pink, pink* call and a repeated penetrating *tsifee* song.

Habits: Inhabits open forest and woodland, plantations and mangroves, favouring paths and clearings and areas of open leaf litter. Feeds singly or in pairs on ground insects, swaying tail and rear body sideways as it walks or runs. Surprisingly inconspicuous and shy. Flies up into trees, calling, when disturbed.

Otto Pfister

Distribution: Locally common winter visitor mainly to hills of northeast and southwest India, Bangladesh and Sri Lanka. Scattered, mainly passage, records from elsewhere in India. Vagrant to Pakistan and Nepal. Breeds very locally in Assam.
Also occurs in N, E and SE Asia.

WHITE WAGTAIL (Hodgon's Pied Wagtail)

Motacilla (alba) alboides 18 cm F: Passeridae

Otto Pfister

Description: Variable, black and white wagtail. Two black-backed races (or incipient species) occur. Black above, on crown, neck, throat to breast. White forehead and round eyes. White underparts. Broad white wing patches, wing and tail edgings. Dark legs. Other race, *leucopsis*, has more extensive white on head including throat. Female have some grey in back.

Voice: Harsh *chizitt*, *zi zee litt*.

Habits: Inhabits upland stream and riversides in breeding season. In winter in open areas often near water but also lawns, fallow, plough and villages. Feeds singly, in pairs or loose groups running over ground after insects and frequently fly-catching. Often perches high.

Distribution: *Alboides* is common breeding summer visitor to northern mountains from Kashmir east to the Myanmar border. Winters widely in northern lowlands. Less common in south. The only black-backed White Wagtail breeding in region. *Leucopsis* is winter visitor to north and north-eastern India, Nepal and the Andamans. Also occurs in Myanmar and China.

WHITE WAGTAIL (Pied, Masked Wagtail)

Motacilla (alba) personata 18 cm F: Passeridae

Otto Pfister

Description: Variable, grey, black and white wagtails. Four grey-backed races (or incipient species) occur. *Personata* similar to *alboides* but with grey mantle. *Ocularis* has black stripe joining eyes to crown. This is a very rare visitor to northern areas. White forecrown separates both of these forms from larger **White-browed.** Female have less black. Juvenile has only a black breast band.

Voice: As *alboides*.

Habits: As *alboides*. The six individual races (or incipient species) have been little studied for variations in behaviour, if any.

Distribution: *Personata* is a common breeding summer visitor to the north-western mountains of Pakistan and India east to Himachal Pradesh. It winters commonly throughout the lowlands south to Sri Lanka. The only grey-backed White Wagtail breeding in region.
Also occurs in W, N and Central Asia.

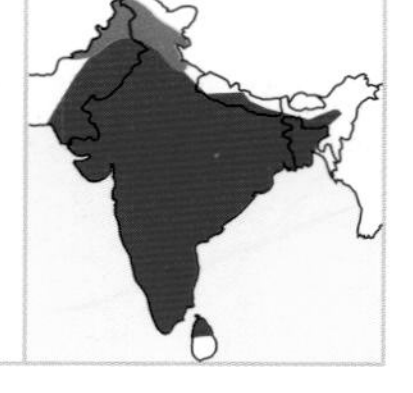

WHITE WAGTAIL (Indian Pied Wagtail)

Motacilla (alba) dukhunensis 18 cm F: Passeridae

Description: The palest of the four grey-backed races (or incipient species) with less black on crown and back. Black of breast extends to chin in breeding plumage. In non-breeding plumage (shown), white chin and throat and very narrow black breast band. Very rare, visiting race *baicalensis* has white chin in all plumages.

Voice: As *alboides*.

Habits: As *alboides*. The six individual races (or incipient species) have been little studied for variations in behaviour, if any.

Otto Pfister

Distribution: *Dukhunensis* winters widely in lowlands south to Andhra Pradesh. Rare further south. Also occurs in W, N and Central Asia.

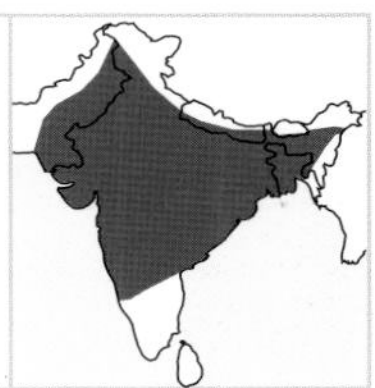

WHITE-BROWED WAGTAIL (Large Pied Wagtail)

Motacilla maderaspatensis 21 cm F: Passeridae

Description: A large, black and white wagtail with distinct, white supercilia. Similar to, but obviously larger than, the **White Wagtails** which never show distinct supercilia. Black above with white wing patches and edgings, white-edged black tail. White under-parts. Female browner above.

Voice: A hard *chezit* call. Jangly, whistling song.

Habits: Inhabits lowland open watersides of all types, particularly favouring irrigation barrages, village tanks and pools. Feeds, usually in pairs, on insects, strutting over ground or fly-catching. Confiding and rather bold. Nests low down close to water.

Otto Pfister

Distribution: Locally common endemic breeding resident throughout lowlands although rare or unknown in parts of the Northwest and Northeast, Bangladesh and Sri Lanka.

CITRINE WAGTAIL (Yellow-headed /hooded Wagtail)

Motacilla (citreola) citreola 17 cm F: Passeridae

Otto Pfister

Description: Variable, yellow and grey wagtail. Breeding plumage male has yellow head and underparts, narrow black collar and grey mantle. Non-breeding plumage male, female and juvenile have variable amounts of yellow on head and breast but all show distinct supercilia (curling right round ear coverts) which in *flava* group are straight. Also shows more prominent, white wing bars. Currently treated as conspecific with *M. c. calcarata*.

Voice: Harsh, wheezy *tzzeep*.

Habits: Inhabits wetlands almost exclusively, including paddy, flooded fields and water hyacinth beds. Feeds in flocks on insects on ground, floating vegetation or in the air. Roosts in reeds, often with **Yellow Wagtails**.

Distribution: Common winter visitor throughout plains, south to Tamil Nadu and Kerala. Rare in Sri Lanka. Includes race *M. c. werae.*
Also occurs in Europe, N, W and Central Asia and China.

CITRINE WAGTAIL (Yellow-headed/hooded Wagtail)

Motacilla (citreola) calcarata 17 cm F: Passeridae

Otto Pfister

♀

Otto Pfister

Description: Variable, black and yellow wagtail. Breeding male has bright yellow head and underparts and black upperparts from hind neck to tail. Prominent white wing bars, flight feather and tail edgings. Non-breeding male, female and juvenile paler than *citreola* but distinctions from *flava* group the same.

Voice: As *citreola*.

Habits: Breeds in montane wet grassland and edges of snow melt. Nests on ground. Otherwise habits as *citreola*. The three **Citrine Wagtail** races are currently treated as as part of one species and there has been little, if any, attempt to study behavioural differences.

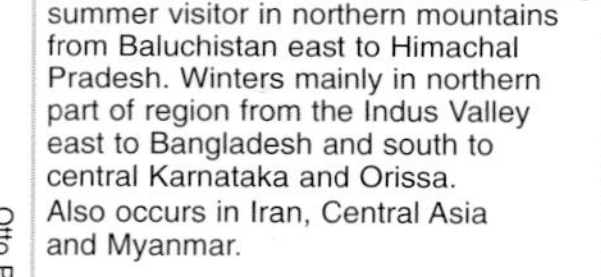

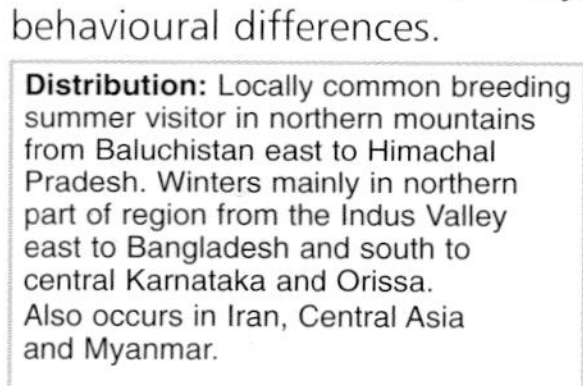

Distribution: Locally common breeding summer visitor in northern mountains from Baluchistan east to Himachal Pradesh. Winters mainly in northern part of region from the Indus Valley east to Bangladesh and south to central Karnataka and Orissa.
Also occurs in Iran, Central Asia and Myanmar.

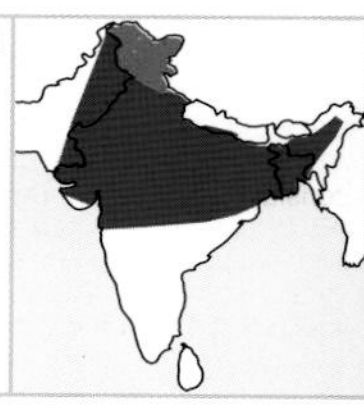

YELLOW WAGTAIL (Blue-headed Wagtail)

Motacilla (flava) thunbergi 17 cm F: Passeridae

Description: Green and yellow wagtails. At least seven races occur, separated by the head colour and pattern on breeding male. *Thunbergi* (shown) is one of the commonest, but *simillina* also has a grey crown. Very common *beema* has pale bluish crown. Scarcer *taivana* has green, *leucocephala* white and *lutea* yellow. All have yellow underparts, green backs and white wing and tail edgings. Females and juveniles duller and less yellow. Straight supercilia separate from **Citrines**.

Voice: A high *tss reep* or *wee zie* call, less harsh than **Citrine**.

Habits: As **Black-headed Yellow**. Different races often seen together, especially on passage.

Distribution: Common winter visitor throughout lowlands of region although the different races vary in abundance and range. *Beema* is also a breeding summer visitor in Ladakh. Various races occur in Europe, Africa and throughout Asia.

Nikhil Devasar

Nikhil Devasar

BLACK-HEADED YELLOW WAGTAIL (Yellow Wagtail)

Motacilla (flava) melanogrisea 17 cm F: Passeridae

Description: Green and yellow wagtail with blackish hood. Breeding male darker green above than other **Yellow** races. White throat then yellow to vent. Juvenile has no yellow and resembles juvenile **Citrine** except for straight supercilia and less white in wings. Specific separation from other **Yellow Wagtails** (perhaps along with more westerly forms) under discussion.

Voice: Slighter harsher *zz ree* call than other **Yellows**, resembling **Citrine** but not as buzzing.

Habits: Inhabits wet grassland, marshes and edges of freshwater bodies. Often accompanying grazing ungulates. Feeds actively in loose flocks on insects, on ground or in air. Mixes with other **Yellows**.

Distribution: Common winter visitor to northern lowlands of region south to northern Kerala and Andhra Pradesh and east to Nepal. Also occurs in Central Asia. Sometimes considered a race of the more western (European, West Asian and African) *feldegg*, and together they are the strongest candidates for separate species status in the Yellow group.

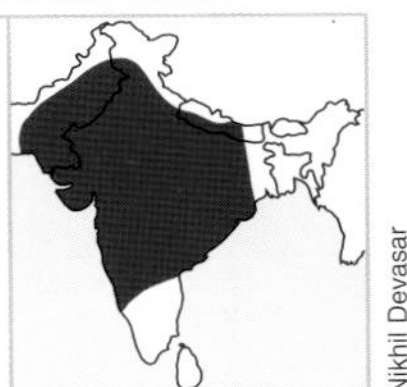

Otto Pfister

Nikhil Devasar

GREY WAGTAIL

Motacilla cinerea 19 cm F: Passeridae

Otto Pfister

Description: A long-tailed, grey and yellow wagtail. Grey above with white-edged black wings and tail. Straight, white supercilia. Yellow below, intense on vent. In some plumages, mainly white with yellow restricted to vent. Short legs pink, unlike all other wagtails except **Forest.** Breeding male has contrasting black throat with white moustache.

Voice: A short, sharp *zit zit* call made commonly in flight.

Habits: Inhabits flowing streams, barrages and adjacent open land. Breeds by fast mountain streams. Usually solitary, feeding actively on invertebrates on ground and in air. Perches on bare branches. Wags long tail constantly. Bounding flight high along water courses. Nests near water.

Distribution: Common breeding summer visitor to northern mountains from Baluchistan east to Bhutan. Winters throughout plains and foothills of region, including Sri Lanka, but always in low concentrations.
Also occurs in Europe, Africa, W, N, E and SE Asia.

RICHARD'S PIPIT

Anthus richardi 18 cm F: Passeridae

Sumit Sen

Description: Large, brown, streaked pipit with a powerful thrush-like bill. Dark brown streaking on sandy or greyish-buff upperparts. Well-streaked breast. Warm buff flanks contrast with white belly. Pale lores give plain faced appearance. Dark moustache bordering white throat. Pointed centres to adult median coverts can separate from **Blyth's**. Long hind-claws, legs and tail and very upright stance.

Voice: Loud grating *shreep* call.

Habits: Inhabits, often wet, grass-land and cultivation. Feeds in loose parties on invertebrates and seeds on ground progressing with fast strutting walk. Powerful bounding flight and often hovers before settling. Generally in lusher habitats than similar species.

Distribution: Fairly common winter visitor to eastern lowlands including Bangladesh and Sri Lanka. Elsewhere scattered records but status uncertain due to identification problems and perhaps only a scarce passage migrant.
Also occurs in N, E and SE Asia.

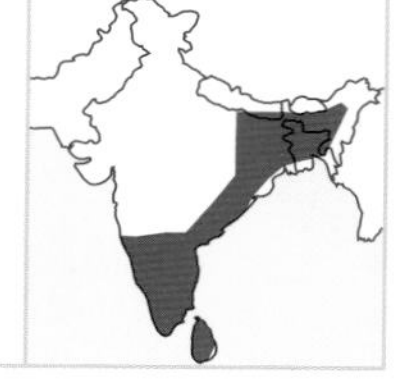

PADDYFIELD PIPIT (Indian Pipit)

Anthus rufulus 15 cm F: Passeridae

Description: A medium-sized, brown, streaked pipit with a fine, short bill. Similar to juvenile **Tawny**, **Blyth's** and **Richard's** but shorter-legged, clawed and tailed and with diffuse, dark lores. Underpart streaking confined to sides of throat and upper breast. Warm buff wash on breast and flanks. White outer tail feathers. Horizontal stance.

Voice: A short weak *tseep tseep* or *chip chip*.

Habits: Inhabits short grassland, marshes, cultivation and fallow. Feeds on ground in pairs or loose groups mainly on invertebrates and seeds. Runs rapidly on ground. Often wags tail. Fluttering flight rarely sustained. Confiding but can be furtive. Parachuting song-flight. Nests on ground.

Sumit Sen

Nikhil Devasar

Distribution: The most widespread pipit. Common breeding resident throughout most of lowlands. Also occurs in SE Asia.

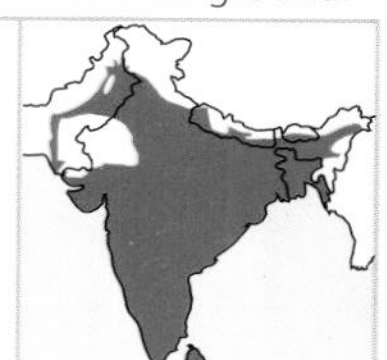

TAWNY PIPIT

Anthus campestris 16 cm F: Passeridae

Description: A medium-sized, rather plain, sandy pipit with dark lores. Adult is plain sandy-brown above with streaking confined to crown and breast sides. Juvenile streaked above and spotted on breast and easily confused with other streaked pipits. Long pale supercilia and indistinct moustache. Obvious white-edged, black median coverts. Outer tail feathers buff. Hind claws short. Horizontal wagtail-like carriage.

Voice: Calls *tseep* or *chulp*.

Habits: Inhabits dry open country including semi-desert, fallow and plough. Feeds singly, in pairs or loose groups on ground on insects and seeds. Runs and walks fast with wagging tail. Often perches on wires or bush tops. Undulating flight.

Otto Pfister

Distribution: Common winter visitor mainly to Pakistan and northwest India but scattered records throughout lowlands. Rare in Nepal and Bangladesh.
Also occurs in Europe, N Africa, W, N and Central Asia.

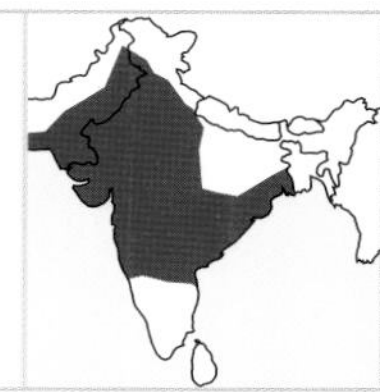

BLYTH'S PIPIT

Anthus godlewskii 17 cm F: Passeridae

Balachandran

Description: A large, streaked pipit with a short, pointed bill. Very similar to **Richard's** but shorter hind-claws, legs, tail and bill and tendency to be more buff below. Adult has squared centres to median coverts. Penultimate outertail feathers have white wedges. Told from immature **Tawny** by pale lores. Horizontal carriage and powerful flight, with no hovering when landing.

Voice: Calls a buzzing *spzeeu* and *chup chup*.

Habits: Inhabits short open grassland, cultivation, fallow and marshes; similar to **Paddyfield**, wetter than **Tawny** and drier than **Richard's** although all four can occur together. Feeds in loose groups on ground on insects and seeds. Can be rather furtive.

Distribution: Probably a widespread and fairly common winter visitor to most of Indian lowlands but identification problems make status uncertain. Very rare in Pakistan and Sri Lanka, rare in Bangladesh and Nepal.
Also occurs in N and Central Asia and Myanmar.

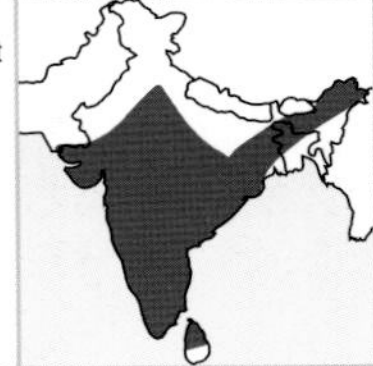

LONG-BILLED PIPIT (Brown Rock Pipit)

Anthus similis 20 cm F: Passeridae

Jan Willem den Besten

Description: Largest pipit; greyish-brown with rufous wing edgings. Dark lores. Powerful thrush-like bill, short legs and upright stance. Northern races, greyest and plainest with barely streaked, whitish-buff or rufous-buff underparts. Southern races more heavily streaked above and more lightly streaked on breast, with rufous-buff underparts. Bounding flight, hovers when landing. Pumps and fans tail upwards.

Voice: A deep *chup* call and a rather slow song.

Habits: Breeds in dry, rocky, grassy uplands. Winters in dry grasslands, open scrub and woodland and fallow cultivation. Feeds, often in loose groups, on ground on invertebrates and seeds. Confiding, often perches upright.

Distribution: Common breeding resident and summer visitor in northern hills from western Pakistan east to west Nepal and some western peninsular hills. Some winter widely in lowlands, mostly from Madhya Pradesh northwestwards to the Indus Valley.
Also occurs in Africa, W and Central Asia and Myanmar.

TREE PIPIT (Eurasian Tree Pipit)

Anthus trivialis 15 cm F: Passeridae

Description: Small, brown, streaked pipit with bold, black breast streaking. Heavily streaked above, especially crown, though rump plain. Buffish edgings to wing feathers and buff washed breast and flanks. White belly. Indistinct whitish supercilia and lores. Sometimes shows white ear covert spots. Pink legs and bill base. Horizontal stance, wags tail.

Voice: Sibilant, piercing *beeze*, usually louder than **Olive-backed**. Loud trilling song in parachuting flight.

Habits: Breeds on grassy mountain slopes. Winters in open wooded areas, cultivation and village margins with trees. Feeds in loose flocks, on invertebrates and seeds. Bouncy flight. Usually flies into trees if disturbed.

Joanna Van Gruisen

Distribution: Local breeding summer visitor to northern mountains from Pakistan to Himachal Pradesh. Winters widely in lowlands except extreme south and east. Commonest in northern and central areas. Vagrant to Bangladesh. Unknown in Sri Lanka. Also occurs in Europe, Africa, W, Central and E Asia.

OLIVE-BACKED PIPIT (Indian, Hodgson's, Olive Tree Pipit)

Anthus hodgsoni 15 cm F: Passeridae

Description: Small, plain-backed pipit with heavily streaked breast and flanks. Like **Tree** but usually shows distinct greenish-olive wash above and mantle barely streaked. Streaked, black-bordered crown, white supercilia with rusty-yellow spot in front of eyes. Often black and white spot on rear ear coverts. Pink legs. Pumps tail constantly when perched and walking.

Voice: Very thin *psee* call. Faster, harder trilling song.

Habits: Breeds in montane open woodland. Winters in open woodland and plantations, almost always beneath trees into which it flies rapidly when disturbed. Feeds furtively in loose parties on ground on invertebrates. Usually shy. Nests on ground.

Sumit Sen

Distribution: Common breeding resident and summer visitor in northern mountains from Himachal east to Arunachal Pradesh. Common winter visitor to hills and plains of northern, eastern, southwestern and central India, Bangladesh and Nepal. Vagrant to Pakistan and Sri Lanka. Also occurs in N, Central and E Asia.

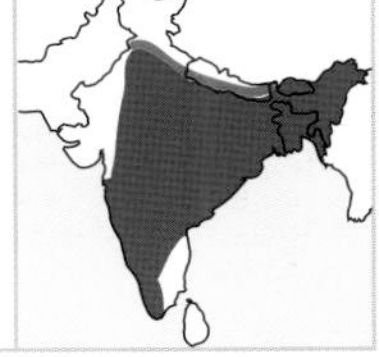

ROSY PIPIT (Vinaceous-breasted, Rose-breasted Pipit)

Anthus roseatus 15 cm F: Passeridae

Otto Pfister

Otto Pfister

Description: Small, heavily streaked pipit. In non-breeding plumage heavily black streaked on greyish-olive above with heavily streaked breast. Prominent buff supercilia, white eye rings and black eyestripes and moustache. Greenish-olive wash to edges to wing feathers. Light pinkish wash on underparts in breeding plumage. Similar **Red-throated** *A. cervinus* brick red on head and breast in breeding plumage with obvious pale braces and white wing edgings. Different call.

Voice: Short, thin *seep seep*.

Habits: Breeds in high altitude meadows. Winters in marshes and damp cultivation. Feeds furtively in loose parties on floating vegetation or wet mud.

Distribution: Common breeding summer visitor to northern mountains from Pakistan east to Arunachal Pradesh. Winters in northern plains south to Rajasthan, Orissa and Bangladesh.
Also occurs in Central and E Asia.

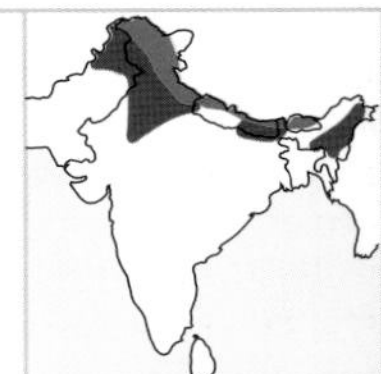

WATER PIPIT

Anthus spinoletta 15 cm F: Passeridae

Otto Pfister

Description: A small, rather grey, dark-legged pipit. Prominent supercilia and usually pale lores. In breeding plumage grey head, browner back and plain pinkish-white underparts. In non-breeding plumage is lightly streaked below and greyer above. Always much less heavily streaked than **Rosy** and **Red-breasted**. Similar **Buff-breasted** *A. rubescens* is darker, less streaked with very pale face and pink legs.

Voice: A thin *tsiip* similar to **Rosy**.

Habits: Inhabits marshes, wet cultivation and the edges of freshwater. Feeds in loose groups on invertebrates on ground. Walks swiftly, often pumping tail. Undulating flight.

Distribution: Scarce winter visitor to lowlands and foothills of Pakistan and north India south to Rajasthan. Scattered records elsewhere in northern India east to Assam and including Nepal.
Also occurs in Europe, N and Central Asia and China.

UPLAND PIPIT

Anthus sylvanus 17 cm F: Passeridae

Description: A large, heavily streaked pipit with a short, broad bill. Dark buff or grey-brown with very heavy, blackish streaking on upperparts and breast. Prominent, buff supercilia, blackish lores and moustache. Pink legs. Thin, pointed tail feathers. Flicks tail.

Voice: Calls a short *chirp* and *zip zip*. Double whistling song *wee chee we chee* and variants.

Habits: Inhabits open woodland, scrubby grassy slopes and abandoned cultivation. Feeds singly or in pairs, on invertebrates and seeds on ground. Often rather furtive and low when feeding but will perch prominently on low eminence or even tree, particularly when singing. Nests on ground.

Jan Willem den Besten

Distribution: Locally common breeding resident in northern foothills from Pakistan east to West Bengal including Nepal. Moves lower down in winter. Also occurs in Central Asia and China.

ALPINE ACCENTOR

Prunella collaris 17 cm F: Passeridae

Description: A large, dark accentor with a prominent yellow bill-base. Grey head, neck and breast with black-barred, white throat. Upperparts streaked brown. Underparts heavily streaked rich chestnut on grey. White tail and wing covert tips. Black wing panels. Sexes alike. Similar, wintering **Altai** *P. himalayana* paler below and white throat centre.

Voice: Call a rolling *chirrirrup* or *chit chitur*.

Habits: Inhabits montane scree slopes, rocky pastures and cliffs. Also near villages in winter. Feeds on ground in pairs or small groups on invertebrates and seeds. Stands upright but runs or walks low. Confiding but unobtrusive. Nests among rocks.

Otto Pfister

Distribution: Fairly common breeding resident above treeline in northern mountains from Pakistan east to Arunachal Pradesh. Moves lower down in winter.
Also occurs in Europe, N Africa, W, Central and E Asia.

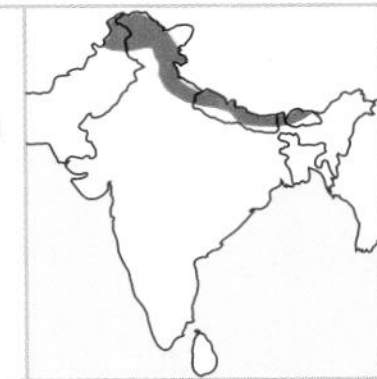

ROBIN ACCENTOR

Prunella rubeculoides 17 cm F: Passeridae

Otto Pfister

Otto Pfister

Description: A large, brown and grey accentor with a broad, unstreaked orange breast band. Grey head and neck and brown-streaked upperparts. Buffish-white belly. White-tipped wing coverts. Sexes alike.

Voice: Calls include *tszi tszi* and a sharp trilling.

Habits: Inhabits damp grassland and scrub, often near water. In winter, stony areas and around villages. Feeds in small groups on invertebrates and seeds on ground. Low lark-like gait. Confiding. Nests low in vegetation. As with all **Accentors** female allows several males to mate with her. Each subsequent male is at great pains to try and remove the sperm of his predecessor from the female's cloaca (as shown).

Distribution: Common breeding resident above treeline of northern mountains from Pakistan east to Bhutan. Moves lower down in winter. Also occurs in Tibet and China.

RUFOUS-BREASTED ACCENTOR

Prunella strophiata 15 cm F: Passeridae

Otto Pfister

Description: A small, chestnut and brown accentor with blackish cheeks and crown edgings. Similar to **Robin Accentor** (apart from head) but throat and belly streaked blackish. Brown-streaked crown and upperparts. Orange supercilia with white in front of eyes and buffy-white moustache. Sexes alike.

Voice: A trilling *tirr r r r t* and *twitt twitt*.

Habits: Inhabits montane scrub. Winters in bushy fallow field margins and scrub forest. Feeds in small groups on invertebrates and seeds on ground. Confiding but unobtrusive as it shuffles along quietly, often in cover. Nests in low bush.

Distribution: Locally common breeding resident at or above treeline in northern mountains from Pakistan east to Arunachal Pradesh. Winters lower down in foothills.
Also occurs in Tibet, Myanmar and China.

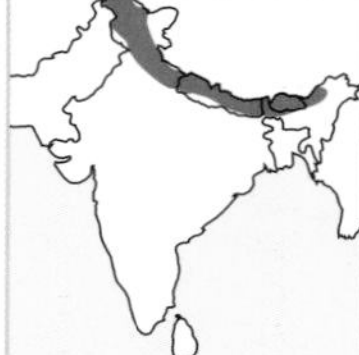

BROWN ACCENTOR

Prunella fulvescens 15 cm F: Passeridae

Description: A small, pale brown and buff accentor with prominent, white supercilia and black cheeks. Lightly brown-streaked crown and upperparts. Rich buff from chin to vent, usually completely unstreaked. Sexes alike.

Voice: Call a thin *zit zit zit*. Also a thin rattle.

Habits: Inhabits montane scrub. Around fallow and villages in winter. Feeds on ground in pairs or small groups on invertebrates and seeds. Unobtrusive. Nests on or near ground.

Otto Pfister

Distribution: Locally common breeding resident at or above treeline in northern mountains from Pakistan east to Nepal. Winters lower down.
Also occurs in Central Asia and China.

BLACK-THROATED ACCENTOR

Prunella atrogularis 15 cm F: Passeridae

Description: A small, streaked accentor with black throat, cheeks and crown sides. Similar to **Brown** but brighter and more heavily streaked above and brighter orange-buff below with some streaking. Supercilia and moustache orange-buff. Note black throat sometimes obscured. Sexes alike.

Voice: Call a soft *trrrt*.

Habits: Inhabits bushes, dry scrub and orchards. Feeds on ground on invertebrates and seeds in loose groups, often with other passerines. Confiding but unobtrusive, although will perch on bush tops.

Jan Willem den Besten

Distribution: Locally common winter visitor to foothills of northern mountains from Pakistan east to western Nepal.
Also occurs in Russia, Iran and Central Asia.

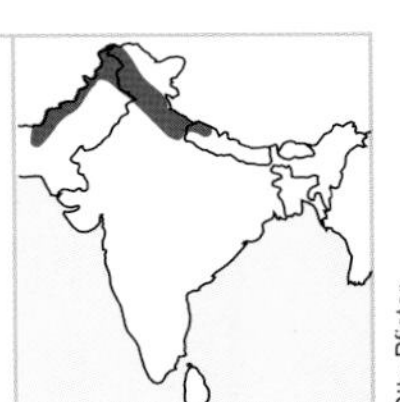

Otto Pfister

MAROON-BACKED ACCENTOR

Prunella immaculata 15 cm F: Passeridae

Tim Loseby

Description: A small, dark accentor with white irises. Uniquely coloured. Grey head, breast and wing panel, white scaling on crown. Maroon-brown upperparts and belly, darker on wings and tail. Pale legs. Sexes alike. Juvenile, heavily streaked.

Voice: Call *zeh dzit*.

Habits: Inhabits the ground beneath rhododendron and conifer forest undergrowth, forest edge and nearby cultivation. Feeds on invertebrates and seeds on ground under cover and difficult to observe. Usually in small parties. Nest not known.

Distribution: Scarce presumed breeding resident in northern mountains from central Nepal east to Arunachal Pradesh. Winters lower down.
Also occurs in Tibet, Myanmar and China.

BLACK-BREASTED WEAVER (Black-throated Weaver)

Ploceus benghalensis 15 cm F: Passeridae

Otto Pfister

Description: A streaked weaver. In breeding plumage male has yellow crown and whitish or dusky cheeks and throat. Solid black breast band, unstreaked whitish underparts and streaked upperparts. In non-breeding plumage, resembles female with dark lines through eyes and round cheeks and dark moustache. Some yellow in supercilia, neck and throat. Smudgy dark breast markings and always unstreaked below.

Voice: Song a quiet repeated *tsik tsik tsik* and call *chit chit*.

Habits: Inhabits tall grass and reed beds along canals and rivers. Feeds there and nearby cultivation on seeds and invertebrates, usually in large flocks with other weavers. Nests in colonies in reedbeds.

Distribution: Locally common endemic breeding resident in parts of Indus Valley and throughout Gangetic Plain and terai east into the Northeast and Bangladesh. Scattered records throughout peninsula.
Also occurs in Europe, N, W and Central Asia and China.

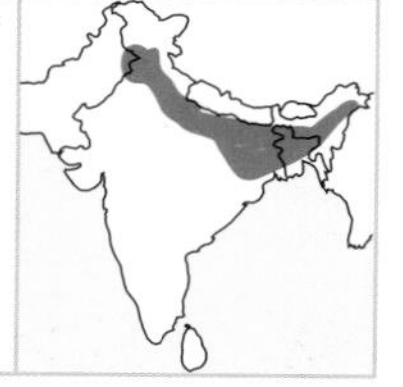

STREAKED WEAVER (Striated Weaver)

Ploceus manyar 15 cm F: Passeridae

Description: A dark, heavily streaked weaver. Breeding male has yellow crown and black cheeks and throat and densely black-streaked breast, flanks and upperparts. Non-breeding male and female are heavily streaked with yellowish supercilia and moustache.

Voice: Noisy. Call *chirt chirt*. Makes complicated wheezing and chattering in colonies.

Habits: Inhabits wet reedbeds and adjoining cultivation, feeding in noisy flocks, often with other weavers, on seeds and invertebrates. Builds woven nest in small colonies in reeds or suspended from overhead wires. Often nests with **Black-breasted** but also uses smaller reedbeds.

Distribution: Locally common breeding resident in Indus and upper Gangetic plains, the Northeast and Sri Lanka. Rare in Nepal, Bangladesh and elsewhere in India.
Also occurs in SE Asia.

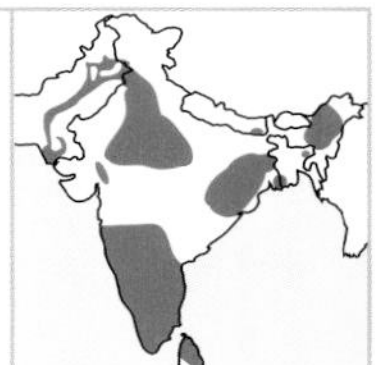

Otto Pfister

BAYA WEAVER (Baya)

Ploceus philippinus 15 cm F: Passeridae

Description: Streaked weaver. Breeding male has yellow crown to nape round to breast (buff breast in eastern race) with black cheeks and throat. Underparts unstreaked. Non-breeding male has yellow supercilia and light breast streaking. Streaked above like female. Rarer and larger **Finn's** *P. megarhynchus* has larger bill and male has all yellow head, underparts and rump. Dark cheeks.

Voice: Noisy. Wheezing and chattering at colonies.

Habits: Commonest weaver. Inhabits arable cultivation with scattered trees often near water. Gregarious. Feeds on seeds and insects. Builds woven nest with spout on end of branch, particularly favouring isolated palms.

Distribution: Common breeding resident in lowlands throughout region. Rare or absent in western Pakistan, northern mountains and parts of the Northeast.
Also occurs in SE Asia.

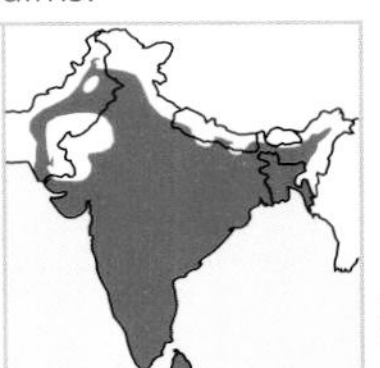

Otto Pfister

♀

Otto Pfister

RED AVADAVAT (Red Munia)

Amandava amandava 10 cm F: Passeridae

Otto Pfister

Tim Loseby

Description: A tiny, brown weaver with variable amounts of red. Breeding male is crimson with brown wings, black mask and belly and white spots on wings and underparts. White-tipped tail. Non-breeding male and female (both shown) brown with dark mask, red rump and white spots on wings. Red bill and pink legs.

Voice: A high *tsee tsee*.

Habits: Inhabits all types of long grass including reedbeds and sugarcane. Also scrub and gardens. Usually in well-watered places. Feeds on seeds, often from grass heads, and invertebrates in small, very active parties, frequently with other seed-eaters. Confiding. Nests low in cover.

Distribution: Locally common breeding resident throughout lowlands of north, central and western India, the Indus Valley and Nepal. Rare or unknown in parts of northeast, east and southeastern India, Bangladesh and Sri Lanka. A popular target as a cage bird so escapes possible anywhere. Also occurs in China and SE Asia.

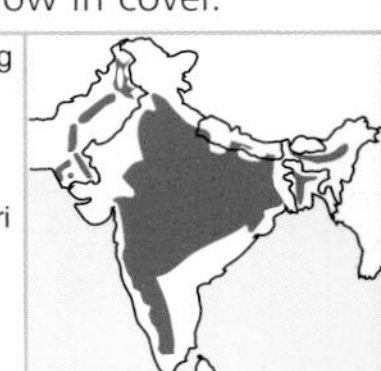

GREEN AVADAVAT (Green Munia)

Amandava formosa 10 cm F: Passeridae

Toby Sinclair

Description: A tiny, green weaver with striped flanks. Green above with black tail. Yellow below with diagnostic zebra-striped flanks and paler throat. Female browner above. Red bill.

Voice: A quiet *swee swee*.

Habits: Inhabits grass, scrub and sugarcane often near water. Feeds on small seeds or invertebrates on ground or plant heads, usually in small parties. Rarely mixes with other seed-eaters. Nests in thick cover.

Distribution: Globally threatened endemic and rare breeding resident known only from a few sites in north-central peninsular India from Rajasthan east to Orissa and south to Andhra Pradesh. Scattered records from elsewhere probably relate to escaped or released cage birds as it is a popular target.

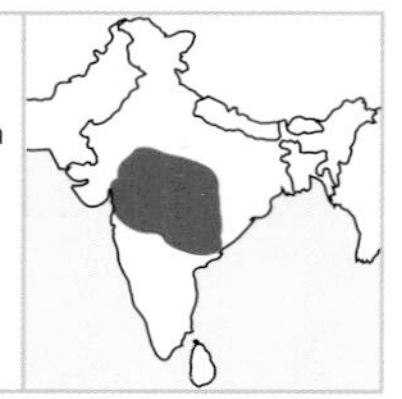

INDIAN SILVERBILL (Common White-throated Munia)

Lonchura malabarica 10 cm F: Passeridae

Description: A tiny, slim, pale weaver with a white rump. Sandy brown with dark brown wings and black, pointed tail. Buffy-white below with faint flank barring. Silvery blue-grey lower mandible. Darker grey upper. Sexes alike.

Voice: Excited but quiet *chip chip* and *see sip* calls.

Habits: Inhabits dry scrub and grass, open woodland and the edges of dry cultivation. Feeds on seeds and invertebrates on ground and on plants, usually in small parties. Often near **Baya Weaver** colonies, where it may roost in their nests. Nests in low tree or bush.

Toby Sinclair

Distribution: Common breeding resident throughout lowlands but rare in parts of south and northeast. A popular target as a cage bird and often artificially coloured. Such birds might escape!
Also occurs in China and SE Asia.

SCALY-BREASTED MUNIA (Spotted, Nutmeg Mannikin)

Lonchura punctulata 10 cm F: Passeridae

Description: A tiny, brown munia with black-scaled white underparts. Warm red-brown head throat and upperparts. Sexes alike. Scarce endemic **Black-throated** *L. kelaarti* is restricted to parts of Western and Eastern Ghats and Sri Lanka. It has a much darker, blackish face and throat, paler neck sides and the Ghats race has rich buff underparts.

Voice: Calls *ki tee ki tee* Also *tee tee* and *seu seu*.

Habits: Inhabits secondary forest growth with grass and scrub, cultivation and gardens. Feeds on seeds and invertebrates on ground and plants in active flocks, sometimes with other seed eaters. Nests low in tree or creeper.

Otto Pfister

Distribution: Common breeding resident throughout most hills and plains of India, Nepal, Bangladesh and Sri Lanka. Very local in extreme north Pakistan.
Also occurs in China and SE Asia.

WHITE-RUMPED MUNIA (White-backed, Striated)

Lonchura striata 10 cm F: Passeridae

Vijay Cavale

Description: A tiny, dark brown weaver with a white rump. Dark brown with white rump and belly and black throat, breast and vent. Northern race, paler and more streaked on head and breast with buffer belly. Sexes alike.

Voice: High pitched *trr trr* and *prrit prrit* calls.

Habits: Inhabits light scrub, grassy forest margins, overgrown cultivation. Feeds on seeds and invertebrates on ground and plants in large active flocks, often with other munias. Nests in tree.

Distribution: Locally common breeding resident in Himalayan foothills from Uttaranchal east to the northeast and Bangladesh. Also in plains and hills of northern peninsula from Orissa west to Gujarat then south to Kerala, southern Tamil Nadu and Sri Lanka.

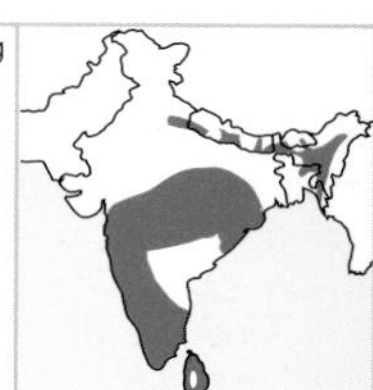

BLACK-HEADED MUNIA (Chestnut Munia)

Lonchura malacca 10 cm F: Passeridae

Otto Pfister

Sumit Sen

Description: A tiny, black and chestnut munia with a striking blue bill. Chestnut above with brighter rump and tail, black head and vent to mid-belly. Southern race has rest of underparts white. In northern race, bright chestnut. They may be different species. Sexes alike. A very popular cage bird (often escaping or deliberately released) and underparts may be dyed red, yellow, blue, or green

Voice: Soft *pee pee* calls. Some racial variation.

Habits: Inhabits reedbeds, paddy, tall grass and grassy scrub. Feeds actively in flocks, often mixed with other seed-eaters, on seeds and invertebrates on ground and on plants. Often very confiding. Nests in reeds.

Distribution: Locally common breeding resident in hills and plains from Uttaranchal east to the north east and Bangladesh. Also in northern, western and southern peninsula and Sri Lanka. Absent from Pakistan and much of northwestern India. Localised pockets probably originate from captive birds. Also occurs in SE Asia.

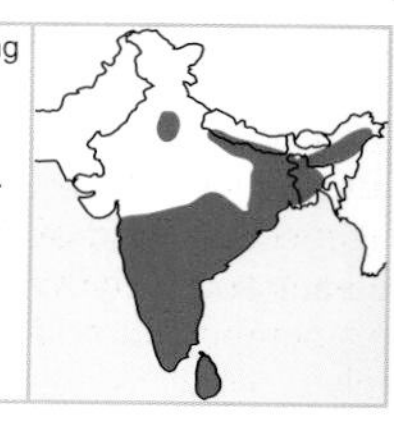

FIRE-FRONTED SERIN (Gold-fronted, Red-fronted Serin)

Serinus pusillus 12 cm F: Fringillidae

Description: A small, dark finch with a bright red forecrown. Heavily streaked black on olive-yellow body, paler below. Olive-yellow rump and wing bars. Whole head, neck and upper breast blackish. Female duller and browner. Tiny, stubby bill.

Voice: Calls a chirpy *chree chree* and a softer *tree tree*. Song is an undulating trill.

Habits: Inhabits bushy, rocky areas at or above the treeline. Also nearby cultivation. Feeds on seeds and berries in lively, noisy flocks on ground, plants and in isolated trees. Bounding, exuberant flight. Can be reasonably confiding, but rarely stays in one place for long. Drinks frequently. Nests in bush or small tree.

Otto Pfister

Distribution: Locally common breeding resident in northern mountains from western Pakistan east to central Nepal. Moves lower down in winter. Also occurs in W Asia and Tibet.

YELLOW-BREASTED GREENFINCH (Himalayan)

Carduelis spinoides 14 cm F: Fringillidae

Description: A medium-sized, black and yellow finch with a stout, pink bill. Blackish brown above with greenish wash to mantle. Yellow supercilia extends round dark cheeks. Yellow underparts with dark moustache. Yellow bars in wings, on rump and tail bases. Female duller.

Voice: Noisy. Calls *tzweee* and *weeee chu*. Also twitters and extends calls into song, often uttered in flight.

Habits: Inhabits forest edge, cultivation, meadows, gardens and scrub. Feeds on seeds, berries and invertebrates in small flocks on ground, plants and trees. Particularly fond of downy seed heads and sunflowers. Active and sometimes confiding. Undulating flight. Nests in tree.

Mohit Aggarwal

Distribution: Locally common breeding resident in northern mountains from Pakistan to the Myanmar border. Moves lower down in winter. Also occurs in SE Asia.

EUROPEAN GOLDFINCH (Goldfinch, Eurasian Goldfinch)

Carduelis carduelis 13 cm F: Fringillidae

Otto Pfister

Otto Pfister

Description: A medium-sized, brown finch with a red face. Greyish-brown above with black wings and broad golden-yellow wing bars. White rump and white-bordered, black tail. Paler buff-grey below. Larger vagrant race *C. c. major* (also shown) is browner and has prominent white border to red face and black crown and neck sides. Sexes alike.

Voice: Calls a jaunty *de di lit* and *chrrilk*. Light, twittering song.

Habits: Inhabits cultivation, orchards, scrub and open coniferous forest. Feeds in small monospecific parties mainly on seed heads, especially thistles and sunflowers. Also on ground. Lively with a bouncy flight. Nests in tree.

Distribution: Locally common breeding resident in northern mountains from western Pakistan east to central Nepal. Winters lower down in foothills and adjacent plains.
Also occurs in Europe, Atlantic Islands, W, N and Central Asia.

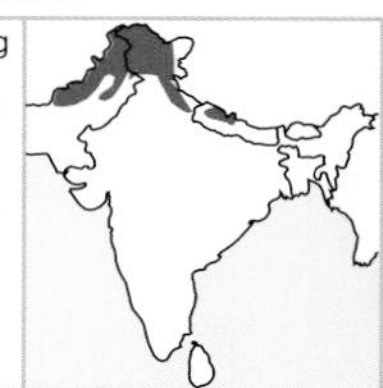

TWITE (Tibetan Twite)

Carduelis flavirostris 13 cm F: Fringillidae

Otto Pfister

Description: A medium-sized, streaked, brown finch with a yellow bill in winter. Streaked brown above and on breast and flanks. Male has pinkish rump. Plain buff supercilia and face. Usually shows distinct whitish wing bars and whitish edges to flight and tail feathers. Stubby bill darkens when breeding. Western race is much paler sandy-brown with less streaking.

Voice: Nasal, drawn out *tweet* and twittering *ditoo di dowit*.

Habits: Inhabits dry, open, montane, stony plateaux and scrubby slopes above treeline. Feeds on ground on seeds and invertebrates, often in flocks mixed with other ground feeding passerines. Undulating, erratic flight. Nests low down.

Distribution: Locally common breeding resident from north Pakistan east to northern Nepal. Moves lower down in winter.
Also occurs in Europe and W, N, Central and E Asia.

EURASIAN LINNET (Linnet, Brown, Common Linnet)

Carduelis cannabina 13 cm F: Fringillidae

Description: A medium-sized, brown and grey finch. Male has grey head and neck with prominent, crimson forehead and breast, most obvious in breeding season. Female streaked brown with greyer head. Whitish rump and edgings to flight and tail feathers. Grey bill.

Voice: Call a dry *teet eet eet eet*.

Habits: Inhabits open stony country and fallow. Feeds on seeds and invertebrates on ground or plants, usually in flocks with other seed-eaters. Will perch prominently. Fast undulating flight.

Otto Pfister

Distribution: An erratic winter visitor to northern Pakistan and Kashmir hills. Vagrant elsewhere in northern India and Nepal.
Also occurs in Europe, N Africa, Atlantic Islands, W and N Asia.

PLAIN MOUNTAIN FINCH (Hodgson's Rosy Finch)

Leucosticte nemoricola 15 cm F: Fringillidae

Description: A streaked, brown finch with two thin, white wing bars. Superficially similar to a female **House Sparrow**. Dark brown streaked above with pale braces and greyer unstreaked rump. Dark centres to wing feathers. Pale supercilia. Greyish-buff below with little or no streaking. Sexes alike.

Voice: Noisy. Call a twittering *chi chi chi*. Also *rik pi vitt* and *diu dip dip dip*.

Habits: Inhabits montane cultivation and meadows, scrub and edges of snowmelt. Winters in cultivation and forest clearings. Feeds on ground on seeds and invertebrates in flocks. Keeps low but will perch on walls, etc. Confiding. Flies in swirling lark-like flocks. Nests on ground.

Otto Pfister

Distribution: Locally common breeding resident in northern mountains from north Pakistan east to Arunachal Pradesh. Moves lower down in winter.
Also occurs in Central Asia, Myanmar and China.

BRANDT'S MOUNTAIN FINCH (Black-headed)

Leucosticte brandti 18 cm F: Fringillidae

Otto Pfister

Distribution: Locally common breeding resident in northern mountains from north Pakistan east to Sikkim. Moves lower down in winter.
Also occurs in N and Central Asia.

Description: A sooty-brown finch. Sooty head and neck. Lightly streaked brown upperparts with pinkish rump in male. Whitish wing panels and tail edgings. Greyish underparts. In non-breeding plumage, dark on head reduced in extent and browner. Juvenile is paler, rather uniform sandy-brown with dark wing edgings.

Voice: Calls *twitt twitt twee ti ti* or *peek peek*. Also churrs.

Habits: Inhabits stony slopes and montane meadows, favouring the edges of snowmelt. Feeds on seeds and invertebrate, in pairs or flocks. Confiding. Behaviour as **Plain**.

MONGOLIAN FINCH (Mongolian Trumpeter Bullfinch)

Bucanetes mongolicus 15 cm F: Fringillidae

Otto Pfister

Distribution: Locally common breeding resident in northern Pakistan and Ladakh. Moves lower down and, rarely, east to Nepal in winter. Rather nomadic.
Also occurs in N, Central and E Asia.

Description: Medium-sized, pale, hill finch with stout, yellow bill. Streaked fawn-brown above with black-edged pink and white in wings and tail. Breeding male has pink supercilia and flanks. Female has pale buff supercilia and underparts. Black eyes and brown legs. North-western **Trumpeter Finch** *B. githagineus* has nasal call, red bill in male (yellow in female), red legs, no supercilia and generally pinker.

Voice: Liquid *djou voud* call. Measured *do mi sol mi* song.

Habits: Inhabits dry stony and gravely hillsides. Feeds on seeds on ground and low growth, usually in pairs or small parties. Confiding. Strong, bounding flight, particularly to water sources. Nests on ground.

DARK-BREASTED ROSEFINCH (Nepal Rosefinch)

Carpodacus nipalensis 15 cm F: Fringillidae

Description: A slim, dark maroon-brown and pink rosefinch with a slender bill. Male maroon-brown above (including rump) and across breast with dark pink forehead, supercilia, throat and belly. Female very dusky grey brown with paler throat, unstreaked underparts and no supercilia.

Voice: A double whistle and a wheezy *chair*.

Habits: Inhabits mixed rhododendron forest and scrub above treeline. In winter in forest clearings and cultivation. Feeds on seeds and berries on ground, in small groups or pairs. Shy and unobtrusive. Nest unknown.

Jan Willem den Besten

Jan Willem den Besten

Distribution: Fairly common breeding resident in northern mountains from Kashmir east to Arunachal Pradesh. Moves lower down in winter.
Also occurs in Tibet, Myanmar and China.

COMMON ROSEFINCH (Scarlet Rosefinch)

Carpodacus erythrinus 15 cm F: Fringillidae

Description: Slim rosefinch with stubby bill and beady, black eyes. Wintering race (shown) has red in male restricted to head, breast and rump with brown stripes through eyes. Whitish belly and streaked brown upperparts. Breeding races are extensively red above and below. Female streaked brown with two obvious whitish wing bars.

Voice: Quiet in winter. *twee* call and whistling repetitive song.

Habits: Inhabits scrub above treeline and open forest. Winters in scrub, open woodland, bushy cultivation, often roosting in reed-beds. Feeds on ground on seeds or in bushes on berries and nectar, often in small flocks. Shy. Will sit inactively in bare branches. Undulating flight. Nests in bush.

Joanna Van Gruisen

Distribution: Locally common breeding resident in northern mountains from north Pakistan east to Arunachal Pradesh. Winters widely in foothills and plains but rare in Bangladesh and unknown in extreme southeastern India and Sri Lanka.
Also occurs in Europe, W, N, Central, E and SE Asia.

BEAUTIFUL ROSEFINCH

Carpodacus pulcherrimus 14 cm F: Fringillidae

Otto Pfister

Description: A small, pale rosefinch with lilac pink supercilia. Male is lilac pink below with faint flank streaking. Pale pinkish-brown above with darker streaking. Rump unstreaked pink. Dark eyestripes and round base of pale bill. Female heavily brown-streaked white below.

Voice: Calls *trip*, *trillip* or *cheet cheet*.

Habits: Inhabits scrub, above treeline in summer and often near cultivation in winter. Feeds on seeds and buds on ground or in low bushes, in small groups or pairs. Frequently sits quietly, especially after being disturbed. Nests low down.

Distribution: Very local breeding resident in northern mountains from Uttaranchal east to Arunachal Pradesh which is where the only recent Indian records come from. Common in Nepal. Moves lower down in winter.
Also occurs in N Asia, Tibet and China.

PINK-BROWED ROSEFINCH

Carpodacus rodochrous 15 cm F: Fringillidae

Goren Ekstrom

Description: A small, bright pink rosefinch with distinct supercilia. Male is dark-streaked pinkish brown above with pink wing bars. Unstreaked red-brown crown and eyestripes and pink rump. Deep pink below. Female has buffy-white supercilia and is warm orange-buff, heavily streaked with brown above and below. Whiter ground colour to throat and unstreaked rump.

Voice: Calls *swe eet perle* and *chew eee*.

Habits: Inhabits open forest and scrub. Winters in oak forest and scrub, sometimes near villages. Feeds on seeds, berries and nectar on ground or in bushes in pairs or small groups. Often mixes with other rosefinches. Unobtrusive. Nests in bush or tree.

Distribution: Locally common breeding resident in northern mountains from north Pakistan east to Nepal. Moves lower down in winter.
Also occurs in Tibet.

DARK-RUMPED ROSEFINCH (Large, Edward's)

Carpodacus edwardsii 17 cm F: Fringillidae

Description: A large, dark rosefinch with distinct supercilia on both sexes. Male dark- streaked brown above with prominent pink supercilia and breast. Indistinct pink wing bars and reddish-brown lower breast band. Rump brown. Female paler and more streaked below with buff wing-bars.

Voice: Calls a sharp *twink* and *chreewee*.

Habits: Inhabits montane scrub, feeding on ground and rhododendrons on seeds and nectar; singly, in pairs or small parties. Rather shy. Nest not known.

Sujan Chatterjee

Distribution: Scarce breeding resident in northern mountains from central Nepal east to Arunachal Pradesh. Moves lower down in winter. Also occurs in Tibet, Myanmar and China.

SPOT-WINGED ROSEFINCH

Carpodacus rodopeplus 16 cm F: Fringillidae

Description: A large, dark rosefinch with very prominent supercilia in both sexes. Male has dark brown upperparts including crown, face and eyestripes. Two pink wing bars and deep pink supercilia, rump and underparts. Female is dark-streaked brown especially on underparts and buffish-white supercilia contrast with blackish cheeks, resembling some accentors. Buff wing bars. Large conical bill.

Voice: Calls *chirp* and *schwee*.

Habits: Inhabits rhododendron shrubs and meadows above treeline in summer and forest ravines and bamboo thickets in winter. Feeding and behaviour as **Beautiful**. Rather shy. Nest not known.

Otto Pfister

Distribution: Very local breeding resident in northern mountains from Himachal Pradesh east to Arunachal Pradesh. Rare in India, less so in Nepal. Winters lower down. Also occurs in Tibet, Myanmar and China.

WHITE-BROWED ROSEFINCH

Carpodacus thura 17 cm F: Fringillidae

Otto Pfister

Description: A large, brown and pink rosefinch with obvious supercilia in both sexes. Male streaked brown above with an unstreaked pink rump and pink supercilia which fade into white at the rear. Lightly white-streaked on pale pink underparts. Female is heavily streaked brown with obvious, white supercilia and distinct, orange ground colour to streaked throat and breast. Long tail.

Voice: Calls include a buzzing *deep deep deep de de de*, a whistling *pwit pwit* and a piping *pupupipipipi*.

Habits: Inhabits montane scrub and open forest. Behaviour as **Spot-winged** but more confiding. Nests low down.

Distribution: Locally common breeding resident in northern mountains from north Pakistan east to Arunachal Pradesh. Winters lower down. Also occurs in Tibet and China.

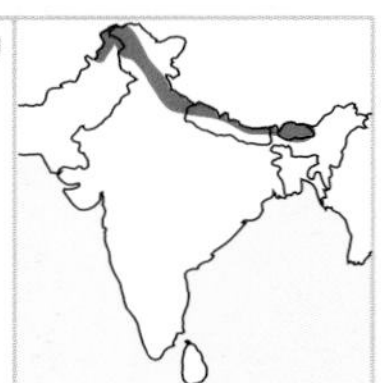

RED-MANTLED ROSEFINCH (Blyth's Rosefinch)

Carpodacus rhodochlamys 18 cm F: Fringillidae

Jan willem den Besten

Description: A large pink and brown rosefinch with a very heavy pale bill. Male rather pale greyish-brown above with a pinkish wash and faint streaking. Rump and supercilia pink. Underparts pink with white streaks. Female pale greyish-brown and heavily streaked above and below.

Voice: Calls a wheezy *quwee* and a sharp *wir*.

Habits: Inhabits dry montane forest and shrubs. Winters also in cultivation. Feeds mainly on or near ground on seeds, berries and buds. Usually singly, in pairs or small parties. Confiding. Powerful undulating flight. Nests low down.

Distribution: Locally common breeding resident in northern mountains from western Pakistan east to Uttaranchal. Moves lower down in winter. Also occurs in Central Asia.

⊙ STREAKED ROSEFINCH (Eastern Great Rosefinch)

Carpodacus rubicilloides 19 cm F: Fringillidae

Description: A large, red and brown rosefinch. Male red, deepest on face, heavily blotched white below and with dark lores. Streaked brown above. Female is greyish-brown streaked above and below.

Voice: Calls *twink*, *sip* and *doid doid*.

Habits: Inhabits dry, stony country with scattered bushes above treeline. Also scrub in winter. Feeds on ground on seeds, usually in pairs or loose parties. Shy. Powerful bounding flight. Nests in bush.

Otto Pfister

Otto Pfister

Distribution: Locally common breeding resident in northern mountains from Ladakh east to Bhutan. Moves lower down in winter.
Also occurs in Central Asia and China.

⊙ GREAT ROSEFINCH

Carpodacus rubicilla 19 cm F: Fringillidae

Description: A large, pale rose-finch with white-spotted, rose pink underparts. Pale sandy-grey above with minimal streaking. Pinkish rump and dark rather long tail. Face darker. Female sandy-grey, finely streaked darker. No supercilia. **Streaked** is darker red and brown and streaked above.

Voice: Varied, including a drawn out *weeep* and *twink*.

Habits: Inhabits dry, rocky and bushy, mountainous country. Feeds singly, in pairs or in small flocks, often with other rose-finches on ground on seeds and berries. Strong, undulating flight. Nests on ground.

Otto Pfister

Distribution: Generally scarce breeding resident in northern mountains from north Pakistan east to Sikkim. Moves lower down in winter.
Also occurs in W and Central Asia east to Mongolia and W China.

RED CROSSBILL (Crossbill)

Loxia curvirostra 16 cm F: Fringillidae

Tim Loseby

Distribution: Scarce and erratic visitor or breeding resident in northern mountains from north Pakistan east to Bhutan. Appearances and nesting dictated by cone crops which vary from year to year.
Also occurs in Europe, N Africa, N, Central, E and SE Asia and N and Central America.

Description: Large-headed, red or greenish finch with crossed mandibles. Male rusty or orange-red with browner wings and tail and lightly streaked, whitish vent. Female dull greenish-brown with yellower rump and darker wings and tail. Dark, crossed bill unique in region. Juvenile, heavily streaked and mandibles not crossed at first.

Voice: Noisy. A distinctive *chip chip* call.

Habits: Inhabits coniferous forest and stands. Favours hemlock. Feeds acrobatically and noisily on the seeds of cones high in canopy. Usually in small flocks. Restless with bounding flight. Drinks frequently. Nest rarely known in region and would be high in coniferous tree.

ORANGE BULLFINCH

Pyrrhula aurantiaca 14 cm F: Fringillidae

Jan Willem den Besten

Distribution: Local endemic breeding resident in northern mountains from north Pakistan east to Uttaranchal. Moves lower down in winter. Brown and Red-headed occurs from Himachal to Arunachal Pradesh while Grey-headed confined to Himalayas from West Bengal east to Arunachal Pradesh.

Description: An orange and black finch with thick, stubby bill. Male all orange with black face, wings and tail, white rump and vent and broad orange wing bars. Female paler orange with grey head. **Red-headed** is larger with grey mantle, red nape and crown and white cheeks and wingbars. **Grey-headed** *P. erythaca* is similar but orange confined to lower breast and belly. **Brown** *P. nipalensis* is dull grey-brown with black wings and tail.

Voice: A rising *tew* or *tewtya*.

Habits: Inhabits open coniferous forest. Feeds singly, in pairs or groups on seeds, berries and buds on ground or in bushes. Confiding. Nests low in tree.

RED-HEADED BULLFINCH

Pyrrhula erythrocephala 17 cm F: Fringillidae

Description: A medium-sized finch with a black face and a stout, black bill. Grey above with mainly black wings and tail. White rump. Male has deep orange crown and nape and paler orange neck and underparts. Female all grey with yellow-green nape. Longer-tailed and more easterly **Grey-headed** has grey head, black rump bar and, in male, reddish breast.

Voice: Calls *phew phew* and *phew flit*. Song a soft *cher per ree*.

Habits: Inhabits montane forest. Feeds in pairs or small parties at all levels on seeds, nectar, fruit and buds. Spends much time sitting quietly.

Sujan Chatterjee

Distribution: Fairly common breeding resident in northern mountains from Himachal Pradesh east to Arunachal Pradesh. Moves lower down in winter. Also occurs in Tibet.

BLACK AND YELLOW GROSBEAK

Mycerobas icterioides 22 cm F: Fringillidae

Description: A large, black and yellow finch with a stout, pale bill. Male has black head, wings, tail and thighs and a pale yellow body. Female is pale grey with a peachy-buff belly and rump.

Voice: Calls *pir riu pir riu* and *chuk*. Song *prr troweet a troweet*.

Habits: Inhabits coniferous forests. Feeds at all levels on seeds, berries, shoots and invertebrates, in pairs or small groups. High undulating flight. Nests in tree.

Otto Pfister

Distribution: Locally common breeding resident in northern mountains from north Pakistan to west Nepal. Moves lower down in winter.
Also occurs in Afghanistan.

COLLARED GROSBEAK (Allied Grosbeak)

Mycerobas affinis 22 cm F: Fringillidae

Joanna Van Gruisen

Description: A large, black and yellow finch with an orange hind-neck. Male very similar to **Black and Yellow** but brighter yellow with orange on hind neck and rump and yellow thighs. Female quite different. Grey head, greenish-grey above and greenish-yellow below.

Voice: Noisy. Ringing *ki ki ki kiw* and *pip pip pip pip ugh* calls and song *high diddle diddle the fiddle*.

Habits: As **Black and Yellow** though also in mixed forest. Nest not known.

Distribution: Locally common breeding resident in northern mountains from Himachal to Arunachal Pradesh. Winters lower down.
Also occurs in Tibet, Myanmar and China.

SPOT-WINGED GROSBEAK (Spotted-winged Grosbeak)

Mycerobas melanozanthos 22 cm F: Fringillidae

Jan Willen den Besten

Description: A large, black and yellow finch with a large, conical, grey bill. Male black on head neck and upper breast with no yellow collar or rump. White wing markings and yellow below. Female yellow, heavily spotted black above and below. **White-winged** *M. carnipes* is duller with more extensive black on breast, yellow rump and in wings. Female grey with same wing markings.

Voice: Call a harsh *krrr.* Also an oriole-like *typo tio.*

Habits: Inhabits mixed forest, feeding in pairs or flocks in trees, bushes or the ground. Eats fruits, particularly cherries, and seeds. Powerful undulating flight in tight flocks. Nests in tree.

Distribution: Uncommon breeding resident in northern mountains from north Pakistan east to the Myanmar border. Moves lower down in winter. White-winged also occurs in western Pakistan hills but no further east than Arunachal Pradesh.
Also occurs in China and SE Asia.

CRESTED BUNTING

Melophus lathami 15 cm F: Fringillidae

Description: A dark bunting with a high, pointed crest. Male blue-black with chestnut wings and tail. May have whitish markings (as shown) when moulting into this plumage. Female streaked brown and buff but also has chestnut in wings and tail.

Voice: Calls a quiet *chip*. Monotonous song.

Habits: Inhabits bushy and grassy rocky slopes. Feeds on seeds on ground in pairs or small groups. Upright stance. Confiding and often perches high. Drinks frequently. Nests low down.

Kamal Sahai

Kamal Sahai

Distribution: Locally common resident in northern foothills from north Pakistan east to the Myanmar border and in northern peninsular hills. Moves lower in winter and there are scattered records throughout northern India. Also occurs in China and SE Asia.

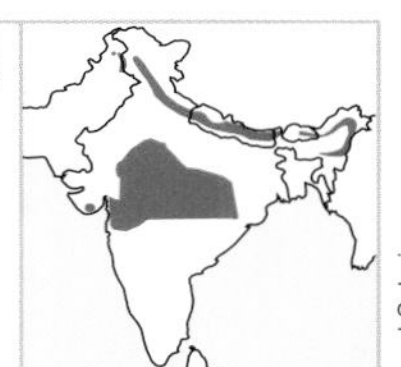

YELLOWHAMMER (Yellow Bunting)

Emberiza citrinella 17 cm F: Fringillidae

Description: A yellow and brown bunting with a rusty rump and white outer tail feathers. Male has yellow head with variable dark markings. Yellow below with rusty breast and flank streaking. Dark-streaked brown above but rump unstreaked. Female streaked brown with yellow on head. Sometimes interbreeds with Pine.

Voice: Call *tzik*. Wheezing song.

Habits: Inhabits upland fallow and stubble fields or scrub. Likely to be with Pine feeding on ground on seeds. Horizontal posture on ground, upright when perched. Rather shy. Undulating flight.

Otto Pfister

Distribution: Vagrant or rare winter visitor to northern India and Nepal. Also occurs in Europe, N Africa, W, N and Central Asia.

PINE BUNTING

Emberiza leucocephalos 17 cm F: Fringillidae

Jan Willem den Besten

Description: A large, brown-streaked bunting with rusty rump. Male in breeding plumage has chestnut head with white crown and cheeks, both edged black. Becomes white-flecked in winter. Female has dark moustache and cheek surrounds and usually a pale spot on rear cheeks. Upper-parts buff-streaked chestnut brown contrasting with streaked rufous rump. Underparts whitish-streaked rufous. No yellow. White outer tail feathers.

Voice: Calls *pit*, *trp* or *dzee*.

Habits: Inhabits hill cultivation, feeding in flocks on seeds on ground. Often mixes with other seed-eaters. Perches upright on bushes but crouches low when feeding. Rather shy.

Distribution: Fairly common but erratic winter visitor to northern hills from Baluchistan east to eastern Nepal. Very scarce in east.
Also occurs in Iraq, N Asia and China.

ROCK BUNTING (Eurasian Rock Bunting)

Emberiza cia 16 cm F: Fringillidae

Otto Pfister

Description: A rich brown bunting with black-marked, grey head. Rufous-brown body, streaked dark above and unstreaked below and on rump. Head pale grey with black crown and cheek surrounds. Northwestern race and female paler. White outer tail feathers. **House** pale whitish-grey on head, mantle and breast.

Voice: Call thin *tsee tsee* and variants.

Habits: Inhabits grassy and rocky hillsides, forest edge. Winters on fallow and stubble fields, often with other seed-eaters. Feeds on ground on seeds, in pairs or small groups. Unobtrusive but not shy. Sometimes perches high.

Distribution: Common breeding resident in northern hills from western Pakistan east to western Nepal. Winters in foothills and sometimes adjacent plains.
Also occurs in Europe, N Africa, N, W and Central Asia and China.

GREY-NECKED BUNTING

Emberiza buchanani 15 cm F: Fringillidae

Description: A grey and rusty-pink bunting with white eye rings. Male has grey head with buffish moustache and throat. Mantle lightly streaked sandy-grey and rump unstreaked grey. Chestnut in wings. Breast rusty-pink. Female paler with streaked crown and breast. Northeastern **Black-faced** *E. spodocephala* has greenish-grey head and yellow underparts. In breeding plumage lores and chin dusky. Otherwise yellow moustache.

Voice: A soft *clik* or *trip trip*.

Habits: Inhabits dry rocky hills with scattered bushes. Feeds on seeds on ground in pairs or small groups, sometimes with other seed-eaters. Inconspicuous. Creeping horizontal stance and low flight.

Clement M Francis

Distribution: Local breeding summer visitor in Baluchistan. Passage migrant through plains of Pakistan and north-west India to winter mainly in western and central peninsula south to Hyderabad. Scattered sightings elsewhere.
Also occurs in W, N and Central Asia.

WHITE-CAPPED BUNTING (Chestnut-breasted Bunting)

Emberiza stewarti 15 cm F: Fringillidae

Description: A chestnut and grey bunting with broad black eye stripes and a black throat. Breeding male has unstreaked chestnut on rump, mantle and across lower breast. Whitish-grey head and upper breast with bold black markings. Pattern obscured and more streaked in winter. Female streaked brown.

Voice: Calls a sharp *tsit* or *chit*.

Habits: Inhabits dry grassy and rocky hill slopes, open woodland and fallow fields. Feeds on seeds on ground in small groups, often with other seed eaters. Crouching posture. Perches quite high. Flies up into trees when disturbed. Nests on ground.

Tim Loseby

Distribution: Fairly common breeding summer visitor to northern mountains from western Pakistan east to Uttaranchal. Winters in plains and hills of north Pakistan and the northern Deccan south to Maharashtra. Passage migrant through northwest India. Rare elsewhere in north.
Also occurs in Central Asia.

HOUSE BUNTING (Striolated Bunting)

Emberiza striolata 14 cm F: Fringillidae

Clement M Francis

Description: A small, pale bunting with a well-marked head. Similar to **Rock** but much paler, particularly below. Streaked greyer on mantle but rufous in wings. Male has finely streaked crown and pale grey cheeks. Dark eye and moustachial stripes. Pale supercilia and diagnostic dark-streaked, grey throat. Female has same pattern but even paler. Yellow lower mandible. No white in rather short tail.

Voice: Calls a nasal *chwer* and *tzwee*. Song *witch witch weech witchy witch*.

Habits: Inhabits scrubby hillsides, semi-desert and ruins, but not usually near human habitation. Feeds on seeds on ground, hopping unobtrusively and low. Usually in pairs or small parties.

Distribution: Locally common breeding resident in lowland Pakistan, western Rajasthan and Gujarat in India. Moves erratically with rare winter records further east and south.
This (the nominate) race also occurs in West Asia while its close relative (*E. (s) sahari*) is a darker bird and closely associated with villages; it is a north and northeastern African species.

CHESTNUT-EARED BUNTING (Grey-headed)

Emberiza fucata 16 cm F: Fringillidae

Jan Willem den Besten

Description: A large russet-brown bunting with chestnut cheeks and lower breast. Male has dark-flecked, grey crown and nape and whitish throat. Black moustache extends down to form broad black gorget across upper breast. All become pale and less distinct in winter. Upperparts buff heavily streaked russet. Rufous rump and white outer tail feathers. Female similarly patterned but very washed out and browner.

Voice: A loud *pzik*, *zii* and *chutt*.

Habits: Inhabits rocky, grassy and bushy hillsides. Often near water in winter when it may roost in reedbeds. Feeds on ground on seeds in pairs or small flocks. Sings from high perch. Nests on ground.

Distribution: Uncommon breeding resident in northern hills from north Pakistan east to central Nepal. Winters in nearby foothills and plains. Most common in northeast India and Bangladesh in winter when more northern birds arrive.
Also occurs in NE and SE Asia and China.

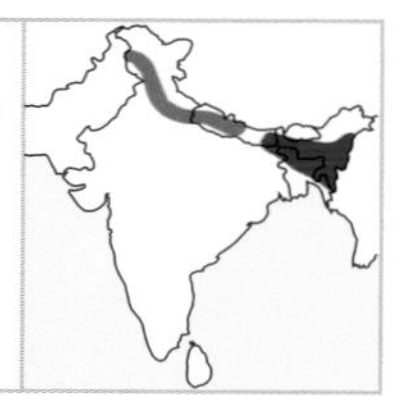

LITTLE BUNTING

Emberiza pusilla 15 cm F: Fringillidae

Description: A small, brown-streaked bunting with chestnut head. Two black crown stripes and black border to rear of cheeks. Distinct, white eye rings. Whitish underparts, heavily streaked on breast and flanks. Sexes alike. Juvenile paler. Legs pink. Small pointed bill. Northwestern **Reed Bunting** *E. schoeniclus* is larger with a larger bill. Male has black (partly obscured in winter) head and throat and white collar and moustache. Female and juvenile have brown cheeks and black moustache. All have grey rumps and lack chestnut.

Voice: Call a clicking *zick*.

Habits: Inhabits stubble, fields and pasture. Feeds on seeds on ground in small parties. Horizontal posture. Unobtrusive.

Otto Pfister

Distribution: Fairly common winter visitor mainly from western Nepal east to the Myanmar border. Rare visitor to Bangladesh and northwest India. Also occurs in N Asia, Myanmar and China.

YELLOW-BREASTED BUNTING

Emberiza aureola 15 cm F: Fringillidae

Description: A dark brown and yellow bunting. Breeding plumage male is dark chestnut-brown above with striking white wing bars, Face and throat black, Yellow underparts with chestnut breast band. Non-breeding male paler and more streaked with dusky cheeks. Female similar but with black-edged, pale cheeks and no breast bar. Plump and rather short-tailed.

Voice: Calls *tzip tzip*.

Habits: Inhabits cultivation and grassland, often near wetlands. Feeds on seeds inconspicuously on ground, usually in flocks which can be very large. Confiding. Perches in trees and bushes when disturbed.

Goren Ekstrom

Distribution: Common winter visitor to plains of Nepal, northeast India and Bangladesh. Rare further west. Also occurs in N, E and SE Asia.

RED-HEADED BUNTING

Emberiza bruniceps 17 cm F: Fringillidae

Otto Pfister

Description: Large, pale bunting. In breeding plumage, male has green-streaked mantle, yellow rump, collar and underparts. Red-brown head and upper breast. In non-breeding plumage, buffer with head mottled chestnut. Female pale and streaked above with yellowish rump and plain head. Larger **Black-headed** *E. melanocephala* male has black cap, yellow collar and rich brown mantle. Non-breeding male and female show shadow of black cap and brown in mantle and on breast sides. Both species have long tail without white edges.

Voice: Calls *chip*, *prit tleep*.

Habits: Inhabits cereal cultivation, often in large flocks. Feeds on cereal ears, often causing damage. Upright stance.

Distribution: Fairly common but sporadic passage migrant through Pakistan and northwest India to winter in western peninsula south to Karnataka and east to Madhya Pradesh. Rare further east and south. Also occurs in Central Asia.

CORN BUNTING

Miliaria calandra 18 cm F: Fringillidae

Otto Pfister

Description: A large, stocky brown-streaked bunting with a large bill. Heavily streaked above and below with no distinctive features, except usually shows a blackish breast patch and ear covert spots. No white in tail. Sexes alike. Flies with dangling legs.

Voice: Calls a thick *tik* and *tsritt*. Jangling song.

Habits: Inhabits cultivation and stubble. Feeds on ground on seeds. Most likely to be found with other seed-eaters.

Distribution: Vagrant to northern Pakistan and India.
Also occurs in Europe, W and Central Asia and China.

Bikram Grewal, a publisher by profession, has written several best-selling guides to Indian birds. He has written more than twenty other books on various subjects, and has a special interest in the history of Indian ornithology. He divides his time between Delhi and his farm in the Himalayan foothills.

Bill Harvey worked with the British Council in the UK, Asia and Africa for 30 years. He has lived in the region for nearly 10 years. His book on the birds of Bangladesh was an instant best seller. He writes widely on birds and conservation matters in scientific journals and popular magazines and newspapers. He now lives in Delhi where he is closely associated with the Northern India Bird Network, which he helped establish.

Otto Pfister, a Swiss national, worked as a civil servant and spent several years in India, where he began his acclaimed bird photography. His work has appeared in several international publications and he is currently working on a book on the wildlife of Ladakh. He now lives in Colombia, when not travelling the world photographing birds.

adult Mature, capable of breeding.

aerial Making use of the open sky.

altitudinal migrant Moving between high mountains and lower foothills.

apical Outer extremities, particularly of the tail.

aquatic Living on or in water.

arboreal Living in trees.

banyan Fig tree (*Ficus bengalensis*).

basal Innermost extremities, particularly of the tail.

bheel Used in Northeast for jheel.

biotope Area of uniform environment, flora and fauna.

buff Yellowish white with a hint of pale brown.

bund Man-made earth embankment.

canopy Leafy foliage of treetops.

cap Upper part of head.

carpal Bend of a closed wing, sometimes called shoulder.

casque Growth above bill of hornbills.

cere Patch of bare skin on upper base of bill of raptors.

colonial Roosting or nesting in groups.

confiding Not shy.

coniferous Forest trees that bear cones.

coverts Small feathers on wings and base of tail.

covey Small group of partridges and allied species.

crepuscular Active at dusk (twilight) and dawn.

crest Extended feathers on head.

crown Upper part of head.

culmen Ridge on upper mandible of bill.

deciduous Forest trees that shed all their leaves seasonally.

diagnostic Sufficient to identify a species or sub-species.

dimorphic Having two forms of plumage.

diurnal Active during daytime.

drumming Rythmic territorial hammering on trees by woodpeckers

duars Forested areas, south of eastern Himalayas.

echolocate To navigate by sound.

eclipse New dull plumage after breeding season, especially in ducks.

endemic Indigenous and confined to one place.

evergreen Forest trees that always have leaves.

eyestripe Stripe through eye.

fallow Cultivated land after harvesting and before ploughing.

family Specified group of genera.

feral Escaped, and living and breeding in the wild.

fledged Having just acquired feathers and ready to leave nest.

fledglings Young birds which have just acquired feathers.

flight feathers Primary and secondary wing feathers.

flushed When disturbed into flight at close quarters.

form Sub-species.

frugivorous Fruit-eating.

fulvous Brownish-yellow.

game birds Pheasants, partridges and allied species.

gape Basal part of the beak (mainly for young birds and raptors).

genera Plural of genus.

genus Group of related species.

ghats Hills parallel to the east and west coasts of India.

gorget Band across upper chest.

granivorous Grain-eating.

gregarious Sociable. Living in communities or flocks.

hackles Long and pointed neck feathers.

hepatic A rust or liver coloured plumage phase, mainly in female cuckoos

hunting group (or party) Group of birds, usually of different species, seeking food.

immature Plumage phases prior to adult.

iris Coloured eye membrane surrounding pupil.

jheel Shallow lake or wetland.

jizz Essence or striking characteristics of a species; from general impression and shape.

uvenile Immature bird immediately after leaving nest.

eading edge Front edge of wing.

ek Sociable courtship gathering.

obe Fleshy extensions to side-edges of toes of some birds.

ocal Unevenly distributed.

oranthus Tropical parasitic plant which grows on tree.

ores Area between eye and bill base.

malar Stripe on side of throat.

mandible Each of the two parts of bill.

mangrove Coastal, salt-tolerant tree or bush living in tidal areas.

mantle Back, between wings.

mask Dark plumage round eye and ear-coverts.

migration Seasonal movement between distant places.

mirror White spots in wing-tips, mainly of gulls.

monsoon Rainy season.

morph One of several distinct types of plumage in the same species.

moult Seasonal shedding of plumage.

nape Back of neck.

necklace Narrow line round neck.

nocturnal Active at night.

nomadic Species without specific territory except when breeding.

nominate First sub-species to be formally named.

nullah Ditch or stream bed, both dry and wet.

orbital ring Narrow ring of skin or feathers round the eye.

order Group of related families.

paddy Wet rice fields.

passerines Perching and song birds.

pectoral Breast area.

pied Black and white.

pipal Tree of the fig family (*Ficus religiosa*).

precocial Young hatched sighted and down-covered, e.g. ducklings.

primaries Outer flight feathers in wing.

ace Sub-species.

aptors Birds of prey and Vultures, excluding owls.

record Published, or otherwise broadcast, occurrence.

resident Non-migratory and breeding in same place.

rufous Reddish brown.

rump Lower back.

sal Dominant tree in North Indian forests (*Shorea robusta*)

salt pans Shallow man-made reservoirs for evaporating out salt.

scalloped Curved markings on edges of feathers.

scapulars Feathers along edge of mantle.

secondaries Inner wing feathers.

sholas Small forests in valleys (mainly in SW India).

speculum Area of colour on secondary feathers of wings.

streamers Long extensions to feathers, usually of tail.

sub-montane Hills below highest mountains.

sub-species Distinct form that does not have specific status.

sub-terminal band Broad band on outer part of feather.

supercilia Plural of supercilium.

supercilium Streak above eye.

teak Dominant tree of South Indian forests (*Tectona grandis*)

terai Alluvial stretch of land, south of the Himalayas.

terminal band Broad band on tip of tail.

terrestrial Ground living.

tertials Innermost wing coverts, often covering secondaries.

trailing edge Rear edge of wing.

vagrant Accidental; irregular

vent Undertail area.

vermiculations Wavy (worm-like) markings.

vinaceous Red wine coloured. Warm pink.

waders Shorebirds. Usually, the smaller, long-legged waterbirds.

water hyacinth Exotic (introduced), pervasive, floating water plant.

wattle Bare skin, often coloured, on part of head.

wildfowl Ducks and geese.

BIBLIOGRAPHY

Abdulali, H. (1973) *Checklist of the Birds of Maharashtra with Notes on their Status around Bombay*. Bombay: BHNS.

Abdulali, H. (1981) *Checklist of the Birds of Borivali National Park with Notes on Their Status*. Bombay: BNHS.

Abdulali, H. & Panday, J. D. (1978) *Checklist of the Birds of Delhi, Agra and Bharatpur*. New Delhi: Privately published.

Agarwal, R. & Bhatnagar, R. (eds.) (1982) *Management of Problem Birds in Aviation and Agriculture*. New Delhi: IARI.

Ahimaz, P. (1990) *Birds of Madras*. Madras: CPR Environmental Education Centre.

Ali, S. (1941) *The Book of Indian Birds*. Bombay: BNHS.

Ali, S. (1945) *The Birds of Kutch*. London: OUP.

Ali, S. (1949) *Indian Hill Birds*. Bombay: OUP.

Ali, S. (1953) *The Birds of Travancore and Cochin*. Bombay: OUP.

Ali, S. (1956) *The Birds of Gujarat*. Bombay: Gujarat Research Society.

Ali, S. (1960) *A Picture Book of Sikkim Birds*. Gangtok: Government of Sikkim.

Ali, S. (1962) *The Birds of Sikkim*. Delhi: OUP.

Ali, S. (1968) *Common Indian Birds, a Picture Album*. New Delhi: NBT.

Ali, S. (1969) *Birds of Kerala*. Madras: OUP.

Ali, S. (1977) *Field Guide to the Birds of the Eastern Himalayas*. Bombay: OUP.

Ali, S. (1979) *Bird Study in India: Its History and Its Importance*. New Delhi: ICCR.

Ali, S. (1980) *Ecological Reconnaissance of Vedaranyam Swamp, Thanjavur District, Tamil Nadu*. Bombay: BNHS.

Ali, S. (1985) *Fall of a Sparrow*. Bombay: OUP.

Ali, S. (1990) *Status and Ecology of the Lesser and Bengal Floricans with Reports on Jerdon's Courser and Mountain Quail*. Bombay: BNHS.

Ali, S. (1996) *The Book of Indian Birds*, 12th edn. New Delhi: BNHS & OUP.

Ali, S. & Abdulali, H. (1941) *The Birds of Bombay and Salsette*. Bombay: Prince of Wales Museum.

Ali, S. & Futehally, L. (1967) *Common Birds*. New Delhi: NBT.

Ali, S. & Futehally, L. (1969) *Hamare Parichat Pakshee*. New Delhi: NBT. (Hindi)

Ali, S. & Grubh, R. (1984) *Potential Problem Birds at Indian Aerodromes*. Bombay: BNHS.

Ali, S. & Grubh, R. (1981–89) *Ecological Study of Bird Hazard at Indian Aerodromes* (2 vols.). Bombay: BNHS.

Ali, S. & Hussain, S. A. (1980–86) *Studies on the Movement and Population of Indian Avifauna*. Annual Reports 1–4. Bombay: BNHS.

Ali, S. & Rahmani, A. (1982–89) *The Great Indian Bustard* (2 vols.). Bombay: BNHS.

Ali, S. & Ripley, D. (1968–74) *Handbook of the Birds of India & Pakistan* (10 vols.). Bombay: OUP.

Ali, S. & Ripley, D. (1983) *A Pictorial Guide to the Birds of the Indian Subcontinent*. Bombay: OUP.

Ali, S. & Ripley, D. (1987) *Handbook of the Birds of India & Pakistan*, compact edn. Bombay: OUP.

Ali, S. & Vijayan, S. (1986) *Keoladeo National Park Ecology Study*. Bombay: BNHS.

Ali, S., Biswas, B. & Ripley, D. (1996) *Birds of Bhutan*. Calcutta: ZSI.

Ali, S., Daniel J. C. & Rahmani, A. (1986) *Study of Ecology of some Endangered Species of Wildlife and Their Habitat*. The Floricans. Bombay: BNHS.

Ali, S. et al. (1981) *Harike Lake Avifauna Project*. Bombay: BNHS.

Ali, S. et al. (1984) *The Lesser Florican in Sailana*. Bombay: BNHS.

Ali, S. et al. (1984) *Strategy for Conservation of Bustards in Maharashtra*. Bombay: BNHS.

Ali, S. et al. (1985) *The Great Indian Bustard in Gujarat*. Bombay: BNHS.

Ambedkar, V. C. (1964) *Some Indian Weaver Birds*. Bombay: University of Bombay.

Anon. (n.d.) *Birds of Delhi & District*.

Anon. (n.d.) *Endemic Birds of Sri Lanka*. Sri Lanka: Department of Wildlife Conservation.

Anon. (1891) *List of Bird's Eggs in the Indian Museum*. Calcutta: Indian Museum.

Anon. (1988) *Rare Birds of India*. Calcutta: Botanical Survey of India.

Anon. (1990) *Wealth of India: Birds*. New Delhi: CSIR.

Anon. (1991) *Pitti Island, Lakshadweep: An Ornithological Study*. Madras: Madras Naturalists' Society.

Anon. (1992) *India's Wetlands, Mangroves and Coral Reefs*. New Delhi: WWF.

Anon. (1993) *Directory of Indian Wetlands*. New Delhi: WWF.

Anon. (1994) *Sri Lanka Avifaunal List.* Colombo: Wildlife & Nature Protection Society of Sri Lanka. (Sinhala–English)

Anon. (1995) *Fauna of Tiger Reserves.* Calcutta: ZSI.

Anon. (1997) *Pocket Guide to Common Birds of South Gujarat.* Surat: Nature Club.

Ara, J. (1970) *Watching Birds.* New Delhi: NBT.

Babault, G. (1920) *Mission Guy Babault dans les Provinces Centrales de l'Inde dans la Region Occidentale de l'Himalaya et Ceylon 1914. Resultats Scientifiques. Oiseaux.* Paris: Impimerie Generale Lahure. (French)

Badshah, M. A. (1968) *Checklist of the Birds of Tamil Nadu.* Madras: Forest Department.

Baker, E. C. S. (1908) *The Indian Ducks and Their Allies.* Bombay: BNHS.

Baker, E. C. S. (1913) *Indian Pigeons and Doves.* London: Witherby & Co.

Baker, E. C. S. (1921–30) *Game-Birds of India, Burma & Ceylon* (3 vols.). London: John Bale.

Baker, E. C. S. (1922–31) *Fauna of British India. Birds* (8 vols.). London: Taylor and Francis.

Baker, E. C. S. (1923) *Handlist of the Birds of the Indian Empire.* Bombay: R. O. Spence.

Baker, E. C. S. (1923) *A Handlist of the Genera and Species of Birds of the Indian Empire.* Bombay: BNHS.

Baker, E. C. S. (1932–35) *The Nidification of Birds of the Indian Empire* (4 vols.). London: Taylor and Francis.

Baker, E. C. S. (1942) *Cuckoo Problems.* London: H. F. & G. Witherby.

Baker, H. R. & Inglis, C. M. (1930) *The Birds of Southern India: Madras,* Malabar, Travancore, Cochin, Coorg and Mysore. Madras: Govt. Press.

Banks, J. (1980) *A Selection of the Birds of Sri Lanka.* London: author's publication.

Baral, H. S. (n.d.) *Birds of Chitwan.* Kathmandu: Victoria Travels.

Barnes, H. E. (1885) *Handbook to the Birds of the Bombay Presidency.* Calcutta: Calcutta Central Press.

Bates, R. S. P. (1931) *Bird Life in India.* Madras: Madras Diocesan Press.

Bates, R. S. P. & Lowther, E. H. N. (1952) *Breeding Birds of Kashmir.* Bombay: OUP.

Beebe, W. (1927, 1994) *Pheasant Jungles.* Reading: WPA.

Bhamburkar, P. M. & Desai, N. (1993) *Study Report of Mansingh Deo Wildlife Sanctuary.* Nagpur: WWF.

Bhushan, B. (1993) *Jerdon's or Double-banded Courser: A Preliminary Survey.* Bombay: BNHS.

BirdLife International (2000) *Threatened Birds of the World.* Barcelona and Cambridge: Lynx Edicions and BirdLife International.

Biswas, B. (1953) *A Checklist of Genera of Indian Birds.* Calcutta: Indian Museum.

Blyth, E. (1849–52) *Catalogue of the Birds in the Museum of Asiatic Society.* Calcutta: J. Thomas Baptist Mission Press.

BNHS Journal (1996) *Dr. Salim Ali Centenary Issue* Vol. 93, No. 3. Bombay: BNHS.

Buckton, S. (1995) *Indian Birding Itineraries.* Sandy UK: OBC.

Bump, G. (1964) *A Study and Review of the Black Francolin and the Gray Francolin.*

Burg, G. et al. (1994) *Ornithology of the Indian Subcontinent 1872–1992: An Annotated Bibliography.* Washington DC: Smithsonian Institution.

Butler, E. A. (1879) *A Catalogue of the Birds of Sind, Katiawar, North Gujarat and Mount Aboo.* Bombay

Butler, E. A. (1880) *A Catalogue of the Birds of the Southern Portion of the Bombay Presidency.* Bombay.

Chakravarthy, A. K. & Tejasvi, K. P. C. C. (1992) *Birds of Hill Region of Karnataka.* Bangalore: Navbharath Enterprises.

Chatrath, K. J. S. (1992) *Wetlands of India.* New Delhi: Ashish Publishing House.

Chaudhari, A. B. & Chakrabarti (n.d.) *Wildlife Biology of the Sundarbans Forests, A Study of the Breeding Biology of Birds.* Calcutta: Office of the Divisional Forest Officer.

Chopra, U. C. (1984) *Our Feathered Friends.* New Delhi: CBT.

Choudhury, A. (1990) *Checklist of the Birds of Assam.* Guwahati: Sofia Press.

Choudhury, A. (1993) *A Naturalist in Karbi-Anglong.* Guwahati: Sofia Press.

Choudhury, A. (1995) *Wildlife Surveys in Bherjan, Borajan, and Podumani Reserve Forests of Tinsukia District, Assam.* Guwahati: Rhino Foundation.

Choudhury, A. (1996) *Surveys of the White-winged Duck and the Bengal Florican in Tinsukia District and Adjacent Areas of Assam and Arunachal Pradesh.* Guwahati: Rhino Foundation.

Choudhury, A. (2001) *Birds of Assam.* Guwahati: Gibbon Books.

Christensen, G. C. et al. (1964) *A Study and Review of the Common Indian Sandgrouse and the Imperial Sandgrouse.* Washington: Department of Fish and Wildlife Services

Cocker, M. & Inskipp, C. (1988) *Hodgson: A Himalayan Ornithologist*. London: OUP.

Cox, J. (1985) *Observation on Falconry and Pakistan*.

Cunningham, D. D. (1903) *Some Indian Friends and Acquaintances*. London: John Murray.

Curson, J. (n.d.) *Birding in Southern India and the Andamans: A Guide to Selected Sites*.

Dani, C. S. (1992) *A Checklist of the Birds of Orissa*. Government of Orissa: Wildlife Wing, Forest Department.

Daniel, J. C. & Rao Y. N. (1991) *Ecology of Point Calimere Sanctuary*. Bombay: BNHS.

Daniel, J. C. et al. (1986–87) *Blacknecked Crane in Ladakh* (2 parts). Bombay: BNHS.

Daniel, J. C. et al. (1991–95) *Ecology and Behavior of Resident Raptors with Special Reference to Endangered Species* (3 parts). Bombay: BNHS.

Daniels, R. (1983) *Birds of Madurai Campus*. Madurai: ACRI.

Daniels, R. (1992) *Birds of Urban South India*. Bangalore: Indian Institute of Science.

Daniels, R. (1996) *Fieldguide to the Birds of Southwest India*. New Delhi: OUP.

Das, A. K. & Dev Roy, M. K. (1989) *A General Account of Mangrove Fauna of Andaman and Nicobar Islands*. Calcutta: ZSI.

Datta, S. (1995) *Birds of the Dhubri District*. Guwahati: Nature's Beckon.

Dave, K. N. (1985) *Birds in Sanskrit Literature*. Delhi: Motilal Banarsidass.

De, R. N. (1990) *The Sundarbans*. Calcutta: OUP.

de Zoysa, N. & Raheem, R. (1990) *Sinharaja: A Rainforest in Sri Lanka*. Colombo: March for Conservation.

de Zylva, T. S. U. (1984) *Birds of Sri Lanka: A Selection of Fifty Six Colour Pictures*. Colombo: Trumpet Publishers.

de Zylva, T. S. U. (1996) *Wings in the Wetlands*. Sri Lanka: Victor Hasselblad Wildlife Trust.

Department of Forests, Punjab (1992) *Waterbirds of Harike Sanctuary*. Chandigarh.

Department of Forests, Punjab (1993) *Punjabi Names of the Birds of Punjab*. Chandigarh.

Dewar, D. (n.d.) *Animals of No Importance*. Calcutta: Thacker, Spink.

Dewar, D. (1906) *Bombay Duck*. London: Bodley Head.

Dewar, D. (1909) *Birds of the Plains*. London: John Lane.

Dewar, D. (1911) *The Indian Crow: His Book*. Madras: Higginbotham.

Dewar, D. (1913) *Glimpses of Indian Birds*. London: John Lane.

Dewar, D. (1915) *Birds of the Indian Hills*. London: Bodley Head.

Dewar, D. (1916) *A Bird Calendar for Northern India*. London: Thacker.

Dewar, D. (1923) *Birds at the Nest*. London: John Lane.

Dewar, D. (1923) *Himalayan and Kashimiri Birds*. London: John Lane.

Dewar, D. (1923) *Indian Birds*. London: Bodley Head.

Dewar, D. (1923–25) *The Common Birds of India* (2 vols.). Calcutta and Simla: Thacker, Spink.

Dewar, D. (1924) *Birds of an Indian Village*. Bombay: OUP.

Dewar, D. (1925) *Indian Bird Life*. London: Bodley Head.

Dewar, D. (1928) *Game Birds*. London: Chapman & Hall.

Dewar, D. (1929) *Indian Birds' Nests*. Bombay: Thacker, Spink.

Dharmakumarsinhji, R. S. (1954) *Birds of Saurashtra*. Bombay: TOI Press.

Dharmakumarsinhji, R. S. & Lavkumar, K. S. (1972) *Sixty Indian Birds*. New Delhi: Ministry of Information

Dhiman, D. K. (1988) *Bird Hit Prevention Programme in IAAI Airports*. New Delhi: Institute of Aviation Management.

Easa, P. S. (1991) *Birds of Peechi Vazhani Wildlife Sanctuary: A Survey Report*. Trichur: Nature Education Society.

EHA (n.d.) *The Common Birds of Bombay*. Bombay: Thacker, Spink.

EHA (E. H. Aitken) (1923) *A Naturalist on the Prowl, or In the Jungle*. London: W. Thacker.

EHA (1947) *The Common Birds of India*. Bombay: Thacker.

Ewans, M. (1989) *Bharatpur, Bird Paradise*. London: Witherby.

Fairbank, H. (1921) *Birds of Mahableshwar*. Ahmadnager: published by the author.

Fairbank, H. (1921) *The Waterfowl of India & Asia*. Published by the author.

Felsinger, C. G. A. (1934) *Birds in the Garden in the Low Country Wet-zone of Ceylon*. Colombo: privately published.

Felsinger, C. G. A. (1972) *It was the Babbler's Nest*. Colombo: Lake House Investments.

Ferguson, W. (1887) *List of the Birds of Ceylon*. Colombo: Ceylon Observer Press.

Finn, F. (1901) *How to Know the Indian Ducks.* Calcutta: Thacker, Spink.

Finn, F. (1901) *Lists of Birds in the Indian Museum.* Calcutta: Indian Museum.

Finn, F. (1916) *Game Birds of India & Asia.* Calcutta: Thacker, Spink.

Finn, F. (1917) *The Birds of Calcutta*. Calcutta: Thacker, Spink.

Finn, F. (1920) *How to Know the Indian Waders*. Calcutta: Thacker, Spink.

Finn, F. (1920) *Indian Sporting Birds.* London: F. Edwards.

Finn, F. (1921) *The Water Fowl of India & Asia.* Calcutta & Simla: Thacker, Spink.

Finn, F. (1950) *Garden & Aviary Birds of India.* Calcutta: Thacker, Spink.

Fleming, R. L. (1977) *Comments on the Endemic Birds of Sri Lanka.* Colombo: Ceylon Bird Club.

Fleming, R. L. (1979) *Birds of the Sagarmatha National Park.* Kathmandu: Avalok.

Fleming, R. L. Sr & Fleming, R. Jr (1970) *Birds of Kathmandu and Surrounding Hills.* Khatmandu: Jore Ganesh Press.

Fleming, R. L. Sr. et al. (1984) *Birds of Nepal, with Reference to Kashmir and Sikkim.* Nepal: Nature Himalayas.

Fletcher, T. & Bainbridge (1924) *Birds of an Indian Garden*. Calcutta & Simla: Thacker, Spink.

Futehally, L. (1959) *About Indian Birds.* India: Blackie.

Gandhi, M. & Husain, O. (1993) *Bird Quiz.* New Delhi: Rupa.

Ganguli, U. (1975) *A Guide to the Birds of the Delhi Area*. New Delhi: ICHR.

Gaur, R. K. (1994) *Indian Birds*. New Delhi: Brijbasi Printers.

Gee, B. (1998) *South India, Sri Lanka and the Andaman Islands.* Sandy: OBC

George, J. (ed.) (1994) *Annotated Checklist of the Birds of Bangalore. Bangalore: Birdwatchers' Field Club.*

Gogoi, C. (1995) *Birds of Greater Guwahati*. Guwahati: Assam Science Society.

Gole, P. (n.d.) *A Guide to the Cranes of India*. Bombay: BNHS.

Gole, P. (1996) *Environment & Ornithology in India.* Jaipur: Rawat Publications.

Gould, J. (1832) *A Century of Birds from the Himalayan Mountains.* London: published by the author.

Gould, J. (1850–73) *Birds of Asia* (6 vols.). London: published by the author.

Green, A. J. (1992) *The Status and Conservation of the White-winged Wood Duck*. Slimbridge: IWRB.

Grewal, B. (ed.) (1988) *Insight Guide to Indian Wildlife.* Singapore: APA.

Grewal, B. (ed.) (1995) *The Avifauna of the Indian Subcontinent Part 1.* Bombay: Sanctuary Vol XV, No 5.

Grewal, B. (ed.) (1995) *The Avifauna of the Indian Subcontinent Part 2.* Bombay: Sanctuary Vol. XV, No 6.

Grewal, B. (1995) *Birds of India & Nepal.* London: New Holland.

Grewal, B. (1995) *Birds of the Indian Subcontinent.* Hong Kong: Guidebook Co. Ltd.

Grewal, B. (1995) *Threatened Birds of India.* New Delhi. Privately published.

Grewal, B. (1997) *Bharatvarsh ke Sankatgrast* Pakshee. New Delhi. (Hindi). Privately published.

Grewal, B. (1997) *Report of the 6th Birdwatching Camp, Sitabani.* Ramnagar: Corbett Birdwatching Programme.

Grewal, B. (2000) *A Photographic Guide to the Birds of the Himalayas, 3rd edn.* London: New Holland.

Grewal, B. & Mahajan, J. (2001) *Splendid Plumage, British Bird Painters in India.* Hong Kong: Local Colour.

Grewal, B. & Sahgal, B. (1995) *Birds of the Corbett Tiger Reserve and Its Environs.* New Delhi. Privately published.

Grimmett, R., Inskipp, T., & Inskipp, C. (1998) *Birds of the Indian Subcontinent.* London: A&C Black.

Grimmett, R., Inskipp, T., & Inskipp, C. (1999) *Field Guide to the Birds of Bhutan.* London: A&C Black.

Grimmett, R., Inskipp C. and Inskipp, T. (1999) *The Pocket Guide to the Birds of the Indian Subcontinent.* London: Chistopher Helm.

Grimmett, R., Inskipp, T., & Inskipp, C. (2000) *Field Guide to the Birds of Nepal.* London: A&C Black.

Haly, A. (1887) *First Report on the Collection of Birds in the Colombo Museum*. Colombo.

Haroun er Rashid (1967) *Systematic List of the Birds of East Pakistan*. Dacca: Asiatic Society of Pakistan.

Harris, P. (1996) *The Birds of Goa: A Complete Checklist*. Lowestoft, UK: Eastern Publications.

Harris, P. (1996) *Goa, The Independent Birder's Guide*. Lowestoft, UK: Eastern Publications.

Harvey, W. G. (1990) *Birds in Bangladesh*. Dhaka: DUP .

Henry, G. M. (1927) *Birds of Ceylon.* London: Ceylon Government.

Henry, G. M. (1953) *A Picture Book of Ceylon Birds*. Ceylon: Department of Information.

Henry, G. M. (1955) *A Guide to the Birds of Ceylon*. London: OUP.

Henry, G. M. & Wait, W. E. (1927–35) *Coloured Plates of the Birds of Ceylon*. Colombo: Ceylon Government.

Hodgson, B. H. (1846) *Catalogue of the Specimens and Drawings of Mammalia and Birds of Nepal & Tibet.* London: British Museum.

Holmer, M. R. N. (1923) *Indian Bird Life*. London: OUP.

Holmer, M. R. N. (1926) *Bird Study in India*. London: OUP.

Horsfield, T. & Moore, F. (1854–58) *Catalogue of the Birds in the Museum of the Hon. East India Company* (2 vols.). London: W. H. Allen.

Hume A. O. (n.d.) *Indian Ornithological Collector's Vade-Mecum*. Calcutta: Central Press.

Hume, A. O. (1869) *My Scrap Book: or Rough Notes on Indian Zoology and Ornithology*. Calcutta: C. B. Lewis, Baptist Mission Press.

Hume, A. O. (1873) *Contributions to Indian Ornithology, with 32 hand coloured plates by Keulemans*. London: L. Reeve.

Hume, A. O. (ed.) (1873–88) *Stray Feathers* (12 vols.). Published by the editor.

Hume, A. O. & Marshall (1879–81) *The Game Birds of India, Burmah, and Ceylon* (3 vols.). London: John Bale.

Hume, A. O. & Oates (1889) *Hume's Nests & Eggs of Indian Birds* (3 vols.). London: R. H. Porter.

Hunter, W. W. (1896) *Life of Brian Houghton Hodgson.* London: John Murray.

Hussain, S. A. (1987–91) *Bird Migration Project* (3 parts). Bombay: BNHS.

Hussain, S. A. & D'Silva, C. (1987) *Waterfowl Indicator.* Bombay: BNHS.

Hussain, S. A. et al. (1984) *Avifaunal Profile of Chilika Lake.* Bombay: BNHS.

Hutson, H. P. W. (1954) *The Birds about Delhi*. DBWS.

Inglis, C. M. (n.d.) *Sixty-eight Indian Birds*. Darjeeling: Natural History Museum.

Inskipp, C. (1988) *A Birdwatchers' Guide to Nepal*. England: Prion.

Inskipp, C. (1989) *Nepal's Forest Birds: Their Status and Conservation*. Cambridge: ICBP.

Inskipp, C. (1989) *A Popular Guide to the Birds and Mammals of the Annapurna Conservation Area.* Nepal: Annapurna Conservation Area Project (ACAP).

Inskipp, C. & Inskipp, T. (n.d.) *An Introduction to Birdwatching in Bhutan*. Thimpu, Bhutan: WWF.

Inskipp, C. & Inskipp, T. (1983) *Report on a Survey of Bengal Floricans*. Cambridge: ICPB.

Inskipp, T. & Inskipp, C. (1991) *A Guide to the Birds of Nepal*. London: Christopher Helm.

Inskipp, T. et al. (1996) *An Annotated Checklist of the Birds of the Oriental Region*. Sandy, UK: OBC.

Jerath, N. (1993) *Harike Wetland: An Avian Paradise*. Punjab State Council for Science & Technology.

Jerdon, T. C. (1845–47) *Illustrations from Indian Ornithology, Containing Fifty Figures of New, Unfigured or Interesting Species of Birds, Chiefly from the South of India.* Madras.

Jerdon, T. C. (1862–64) *The Birds of India: A Natural History* (3 vols.). Calcutta: The Military Orphan Press.

Jerdon, T. C. (1864) *The Game Birds and Wildfowl of India*. London: Military Orphan Press.

Jonathan, J. K. & Kulkarni (1989) *Beginners' Guide to Field Ornithology*. Calcutta: ZSI.

Jones, A. E. (n.d.) *The Common Birds of Simla*. Simla: Liddells Printing Works.

Kalpavriksh (1991) *What's That Bird? A Guide to Birdwatching, with Special Reference to Delhi*. New Delhi: Kalpavrksh.

Kalpavriksh (1996) *Small & Beautiful: Sultanpur National Park.* New Delhi: Kalpavriksh.

Karoor, J. J. (1986) *List of Birds That May Be Found in Periyar Wildlife Sanctuary.* Peermade Wildlife Preservation Society.

Kaul, S. C. (1939) *Birds of Kashmir*. Srinagar: The Normal Press.

Kazmierczak, K. (1996) *A Checklist of Indian Birds*. Published by the author.

Kazmierczak, K. & Singh, R. (2001) *A Bird Watchers Guide to India.* Delhi. OUP.

Kazmierczak, K. & van Perlo, B. (2000) *A Field-Guide to the Birds of the Indian Subcontinent*. UK: Pica Press

Kelaart, E.F. (1853) *Prodromus Faunae Zeylanicae*. Colombo: Observer Press.

Kershaw, C. (1925) *Familiar Birds of Ceylon*. Colombo: H. W. Cave.

Kershaw, C. (1949) *Bird Life in Ceylon*. Colombo: Times of Ceylon.

Khacher, L. (n.d.) *Birds of the Indian Wetland*. New Delhi: Department of Environment.

King, B. et al. (1991) *A Field Guide to the Birds of South East Asia*. London: Collins.

Kotagama, S. (1986) *Kurullan Narambamu*. Colombo: Pubudu Publishers. (Sinhala)

Kotagama, S. & Fernando, P. (1994) *A Field Guide to the Birds of Sri Lanka*. Colombo: Wildlife Heritage Trust of Sri Lanka.

Kothari, A. & Chhapgar, B. F. (eds.) (1996) *Salim Ali's India*. New Delhi: OUP & BNHS.

Krishnan, M. (1986) *Vedanthangal Water-Bird Sanctuary*. Chennai: Government of Tamil Nadu.

Kumar, G. & Lamba, B. (1985) *Studies on Migratory Birds and Their Behavior in Corbett National Park*. Calcutta: ZSI.

Kumar, K. (1992) *Keoladeo National Park, Bharatpur: Flora & Fauna*. Bharatpur.

Lamba, B. (1987) *Fauna of Corbett National Park*. Calcutta: ZSI.

Lamba, B. (1987) *Status Survey of Fauna, Nanda Devi National Park*. Calcutta: ZSI.

Lamfuss, G. (1994) *The Birds of Sri Lanka: A Complete Birdwatcher's Checklist*. Colombo: Ceylon Bird Club.

Law, S. C. (1923) *Pet Birds of Bengal*. Calcutta & Simla: Thacker, Spink.

Law, S. C. (1934) *Kalidaser Pakhi*. Calcutta: Gurudas Chattopadhayay. (Bengali)

Law, S. C. (1935) *Jalachari*. Calcutta: Gurudas Chattopadhyay. (Bengali)

Le Messurier, A. (1904) *Game, Shore and Water Birds of India*. London: W. Thacker.

Legge, W. V. A. (1867) *History of the Birds of Ceylon*. London: Taylor & Francis.

Lester, C. D. (1904) *The Birds of Kutch*. Bhuj: Kutch Darbar

Lister, M. D. (1954) *A Contribution to the Ornithology of the Darjeeling Area*. Bombay: BNHS.

Lowther, E. H. N. (1949) *A Bird Photographer in India*. London: OUP.

Lushington, C. (1949) *Bird Life in Ceylon*. Colombo: Times of Ceylon.

MacDonald, M. (1960) *Birds in My Indian Garden*. London: Jonathan Cape.

MacDonald, M. (1962) *Birds in the Sun: Some Beautiful Birds of India*. London: D. B. Taraporevala Sons.

Mackintosh, L. J. (1915) *Birds of Darjeeling and India: Part I*. Calcutta: J. N. Banerjee.

Mahabal, A. & Lamba, B. (1987) *On the Birds of Poona and Vicinity*. Calcutta: ZSI.

Majumdar, N. (1977) *Bird Migration*. New Delhi: NCERT.

Majumdar, N. (1984) *On a Collection of Birds from Adilabad District*. Andhra Pradesh. Calcutta: ZSI.

Majumdar, N. (1984) *On a Collection of Birds from Bastar District*. Calcutta: ZSI.

Majumdar, N. (1988) *On a Collection of Birds from Koraput District. Orissa*. Calcutta: ZSI.

Malik, D. & Malik, M. S. (1992) *Checklist of Birds of Zainabad*. Gujarat: Desert Courser Camp.

Mandal, A. K. & Nandi, N. C. (1989) *Fauna of the Sundarban Mangrove Eco-system*. Calcutta: ZSI.

Marshall, G. F. (1877) *Birds Nesting in India*. Calcutta: Central Press.

Martens, J. (1980) *Vocalization, Relationships and Distributional History of the Asian Phylloscopine Warblers*. Berlin: Paul Parey.

Mason, C. W. & Maxwell-Lefroy, H. (1912) *The Food of Birds in India*. Calcutta: Thacker, Spink.

Mathews, W. H. & Edwards, V. S. (1944) *A List of Birds of Darjeeling*. Published by the authors.

Mathur, H. N. et al. (1993) *Birds of Tripura: A Checklist*. Agartala: Tripura State Council for Science and Technology.

Mehrotra, K. N. & Bhatnagar R. K. (1979) *Status of Economic Ornithology in India*. New Delhi: ICAR.

Mierow, D. (1988). *Birds of the Central Himalayas*. Bangkok: Craftsman Press.

Mookherjee, K. (1995) *Birds and Trees of Tolly*. Calcutta: Tollygunge Club.

Mukherjee, A. K. (1979) *Peacock, Our National Bird*. New Delhi: Publications Division.

Mukherjee, A. K. (1992) *Birds of Goa*. Calcutta: ZSI.

Mukherjee, A. K. (1995) *Birds of Arid & Semi Arid Tracts*. Calcutta: ZSI.

Mukherjee, A. K. & Dasgupta, J. M. (1986) Catalogue of Birds in the Zoological Survey of India. Calcutta: ZSI.

Mukherjee, K. (1994) *Narendrapur Wildlife Sanctuary*. Calcutta: K. Dey.

Murray, J. A. (1888) *Indian Birds or the Avifauna of British India* (2 vols.). London: Trubner.

Murray, J. A. (1889) *The Edible and Game Birds of British India with its Dependencies and Ceylon*. London: Trubner.

Murray, J. A. (1890) *The Avifauna of the Island of Ceylon*. Bombay: Educational Society Press.

Musavi, A.H. & Urfi, A. J. (1987) *Avifauna of the Aligarh Region*. Nature Conservation Society of India.

Nair, S. M. (1992) *Endangered Animals of India and their Conservation*. New Delhi: NBT.

Nameer, P. O. (1992) *Birds of Chimmoni Wildlife Sanctuary*. Trichur: Kerala FRI & Forest Dept.

Nameer, P. O. (1992/93) *Birds of Kole Wetlands* (2 parts). Trichur: FRI & Kerala Forest Department.

Neelakantan, K. K. (1958) *Keralathile Pakshikal*. Trichur: Kerala Sahitya Academy. (Malayalam)

Neelakantan, K. K., Sashikumar & Venugopalan (1993) *A Book of Kerala Birds*. Trivandrum: WWF.

Nugent, R. (1991) *The Search for the Pink-headed Duck*. Boston: Houghton Mifflin.

Oates, E. W. (1883) *Handbook of the Birds of British Burma* (2 vols.). London: R. H. Porter.

Oates, E. W. (1898) *A Manual of the Game Birds of India, Part 1: Land Birds.* Bombay: A. J. Combridge.

Oates, E. W. (1899) *A Manual of the Game Birds of India, Part 2: Game Birds.* Bombay: A. J. Combridge.

Oates, E. W. & Blandford (1889–98) *Fauna of British India Birds* (4 vols.). London: Taylor & Francis.

Oriental Bird Club (1995) *Bulletin 22 Special India Issue*. Sandy: OBC.

Osman, S. M. (1991) *Hunters of the Air: A Falconer's Notes*. New Delhi: WWF.

Osmaston, B. B. & Sale, J. B. (1989) *Wildlife of Dehra Dun and Adjacent Hills*. Dehra Dun: Natraj Publishers.

Palin, H. & Lester, C. D. (1904) *The Birds of Cutch*. Bombay: Times Press.

Panani, D. (1996) *Rethinking Wetland Laws: A Case Study of Pulicat Bird Sanctuary*. New Delhi: WWF.

Perera, D. G. A. & Kotagama, S. (1983) *A Systematic Nomenclature for the Birds of Sri Lanka*. Dehiwela: Tisara Prakashana.

Phillips, W. W. A. (1949–61) *Birds of Ceylon* (4 vols.). Colombo: Ceylon Daily News Press.

Phillips, W. W.A. (1953) *Revised Checklist of the Birds of Ceylon*. National Museums of Ceylon.

Phillips, W. W. A. (1978) *Annotated Checklist of the Birds of Ceylon.* Colombo: Ceylon Bird Club.

Phukan, H. P. (1987) *Death at Jatinga: An Enquiry into the Jatinga Bird Mystery*. Guwahati: Forest Department Assam.

Pinn, F. (1985) *L. Mandelli, Darjeeling Ornithologist*. London: Privately published.

Pittie, A. (1995) *A Bibliographic Index to the Orinthology of the Indian Region. Part 1*. Hyderabad: Privately published.

Pittie, A. & Robertson, A. (1993) *A Nomenclature of Birds of the Indian Subcontinent.* Bangalore: OSI.

Rahmani, A. (1986) *Status of the Great Indian Bustard in Rajasthan*. Bombay: BNHS.

Rahmani, A. (1987) *Dihaila Jheel: Conservation Strategies*. Bombay: BNHS.

Rahmani, A. (1987) *The Great Indian Bustard*. Bombay: BNHS.

Rahmani, A. & Manakadan, R. (1988) *Bustard Sanctuaries in India*. Bombay: BNHS.

Rahmani, A. & Sankaran, R. (1986) *The Lesser Florican*. Bombay: BNHS.

Rahmani, A. et al. (1985) *Threats to the Karera Bustard Sanctuary*. Bombay: BNHS.

Rahmani, A. et al. (1985, 1989) *The Floricans. Annual Report 1984–85, 1988–89.* Bombay: BNHS.

Rahmani, A. et al. (1988) *The Bengal Florican.* Bombay: BNHS.

Rahmani, A. et al. (1996) *A Study of the Ecology of the Grasslands of the Indian Plains with Particular Reference to Their Endangered Fauna.* Bombay: BNHS.

Rai, Y .M. (1983) *Birds of the Meerut Region.* Meerut: Vardhaman Printers.

Raju, K. (1985) *A Checklist of Birds of Vishakapatnam Region*. Vishakapatnam: APNHS.

Ranasinghe, D. (1976) *Asirimath Kurulu Lokaya*. Colombo: Wildlife and Nature Society. (Sinhala)

Ranasinghe, D. (1977) *A Guide to Bird Watching in Sri Lanka*. Colombo.

Rangaswami, S. & Sridhar, S. (1993) *Birds of Rishi Valley*. Andhra Pradesh.

Ranasinghe, D. & De Zylva, T. S. U. (1978) *Sri Lanka Avifaunal List*. Colombo: Ceylon Bird Club.

Reid, G. (1886) *Catalogue of the Birds in the Provincial Museum, N-W P and Oudh, Lucknow.* Calcutta.

Ripley, D. (1952) *Search for the Spiny Babbler*. Boston: Houghton Mifflin.

Ripley, D. (1978) *A Bundle of Feathers*. London: OUP.

Ripley, D. (1982) *A Synopsis of the Birds of India and Pakistan*, 2nd edn. Bombay: BNHS.

Roberts, T. J. (1991–92) *The Birds of Pakistan* (2 vols.). Karachi: OUP.

Roberts,T. J. et al. (n.d.) *A Checklist of Birds of Karachi & Lower Sindh*. Pakistan: WWF.

Robertson, A. & Jackson, M. (1992) *Birds of Periyar: An Aid to Birdwatching in the Periyar Sanctuary.* Jaipur.

Saha, B. & Dasgupta, J. (1992) *Birds of Goa.* Calcutta: ZSI.

Samsad, P. (1984) *Chilika: A Report*. Calcutta.

Sankala, K. (1990) *Gardens of Eden: The Waterbird Sanctuary of Bharatpur.* New Delhi: Vikas.

Sankaran, R. (1993) *The Status and Conservation of the Nicobar Scrubfowl*. Coimbatore: Sacon.

Sankaran, R. (1995) *Impact Assessment of Nest Collection on the Edible Swiftlet in the Nicobar Islands.* Coimbatore: Sacon.

Sankaran, R. (1995) *The Nicobar Megapode and other Endemic Avifauna of the Nicobar Islands*. Coimbatore: Sacon.

Sargeant, D. (1994) *A Birders' Checklist of the Birds of Nepal.* Norfolk, UK: private publication.

Sarmah, N. C. (1996) *Checklist, Dibru-Saikhowa Wildlife Sanctuary.* Tinsukia: Muniruddin Ahmed.

Satyamurti, S. T (1970) *Catalogue of the Bird Gallery Museum*. Madras.

Satyamurti, S. T. (1979) *Bird's Eggs and Nests.* Madras.

Sawhney, J. C. (n.d.) *The Cranes.* Bombay: WWF.

Saxena, V. S. (1969) *Birds of Bharatpur: A Field Checklist*. Jaipur: Govt. Central Press.

Saxena, V. S. (1975) *A Study of the Flora and Fauna of Bharatpur Bird Sanctuary*. Rajasthan: Department of Tourism.

Sen Gupta, P. K. (1955) *Birds around Shantiniketan*. West Bengal.

Sharma, B. (1991) *The Book of Indian Birds*. New Delhi: Harper Collins.

Sharma, S. C. (1985) *Birds of Sultanpur Bird Sanctuary: A Checklist*. New Delhi: WWF.

Sharma, S. C. (1989) *Ten Thousand Ducks in Five Acres*. New Delhi: Kalpavrikish.

Sharma, S. C. & Kothari, A. (1991) *Save the Bhindawas Lake Bird Sanctuary.* New Delhi: Kalpavrikish.

Sharma, S. K. (n.d.) *Ornithology of Indian Weaver Birds.* Delhi: Himanshu Publishers.

Sharma, S. K. (1995) *Ornithobotany.* Delhi: Himanshu Publishers.

Sharpe, R. B. (1891) *Scientific Results of the Second Yarkhand Mission*. London: Taylor & Francis.

Shepard, M. (1978) *Let's Look at Sri Lanka. An Orintholidays' Guide.* UK.

Shepard, M. (1987) *Let's Look at North India. An Orintholidays' Guide.* UK.

Singh, A. & Singh, N. (1985) *Birds of Dudhwa National Park*. Uttar Pradesh.

Singh, B. (1996) *Siberian Cranes*. New Delhi: WWF.

Singh, G. (1993) *A Checklist of Birds of Punjab*. Punjab: Govt. Press.

Singh, K. R. (1994) *Birds of Rajasthan*. Jaipur: Rajasthan Tourism.

Singh, R. (n.d.) *Bird and Wildlife Sanctuaries of India, Nepal & Bhutan.* New Delhi.

Singh, R. & Singh, K. S. (n.d.) *A Pocket-book of Indian Pheasants*. Dehradun: WII.

Singh, R. & Singh, K. S. (1995) *Pheasants of India and Their Aviculture*. New Delhi: WWF.

Singh, R. N. (1962) *Harmare Jal Pakshi.* New Delhi: Publications Division. (Hindi)

Singh, R. N. (1977) *Our Birds.* New Delhi: Publications Division.

Siromoney, G. (1971) *Birds of Tambaram Area and Water-birds of Vedanthangal.* Madras: MCC.

Smetackek, F. (n.d.) *Birds of the Lake Region, Nainital-Naukotchia Tal.* Bhimtal: Society of Appeal for Vanishing Environments.

Smith, C. W. & Doyly, Sir C. (1828) *The Feathered Game of Hindostan.* Patna: Behar Amateur Lithographic Press.

Smith, C. W. & Doyly, Sir C. (1829) *Oriental Ornithology.* Patna: Behar Amateur Lithographic Press.

Smythies, B. E. (1953) *The Birds of Burma*. London: Oliver and Boyd.

Snilloc (1945) *Mystery Birds of India.* Bombay: Thacker.

Soni, V. C. (1988) *Ecology and Behavior of the Indian Black Ibis.* Gujarat: University of Saurasthra.

Sonobe, K. (ed) (1993) *A Field Guide to the Waterbirds of Asia*. Tokyo: Wildbird Society of Japan.

Stanford, J. K. & Mayr, E. (1940) *The Vernay-Cutting Expedition to Southern Burma*.

Stefee, N. D. (1981) *Field Checklist of the Birds of Nepal, Kashmir, Garwal and Sikkim*. Russ Mason's Natural History Tours.

Stefee, N. D. (1981) *Field Checklist of the Birds of Peninsular India.* Russ Mason's Natural History Tours.

Sugathan, R. et al. (1985) *Studies on the Movement and Population Structure of Indian Avifauna.* Bombay: BNHS.

Taher, S. & Pittie, A. (1989) *A Checklist of the Birds of Andhra Pradesh*. Published by the author.

Talukdar, B. N. & Mahanta, R. (1991) *Pabitora Wildlife Sanctuary (Checklist of Birds).* Guwahati.

Talukdar, B. N. & Sharma, P. (1995) *Checklist of Birds of Orang Wildlife Sanctuary*. Guwahati.

Tennent, J. E. (1861) *Sketches of the Natural History of Ceylon.* London: Longman Green.

Thompson, P .M. et al. (1998) "Recent Notable Bird Records from Bangaldesh". *Forktail* 9: 13–44.

Tikader, B. K. (1983) *Threatened Animals of India.* Calcutta: ZSI.

Tikader, B. K. (1984) *Birds of Andaman & Nicobar Islands*. Calcutta: ZSI.

Tikader, B. K. & Das, A. K. (1984) *Glimpses of Animal Life of Andaman & Nicobar Islands*. Calcutta: ZSI.

Toor, H. S. & Sandhu, P. S. (1981) *Harmful Birds and Their Control*. Ludhiana: Punjab Agriculture University.

Torfrida (1944) *Nurseries of Heaven: Birds. Nilgiris. Kotagiri: Mrs May Dart.*

Torfrida. (1944) *Nurseries of Heaven: More Birds. Nilgiris.* Wellington: Mrs May Dart.

Tyabji, H. (1994) *The Birds of Bandavgarh National Park.* New Delhi.

Tyagi, A. K. & Lamba, B. S. (1984) *A Contribution to the Breeding Biology of Two Indian Mynas*. Calcutta: ZSI.

"Vagrant" (1868) *Random Notes on Indian and Burman Ornithology.* Bangalore: Regimental Press.

Vaurie, Charles. (1972) *Tibet and Its Birds*. London: Witherby.

Verghese, et al. (1993) *Bird Conservation: Strategies for the Nineties and Beyond.* Bangalore: OSI.

Verghese, et al. (1995) *Bird Diversity & Conservation Thrusts for the Nineties and Beyond.* Bangalore: OSI.

Vijayan, L. (ed) (1995) *Avian Conservation in India*. Coimbatore: SACON.

Vira, R. & Dave, K. N. (1943) *Scientific Nomenclature of Birds of India, Burma and Ceylon*. Nagpur: Indian Academy of Indian Culture.

Wait, W. E. (1925) *Birds of Ceylon*. London: Dualu.

Wan Tho Loke (1958) *A Company of Birds*. London: Michael Joseph.

Ward, Geoff. (1994) *Islamabad Birds*. Islamabad: Asian Study Group.

Wedderburn, Sir William (1912) *Allan Octavian Hume.* London: T. Fisher Unwin.

Wheatley, N. (1996) *Where to Watch Birds in Asia.* London: Christopher Helm.

Whistler, H. (1928) *Popular Handbook of Indian Birds.* London: Gurney and Jackson.

Whistler, H., Kinnear, N. and Ali, S. (n.d.) *The Vernay Scientific Survey of the Eastern Ghat; Ornithological Section, Together with the Hyderabad State Ornithological Survey 1930–38.*

Wijesinghe, D. P. (1994) *Checklist of the Birds of Sri Lanka*. Colombo: Ceylon Bird Club.

Wikramanayake, E. B. (1997) *Go to the Birds.* Colombo: Ceylon Bird Club.

Willoughby, Paul. (1996) *Goa: A Birder's Guide.*

Woodcock, Martin. (1980) *Collins Handguide to the Birds of the Indian Subcontinent.* London: Collins.

Wright, R. C and Dewar, D. (1925) *The Ducks of India*. London: H.F. & G. Witherby.

WWF India (1994) *Ramsar Sites; Chilika, Sambhar Lake, Keoladeo National Park, Loktak Lake, Wular Lake, Harike Lake* (6 booklets). WWF.

Ziddi, S. & Bhagava, S. (1995) *Bharatpur*. New Delhi: Indus.

ZSI. (1992) *Fauna of West Bengal Part 1.* Calcutta: ZSI.

Sound Guides:

Barucha, E. (1999) *Indian Bird Calls,* Bombay, BNHS

Connop, S. (1993) *Birdsongs of Nepal.* New York: Turaco.

Connop, S. (1995) *Birdsongs of the Himalayas.* Toronto: Turaco.

Sivaprasad, P. S. (1994) *An Audio Guide to the Birds of South India.* London: OBC.

Warakagoda, D. (1997) *Birdsounds of Sri Lanka Nugegoda,* DHW Library

White, T. (1984) *A Field Guide to the Bird Songs of South East Asia.* London: British Library.

USEFUL ADDRESSES

Many organisations (governmental, NGOs and informal groups) work with birds and their conservation in this region. Birders who are interested in the region are encouraged to join the Oriental Bird Club (OBC). Most big cities have an informal group. Some of the important postal addresses, e-mail addresses (and websites where available) are given below:

INTERNATIONAL

BirdLife International (formerly ICBP)
Asia Division, Wellbrook Court
Girton Road, Cambridge
CB3 ONA, UK
Tel: 01223-277318
Publishes "World Birdwatch"

Oriental Bird Club (OBC)
c/o The Lodge
Sandy, Bedfordshire
SG19 2DL, UK
E-mail: mail@orientalbirdclub.org.
Website: www.orientalbirdclub.org.
Publishes "OBC Bulletin" twice a year and a journal, "Forktail", once a year

Oriental Birding is an e-mail discussion group. Membership is free. To join, send an e-mail to orientalbirding-subscribe@egroups.com. The e-mail can be blank or with the subject "subscribe".

South Asian Natural History Discussion Group
To join, e-mail Listproc@lists.princeton.edu leaving the "subject" heading blank.

Wetlands International
Asia Pacific Regional Office
E-mail: wiap@wiap.nasionet.net
Website: http://ngo.asiapac.net/wetlands

World Pheasant Association (WPA).
E-mail: wpa@gn.apc.org
Publishes "WPA News"

BANGLADESH

Nature Conservation Movement (NACON)
125–127 (2nd Floor)
Mohammedia Super Market
Sobhanbag, Dhaka 1207

Wildlife Society of Bangladesh
c/o Department of Zoology
University of Dhaka, Dhaka

BHUTAN

Royal Society for the Protection of Nature (RSPN)
P.O. Box 325, Thimphu
Members receive two newsletters "Thrung Thrung" annually & "Rangzbin" quarterly

World Wildlife Fund
P.O. Box 210, Thimphu

INDIA

Bombay Natural History Society (BNHS)
Hornbill House
Dr Salim Ali Chowk
Opp. Lion Gate, Shahid Bhagat
Singh Road, Bombay 400 023
Tel: 022-2843869/2843421
E-mail: bnhs@bom4.vsnl.net.in.
Publishes "Journal of the Bombay Natural Society" and the "Hornbill Magazine"

Newsletter for Birdwatchers
C/o Navbharat Enterprises
No.10 Sirur Park 'B' Street
Seshadripuram
Bangalore 560 020
Tel: 080-336 4142/336 4682

Salim Ali Centre for Ornithology and Natural History
(SACON) Kalampalayam PL
Coimbatore 641 010, Tamil Nadu
Tel: 0422-807973/807083
E-mail: lv@sacon.ernet.in or centre@sacon.ernet.in.
Publishes "Sacon Newsletter"

Wildlife Institute of India
Chandrabani
Dehradun 248 001

Wildlife Protection Society of India (WPSI)
Thapar House,
124 Janpath
New Delhi 110 001
Tel: 011-6213864
E-mail: wpsi@vsnl.com.

World Wildlife Fund India
172B – Lodi Estate
New Delhi 110003
Tel: 011-4633473/4627586

Zoological Survey of India (ZSI)
Prani Vigyan Bhawan
M Block, New Alipore
Calcutta 700 053

Northern India Bird Network
E-mail: www.delhibird.org (includes e-mail group)

Birds and Wildlife of Bangalore
E-mail: bngbirds@yahoogroups.com

Birdwatcher's Society of Andhra Pradesh
P.O. Box 45, Banjara Hills, Hyderabad 500034.
E-mail: bsap@bsaponline.org
Publishes "Pitta".

Birds of Bombay
E-mail:birdsofbombay@yahoogroups.com

Maharashtra Pakshimrita Sanghatana (Birds of Maharashtra)
E-mail: maharashtrapakshimita @yahoogroups.com

Birds of Kerala
E-mail: Keralabirder@yahoogroups.com

Madras Naturalist Society
No. 8. Janaki Avenue
Abhiramapuram
Chennai 600018
E-mail: MNS@dmiaetine.com

Birds of Calcutta
Website: www.kolkatabirds.com

Sanctuary Asia Magazine
Website: www.sanctuaryasia.com

NEPAL

Bird Conservation Nepal (BCN)
(formerly the Nepal Birdwatching Club)
Post Box 12465
Kathmandu
Members receive the quarterly newsletter "Danphe"

World Wildlife Fund
P.O. Box 7660
Lal Durbar
Kathmandu

PAKISTAN

Ornithological Society of Pakistan
Near Chowk Fara
Block 'D', P.O. Box 73
Dera Ghazi Khan 32200
Publishes the "Pakistan Journal of Ornithology"

World Wildlife Fund
Ferozepur Road
54600 Lahore
Postal address: P.O. Box 5180
54600 Lahore
Tel: +92-2-86-360/586-429

SRI LANKA

Ceylon Bird Club (CBC)
39 Chatham Street
Colombo 1
Tel: ++94-1-328625/328627
E-mail: birdclub@sltnet.lk.
Publishes "Ceylon Bird Notes"

Field Ornithology Group of Sri Lanka (FOGSL)
c/o Department of Zoology
University of Colombo
Colombo 3
E-mail: fogsl@slt.lk.
Publishes "Malkoba"

SYSTEMATIC LIST OF FAMILIES AND SPECIES

R	widespread resident	I	introduced resident
r	very local resident	Ex	extinct
W	widespread winter visitor	C	critically endangered
w	sparse winter visitor	E	endangered
P	widespread migrant	V	vulnerable
p	sparse migrant	D	conservation-dependent
V	vagrant or irregular visitor	N	near-threatened

ORDER: CRACIFORMES

Family: Megapodiidae

Nicobar Scrubfowl	V	r	*Megapodius nicobariensis* (Spurfowl)

ORDER: GALLIFORMES

Family: Phasianidae

Snow Partridge		r	*Lerwa lerwa*
See-See Partridge		r	*Ammoperdix griseogularis*
Tibetan Snowcock		r	*Tetraogallus tibetanus*
Himalayan Snowcock		r	*Tetraogallus himalayensis*
Buff-throated Partridge		r	*Tetraophasis szechenyii*
Chukar		R	*Alectoris chukar*
Black Francolin		R	*Francolinus francolinus* (Black Partridge)
Painted Francolin		R	*Francolinus pictus* (Painted Partridge)
Chinese Francolin		r	*Francolinus pintadeanus*
Grey Francolin		R	*Francolinus pondicerianus* (Grey Partridge)
Swamp Francolin	V	r	*Francolinus gularis* (Swamp Partridge)
Tibetan Partridge		r	*Perdix hodgsoniae*
Common Quail		rw	*Coturnix coturnix*
Japanese Quail		w	*Coturnix japonica*
Rain Quail		r	*Coturnix coromandelica* (Black-breasted Quail)
Blue-breasted Quail		r	*Coturnix chinensis*
Jungle Bush Quail		R	*Perdicula asiatica*
Rock Bush Quail		R	*Perdicula argoondah*
Painted Bush Quail		r	*Perdicula erythrorhyncha*
Manipur Bush Quail	V	r	*Perdicula manipurensis*
Hill Partridge		r	*Arborophila torqueola* (Common Hill Partridge)
Rufous-throated Partridge		r	*Arborophila rufogularis* (Rufous-throated Hill Partridge)
White-cheeked Partridge	N	r	*Arborophila atrogularis* (White-cheeked Hill Partridge)
Chestnut-breasted Partridge	V	r	*Arborophila mandellii* (Red-breasted Hill Partridge)
Mountain Bamboo Partridge		r	*Bambusicola fytchii* (Bamboo Partridge)
Red Spurfowl		r	*Galloperdix spadicea*
Painted Spurfowl		r	*Galloperdix lunulata*
Sri Lanka Spurfowl		r	*Galloperdix bicalcarata* (Ceylon Spurfowl)
Himalayan Quail	C	Ex?	*Ophrysia superciliosa* (Mountain Quail)

Blood Pheasant		r	*Ithaginis cruentus*
Western Tragopan	V	r	*Tragopan melanocephalus*
Satyr Tragopan	N	r	*Tragopan satyra* (Crimson Tragopan, Crimson Horned Pheasant)
Blyth's Tragopan	V	r	*Tragopan blythii*
Temminck's Tragopan		r	*Tragopan temminckii*
Koklass Pheasant		r	*Pucrasia macrolopha* (Koklas Pheasant)
Himalayan Monal		r	*Lophophorus impejanus* (Impeyan Pheasant)
Sclater's Monal	V	r	*Lophophorus sclateri*
Red Junglefowl		R	*Gallus gallus*
Grey Junglefowl		R	*Gallus sonneratii*
Sri Lanka Junglefowl		R	*Gallus lafayetii* (Ceylon Junglefowl, Lafayetti's Junglefowl)
Kalij Pheasant		R	*Lophura leucomelanos*
Tibetan Eared Pheasant	N	V	*Croosoptilon harmani*
Cheer Pheasant	V	r	*Catreus wallichii* (Chir Pheasant)
Mrs Hume's Pheasant	V	r	*Syrmaticus humiae* (Mrs Hume's Barred-back Pheasant)
Grey Peacock Pheasant		r	*Polyplectron bicalcaratum* (Peacock Pheasant)
Indian Peafowl		R	*Pavo crista* (Common Peafowl, Blue Peafowl)
Green Peafowl		r	*Pavo muticus* (Burmese Peafowl)

ORDER: ANSERIFORMES

Family: Dendrocygnidae

Fulvous Whistling Duck		r	*Dendrocygna bicolor* (Large Whistling-teal, Fulvous Treeduck)
Lesser Whistling Duck		R	*Dendrocygna javanica* (Lesser Whistling-teal, Lesser Treeduck)

Family: Anatidae

Oxyurinae

White-headed Duck	E	V	*Oxyura leucocephala* (White-headed Stiff-tailed Duck)

Cygninae

Mute Swan		V	*Cygnus olor*
Whooper Swan		V	*Cygnus cygnus*
Tundra Swan		V	*Cygnus columbianus* (Bewick's Swan)

Anatinae

Anserini

Bean Goose		V	*Anser fabalis*
Greater White-fronted Goose		V	*Anser albifrons* (White-fronted Goose)
Lesser White-fronted Goose	V	V	*Anser erythropus*
Grey-lag Goose		W	*Anser anser*
Bar-headed Goose		rW	*Anser indicus*
Snow Goose		V	*Anser caerulescens*
Red-breasted Goose		V	*Branta ruficollis* (Siberian Red-breasted Goose)
Ruddy Shelduck		RW	*Tadorna ferruginea* (Ruddy Sheldrake, Brahminy Duck)

Common Shelduck		w	*Tadorna tadorna*
White-winged Duck	E	r	*Cairina scutulata* (White-winged Wood Duck)
Comb Duck		r	*Sarkidiornis melanotos*
Cotton Pygmy-goose		r	*Nettapus coromandelianus* (Cotton Teal, Quacky Duck)

Anatini

Mandarin Duck		V	*Aix galericulata*
Gadwall		W	*Anas strepera*
Falcated Duck		V	*Anas falcata* (Falcated Teal)
Eurasian Wigeon		W	*Anas penelope* (Wigeon)
Mallard		rW	*Anas platyrhynchos*
Spot-billed Duck		R	*Anas poecilorhyncha* (Spotbill Duck, Spotbill)
Northern Shoveler		W	*Anas clypeata* (Shoveller)
Sunda Teal		r	*Anas gibberifrons* (Grey Teal)
Northern Pintail		W	*Anas acuta* (Common Pintail, Pintail)
Garganey		W	*Anas querquedula*
Baikal Teal	V	V	*Anas formosa*
Common Teal		W	*Anas crecca*
Marbled Duck	V	V	*Marmaronetta angustirostris* (Marbled Teal)
Pink-headed Duck	C	Ex?	*Rhodonessa caryophyllacea*
Red-crested Pochard		w	*Rhodonessa rufina*
Common Pochard		W	*Aythya ferina*
Ferruginous Pochard	N	w	*Aythya nyroca* (Ferruginous Duck, White-eyed Pochard)
Baer's Pochard	V	w	*Aythya baeri*
Tufted Duck		W	*Aythya fuligula* (Tufted Pochard)
Greater Scaup		V	*Aythya marila* (Scaup Duck)
Long-tailed Duck		V	*Clangula hyemalis* (Long-tail Duck, Old Squaw)
White-winged Scoter		V	*Meanitta fusca* (Velvet Scoter)
Common Goldeneye		V	*Bucephala clangula* (Goldeneye Duck, Goldeneye)
Smew		w	*Mergellus albellus*
Red-breasted Merganser		V	*Mergus serrator*
Common Merganser		RW	*Mergus merganser* (Goosander, Merganser)

ORDER: TURNICIFORMES

Family: Turnicidae

Small Buttonquail		R	*Turnix sylvatica* (Little Bustardquail, Little Buttonquail)
Yellow-legged Buttonquail		R	*Turnix tanki* (Button Quail)
Barred Buttonquail		R	*Turnix suscitator* (Common Bustardquail, Barred Bustard-Quail)

ORDER: PICIFORMES

Family: Indicatoridae

Yellow-rumped Honeyguide	N	r	*Indicator xanthonotus* (Honeyguide, Orange-rumped Honeyguide, Himalayan Honeyguide)

Family: Picidae

Eurasian Wryneck		sw	*Jynx torquilla*
Speckled Piculet		r	*Picumnus innominatus* (Spotted Piculet)
White-browed Piculet		r	*Sasia ochracea* (Rufous Piculet)
Brown-capped Pygmy Woodpecker		R	*Dendrocopos nanus* (Pygmy Woodpecker, Brown-capped Woodpecker, Brown-crowned Pigmy Woodpecker)
Grey-capped Pygmy Woodpecker		R	*Dendrocopos canicapillus* (Grey-crowned Pigmy Woodpecker, Grey-capped Woodpecker)
Brown-fronted Woodpecker		R	*Dendrocopos auriceps* (Brown-fronted Pied Woodpecker)
Fulvous-breasted Woodpecker		R	*Dendrocopos macei* (Fulvous-breasted Pied Woodpecker)
Stripe-breasted Woodpecker		r	*Dendrocopos atratus* (Stripe-breasted Pied Woodpecker)
Yellow-crowned Woodpecker		R	*Dendrocopos mahrattensis* (Yellow-fronted Pied Woodpecker, Yellow-fronted Woodpecker)
Rufous-bellied Woodpecker		r	*Dendrocopos hyperythrus* (Rufous-bellied Sapsucker)
Crimson-breasted Woodpecker		r	*Dendrocopos cathpharius* (Crimson-breasted Pied Woodpecker, Small Crimson-breasted Pied Woodpecker)
Darjeeling Woodpecker		R	*Dendrocopos darjellensis* (Darjeeling Pied Woodpecker)
Great Spotted Woodpecker		R	*Dendrocopos major*
Sind Woodpecker		r	*Dendrocopos assimilis* (Sind Pied Woodpecker)
Himalayan Woodpecker		R	*Dendrocopos himalayensis* (Himalayan Pied Woodpecker)
Rufous Woodpecker		R	*Celeus brachyurus* (Brown Woodpecker)
White-bellied Woodpecker		r	*Dryocopus javensis* (Indian Great Black Woodpecker)
Andaman Woodpecker	N	r	*Dryocopus hodgei*
Lesser Yellownape		R	*Picus chlorolophus* (Small Yellow-naped Woodpecker, Lesser Yellow-naped Woodpecker)
Greater Yellownape		R	*Picus flavinucha* (Large Yellow-naped Woodpecker, Greater Yellow-naped Woodpecker)
Streak-breasted Woodpecker		r	*Picus viridanus*
Streak-throated Woodpecker		R	*Picus xanthopygaeus* (Little Scaly-bellied Green Woodpecker, Small Scaly-bellied Woodpecker)
Scaly-bellied Woodpecker		R	*Picus squamatus* (Scaly-bellied Green Woodpecker, Large Scaly-bellied Woodpecker)

Grey-headed Woodpecker		R	*Picus canus* (Black-naped Green Woodpecker, Grey-faced Woodpecker, Black-naped Wooker)
Himalayan Flameback		R	*Dinopium shorii* (Himalayan Golden-backed Three-toed Woodpecker, Himalayan Goldenback, Three-toed Golden-backed Woodpecker)
Common Flameback		R	*Dinopium javanense* (Indian Golden-backed Three-toed Woodpecker, Common Goldenback)
Black-rumped Flameback		R	*Dinopium benghalense* (Lesser Golden-backed Woodpecker, Black-rumped Goldenback, Golden-back Woodpecker)
Greater Flameback		R	*Chrysocolaptes lucidus* (Larger Golden-backed Woodpecker, Greater Goldenback, Crimson-backed Woodpecker, Golden-backed Woodpecker)
White-naped Woodpecker		r	*Chrysocolaptes festivus* (Black-backed Woodpecker, Black-backed Yellow Woodpecker)
Pale-headed Woodpecker		r	*Gecinulus grantia*
Bay Woodpecker		r	*Blythipicus pyrrhotis* (Red-eared Bay Woodpecker, Red-eared Rufous Woodpecker)
Heart-spotted Woodpecker		r	*Hemicircus canente*
Great Slaty Woodpecker		r	*Mulleripicus pulverulentus* (Himalayan Great Slaty Woodpecker)

Family: Megalaimidae

Great Barbet		R	*Megalaima virens* (Great Hill Barbet, Great Himalayan Barbet)
Brown-headed Barbet		R	*Megalaima zeylanica* (Green Barbet)
Lineated Barbet		R	*Megalaima lineata*
White-cheeked Barbet		R	*Megalaima viridis* (Small Green Barbet)
Yellow-fronted Barbet		r	*Megalaima flavifrons*
Golden-throated Barbet		r	*Megalaima franklinii*
Blue-throated Barbet		R	*Megalaima asiatica*
Blue-eared Barbet		r	*Megalaima australis*
Crimson-fronted Barbet		r	*Megalaima rubricapilla* (Crimson-throated Barbet, Ceylon Small Barbet)
Coppersmith Barbet		R	*Megalaima haemacephala* (Crimson-breasted Barbet)

ORDER: BUCEROTIFORMES

Family: Bucerotidae

Indian Grey Hornbill		R	*Ocyceros birostris* (Common Grey Hornbill, Gray Hornbill)
Malabar Grey Hornbill		R	*Ocyceros griseus*
Sri Lanka Grey Hornbill		R	*Ocros gingalensis* (Ceylon Grey-Hornbill)
Great Hornbill	N	R	*Buceros bicornis* (Great Pied Hornbill, Giant Hornbill)

Brown Hornbill	N	r	*Anorrhinus tickelli* (White-throated Brown Hornbill)
Rufous-necked Hornbill	V	r	*Aceros nipalensis*
Wreathed Hornbill		r	*Aceros undulatus*
Narcondam Hornbill	V	r	*Aceros narcondami*
Oriental Pied Hornbill		r	*Anthracoceros albirostris* (Indian Pied Hornbill, Pied Hornbill)
Malabar Pied Hornbill	N	r	*Anthracoceros coronatus*

ORDER: UPUPIFORMES

Family: Upupidae

Common Hoopoe		RW	*Upupa epops* (Hoopoe)

ORDER: TROGONIFORMES

Family: Trogonidae

Malabar Trogon		r	*Harpactes fasciatus* (Ceylon Trogon)
Red-headed Trogon		r	*Harpactes erythrocephalus*
Ward's Trogon	N	r	*Harpactes wardi*

ORDER: CORACIIFORMES

Family: Coraciidae

European Roller		rp	*Coracias garrulus*
Indian Roller		R	*Coracias benghalensis*
Dollarbird		r	*Eurystomus orientalis* (Broad-billed Roller, Dark Roller, Asian Broad-billed Roller)

Family: Alcedinidae

Blyth's Kingfisher	N	r	*Alcedo hercules* (Great Blue Kingfisher)
Common Kingfisher		R	*Alcedo atthis* (Small Blue Kingfisher, Eurasian Kingfisher)
Blue-eared Kingfisher		r	*Alcedo meninting*
Oriental Dwarf Kingfisher		r	*Ceyx erithacus* (Three-toed Kingfisher, Black-backed Kingfisher, Oriental Kingfisher, Indian Three-toed Kingfisher)

Family: Halcyonidae

Brown-winged Kingfisher	N	r	*Halcyon amauroptera*
Stork-billed Kingfisher		R	*Halcyon capensis*
Ruddy Kingfisher		r	*Halcyon coromanda*
White-throated Kingfisher		R	*Halcyon smyrnensis* (White-breasted Kingfisher, Smyrna Kingfisher)
Black-capped Kingfisher		R	*Halcyon pileata* (Black-capped Purple Kingfisher)
Collared Kingfisher		r	*Todiramphus chloris* (White-collared Kingfisher)

Family: Cerylidae

Crested Kingfisher		R	*Megaceryle lugubris* (Greater Pied Kingfisher, Large-pied Kingfisher)

Pied Kingfisher		R	*Ceryle rudis* (Lesser Pied Kingfisher, Small-pied Kingfisher)

Family: Meropidae

Blue-bearded Bee-eater		r	*Nyctyornis athertoni*
Green Bee-eater		R	*Merops orientalis* (Little Green Bee-eater)
Blue-cheeked Bee-eater		PS	*Merops persicus*
Blue-tailed Bee-eater		R	*Merops philippinus*
European Bee-eater		sP	*Merops apiaster*
Chestnut-headed Bee-eater		R	*Merops leschenaulti*

ORDER: CUCULIFORMES

Family: Cuculidae

Pied Cuckoo		rS	*Clamator jacobinus* (Pied Crested Cuckoo, Jacobin Cuckoo)
Chestnut-winged Cuckoo		r	*Clamator coromandus* (Red-winged Creasted Cuckoo)
Large Hawk Cuckoo		r	*Hierococcyx sparverioides*
Common Hawk Cuckoo		R	*Hierococcyx varius*
Hodgson's Hawk Cuckoo		r	*Hierococcyx fugax*
Indian Cuckoo		R	*Cuculus micropterus*
Eurasian Cuckoo		R	*Cuculus canorus* (The Cuckoo, Common Cuckoo)
Oriental Cuckoo		r	*Cuculus saturatus* (Himalayan Cuckoo)
Lesser Cuckoo		r	*Cuculus poliocephalus* (Small Cuckoo)
Banded Bay Cuckoo		r	*Cacomantis sonneratii* (Indian Banded Bay Cuckoo, Bay-banded Cuckoo)
Grey-bellied Cuckoo		r	*Cacomantis passerinus* (Indian Plaintive Cuckoo, Grey-bellied Plaintive Cuckoo)
Plaintive Cuckoo		r	*Cacomantis merulinus* (Rufous-bellied Plaintive Cuckoo)
Asian Emerald Cuckoo		r	*Chrysococcyx maculatus* (Emerald Cuckoo)
Violet Cuckoo		r	*Chrysococcyx xanthorhynchus*
Drongo Cuckoo		r	*Surniculus lugubris*
Asian Koel		R	*Eudynamys scolopacea* (Koel, Common Koel, Koel Cuckoo, Indian Koel)
Green-billed Malkoha	V	r	*Phaenicophaeus tristis* (Large Green-billed Malkoha)
Blue-faced Malkoha		r	*Phaenicophaeus viridirostris* (Small Green-billed Malkoha)
Sirkeer Malkoha		r	*Phaenicophaeus leschenaultii* (Sirkeer Cuckoo, Sirkeer)
Red-faced Malkoha	V	r	*Phaenicophaeus pyrrhocephalus*

Family: Centropodidae

Greater Coucal		R	*Centropus sinensis* (Crow-pheasant, Common Coucal, Large Coucal)
Brown Coucal		r	*Centropus andamanensis* (Andaman Coucal)
Lesser Coucal		r	*Centropus bengalensis* (Small Coucal)
Green-billed Coucal	V	r	*Centropus chlororhynchus* (Ceylon Coucal)

ORDER: PSITTACIFORMES

Family: Psittacidae

Vernal Hanging Parrot		R	*Loriculus vernalis* (Indian Lorikeet)
Sri Lanka Hanging Parrot		R	*Loriculus beryllinus* (Ceylon Lorikeet, Ceylon Hanging Parrot)
Alexandrine Parakeet		R	*Psittacula eupatria* (Large Indian Parakeet, Large Parakeet)
Rose-ringed Parakeet		R	*Psittacula krameri*
Slaty-headed Parakeet		R	*Psittacula himalayana*
Grey-headed Parakeet		R	*Psittacula finschii* (Eastern Slaty-headed Parakeet)
Intermediate Parakeet		r?	*Psittacula intermedia* (Rothschild's Parakeet)
Plum-headed Parakeet		R	*Psittacula cyanocephala* (Blossom-headed Parakeet)
Blossom-headed Parakeet		R	*Psittacula roseata* (Eastern Blossom-headed)
Malabar Parakeet		R	*Psittacula columboides* (Blue-winged Parakeet)
Layard's Parakeet		r	*Psittacula calthropae*
Derbyan Parakeet		r	*Psittacula derbiana* (Lord Derby's Parakeet)
Red-breasted Parakeet		R	*Psittacula alexandri* (Moustached Parakeet, Rose-breasted Parakeet)
Nicobar Parakeet		r	*Psittacula caniceps*
Long-tailed Parakeet	N	R	*Psittacula longicauda* (Red-cheeked Parakeet)

ORDER: APODIFORMES

Family: Apodidae

Glossy Swiftlet		R	*Collocalia esculenta* (White-bellied Swiftlet)
Indian Swiftlet		R	*Collocalia unicolor* (Indian Edible-nest Swiftlet)
Himalayan Swiftlet		R	*Collocalia brevirostris* (Edible-nest Swiftlet)
Black-nest Swiftlet			*Collocalia maxima*
Edible-nest Swiftlet		R	*Collocalia fuciphaga* (Andaman Grey-rumped Swiftlet)
White-rumped Needletail		R	*Zoonavena sylvatica* (White-rumped Spinetail)
White-throated Needletail		s	*Hirundapus caudacutus* (White-throated Spinetail Swift)
Silver-backed Needletail		r	*Hirundapus cochinchinensis* (Cochinchina Spinetail Swift, White-vented Needletail)
Brown-backed Needletail		R	*Hirundapus giganteus* (Large Brown-throated Spinetail Swift, Brown Needletail, Brown-throated SpineSwift)
Asian Palm Swift		R	*Cypsiurus balasiensis* (Palm Swift)
Alpine Swift		r	*Tachymarptis melba*
Common Swift		W	*Apus apus* (Swift, Black Swift)
Pallid Swift		r	*Apus pallidus*
Fork-tailed Swift		r	*Apus pacificus* (Large White-rumped Swift)
Dark-rumped Swift	V	r	*Apus acuticauda* (Dark-backed Swift, Khasi Hills Swift)
House Swift		R	*Apus affinis* (Little Swift)

Family: Hemiprocnidae

Crested Treeswift		R	*Hemiprocne coronata* (Crested Swift)

ORDER: STRIGIFORMES

Family: Tytonidae

Barn Owl		r	*Tyto alba*
Grass Owl		r	*Tyto capensis* (Eastern Grass Owl)
Oriental Bay Owl		r	*Phodilus badius* (Bay Owl)

Family: Strigidae

Andaman Scops Owl	N	r	*Otus balli*
Mountain Scops Owl		r	*Otus spilocephalus* (Spotted Scops Owl)
Pallid Scops Owl		r	*Otus brucei* (Striated Scops Owl)
Eurasian Scops Owl		w	*Otus scops* (European Scops Owl)
Nicobar Scops Owl		r?	*Otus alius*
Oriental Scops Owl		R	*Otus sunia* (Scops Owl, Little Scops Owl)
Collared Scops Owl		R	*Otus bakkamoena*
Eurasian Eagle Owl		R	*Bubo bubo* (Great Horned Owl, Eagle Owl, Northern Eagle Owl)
Rock Eagle Owl		R	*Bubo bengalensis* (Eurasian Eagle Owl)
Spot-bellied Eagle Owl		r	*Bubo nipalensis* (Forest Eagle Owl)
Dusky Eagle Owl		R	*Bubo coromandus* (Dusky Horned Owl)
Brown Fish Owl		r	*Ketupa zeylonensis*
Tawny Fish Owl		r	*Ketupa flavipes*
Buffy Fish Owl		r?	*Ketupa ketupu* (Malay Fish Owl)
Snowy Owl		V	*Nyctia scandiaca*
Mottled Wood Owl		r	*Strix ocellata*
Brown Wood Owl		r	*Strix leptogrammica*
Tawny Owl		r	*Strix aluco* (Tawny Wood Owl)
Hume's Owl			*Strix butleri* (Hume's Wood Owl)
Collared Owlet		r	*Glaucidium brodiei* (Collared Pygmy Owl)
Asian Barred Owlet		r	*Glaucidium cuculoides* (Barred Owlet)
Jungle Owlet		R	*Glaucidium radiatum* (Barred Jungle Owlet)
Chestnut-backed Owlet	N	r	*Glaucidium castanonotum*
Little Owl		r	*Athene noctua*
Spotted Owlet		R	*Athene brama*
Forest Owlet	C	r	*Athene blewitti* (Forest Spotted Owlet
Boreal Owl		V	*Aegolius funereus* (Tengmalm's Owl)
Brown Hawk Owl		r	*Ninox scutulata* (Brown Boobook)
Andaman Hawk Owl	N	r	*Ninox affinis* (Andaman Brown Hawk Owl)
Long-eared Owl		rw	*Asio otus*
Short-eared Owl		w	*Asio flammeus*

Family: Batrachostomidae

Sri Lanka Frogmouth		r	*Batrachostomus moniliger* (Ceylon Frogmouth)
Hodgson's Frogmouth		r	*Batrachostomus hodgsoni*

Family: Eurostopodidae

Great Eared Nightjar		r	*Eurostopodus macrotis*

Family: Caprimulgidae

Grey Nightjar		R	*Caprimulgus indicus* (Indian Jungle Nightjar, Jungle Nightjar, Highland Nightjar)
Eurasian Nightjar		rp	*Caprimulgus europaeus* (European Nightjar)
Egyptian Nightjar		V	*Caprimulgus aegyptius*
Sykes's Nightjar		r	*Caprimulgus mahrattensis*
Large-tailed Nightjar		R	*Caprimulgus macrurus* (Long-tailed Nightjar)
Jerdon's Nightjar		R	*Caprimulgus atripennis* (Long-tailed Nightjar, Horsfield's Jungle Nightjar)
Indian Nightjar		R	*Caprimulgus asiaticus* (Common Indian Nightjar, Common Nightjar, Little Nightjar)
Savannah Nightjar		r	*Caprimulgus affinis* (Franklin's Nightjar)

ORDER:COLUMBIFORMES

Family:Columbidae

Rock Pigeon		R	*Columba livia* (Blue Rock Pigeon)
Hill Pigeon		R	*Columba rupestris*
Snow Pigeon		R	*Columba leuconota*
Yellow-eyed Pigeon	V	w	*Columba eversmanni* (Eastern Stock Pigeon, Pale-backed Pigeon, Pale-backed Eastern Stock Dove)
Common Wood Pigeon		w	*Columba palumbus* (Wood Pigeon)
Speckled Wood Pigeon		r	*Columba hodgsonii*
Ashy Wood Pigeon		r	*Columba pulchricollis*
Nilgiri Wood Pigeon	V	r	*Columba elphinstonii*
Sri Lanka Wood Pigeon	V	r	*Columba torringtoni* (Ceylon Wood Pigeon)
Pale-capped Pigeon	V	w	*Columba punicea* (Purple Wood)
Andaman Wood Pigeon	N	r	*Columba palumboides*
European Turtle Dove		V	*Streptopelia turtur* (Turtle Dove, Western Turtle Dove)
Oriental Turtle Dove		RW	*Streptopelia orientalis* (Rufous Turtle Dove, Eastern Rufous Turtle Dove)
Laughing Dove		R	*Streptopelia senegalensis* (Little Brown Dove)
Spotted Dove		R	*Streptopelia chinensis*
Red Collared Dove		R	*Streptopelia tranquebarica* (Red Turtle Dove)
Eurasian Collared Dove		R	*Streptopelia decaocto* (Indian Ring Dove, Collared Dove)
Barred Cuckoo Dove		r	*Macropygia unchall* (Bar-tailed Cuckoo Dove, Long-tailed Cuckoo Dove)
Andaman Cuckoo Dove	N	r	*Macropygia rufipennis*
Emerald Dove		R	*Chalcophaps indica* (Green-winged Pigeon, Bronze-winged Dove)
Nicobar Pigeon	N	r	*Caloenas nicobarica*
Orange-breasted Green Pigeon		r	*Treron bicincta* (Orange-breasted Pigeon)
Pompadour Green Pigeon		r	*Treron pompadora* (Pompadour Pigeon, Gray-fronted Green Pigeon)
Thick-billed Green Pigeon		r	*Treron curvirostra* (Thick-billed Pigeon)
Yellow-footed Green Pigeon		R	*Treron phoenicoptera* (Yellow-legged, Bengal Green Pigeon)
Pin-tailed Green Pigeon		r	*Treron apicauda* (Pin-tailed Pigeon)

Wedge-tailed Green Pigeon		r	*Treron sphenura* (Wedge-tailed Pigeon)
Green Imperial Pigeon		r	*Ducula aenea*
Mountain Imperial Pigeon		r	*Ducula badia* (Imperial Pigeon)
Pied Imperial Pigeon		r	*Ducula bicolor*

ORDER: GRUIFORMES

Family: Otididae

Little Bustard	N	V	*Tetrax tetrax*
Great Bustard	V	V	*Otis tarda*
Indian Bustard	E	r	*Ardeotis nigriceps* (Great Indian Bustard)
MacQueen's Bustard	N	w	*Chlamydotis macqueeni* (Houbara Bustard)
Bengal Florican	E	r	*Houbaropsis bengalensis* (Bengal Likh)
Lesser Florican	E	r	*Sypheotides indica* (Likh)

Family: Gruidae

Siberian Crane	C	w	*Grus leucogeranus*
Sarus Crane	V	r	*Grus antigone*
Demoiselle Crane		w	*Grus virgo*
Common Crane		w	*Grus grus*
Hooded Crane		V	*Grus monacha*
Black-necked Crane	V	r	*Grus nigricollis*

Family: Heliornithidae

Masked Finfoot	V	r	*Heliopais personata*

Family: Rallidae

Andaman Crake		r	*Rallina canningi* (Andaman Banded Rail)
Red-legged Crake		V?	*Rallina fasciata* (Red-legged Banded Crake)
Slaty-legged Crake		r	*Rallina eurizonoides* (Banded Crake, Slaty-legged Banded Crake)
Slaty-breasted Rail		r	*Gallirallus striatus* (Blue-breasted Banded Rail)
Water Rail		rw	*Rallus aquaticus*
Corn Crake		V	*Crex crex*
Brown Crake		r	*Amaurornis akool*
White-breasted Waterhen		R	*Amaurornis phoenicurus*
Black-tailed Crake		r	*Porzana bicolor* (Elwes's Crake)
Little Crake		V	*Porzana parva*
Baillon's Crake		rw	*Porzana pusilla*
Spotted Crake		V	*Porzana porzana*
Ruddy-breasted Crake		r	*Porzana fusca* (Ruddy Crake)
Watercock		r	*Gallicrex cinerea*
Purple Swamphen		R	*Porphyrio porphyrio* (Purple Moorhen, Gallinule)
Common Moorhen		R	*Gallinula chloropus* (Moorhen, Waterhen, Indian Gallinule)
Common Coot		rW	*Fulica atra* (Coot, Eurasian Coot, Black Coot)

ORDER: CICONIIFORMES

Family: Pteroclidae

Tibetan Sandgrouse		r	*Syrrhaptes tibetanus*
Pallas's Sandgrouse		V	*Syrrhaptes paradoxus*
Pin-tailed Sandgrouse		w	*Pterocles alchata* (Large Pintail Sandgrouse)
Chestnut-bellied Sandgrouse		r	*Pterocles exustus* (Indian Sandgrouse)
Spotted Sandgrouse		w	*Pterocles senegallus*
Black-bellied Sandgrouse		rw	*Pterocles orientalis* (Imperial Sandgrouse)
Crowned Sandgrouse		r	*Pterocles coronatus* (Coronetted Sandgrouse)
Painted Sandgrouse		r	*Pterocles indicus* (Close-barred Sandgrouse)
Lichtenstein's Sandgrouse		r	*Pterocles lichtensteinii* (Close-barred Sandgrouse)

Family: Scolopacidae

Scolopacinae

Eurasian Woodcock		rw	*Scolopax rusticola* (Woodcock)
Solitary Snipe		r	*Gallinago solitaria*
Wood Snipe	V	r	*Gallinago nemoricola*
Pintail Snipe		W	*Gallinago stenura*
Swinhoe's Snipe		w	*Gallinago megala*
Great Snipe		V	*Gallinago media*
Common Snipe		rW	*Gallinago gallinago* (Fantail Snipe)
Jack Snipe		w	*Lymnocryptes minimus*

Tringinae

Black-tailed Godwit		W	*Limosa limosa*
Bar-tailed Godwit		W	*Limosa lapponica*
Whimbrel		W	*Numenius phaeopus*
Eurasian Curlew		W	*Numenius arquata* (Curlew)
Eastern Curlew		V	*Numenius madagascariensis* (Far Eastern Curlew)
Spotted Redshank		W	*Tringa erythropus* (Dusky Redshank)
Common Redshank		sW	*Tringa totanus* (Redshank)
Marsh Sandpiper		W	*Tringa stagnatilis*
Common Greenshank		W	*Tringa nebularia* (Greenshank)
Nordmann's Greenshank	E	V	*Tringa guttifer* (Spotted Greenshank)
Green Sandpiper		W	*Tringa ochropus*
Wood Sandpiper		W	*Tringa glareola*
Terek Sandpiper		w	*Xenus cinereus*
Common Sandpiper		sW	*Actitis hypoleucos*
Grey-tailed Tattler		V	*Heteroscelus brevipes*
Ruddy Turnstone		w	*Arenaria interpres* (Turnstone)
Long-billed Dowitcher		V	*Limnodromus scolopaceus*
Asian Dowitcher	N	w	*Limnodromus semipalmatus* (Snipe-billed Godwit)
Great Knot		w	*Calidris tenuirostris* (Eastern Knot)
Red Knot		w	*Calidris canutus* (Knot)
Sanderling		w	*Calidris alba*
Spoon-billed Sandpiper	V	w	*Calidris pygmea* (Spoonbill Sandpiper)
Little Stint		W	*Calidris minuta*

Red-necked Stint		w	*Calidris ruficollis* (Eastern Little Stint, Rufous-necked Stint)
Temminck's Stint		W	*Calidris temminckii*
Long-toed Stint		w	*Calidris subminuta*
Sharp-tailed Sandpiper		V	*Calidris acuminata* (Asian Pectoral Sandpiper)
Pectoral Sandpiper		V	*Calidris melanotos*
Dunlin		w	*Calidris alpina*
Curlew Sandpiper		W	*Calidris ferruginea*
Buff-breasted Sandpiper		V	*Tryngites subruficollis*
Broad-billed Sandpiper		w	*Limicola falcinellus*
Ruff		W	*Philomachus pugnax* (Reeve for female)
Red-necked Phalarope		w	*Phalaropus lobatus*
Red Phalarope		V	*Phalaropus fulicaria* (Grey Phalarope)
Family: Rostratulidae			
Greater Painted-snipe		r	*Rostratula benghalensis* (Painted Snipe)
Family: Jacanidae			
Pheasant-tailed Jacana		R	*Hydrophasianus chirurgus*
Bronze-winged Jacana		R	*Metopidius indicus*
Family: Burhinidae			
Eurasian Thick-knee		R	*Burhinus oedicnemus* (Stone-curlew, Stone-Plover)
Great Thick-knee		r	*Esacus recurvirostris* (Great Stone Plover)
Beach Thick-knee	N	r	*Esacus neglectus*
Family: Charadriidae			
Recurvirostrinae			
Haematopodini			
Eurasian Oystercatcher		w	*Haematopus ostralegus* (Oystercatcher. Common Oystercatcher)
Recurvirostrini			
Ibisbill		r	*Ibidorhyncha struthersii*
Black-winged Stilt		RW	*Himantopus himantopus*
Pied Avocet		rW	*Recurvirostra avosetta* (Avocet)
Charadriinae			
European Golden Plover		V	*Pluvialis apricaria* (Golden Plover, Eurasian Golden Plover, Greater Golden Plover)
Pacific Golden Plover		W	*Pluvialis fulva* (Eastern Golden Plover, Lesser Golden Plover, Asiatic Golden Plover)
Grey Plover		w	*Pluvialis squatarola* (Black-bellied Plover)
Common Ringed Plover		w	*Charadrius hiaticula* (Ringed Plover)
Long-billed Plover		w	*Charadrius placidus* (Long-billed Ringed Plover, Long-billed Ring Plover)
Little Ringed Plover		RW	*Charadrius dubius* (Little Plover, Little Ring Plover)
Kentish Plover		RW	*Charadrius alexandrinus*
Lesser Sand Plover		sW	*Charadrius mongolus* (Mongolian Plover)

Greater Sand Plover		w	*Charadrius leschenaultii* (Large Sand Plover)
Caspian Plover		V	*Charadrius asiaticus* (Sand Plover)
Oriental Plover		V	*Charadrius veredus* (Sand Plover)
Black-fronted Dotterel		V	*Elseyornis melanops* (Australian Black-fronted Plover)
Northern Lapwing		w	*Vanellus vanellus* (Eurasian Lapwing, Lapwing, Green Plover, Peewit)
Yellow-wattled Lapwing		r	*Vanellus malarbaricus*
River Lapwing		R	*Vanellus duvaucelii* (Spur-winged Plover)
Grey-headed Lapwing		w	*Vanellus cinereus*
Red-wattled Lapwing		R	*Vanellus indicus*
Sociable Lapwing	V	w	*Vanellus gregarius* (Sociable Plover)
White-tailed Lapwing		W	*Vanellus leucurus*

Family: Glareolidae

Dromadinae

Crab-plover		w	*Dromas ardeola*

Glareolinae

Jerdon's Courser	C	r	*Rhinoptilus bitorquatus*
Cream-colored Courser		rw	*Cursorius cursor*
Indian Courser		r	*Cursorius coromandelicus*
Collared Pratincole		rw	*Glareola pratincola* (Common Pratincole, European Pratincole)
Oriental Pratincole		R	*Glareola maldivarum* (Indian Large, Large Indian Pratincole)
Small Pratincole		R	*Glareola lactea* (Small Indian Pratincole, Little Pratincole, Milky Pratincole)

Family: Laridae

Larinae

Stercorariini

Brown Skua		sp	*Catharacta antarctica* (Antarctic Skua, Catharacta Skua)
South-Polar Skua		V	*Catharacta maccormicki* (MacCormick's Skua, Catharacta Skua)
Pomarine Jaeger		W	*Stercorarius pomarinus* (Pomatorhine Skua)
Parasitic Jaegar		W	*Stercorarius parasiticus* (Parasitic Skua, Arctic Skua)

Rynchopini

Indian Skimmer	V	r	*Rynchops albicollis*

Larini

White-eyed Gull		V	*Larus leucophthalmus*
Sooty Gull		V	*Larus hemprichii* (Hemprich's Gull)
Mew Gull		V	*Larus canus* (Common Gull)
Yellow-legged Gull		w	*Larus cachinnans* (Herring Gull, Caspian Gull)
Armenian Gull		w	*Larus armenicus* (Yellow-legged Gull)

Heuglin's Gull		w	*Larus heuglini* (Herring Gull, Lesser Black-backed Gull)
Pallas's Gull		w	*Larus ichthyaetus* (Great Black-headed Gull)
Brown-headed Gull		sW	*Larus brunnicephalus* (Indian Black-headed Gull)
Black-headed Gull		W	*Larus ridibundus* (Common Black-headed Gull)
Slender-billed Gull		w	*Larus genei*
Little Gull		V	*Larus minutus*

Sternini

Gull-billed Tern		rW	*Gelochelidon nilotica*
Caspian Tern		W	*Sterna caspia*
River Tern		R	*Sterna aurantia* (Indian River Tern)
Lesser Crested Tern		r	*Sterna bengalensis* (Indian Lesser Crested Tern, Small Crested Tern)
Great Crested Tern		r	*Sterna bergii* (Large Crested Tern, Swift Tern)
Sandwich Tern		w	*Sterna sandvicensis*
Roseate Tern		r	*Sterna dougallii* (Rosy Tern)
Black-naped Tern		r	*Sterna sumatrana*
Common Tern		sW	*Sterna hirundo*
Arctic Tern		V	*Sterna paradisaea*
Little Tern		r	*Sterna albifrons*
Saunders's Tern		r	*Sterna saundersi* (Saunders's Little Tern)
White-cheeked Tern		p	*Sterna repressa*
Black-bellied Tern	N	r	*Sterna acuticauda*
Bridled Tern		s	*Sterna anaethetus* (Brown-winged Tern)
Sooty Tern		s	*Sterna fuscata*
Whiskered Tern		RW	*Chlidonias hybridus*
White-winged Tern		w	*Chlidonias leucopterus* (White-winged Black Tern)
Black Tern		V	*Chlidonias niger*
Brown Noddy		V	*Anous stolidus* (Noddy Tern, Common Noddy)
Black Noddy		V	*Anous minutus* (White-capped Noddy)
Lesser Noddy		V	*Anous tenuirostris*
White Tern		r	*Gygis alba* (Indian Ocean White Tern, Common White Tern, Fairy Tern)

Family: Accipitridae

Pandioninae

Osprey		rW	*Pandion haliaetus*

Accipitrinae

Jerdon's Baza		r	*Aviceda jerdoni* (Blyth's Baza, Brown Baza)
Black Baza		r	*Aviceda leuphotes* (Indian Black-crested Baza, Black-crested Baza)
Oriental Honey-buzzard		RW	*Pernis ptilorhyncus* (Honey Buzzard, Eurasian Honey Buzzard, Crested Honey Buzzard, Honey Kite)
Black-shouldered Kite		R	*Elanus caeruleus* (Black-winged Kite)

Red Kite		V	*Milvus milvus* (Kite)
Black Kite		RW	*Milvus migrans* (Pariah Kite, Dark Kite, Black-eared Kite)
Brahminy Kite		R	*Haliastur indus* (Garuda)
White-bellied Sea Eagle		R	*Haliaeetus leucogaster* (White-bellied Fish-Eagle)
Pallas's Fish Eagle	V	r	*Haliaeetus leucoryphus* (Pallas's Fishing Eagle)
White-tailed Eagle	N	w	*Haliaeetus albicilla* (White-tailed Sea Eagle, Sea Eagle)
Lesser Fish Eagle	N	r	*Ichthyophaga humilis* (Himalyan Grey-headed Fishing Eagle, Lesser Fishing Eagle)
Grey-headed Fish Eagle	N	r	*Ichthyophaga ichthyaetus* (Himalayan Fishing Eagle, Grey-headed Fishing Eagle)
Lammergeier		r	*Gypaetus barbatus* (Bearded Vulture)
Egyptian Vulture		R	*Neophron percnopterus* (Small White Vulture, Scavenger Vulture)
White-rumped Vulture	C	r	*Gyps bengalensis* (Indian White-backed Vulture, Oriental White-backed Vulture, White-backed Vulture)
Indian Vulture	C	r	*Gyps indicus* (Indian Long-billed Vulture, Indian Griffon Vulture)
Slender-billed Vulture	?	r	*Gyps tenvirostris* (Long-billed Vulture)
Himalayan Griffon		r	*Gyps himalayensis* (Himalayan Griffon Vulture)
Eurasian Griffon		r	*Gyps fulvus* (Griffon Vulture, Eurasian Griffon Vulture)
Cinereous Vulture	N	rw	*Aegypius monachus* (Black Vulture)
Red-headed Vulture	N	r	*Sarcogyps calvus* (King Vulture, Black Vulture)
Short-toed Snake Eagle		r	*Circaetus gallicus* (Short-toed Eagle)
Crested Serpent Eagle		R	*Spilornis cheela*
Nicobar Serpent Eagle	N	r	*Spilornis minimus* (Great Nicobar Serpent Eagle)
Andaman Serpent Eagle	N	r	*Spilornis elgini* (Andaman Dark Serpent Eagle)
Eurasian Marsh Harrier		W	*Circus aeruginosus* (Western Marsh Harrier, Eastern Marsh Harrier, Marsh Harrier)
Hen Harrier		w	*Circus cyaneus* (Northern Harrier)
Pallid Harrier	N	w	*Circus macrourus* (Pale Harrier)
Pied Harrier		rw	*Circus melanoleucos*
Montagu's Harrier		w	*Circus pygargus*
Crested Goshawk		r	*Accipiter trivirgatus*
Shikra		R	*Accipiter badius* (Shikra Hawk)
Nicobar Sparrowhawk	N	r	*Accipiter butleri*
Chinese Sparrowhawk		W	*Accipiter soloensis* (Horsfield's Goshawk, Chinese Goshawk)
Japanese Sparrowhawk		w	*Accipiter gularis*
Besra		r	*Accipiter virgatus* (Besra Sparrow-hawk)
Eurasian Sparrowhawk		rw	*Accipiter nisus* (Sparrow Hawk, Northern Sparrowhawk)
Northern Goshawk		rw	*Accipiter gentilis* (Goshawk)
White-eyed Buzzard		R	*Butastur teesa* (White-eyed Buzzard Eagle, White-eyed Hawk)

Grey-faced Buzzard		V	*Bustatur indicus*
Common Buzzard		rw	*Buteo buteo* (Buzzard, Eurasian Buteo)
Long-legged Buzzard		rW	*Buteo rufinus* (Long-legged Buteo)
Upland Buzzard		w	*Buteo hemilasius* (Upland Buteo)
Rough-legged Buzzard		V	*Buteo lagopus* (Rough-legged Buteo)
Black Eagle		r	*Ictinaetus malayensis*
Lesser Spotted Eagle		r	*Aquila pomarina*
Greater Spotted Eagle	V	W	*Aquila clanga* (Spotted Eagle)
Tawny Eagle		R	*Aquila rapax* (Eurasian Tawny Eagle)
Steppe Eagle		W	*Aquila nipalensis*
Imperial Eagle	V	w	*Aquila heliaca* (Eastern Imperial Eagle)
Golden Eagle		r	*Aquila chrysaetos*
Bonelli's Eagle		r	*Hieraaetus fasciatus* (Bonelli's Hawk Eagle)
Booted Eagle		w	*Hieraaetus pennatus* (Booted Hawk Eagle)
Rufous-bellied Eagle		r	*Hieraaetus kienerii* (Rufous-bellied Hawk Eagle)
Changeable Hawk Eagle		R	*Spizaetus cirrhatus* (Crested Hawk Eagle)
Mountain Hawk Eagle		r	*Spizaetus nipalensis* (Hodgson's Hawk Eagle, Legge's Hawk Eagle)

Family: Falconidae

Collared Falconet		r	*Microhierax caerulescens* (Red-breasted Falconet, Red-thighed Falconet)
Pied Falconet		r	*Microhierax melanoleucus* (White-legged Falconet)
Lesser Kestrel	V	p	*Falco naumanni*
Common Kestrel		RW	*Falco tinnunculus* (Kestrel, Eurasian Kestrel)
Red-necked Falcon		r	*Falco chicquera* (Red-headed Merlin, Red-headed Falcon)
Amur Falcon		p	*Falco amurensis* (Red-legged Falcon, Eastern Red-legged Falcon)
Sooty Falcon			*Falco concolor*
Merlin		w	*Falco columbarius*
Eurasian Hobby		rp	*Falco subbuteo* (Hobby, Northern Hobby)
Oriental Hobby		r	*Falco severus* (Indian Hobby)
Laggar Falcon		r	*Falco jugger*
Saker Falcon		w	*Falco cherrug* (Lanner, Saker, Shangar)
Peregrine Falcon		rw	*Falco peregrinus* (Red-capped Falcon, Barbary Falcon, Shaheen Falcon)

Family: Podicipedidae

Little Grebe		R	*Tachybaptus ruficollis* (Dabchick)
Red-necked Grebe		W	*Podiceps grisegena*
Great Crested Grebe		rw	*Podiceps cristatus*
Horned Grebe		w	*Podiceps auritus* (Slavonian Grebe)
Black-necked Grebe		rw	*Podiceps nigricollis* (Eared Grebe)

Family: Phaethontidae

Red-billed Tropicbird		w	*Phaethon aethereus* (Short-tailed Tropicbird)
Red-tailed Tropicbird		r	*Phaethon rubricauda*

White-tailed Tropicbird		r	*Phaethon lepturus* (Long-tailed Tropicbird, Yellow-billed Tropicbird)

Family: Sulidae

Masked Booby		r	*Sula dactylatra*
Red-footed Booby		r	*Sula sula*
Brown Booby		r	*Sula leucogaster*

Family: Anhingidae

Darter	N	R	*Anhinga melanogaster* (Indian Darter, Oriental Darter, Snakebird)

Family: Phalacrocoracidae

Pygmy Cormorant		V	*Phalacrocorax pygmeus*
Little Cormorant		R	*Phalacrocorax niger*
Indian Cormorant		R	*Phalacrocorax fuscicollis* (Shag)
Great Cormorant		RW	*Phalacrocorax carbo* (Cormorant, Large Cormorant)

Family: Ardeidae

Little Egret		R	*Egretta garzetta*
Western Reef Egret		R	*Egretta gularis* (Indian Reef Heron)
Pacific Reef Egret		r	*Egretta sacra* (Reef Heron)
Grey Heron		RW	*Ardea cinerea* (Heron)
Goliath Heron		r?	*Ardea goliath* (Giant Heron)
White-bellied Heron	E	r	*Ardea insignis* (Great White-bellied Heron, Imperial Heron)
Great-billed Heron		V	*Ardea sumatrana*
Purple Heron		R	*Ardea purpurea*
Great Egret		RW	*Casmerodius albus* (Large Egret)
Intermediate Egret		R	*Mesophoyx intermedia* (Smaller Egret, Plumed Egret, Median Egret, Yellow-billed Egret)
Cattle Egret		R	*Bubulcus ibis*
Indian Pond Heron		R	*Ardeola grayii* (Pond Heron)
Chinese Pond Heron		r	*Ardeola bacchus*
Little Heron		r	*Butorides striatus* (Little Green Heron, Striated Heron)
Black-crowned Night Heron		R	*Nycticorax nycticorax* (Night Heron)
Malayan Night Heron		r	*Gorsachius melanolophus* (Malay Bittern, Malay Night Heron, Tiger Bittern)
Little Bittern		r	*Ixobrychus minutus*
Yellow Bittern		r	*Ixobrychus sinensis*
Cinnamon Bittern		r	*Ixobrychus cinnamomeus* (Chestnut Bittern)
Black Bittern		r	*Dupetor flavicollis*
Great Bittern		w	*Botaurus stellaris* (Bittern, Eurasian Bittern)

Family: Phoenicopteridae

Greater Flamingo		rW	*Phoenicopterus ruber* (Flamingo)
Lesser Flamingo	N	r	*Phoenicopterus minor*

Family: Threskiornithidae

Glossy Ibis		RW	*Plegadis falcinellus*
Black-headed Ibis	N	R	*Threskiornis melanocephalus* (White Ibis, Oriental White Ibis)
Black Ibis		R	*Pseudibis papillosa* (Red-naped Ibis)
Eurasian Spoonbill		RW	*Platalea leucorodia* (Spoonbill, White Spoonbill)

Family: Pelecanidae

Great White Pelican		rW	*Pelecanus onocrotalus* (Rosy Pelican, White Pelican, Eastern White Pelican)
Dalmatian Pelican	D	w	*Pelecanus crispus*
Spot-billed Pelican	V	R	*Pelecanus philippensis* (Grey Pelican, Dalmatian Pelican)

Family: Ciconiidae

Painted Stork	N	R	*Mycteria leucocephala*
Asian Openbill		R	*Anastomus oscitans* (Asian Openbill Stork, Openbill, Openbill Stork, Open-billed Stork)
Black Stork		w	*Ciconia nigra*
Woolly-necked Stork		R	*Ciconia episcopus* (White-necked Stork)
White Stork		w	*Ciconia ciconia*
Oriental Stork		V	*Ciconia boyciana* (Oriental White Stork, White Stork)
Black-necked Stork	N	r	*Ephippiorhynchus asiaticus*
Lesser Adjutant	V	r	*Leptoptilos javanicus* (Lesser Adjutant Stork)
Greater Adjutant	E	r	*Leptoptilos dubius* (Adjutant, Greater Adjutant Stork, Adjutant Stork)

Family: Fregatidae

Great Frigatebird		P	*Fregata minor* (Lesser)
Lesser Frigatebird		r	*Fregata ariel* (Least Frigate Bird, Small Frigatebird)
Christmas Island Frigatebird	C	V	*Fregata andrewsi* (Christmas Frigatebird)

Family: Gaviidae

Red-throated Loon		V	*Gavia stellata* (Red-throated Diver)
Black-throated Loon		V	*Gavia arctica* (Black-throated Diver)

Family: Procellariidae

Procellariinae

Cape Petrel		V	*Daption capense*
Barau's Petrel		V	*Pterodroma baraui*
Bulwer's Petrel		V	*Bulweria bulwerii* (Bulwer's Gadfly Petrel)
Jouanin's Petrel		V	*Bulweria fallax* (Jouanin's Gadfly Petrel)
Streaked Shearwater		V	*Calonectris leucomelas* (White-fronted Shearwater)
Wedge-tailed Shearwater		s	*Puffinus pacificus* (Green-billed Shearwater)
Flesh-footed Shearwater		s	*Puffinus carneipes* (Pink-footed Shearwater, Pale-footed Shearwater)

Sooty Shearwater		V	*Puffinus griseus*
Short-tailed Shearwater		V	*Puffinus tenuirostris* (Slender-billed Shearwater)
Audubon's Shearwater		r	*Puffinus lherminieri*
Persian Shearwater	N	P	*Puffinus persicus*

Hydrobatinae

Wilson's Storm-petrel		P	*Oceanites oceanicus*
White-faced Storm-petrel		V	*Pelagodroma marina*
Black-bellied Storm-petrel		V	*Fregetta tropica* (Dusky-vented Storm-petrel)
White-bellied Storm-petrel		V	*Fregetta grallaria*
Swinhoe's Storm-petrel		p	*Oceanodroma monorhis* (Ashy Storm Petrel, Fork-tailed Storm Petrel)

ORDER: PASSERIFORMES

Family: Pittidae

Eared Pitta		V	*Pitta phayeri*
Blue-naped Pitta		r	*Pitta nipalensis*
Blue Pitta		r	*Pitta cyanea*
Hooded Pitta		r	*Pitta sordida* (Green-breasted Pitta)
Indian Pitta		R	*Pitta brachyura* (Pitta)
Mangrove Pitta	N	r	*Pitta megarhyncha*

Family: Eurylaimidae

Silver-breasted Broadbill		r	*Serilophus lunatus* (Collared, Hodgson's Broadbill)
Long-tailed Broadbill		r	*Psarisomus dalhousiae*

Family: Irenidae

Asian Fairy Bluebird		r	*Irena puella* (Fairy Bluebird)
Blue-winged Leafbird		R	*Chloropsis cochinchinensis* (Gold-mantled Leafbird, Jerdon's Chloropsis)
Golden-fronted Leafbird		R	*Chloropsis aurifrons* (Golden-fronted Chloropsis)
Orange-bellied Leafbird		r	*Chloropsis hardwickii* (Orange-bellied Chloropsis)

Family: Laniidae

Red-backed Shrike		p	*Lanius collurio*
Rufous-tailed Shrike		W	*Lanius isabellinus* (Isabelline Shrike)
Brown Shrike		W	*Lanius cristatus* (Philippine Shrike)
Burmese Shrike		p	*Lanius collurioides*
Bay-backed Shrike		R	*Lanius vittatus*
Long-tailed Shrike		R	*Lanius schach* (Rufous-backed Shrike, Rufous-rumped Shrike, Black-headed Shrike)
Grey-backed Shrike		rW	*Lanius tephronotus* (Tibetan Shrike)
Lesser Grey Shrike		V	*Lanius minor*
Great Grey Shrike		V	*Lanius excubitor*
Southern Grey Shrike		R	*Lanius meridionalis* (Grey Shrike)

Family: Corvidae

Pachycephalinae

Mangrove Whistler		r	*Pachycephala grisola* (Grey Thickhead)

Corvinae

Corvini

Eurasian Jay		R	*Garrulus glandarius* (Jay)
Black-headed Jay		R	*Garrulus lanceolatus* (Black-throated Jay, Lanceolated Jay)
Sri Lanka Blue Magpie	V	r	*Urocissa ornata* (Ceylon Magpie, Ceylon Blue Magpie, Sri Lanka Magpie)
Yellow-billed Blue Magpie		R	*Urocissa flavirostris* (Gold-billed Magpie)
Red-billed Blue Magpie		R	*Urocissa erythrorhyncha* (Red-billed Magpie)
Common Green Magpie		r	*Cissa chinensis*
Rufous Treepie		R	*Dendrocitta vagabunda* (Indian Treepie)
Grey Treepie		R	*Dendrocitta formosae* (Himalayan Treepie)
White-bellied Treepie		r	*Dendrocitta leucogastra* (Southern Tree Pie)
Collared Treepie		r	*Dendrocitta frontalis* (Black-browed Treepie)
Andaman Treepie	N	R	*Dendrocitta bayleyi*
Black-billed Magpie		r	*Pica pica* (Eurasian Magpie, Magpie)
Hume's Groundpecker		r	*Pseudopodoces humilis* (Hume's Ground Jay, Hume's Ground-chough, Tibetan Ground Jay)
Spotted Nutcracker		r	*Nucifraga caryocatactes* (Nutcracker, Eurasian Nutcraker)
Red-billed Chough		R	*Pyrrhocorax pyrrhocorax*
Yellow-billed Chough		R	*Pyrrhocorax graculus* (Alpine Chough)
Eurasian Jackdaw		rw	*Corvus monedula* (Jackdaw, Common Jackdaw)
House Crow		R	*Corvus splendens*
Rook		w	*Corvus frugilegus*
Carrion Crow		rw	*Corvus corone* (Hooded Crow)
Large-billed Crow		R	*Corvus macrorhynchos* (Jungle Crow, Black Crow)
Brown-necked Raven		r	*Corvus ruficollis*
Common Raven		r	Corvus corax (Raven)

Artamini

Ashy Woodswallow		R	*Artamus fuscus* (Ashy Swallow-shrike)
White-breasted Woodswallow		r	*Artamus leucorynchus* (White-rumped Woodswallow)

Oriolini

Eurasian Golden Oriole		R	*Oriolus oriolus* (Golden Oriole)
Black-naped Oriole		rw	*Oriolus chinensis*
Slender-billed Oriole		r	*Oriolus tenuirostris*
Black-hooded Oriole		R	*Oriolus xanthornus* (Black-headed Oriole)
Maroon Oriole		r	*Oriolus traillii*
Large Cuckooshrike		r	*Coracina macei*
Bar-bellied Cuckooshrike		r	*Coracina striata* (Barred Cuckooshrike)

Black-winged Cuckooshrike		r	*Coracina melaschistos* (Dark Cuckooshrike, Smaller Grey Cuckooshrike)
Black-headed Cuckooshrike		r	*Coracina melanoptera*
Pied Triller		r	*Lalage nigra* (Pied Cuckooshrike)
Rosy Minivet		rw	*Pericrocotus roseus*
Swinhoe's Minivet		V	*Pericrocotus cantonensis*
Ashy Minivet		w	*Pericrocotus divaricatus*
Small Minivet		R	*Pericrocotus cinnamomeus* (Little Minivet)
White-bellied Minivet		r	*Pericrocotus erythropygius*
Grey-chinned Minivet		r	*Pericrocotus solaris* (Yellow-throated Minivet)
Long-tailed Minivet		R	*Pericrocotus ethologus*
Short-billed Minivet		r	*Pericrocotus brevirostris*
Scarlet Minivet		R	*Pericrocotus flammeus* (Orange Minivet)
Bar-winged Flycatcher-shrike		R	*Hemipus picatus* (Pied Flycatcher-shrike, Pied Shrike, Pied Wood-Shrike)
Dicrurinae			
Rhipidurini			
Yellow-bellied Fantail		R	*Rhipidura hypoxantha* (Yellow-bellied Fantail Flycatcher)
White-throated Fantail		R	*Rhipidura albicollis* (White-throated Fantail Flycatcher)
White-browed Fantail Flycatcher		R	*Rhipidura aureola* (White-breasted Fantail Flycatcher)
Dicrurini			
Black Drongo		R	*Dicrurus macrocercus*
Ashy Drongo		R	*Dicrurus leucophaeus* (Grey Drongo)
White-bellied Drongo		r	*Dicrurus caerulescens* (White-vented Drongo)
Crow-billed Drongo		r	*Dicrurus annectans*
Bronzed Drongo		r	*Dicrurus aeneus* (Little Bronzed Drongo)
Lesser Racket-tailed Drongo		r	*Dicrurus remifer* (Small Racket-tailed Drongo)
Spangled Drongo		R	*Dicrurus hottentottus* (Hair-crested Drongo)
Andaman Drongo	N	R	*Dicrurus andamanensis*
Greater Racket-tailed Drongo		r	*Dicrurus paradiseus* (Racket-tailed Drongo, Ceylon Crested Drongo, Large Racquet-tailed Drongo)
Monarchini			
Black-naped Monarch		r	*Hypothymis azurea* (Black-naped Flycatcher, Azure Flycatcher)
Asian Paradise Flycatcher		R	*Terpsiphone paradisi* (Paradise-flycatcher)
Aegithininae			
Common Iora		R	*Aegithina tiphia* (Iora)
Marshall's Iora		r	*Aegithina nigrolutea* (White-tailed Iora)
Malaconotinae			
Large Woodshrike		r	*Tephrodornis gularis*

Common Woodshrike		R	*Tephrodornis pondicerianus* (Woodshrike, Lesser Wood-shrike)

Family: Bombycillidae

Bohemian Waxwing		V	*Bombycilla garrulus* (Waxwing)

Family: Cinclidae

White-throated Dipper		R	*Cinclus cinclus* (White-breasted Dipper, Dipper)
Brown Dipper		R	*Cinclus pallasii*

Family: Muscicapidae
Turdinae

Rufous-tailed Rock Thrush		sp	*Monticola saxatilis* (Rock Thrush)
Blue-capped Rock Thrush		R	*Monticola cinclorhynchus* (Blue-headed Rock Thrush)
Chestnut-bellied Rock Thrush		r	*Monticola rufiventris*
Blue Rock Thrush		rW	*Monticola solitarius*
Sri Lanka Whistling Thrush	E	R	*Myophonus blighi* (Ceylon Whistling Thrush)
Malabar Whistling Thrush		R	*Myophonus horsfieldii*
Blue Whistling Thrush		R	*Myophonus caeruleus* (Whistling Thrush)
Pied Thrush		sp	*Zoothera wardii* (Pied Ground Thrush)
Orange-headed Thrush		R	*Zoothera citrina* (White-throated Thrush, Orange-headed Grand Thrush)
Siberian Thrush		w	*Zoothera sibirica* (Siberian Ground Thrush)
Spot-winged Thrush	N	r	*Zoothera spiloptera* (Spotted-winged Thrush, Spotted-winged Ground Thrush)
Plain-backed Thrush		r	*Zoothera mollissima* (Plain-backed Mountain Thrush)
Long-tailed Thrush		r	*Zoothera dixoni* (Long-tailed Mountain Thrush)
Scaly Thrush		r	*Zoothera dauma* (Speckled Mountain Thrush, Golden Mountain Thrush, White's Thrush)
Long-billed Thrush		r	*Zoothera monticola* (Large Brown Thrush, Large Long-billed Thrush)
Dark-sided Thrush		r	*Zoothera marginata* (Lesser Brown Thrush, Lesser Long-billed Thrush)
Tickell's Thrush		R	*Turdus unicolor* (Indian Grey Thrush)
Black-breasted Thrush		w	*Turdus dissimilis*
White-collared Blackbird		r	*Turdus albocinctus*
Grey-winged Blackbird		r	*Turdus boulboul*
Eurasian Blackbird		r	*Turdus merula* (Blackbird)
Chestnut Thrush		r	*Turdus rubrocanus* (Grey-headed Thrush)
Kessler's Thrush		V	*Turdus kessleri* (White-backed)
Grey-sided Thrush	V	w	*Turdus feae* (Fea's Thrush)
Eye-browed Thrush		w	*Turdus obscurus* (Dark Thrush)
Dark-throated Thrush		w	*Turdus ruficollis* (Red-throated Thrush, Black-throated Thrush)
Dusky Thrush		w	*Turdus naumanni* (Rufous-tailed Thrush)
Fieldfare		V	*Turdus pilaris*
Song Thrush		V	*Turdus philomelos*
Mistle Thrush		r	*Turdus viscivorus*

Gould's Shortwing		r	*Brachypteryx stellata*
Rusty-bellied Shortwing	V	r	*Brachypteryx hyperythra*
White-bellied Shortwing	V	r	*Brachypteryx major* (Rufous-bellied Shortwing)
Lesser Shortwing		r	*Brachypteryx leucophrys*
White-browed Shortwing		r	*Brachypteryx montana*

Muscicapinae
Muscicapini

Brown-chested Jungle Flycatcher		wr	*Rhinomyias brunneata* (Olive Flycatcher, Brown-chested Flycatcer)
Spotted Flycatcher		p	*Muscicapa striata*
Dark-sided Flycatcher		r	*Muscicapa sibirica* (Sooty Flycatcher, Asian Sooty Flycatcher)
Asian Brown Flycatcher		rw	*Muscicapa dauurica* (Brown Flycatcher)
Rusty-tailed Flycatcher		r	*Muscicapa ruficauda* (Rufous-tailed Flycatcher)
Brown-breasted Flycatcher		r	*Muscicapa muttui* (Layard's Flycatcher)
Ferruginous Flycatcher		r	*Muscicapa ferruginea*
Yellow-rumped Flycatcher		V	*Ficedula zanthopygia*
Slaty-backed Flycatcher		r	*Ficedula hodgsonii* (Rusty-breasted Blue Flycatcher)
Rufous-gorgeted Flycatcher		r	*Ficedula strophiata* (Orange-gorgeted Flycatcher)
Red-throated Flycatcher		W	*Ficedula parva* (Red-breasted Flycatcher, Taiga Flycatcher)
Kashmir Flycatcher	V	r	*Ficedula subrubra* (Kashmir Red-breasted Flycatcher)
White-gorgeted Flycatcher		r	*Ficedula monileger*
Snowy-browed Flycatcher		r	*Ficedula hyperythra* (Rufous-breasted Blue Flycatcher)
Little Pied Flycatcher		r	*Ficedula westermanni* (Westermann's Flycatcher)
Ultramarine Flycatcher		r	*Ficedula superciliaris* (White-browed Blue Flycatcher)
Slaty-blue Flycatcher		r	*Ficedula tricolor*
Sapphire Flycatcher		r	*Ficedula sapphira* (Sapphire-headed Flycatcher)
Black-and-orange Flycatcher	N	r	*Ficedula nigrorufa* (Black-and-rufous Flycatcher)
Verditer Flycatcher		R	*Eumyias thalassina*
Dull-blue Flycatcher	N	r	*Eumyias sordida* (Dusky Blue Flycatcher)
Nilgiri Flycatcher	N	r	*Eumyias albicaudata*
Large Niltava		r	*Niltava grandis*
Small Niltava		r	*Niltava macgrigoriae*
Rufous-bellied Niltava		r	*Niltava sundara* (Beautiful Niltava)
Vivid Niltava		r	*Niltava vivida* (Rufous-bellied Blue Flycatcher)
White-tailed Flycatcher		r	*Cyornis concretus* (White-tailed Blue Flycatcher)
White-bellied Blue Flycatcher		r	*Cyornis pallipes*
Pale-chinned Flycatcher		r	*Cyornis poliogenys* (Brook's Flycatcher, Pale-chinned Blue Flycatcher)

Pale Blue Flycatcher		r	*Cyornis unicolor*
Blue-throated Flycatcher		r	*Cyornis rubeculoides*
Hill Blue Flycatcher		r	*Cyornis banyumas* (Large-billed Blue Flycatcher)
Tickell's Blue Flycatcher		R	*Cyornis tickelliae* (Orange-breasted Blue Flycatcher)
Pygmy Blue Flycatcher		r	*Muscicapella hodgsoni*
Grey-headed Canary Flycatcher		R	*Culicicapa ceylonensis* (Grey-headed Flycatcher)

Saxicolini

Common Nightingale		V	*Luscinia megarhynchos* (Nightingale)
Siberian Rubythroat		w	*Luscinia calliope* (Rubythroat, Eurasian Rubythroat)
White-tailed Rubythroat		rW	*Luscinia pectoralis* (Himalayan Rubythroat)
Bluethroat		sW	*Luscinia svecica*
Firethroat	N	V	*Luscinia pectardens*
Indian Blue Robin		r	*Luscinia brunnea* (Blue Chat)
Siberian Blue Robin		V	*Luscinia cyane* (Siberian Blue Chat)
Orange-flanked Bush Robin		r	*Tarsiger cyanurus* (Red-flanked Bluetail)
Golden Bush Robin		r	*Tarsiger chrysaeus*
White-browed Bush Robin		r	*Tarsiger indicus*
Rufous-breasted Bush Robin		r	*Tarsiger hyperythrus* (Rufous-bellied Bush Robin)
Rufous-tailed Scrub Robin		p	*Cercotrichas galactotes* (Rufous Chat, Rufous-tailed Bush Chat, Rufous Bush Robin)
Oriental Magpie Robin		R	*Copsychus saularis* (Magpie Robin, Asian Magpie Robin, Robin Dayal)
White-rumped Shama		R	*Copsychus malabaricus* (Shama)
Indian Robin		R	*Saxicoloides fulicata* (Black Robin, Indian Chat)
Rufous-backed Redstart		w	*Phoenicurus erythronota* (Eversmann's Redstart)
Blue-capped Redstart		r	*Phoenicurus coeruleocephalus* (Blue-headed Redstart)
Black Redstart		rW	*Phoenicurus ochruros*
Common Redstart		P	*Phoenicurus phoenicurus* (Redstart)
Hodgson's Redstart		w	*Phoenicurus hodgsoni*
White-throated Redstart		r	*Phoenicurus schisticeps*
Daurian Redstart		*rw*	*Phoenicurus auroreus*
White-winged Redstart		r	*Phoenicurus erythrogaster* (Guldenstadt's Redstart)
Blue-fronted Redstart		r	*Phoenicurus frontalis*
White-capped Water Redstart		r	*Chaimarrornis leucocephalus* (White-capped Redstart, River Chat, White-capped River Chat)
Plumbeous Water Redstart		r	*Rhyacornis fuliginosus* (Plumbeous Redstart)
White-bellied Redstart		r	*Hodgsonius phaenicuroides* (Hodgson's Shortwing)
White-tailed Robin		r	*Myiomela leucura* (White-tailed Blue Robin)
Blue-fronted Robin		r	*Cinclidium frontale* (Blue-fronted Long-tailed Robin)
Grandala		r	*Grandala coelicolor* (Hodgson's Grandala)
Little Forktail		r	*Enicurus scouleri*
Black-backed Forktail		r	*Enicurus immaculatus*
Slaty-backed Forktail		r	*Enicurus schistaceus*

White-crowned Forktail		r	*Enicurus leschenaulti* (Leschenault's Forktail)
Spotted Forktail		r	*Enicurus maculatus*
Purple Cochoa		r	*Cochoa purpurea*
Green Cochoa		r	*Cochoa viridis*
Stoliczka's Bushchat	V	r	*Saxicola macrorhyncha* (White-browed Bushchat, Stoliczka's Whinchat)
Hodgson's Bushchat	V	w	*Saxicola insignis* (White-throated Bushchat)
Common Stonechat		R	*Saxicola torquata* (Collared Bushchat, Siberian Stonechat, Stonechat)
White-tailed Stonechat		r	*Saxicola leucura* (White-tailed Bushchat)
Pied Bushchat		R	*Saxicola caprata* (Pied Stone-chat)
Jerdon's Bushchat		r	*Saxicola jerdoni*
Grey Bushchat		R	*Saxicola ferrea* (Dark-grey Bushchat)
Hooded Wheatear			*Oenanthe monacha* (Hooded Chat)
Hume's Wheatear		r	*Oenanthe alboniger* (Hume's Chat)
Northern Wheatear		P	*Oenanthe oenanthe* (Wheatear)
Finsch's Wheatear		w	*Oenanthe finschii* (Barnes's Chat)
Variable Wheatear		rw	*Oenanthe picata* (Pied Chat, Pied Wheatear)
Pied Wheatear		rw	*Oenanthe pleschanka* (Pleschanka's Pied Chat)
Rufous-tailed Wheatear		rw	*Oenanthe xanthoprymna* (Red-tailed Chat, Red-tailed Wheatear)
Desert Wheatear		rw	*Oenanthe deserti*
Isabelline Wheatear		rw	*Oenanthe isabellina* (Isabelline Chat)
Brown Rock-chat		R	*Cercomela fusca* (Indian Chat)

Family: Sturnidae

Asian Glossy Starling		r	Aplonis panayensis (Glossy Stare, Philippine Glossy Starling)
Spot-winged Starling		r	Saroglossa spiloptera (Spotted-winged Stare, Spot-winged Stare)
White-faced Starling	V	r	Sturnus senex (Ceylon White-headed Myna/Starling)
Chestnut-tailed Starling		R	Sturnus malabaricus (Grey-headed Myna)
White-headed Starling		r	Sturnus erythropygius (White-headed Mynah)
Brahminy Starling		R	Sturnus pagodarum (Black-headed /Brahminy Myna)
Purple-backed Starling		V	Sturnus sturninus (Daurian Myna, Daurian Staring)
White-shouldered Starling		V	Sturnus sinensis (Chinese Myna)
Rosy Starling		WP	Sturnus roseus (Rosy Pastor, Rose-coloured Starling)
Common Starling		wp	Sturnus vulgaris (Starling, Eurasian Starling)
Asian Pied Starling		R	Sturnus contra (Pied Myna)
Common Myna		R	Acridotheres tristis (Mynah)
Bank Myna		R	Acridotheres ginginianus
Jungle Myna		R	Acridotheres fuscus
White-vented Myna		r	*Acridotheres cinereus* (Orange-billed Jungle Myna, Great Myna)
Collared Myna		r	*Acridotheres albocinctus*

Golden-crested Myna		r	*Ampeliceps coronatus* (Gold-crested Myna)
Sri Lanka Myna	N	r	*Gracula ptilogenys* (Ceylon Mynah)
Hill Myna		r	*Gracula religiosa* (Grackle, Common Hill Myna, Talking Myna)

Family: Sittidae

Sittinae

Chestnut-vented Nuthatch		r	*Sitta nagaensis* (Common Nuthatch, Eurasian Nuthatch)
Kashmir Nuthatch		r	*Sitta cashmirensis* (European Nuthatch, Common Nuthatch)
Chestnut-bellied Nuthatch		R	*Sitta castanea*
White-tailed Nuthatch		r	*Sitta himalayensis*
White-cheeked Nuthatch		r	*Sitta leucopsis*
Eastern Rock Nuthatch		r	*Sitta tephronota*
Velvet-fronted Nuthatch		R	*Sitta frontalis* (Velvet-fronted Blue Nuthatch)
Beautiful Nuthatch	V	r	*Sitta formosa*

Tichodrominae

Wallcreeper		rw	*Tichodroma muraria*

Family: Certhiidae

Certhiinae

Certhinii

Eurasian Treecreeper		R	*Certhia familiaris* (Tree Creeper, Common Treecreeper, Northern Tree Creeper)
Bar-tailed Treecreeper		R	*Certhia himalayana* (Himalayan Treecreeper)
Rusty-flanked Treecreeper		r	*Certhia nipalensis* (Nepal Treecreeper)
Brown-throated Treecreeper		r	*Certhia discolor* (Sikkim Treecreeper)

Salpornithini

Spotted Creeper		r	*Salpornis spilonotus* (Grey Creeper)

Troglodytinae

Winter Wren		r	*Troglodytes troglodytes* (Wren, Northern Wren)

Family: Paridae

Remizinae

White-crowned Penduline Tit		r	*Remiz coronatus* (Penduline Tit, Eurasian Penduline Tit)
Fire-capped Tit		r	*Cephalopyrus flammiceps*

Parinae

Rufous-naped Tit		r	*Parus rufonuchalis* (Simla Black, Dark-grey Tit)
Rufous-vented Tit		r	*Parus rubidiventris* (Rufous-bellied Crested Tit, Rufous-breasted Black Tit)
Spot-winged Tit		r	*Parus melanolophus* (Crested Black Tit, Black-crested Tit, Spot-winged Black Tit)
Coal Tit		r	*Parus ater*

Grey-crested Tit		r	*Parus dichrous* (Brown Crested Tit, Crested Brown Tit)
Great Tit		R	*Parus major* (Grey Tit)
Green-backed Tit		R	*Parus monticolus*
White-naped Tit	V	r	*Parus nuchalis* (White-winged Black Tit)
Black-lored Tit		r	*Parus xanthogenys* (Yellow-cheeked Tit)
Yellow-cheeked Tit		r	*Parus spilonotus* (Black-spotted Yellow Tit)
Azure Tit		V	*Parus cyanus*
Yellow-breasted Tit		V	*Parus flavipectus* (Yellow-breasted Blue Tit)
Yellow-browed Tit		r	*Sylviparus modestus*
Sultan Tit		r	*Melanochlora sultanea*

Family: Aegithalidae

White-cheeked Tit	r	*Aegithalos leucogenys*
Black-throated Tit	R	*Aegithalos concinnus* (Red-headed Tit)
White-throated Tit	r	*Aegithalos niveogularis*
Rufous-fronted Tit	r	*Aegithalos iouschistos* (Black-browed Tit)

Family: Hirundinidae

Pale Martin	pr?	*Riparia diluta* (Collared Sand Martin)
Sand Martin	r	*Riparia riparia* (Collared Sand Martin)
Plain Martin	R	*Riparia paludicola* (Plain Sand Martin, Brown-throated Sand Martin, Sand Martin)
Eurasian Crag Martin	r	*Hirundo rupestris* (Crag Martin, Northern Crag Martin)
Rock Martin	V	*Hirundo fuligula* (Pale Crag Martin)
Dusky Crag Martin	R	*Hirundo concolor*
Barn Swallow	RW	*Hirundo rustica* (Swallow, Common Swallow)
Pacific Swallow	r	*Hirundo tahitica* (House Swallow, Hill Swallow)
Wire-tailed Swallow	R	*Hirundo smithii*
Red-rumped Swallow	RW	*Hirundo daurica* (Striated Swallow)
Striated Swallow	r	*Hirundo striolata* (Larger Striated Swallow)
Streak-throated Swallow	R	*Hirundo fluvicola* (Indian Cliff Swallow)
Northern House Martin	rw	*Delichon urbica* (House Martin, Common House Martin, Eurasian House Martin)
Asian House Martin	r	*Delichon dasypus* (Asian Martin)
Nepal House Martin	r	*Delichon nipalensis* (Nepal Martin)

Family: Regulidae

Goldcrest	r	*Regulus regulus*

Family: Pycnonotidae

Crested Finchbill	r	*Spizixos canifrons* (Finch-billed Bulbul)
Striated Bulbul	r	*Pycnonotus striatus* (Striated Green Bulbul)
Grey-headed Bulbul	r	*Pycnonotus priocephalus*
Black-headed Bulbul	r	*Pycnonotus atriceps*
Black-crested Bulbul	R	*Pycnonotus melanicterus*
Red-whiskered Bulbul	R	*Pycnonotus jocosus*
White-eared Bulbul	R	*Pycnonotus leucotis* (White-cheeked Bulbul)

Himalayan Bulbul		R	*Pycnonotus leucogenys* (White-cheeked Bulbul)
Red-vented Bulbul		R	*Pycnonotus cafer*
Yellow-throated Bulbul	V	r	*Pycnonotus xantholaemus*
Yellow-eared Bulbul	N	r	*Pynconotus penicillatus*
Flavescent Bulbul		r	*Pycnonotus flavescens* (Blyth's Bulbul)
White-browed Bulbul		r	*Pycnonotus luteolus*
White-throated Bulbul		r	*Alophoixus flaveolus*
Olive Bulbul		r	*Iole virescens*
Yellow-browed Bulbul		r	*Iole indica*
Ashy Bulbul		r	*Hemixos flavala* (Brown-eared Bulbul)
Mountain Bulbul		r	*Hypsipetes mcclellandii* (Rufous-bellied Bulbul)
Black Bulbul		R	*Hypsipetes leucocephalus* (Gray Bulbul)
Nicobar Bulbul	V	r	*Hypsipetes nicobariensis*

Family Hypocoliidae

Grey Hypocolius		w	*Hypocolius ampelinus*

Family: Cisticolidae

Zitting Cisticola		R	*Cisticola juncidis* (Streaked Fantail Warbler)
Bright-headed Cisticola		r	*Cisticola exilis* (Fantail Warbler, Bright-capped Cisticola)
Streaked Scrub Warbler		r	*Scotocerca inquieta* (Scrub Warbler)
Rufous-vented Prinia	N	r	*Prinia burnesii* (Long-tailed Grass Warbler, Long-tailed Prinia)
Striated Prinia		R	*Prinia criniger* (Brown Hill Warbler, Brown Hill Prinia)
Hill Prinia		r	*Prinia atrogularis* (Black-throated Hill Warbler/Prinia)
Grey-crowned Prinia	V	r	*Prinia cinereocapilla* (Hodgson's Wren-Warbler, Grey-capped Prinia)
Rufous-fronted Prinia		R	*Prinia buchanani* (Rufous-fronted Wren-Warbler)
Rufescent Prinia		r	*Prinia rufescens* (Beavan's Wren-Warbler)
Grey-breasted Prinia		R	*Prinia hodgsonii* (Franklin's Wren-Warbler, Franklin's Prinia, Hodgson's Prinia)
Graceful Prinia		R	*Prinia gracilis* (Streaked Wren-Warbler, Graceful Stripe-backed Prinia, Fulvous-streaked Prinia)
Jungle Prinia		R	*Prinia sylvatica* (Jungle Wren-Warbler, Large Prinia)
Yellow-bellied Prinia		R	*Prinia flaviventris* (Yellow-bellied Wren-Warbler)
Ashy Prinia		R	*Prinia socialis* (Ashy Wren-Warbler)
Plain Prinia		R	*Prinia inornata* (Plain Wren-Warbler, White-browed Prinia)

Family: Zosteropidae

Sri Lanka White-eye		R	*Zosterops ceylonensis* (Ceylon White-eye, Ceylon Hill White-eye)

Oriental White-eye		R	*Zosterops palpebrosus* (White-eye, Small White-eye)

Family: Sylviidae
Acrocephalinae

Chestnut-headed Tesia		r	*Tesia castaneocoronata* (Chestnut-headed Ground Warbler)
Slaty-bellied Tesia		r	*Tesia olivea* (Slaty-bellied Ground Warbler)
Grey-bellied Tesia		r	*Tesia cyaniventer* (Dull Slaty-bellied Ground Warbler, Slaty-bellied Ground Warbler)
Asian Stubtail			*Urospena squameiceps* (Stub-tailed Bush-Warbler)
Pale-footed Bush Warbler		r	*Cettia pallidipes* (Blanford's Bush Warbler)
Japanese Bush Warbler		V	*Cettia diphone* (Chinese Bush Warbler)
Brownish-flanked Bush Warbler		r	*Cettia fortipes* (Strong-footed Bush Warbler, Brown-flanked Bush Warbler)
Chestnut-crowned Bush Warbler		r	*Cettia major* (Large Bush Warbler)
Aberrant Bush Warbler		r	*Cettia flavolivacea*
Yellowish-bellied Bush Warbler		r	*Cettia acanthizoides* (Yellow-bellied Bush Warbler, Hume's Bush Warbler, Verreaux's Bush Warbler)
Grey-sided Bush Warbler		r	*Cettia brunnifrons* (Rufous-capped Bush Warbler)
Cetti's Bush Warbler		w	*Cettia cetti* (Cetti's Warbler)
Spotted Bush Warbler		r	*Bradypterus thoracicus*
Long-billed Bush Warbler	N	r	*Bradypterus major* (Large Bush Warbler)
Chinese Bush Warbler		V	*Bradypterus tacsanowskius*
Brown Bush Warbler		r	*Bradypterus luteoventris*
Russet Bush Warbler		V	*Bradypterus seebohmi*
Sri Lanka Bush Warbler	N	r	*Bradypterus palliseri* (Palliser's Warbler, Ceylon Bush Warbler, Ceylon Warbler)
Lanceolated Warbler		w	*Locustella lanceolata* (Streaked Grasshopper Warbler)
Grasshopper Warbler		w	*Locustella naevia* (Common Grasshopper Warbler)
Rusty-rumped Warbler		w	*Locustella certhiola* (Pallas's Warbler, Pallas's Grasshopper Warbler)
Moustached Warbler		rw	*Acrocephalus melanopogon* (Moustached Sedge Warbler)
Sedge Warbler		V	*Acrocephalus schoenobaenus*
Black-browed Reed Warbler		w	*Acrocephalus bistrigiceps*
Paddyfield Warbler		w	*Acrocephalus agricola*
Blunt-winged Warbler		r	*Acrocephalus concinens* (Blunt-winged Paddyfield Warbler)
Blyth's Reed Warbler		W	*Acrocephalus dumetorum*
Large-billed Reed Warbler		r?	*Acrocephalus orinus*
Great Reed Warbler		w	*Acrocephalus arundinaceus* (Eurasian Great Reed Warbler)
Oriental Reed Warbler		w	*Acrocephalus orientalis* (Eastern Great Reed Warbler)

Clamorous Reed Warbler	R	*Acrocephalus stentoreus* (Indian Great Reed Warbler, Great Reed Warbler, Large-billed Reed Warbler)
Thick-billed Warbler	w	*Acrocephalus aedon*
Booted Warbler	rw	*Hippolais caligata* (Booted Tree Warbler)
Sykes's Warbler	w	*Hippolais rama* (Booted Tree, Booted Warbler)
Upcher's Warbler	r	*Hippolais languida*
Mountain Tailorbird	r	*Orthotomus cuculatus* (Golden-headed Tailorbird)
Common Tailorbird	R	*Orthotomus sutorius* (Tailorbird)
Dark-necked Tailorbird	r	*Orthotomus atrogularis* (Black-necked Tailorbird)
White-browed Tit Warbler	r	*Leptopoecile sophiae* (Stoliczka's Tit Warbler)
Common Chiffchaff	sW	*Phylloscopus collybita* (Brown Leaf Warbler, Chiffchaff)
Mountain Chiffchaff	Rw	*Phylloscopus sindianus*
Plain Leaf Warbler	w	*Phylloscopus neglectus*
Dusky Warbler	w	*Phylloscopus fuscatus* (Dusky Leaf Warbler)
Smoky Warbler	sw	*Phylloscopus fuligiventer* (Smoky Willow Warbler, Smoky Leaf Warbler)
Tickell's Leaf Warbler	sW	*Phylloscopus affinis* (Tickell's Warbler)
Buff-throated Warbler	V	*Phylloscopus subaffinis*
Sulphur-bellied Warbler	sw	*Phylloscopus griseolus* (Olivaceous Leaf Warbler)
Radde's Warbler	V	*Phylloscopus schwarzi* (Radde's Leaf Warbler)
Buff-barred Warbler	r	*Phylloscopus pulcher* (Orange-barred Leaf Warbler)
Ashy-throated Warbler	r	*Phylloscopus maculipennis* (Grey-faced Leaf Warbler)
Lemon-rumped Warbler	rW	*Phylloscopus chloronotus* (Pallas's Leaf Warbler, Yellow-rumped Leaf Warbler, Pale-rumped Warbler)
Chinese Leaf Warbler	V	*Phylloscopus sichuanensis*
Brooks's Leaf Warbler	w	*Phylloscopus subviridis*
Yellow-browed Warbler	rW	*Phylloscopus inornatus* (Inornate Leaf Warbler, Plain Leaf Warbler)
Hume's Warbler	W	*Phylloscopus humei* (Plain Leaf Warbler, Yellow-browed Warbler)
Arctic Warbler	V	*Phylloscopus borealis* (Arctic Leaf Warbler)
Greenish Warbler	rW	*Phylloscopus trochiloides* (Dull Green Leaf Warbler, Greenish Tree Warbler, Green Leaf, Green Tree Warbler)
Pale-legged Leaf Warbler	V	*Phylloscopus tenellipes*
Large-billed Leaf Warbler	rw	*Phylloscopus magnirostris* (Large-billed Tree Warbler)
Tytler's Leaf Warbler	r	*Phylloscopus tytleri* (Slender-billed Warbler)
Western Crowned Warbler	r	*Phylloscopus occipitalis* (Large Crowned Leaf Warbler)
Eastern Crowned Warbler	w	*Phylloscopus coronatus*
Blyth's Leaf Warbler	r	*Phylloscopus reguloides* (Crowned Leaf Warbler)

Yellow-vented Warbler		r	*Phylloscopus cantator* (Black-browed Leaf Warbler, Yellow-throated Leaf Warbler)
Golden-spectacled Warbler		R	*Seicercus burkii* (Black-browed Flycatcher Warbler, Yellow-eyed Warbler)
Grey-hooded Warbler		R	*Seicercus xanthoschistos* (Grey-headed Flycatcher Warbler, Grey-headed Warbler)
White-spectacled Warbler		r	*Seicercus affinis* (Allied Flycatcher Warbler, Allied Warbler)
Grey-cheeked Warbler		r	*Seicercus poliogenys* (Grey-cheeked Flycatcher Warbler)
Chestnut-crowned Warbler		r	*Seicercus castaniceps* (Chestnut-headed Flycatcher Warbler)
Broad-billed Warbler		r	*Tickellia hodgsoni* (Broad-billed Flycatcher Warbler)
Rufous-faced Warbler		r	*Abroscopus albogularis* (White-throated Flycatcher Warbler, White-throated Warbler)
Black-faced Warbler		r	*Abroscopus schisticeps* (Black-faced Flycatcher Warbler)
Yellow-bellied Warbler		r	*Abroscopus superciliaris* (Yellow-bellied Flycatcher Warbler)
Megalurinae			
Striated Grassbird		R	*Megalurus palustris* (Striated Marsh Warbler, Striated Warbler)
Bristled Grassbird	V	r	*Chaetornis striatus* (Bristled Grass Warbler)
Rufous-rumped Grassbird	N	r	*Graminicola bengalensis* (Large Grass Warbler)
Broad-tailed Grassbird	V	r	*Schoenicola platyura* (Broad-tailed Grass Warbler)
Garrulacinae			
Ashy-headed Laughingthrush	V	r	*Garrulax cinereifrons* (Ashy-headed Babbler)
White-throated Laughingthrush		R	*Garrulax albogularis*
White-crested Laughingthrush		R	*Garrulax leucolophus*
Lesser Necklaced Laughingthrush		r	*Garrulax monileger* (Necklaced Laughing Thrush)
Greater Necklaced Laughingthrush		r	*Garrulax pectoralis* (Black-gorgeted Laughing Thrush, Large Necklaced Laughing Thrush)
Striated Laughingthrush		r	*Garrulax striatus*
Rufous-necked Laughingthrush		r	*Garrulax ruficollis*
Chestnut-backed Laughingthrush	N	r	*Garrulax nuchalis*
Yellow-throated Laughingthrush		r	*Garrulax galbanus*
Wynaad Laughingthrush		r	*Garrulax delesserti*
Rufous-vented Laughingthrush		r	*Garrulax gularis*
Moustached Laughingthrush		r	*Garrulax cineraceus* (Ashy Laughing Thrush)
Rufous-chinned Laughingthrush		r	*Garrulax rufogularis*
Spotted Laughingthrush		r	*Garrulax ocellatus* (White-spotted Laughing Thrush)
Grey-sided Laughingthrush		r	*Garrulax caerulatus*
Spot-breasted Laughingthrush		r	*Garrulax merulinus* (Spotted-breasted)
White-browed Laughingthrush		r	*Garrulax sannio*

Nilgiri Laughingthrush	E	r	*Garrulax cachinnans* (Rufous-breasted Laughing Thrush)
Grey-breasted Laughingthrush	N	r	*Garrulax jerdoni* (White-breasted Laughing Thrush)
Streaked Laughingthrush		R	*Garrulax lineatus*
Striped Laughingthrush		r	*Garrulax virgatus* (Manipur Streaked Laughing Thrush)
Brown-capped Laughingthrush		r	*Garrulax austeni*
Blue-winged Laughingthrush		r	*Garrulax squamatus*
Scaly Laughingthrush		r	*Garrulax subunicolor* (Plain-colored Laughing Thrush)
Elliot's Laughingthrush		r	*Garrulax elliotii*
Variegated Laughingthrush		r	*Garrulax variegatus*
Brown-cheeked Laughingthrush		r	*Garrulax henrici* (Prince Henry's Laughing Thrush)
Black-faced Laughingthrush		r	*Garrulax affinis*
Chestnut-crowned Laughingthrush		r	*Garrulax erythrocephalus* (Red-headed Laughing Thrush)
Red-faced Liocichla		r	*Liocichla phoenicea* (Crimson-winged Laughing Thrush)

Sylviinae

Timaliini

Abbott's Babbler		r	*Malacocincla abbotti*
Buff-breasted Babbler		r	*Pellorneum tickelli* (Tickell's Babbler)
Spot-throated Babbler		r	*Pellorneum albiventre* (Brown Babbler)
Marsh Babbler	V	r	*Pellorneum palustre* (Marsh Spotted Babbler)
Puff-throated Babbler		R	*Pellorneum ruficeps* (Spotted Babbler)
Brown-capped Babbler		r	*Pellorneum fuscocapillum*
Large Scimitar Babbler		r	*Pomatorhinus hypoleucos* (Long-billed Scimitar Babbler)
Spot-breasted Scimitar Babbler		r	*Pomatorhinus erythrocnemis*
Rusty-cheeked Scimitar Babbler		r	*Pomatorhinus erythrogenys*
White-browed Scimitar Babbler		r	*Pomatorhinus schisticeps* (Slaty-headed Scimitar Babbler)
Indian Scimitar Babbler		r	*Pomatorhinus horsfieldii*
Streak-breasted Scimitar Babbler		r	*Pomatorhinus ruficollis* (Rufous-necked Scimitar Babbler)
Red-billed Scimitar Babbler		r	*Pomatorhinus ochraciceps* (Lloyd's Scimitar Babbler)
Coral-billed Scimitar Babbler		r	*Pomatorhinus ferruginosus*
Slender-billed Scimitar Babbler		r	*Xiphirhynchus superciliaris*
Long-billed Wren Babbler		r	*Rimator malacoptilus*
Streaked Wren Babbler		r	*Napothera brevicaudata* (Short-tailed Wren Babbler)
Eyebrowed Wren Babbler		r	*Napothera epilepidota* (Small Wren Babbler)
Scaly-breasted Wren Babbler		r	*Pnoepyga albiventer* (Greater Scaly-breasted Wren Babbler)
Nepal Wren Babbler		r	*Pnoepyga immaculata*
Pygmy Wren Babbler		r	*Pnoepyga pusilla* (Brown Wren Babbler)

Rufous-throated Wren Babbler	N	r	*Spelaeornis caudatus* (Tailed Wren Babbler)
Rusty-throated Wren Babbler	V	r	*Spelaeornis badeigularis* (Mishmi Wren Babbler)
Bar-winged Wren Babbler		r	*Spelaeornis troglodytoides* (Long-tailed Spotted Wren Babbler)
Spotted Wren Babbler		r	*Spelaeornis formosus*
Long-tailed Wren Babbler		r	*Spelaeornis chocolatinus* (Streaked Long-tailed Wren Babbler)
Tawny-breasted Wren Babbler	V	r	*Spelaeornis longicaudatus* (Long-tailed Wren Babbler)
Wedge-billed Wren Babbler	N	r	*Sphenocichla humei* (Wedge-billed Wren)
Rufous-fronted Babbler		r	*Stachyris rufifrons* (Red-fronted /Buff-chested Babbler)
Rufous-capped Babbler		r	*Stachyris ruficeps* (Red-headed Babbler)
Black-chinned Babbler		R	*Stachyris pyrrhops* (Red-billed Babbler)
Golden Babbler		r	*Stachyris chrysaea* (Golden/Gold-headed Babbler)
Grey-throated Babbler		r	*Stachyris nigriceps* (Black-throated Babbler)
Snowy-throated Babbler	V	r	*Stachyris oglei* (Austen's Spotted Babbler)
Tawny-bellied Babbler		R	*Dumetia hyperythra* (Rufous-bellied Babbler, White-throated Babbler)
Dark-fronted Babbler		r	*Rhopocichla atriceps* (Black-headed Babbler, Black-fronted Babbler)
Striped Tit Babbler		R	*Macronous gularis* (Yellow-breasted Babbler)
Chestnut-capped Babbler		R	*Timalia pileata* (Red-capped Babbler)
Yellow-eyed Babbler		R	*Chrysomma sinense*
Jerdon's Babbler	V	r	*Chrysomma altirostre* (Jerdon's Moupinia)
Spiny Babbler		r	*Turdoides nipalensis*
Common Babbler		R	*Turdoides caudatus*
Striated Babbler		R	*Turdoides earlei*
Slender-billed Babbler	V	r	*Turdoides longirostris*
Large Grey Babbler		R	*Turdoides malcolmi*
Rufous Babbler		r	*Turdoides subrufus*
Jungle Babbler		R	*Turdoides striatus* (Seven Sisters)
Orange-billed Babbler	N	r	*Turdoides rufescens* (Ceylon Rufous Babbler)
Yellow-billed Babbler		R	*Turdoides affinis* (White-headed Babbler)
Chinese Babax		V	*Babax lanceolatus*
Giant Babax	N	r	*Babax waddelli* (Giant Tibetan Babax)
Silver-eared Mesia		r	*Leiothrix argentauris*
Red-billed Leiothrix		r	*Leiothrix lutea* (Pekin Robin)
Cutia		r	*Cutia nipalensis* (Nepal Cutia)
Black-headed Shrike Babbler		r	*Pteruthius rufiventer* (Rufous-bellied Shrike Babbler)
White-browed Shrike Babbler		r	*Pteruthius flaviscapis* (Red-winged Shrike Babbler)
Green Shrike Babbler		r	*Pteruthius xanthochlorus*
Black-eared Shrike Babbler		r	*Pteruthius melanotis* (Chestnut-throated Shrike Babbler)
Chestnut-fronted Shrike Babbler		r	*Pteruthius aenobarbus*

White-hooded Babbler		r	*Gampsorhynchus rufulus* (White-headed Shrike Babbler)
Rusty-fronted Barwing		r	*Actinodura egertoni* (Spectacled Barwing)
Hoary-throated Barwing		r	*Actinodura nipalensis* (Hoary Barwing)
Streak-throated Barwing		r	*Actinodura waldeni* (Austen's Barwing)
Blue-winged Minla		r	*Minla cyanouroptera* (Blue-winged Siva)
Chestnut-tailed Minla		r	*Minla strigula* (Bar-throated Siva, Bar-throated Minla)
Red-tailed Minla		r	*Minla ignotincta*
Golden-breasted Fulvetta		r	*Alcippe chrysotis* (Golden-breasted Tit Babbler)
Yellow-throated Fulvetta		r	*Alcippe cinerea* (Dusky-green Tit Babbler)
Rufous-winged Fulvetta		r	*Alcippe castaneceps* (Chestnut-headed Tit Babbler)
White-browed Fulvetta		r	*Alcippe vinipectus* (White-browed Tit Babbler)
Streak-throated Fulvetta		r	*Alcippe cinereiceps* (Brown-headed Tit Babbler)
Brown-throated Fulvetta		r	*Alcippe ludlowi* (Ludlow's Fulvetta, Brown-headed Tit Babbler)
Rufous-throated Fulvetta		r	*Alcippe rufogularis* (Red-throated Tit Babbler)
Rusty-capped Fulvetta		r	*Alcippe dubia* (Rufous-headed Tit Babbler)
Brown-cheeked Fulvetta		R	*Alcippe poioicephala* (Quaker Babbler)
Nepal Fulvetta		r	*Alcippe nipalensis* (Nepal Babbler)
Rufous-backed Sibia		r	*Heterophasia annectans* (Chestnut-backed Sibia)
Rufous Sibia		R	*Heterophasia capistrata* (Black-capped Sibia)
Grey Sibia		r	*Heterophasia gracilis*
Beautiful Sibia		r	*Heterophasia pulchella*
Long-tailed Sibia		r	*Heterophasia picaoides*
Striated Yuhina		r	*Yuhina castaniceps* (White-browed Yuhina)
White-naped Yuhina		r	*Yuhina bakeri*
Whiskered Yuhina		R	*Yuhina flavicollis* (Yellow-naped Yuhina)
Stripe-throated Yuhina		R	*Yuhina gularis*
Rufous-vented Yuhina		r	*Yuhina occipitalis*
Black-chinned Yuhina		R	*Yuhina nigrimenta*
White-bellied Yuhina		r	*Yuhina zantholeuca*
Fire-tailed Myzornis		r	*Myzornis pyrrhoura*
Bearded Parrotbill		V	*Panurus biarmicus* (Bearded Tit Babbler, Bearded Reedling, Bearded Tit)
Great Parrotbill		r	*Conostoma oemodium*
Brown Parrotbill		r	*Paradoxornis unicolor* (Brown Suthora)
Grey-headed Parrotbill		r	*Paradoxornis gularis*
Black-breasted Parrotbill	V	r	*Paradoxornis flavirostris* (Gould's Parrotbill)
Spot-breasted Parrotbill		r	*Paradoxornis guttaticollis* (White-throated Parrotbill)
Fulvous Parrotbill		r	*Paradoxornis fulvifrons* (Fulvous-fronted Suthora/Parrotbill)
Black-throated Parrotbill		r	*Paradoxornis nipalensis* (Orange Suthora, Nepal Parrotbill)
Lesser Rufous-headed Parrotbill		r	*Paradoxornis atrosuperciliaris* (Lesser Red-headed Suthora, Black-browed Parrotbill)

Greater Rufous-headed Parrotbill	r	*Paradoxornis ruficeps* (Greater Red-headed Parrotbill, Rufous-headed Parrotbill)
Sylviini		
Garden Warbler	V	*Sylvia borin*
Greater Whitethroat	p	*Sylvia communis* (Common Whitethroat)
Lesser Whitethroat	W	*Sylvia curruca blythi*
Hume's Lesser Whitethroat	W	*Sylvia althaea* (Lesser Whitethroat, Hume's Whitethroat)
Small Whitethroat	w	*Sylvia minula* (Lesser Whitethroat)
Menetries Warbler	s	*Sylvia mystacea*
Eastern Desert Warbler	w	*Sylvia nana* (Desert Warbler)
Barred Warbler	V	*Sylvia nisoria*
Eastern Orphean Warbler	rW	*Sylvia crassirostris* (Orphean Warbler)
Family: Alaudidae		
Singing Bushlark	r	*Mirafra cantillans* (Singing Lark)
Indian Bushlark	R	*Mirafra erythroptera* (Red-winged Bushlark, Indian Lark)
Bengal Bushlark	R	*Mirafra assamica*
Rufous-winged Bushlark	R	*Mirafra affinis* (Rufous-winged Lark, Bush Lark)
Black-crowned Sparrow Lark	r	*Eremopterix nigriceps* (Black-crowned Finch Lark)
Ashy-crowned Sparrow Lark	R	*Eremopterix grisea* (Ashy-crowned Finch Lark)
Bar-tailed Lark	r	*Ammomanes cincturus* (Bar-tailed Desert Lark)
Rufous-tailed Lark	R	*Ammomanes phoenicurus* (Rufous-tailed Finch Lark)
Desert Lark	r	*Ammomanes deserti* (Finch Lark)
Greater Hoopoe Lark	r	*Alaemon alaudipes* (Bisfasciated Lark, Hoopoe Lark)
Bimaculated Lark	w	*Melanocorypha bimaculata* (Eastern Calandra Lark)
Tibetan Lark	r	*Melanocorypha maxima* (Long-billed Calandra Lark)
Greater Short-toed Lark	W	*Calandrella brachydactyla* (Short-toed lark)
Hume's Short-toed Lark	rw	*Calandrella acutirostris* (Hume's Lark)
Lesser Short-toed Lark	W	*Calandrella rufescens*
Asian Short-toed Lark	V	*Calandrella cheleensis*
Sand Lark	R	*Calandrella raytal* (Indian Short-toed Lark, Indian Sandlark)
Crested Lark	R	*Galerida cristata*
Malabar Lark	r	*Galerida malabarica* (Malabar Crested Lark)
Sykes's Lark	r	*Galerida deva* (Syke's Crested Lark, Tawny Lark)
Eurasian Skylark	w	*Alauda arvensis* (Skylark, Common Skylark)
Oriental Skylark	R	*Alauda gulgula* (Eastern Skylark, Oriental Lark, Indian Lark, Little Skylark)
Horned Lark	r	*Eremophila alpestris*

Family: Nectariniidae
Nectariniinae
Dicaeini

Thick-billed Flowerpecker		R	*Dicaeum agile*
Yellow-vented Flowerpecker		r	*Dicaeum chrysorrheum*
Yellow-bellied Flowerpecker		r	*Dicaeum melanoxanthum*
Legge's Flowerpecker	N	r	*Dicaeum vincens* (White-throated Flowerpecker)
Orange-bellied Flowerpecker		r	*Dicaeum trigonostigma*
Pale-billed Flowerpecker		R	*Dicaeum erythrorynchos* (Tickell's Flowerpecker, Small Flowerpecker)
Plain Flowerpecker		r	*Dicaeum concolor* (Plain-coloured Flowerpecker)
Fire-breasted Flowerpecker		r	*Dicaeum ignipectus* (Buff–bellied)
Scarlet-backed Flowerpecker		r	*Dicaeum cruentatum*

Nectariniini

Ruby-cheeked Sunbird		r	*Anthreptes singalensis* (Rubycheek)
Purple-rumped Sunbird		R	*Nectarinia zeylonica*
Crimson-backed Sunbird		r	*Nectarinia minima* (Small Sunbird)
Purple-throated Sunbird		r	*Nectarinia sperata* (Van Hasselt's Sunbird)
Olive-backed Sunbird		r	*Nectarinia jugularis*
Purple Sunbird		R	*Nectarinia asiatica*
Loten's Sunbird		R	*Nectarinia lotenia* (Long-billed Sunbird)
Mrs Gould's Sunbird		r	*Aethopyga gouldiae* (Gould's Sunbird)
Green-tailed Sunbird		r	*Aethopyga nipalensis* (Nepal Yellow-backed Sunbird, Nepal Sunbird)
Black-throated Sunbird		r	*Aethopyga saturata* (Black-breasted Sunbird)
Crimson Sunbird		R	*Aethopyga siparaja* (Yellow-backed Sunbird, Scarlet-breasted Sunbird)
Fire-tailed Sunbird		r	*Aethopyga ignicauda*
Little Spiderhunter		r	*Arachnothera longirostra*
Streaked Spiderhunter		r	*Arachnothera magna*

Family: Passeridae
Passerinae

House Sparrow		R	*Passer domesticus*
Spanish Sparrow		w	*Passer hispaniolensis*
Sind Sparrow		r	*Passer pyrrhonotus* (Sind Jungle Sparrow)
Russet Sparrow		R	*Passer rutilans* (Cinnamon Tree Sparrow, Cinnamon Sparrow)
Dead Sea Sparrow		V	*Passer moabiticus* (Scrub Sparrow)
Eurasian Tree Sparrow		R	*Passer montanus* (Tree Sparrow)
Chestnut-shouldered Petronia		R	*Petronia xanthocollis* (Yellow-throated Sparrow, Chestnut-shouldered Rock Sparrow)
Rock Sparrow		w	*Petronia petronia* (Streaked Rock Sparrow)
Tibetan Snowfinch		r	*Montifringilla adamsi* (Tibet Snow Finch, Black-winged Snowfinch)
White-rumped Snowfinch		r	*Pyrgilauda taczanowskii* (Mandelli's Snowfinch)
Rufous-necked Snowfinch		w	*Pyrgilauda ruficollis* (Red-necked Snowfinch)
Plain-backed Snowfinch		w	*Pyrgilauda blanfordi* (Blanford's Snowfinch)

Motacillinae

Forest Wagtail		rW	*Dendronanthus indicus*
White Wagtail		rW	*Motacilla alba* (Pied Wagtail)
White-browed Wagtail		R	*Motacilla maderaspatensis* (Large Pied Wagtail)
Citrine Wagtail		rW	*Motacilla citreola* (Yellow-headed, Yellow-hooded Wagtail)
Yellow Wagtail		W	*Motacilla flava* (Black-headed Wagtail)
Grey Wagtail		rW	*Motacilla cinerea*
Richard's Pipit		W	*Anthus richardi*
Paddyfield Pipit		R	*Anthus rufulus* (Indian Pipit)
Tawny Pipit		W	*Anthus campestris*
Blyth's Pipit		w	*Anthus godlewskii*
Long-billed Pipit		Rw	*Anthus similis* (Brown Rock Pipit)
Tree Pipit		rW	*Anthus trivialis* (Eurasian Tree Pipit)
Olive-backed Pipit		RW	*Anthus hodgsoni* (Indian Tree Pipit, Olive Tree Pipit, Hodgson's Tree Pipit)
Meadow Pipit		V	*Anthus pratensis*
Red-throated Pipit		p	*Anthus cervinus*
Rosy Pipit		r	*Anthus roseatus* (Vinaceous-breasted Pipit, Rose-breasted Pipit)
Water Pipit		w	*Anthus spinoletta*
Buff-bellied Pipit		w	*Anthus rubescens* (Water Pipit)
Upland Pipit		r	*Anthus sylvanus*
Nilgiri Pipit	N	r	*Anthus nilghiriensis*

Prunellinae

Alpine Accentor		r	*Prunella collaris*
Altai Accentor		w	*Prunella himalayana* (Rufous-streaked Accentor)
Robin Accentor		R	*Prunella rubeculoides*
Rufous-breasted Accentor		r	*Prunella strophiata*
Radde's Accentor		V	*Prunella ocularis* (Spot-throated Accentor)
Brown Accentor		r	*Prunella fulvescens*
Black-throated Accentor		w	*Prunella atrogularis*
Maroon-backed Accentor		r	*Prunella immaculata*

Ploceinae

Black-breasted Weaver		R	*Ploceus benghalensis* (Black-throated Weaver)
Streaked Weaver		R	*Ploceus manyar* (Striated Weaver)
Baya Weaver		R	*Ploceus philippinus* (Baya)
Finn's Weaver	V	r	*Ploceus megarhynchus* (Finn's Baya, Yellow Weaver)

Estrildinae

Red Avadavat		R	*Amandava amandava* (Red Munia)
Green Avadavat	V	r	*Amandava formosa* (Green Munia)
Indian Silverbill		R	*Lonchura malabarica* (White-throated Munia, Silverbill, Common Silverbill)
White-rumped Munia		R	*Lonchura striata* (White-backed Munia, Striated Munia, Sharp-tailed Munia)

Black-throated Munia	r	*Lonchura kelaarti* (Rufous-bellied Munia, Ceylon Hill Munia)
Scaly-breasted Munia	R	*Lonchura punctulata* (Spotted Munia, Nutmeg Mannikin)
Black-headed Munia	R	*Lonchura malacca* (Chestnut Munia)
Java Sparrow	Ir	*Lonchura oryzivora*

Family: Fringillidae
Fringillinae
Fringillini

Chaffinch	V	*Fringilla coelebs* (Common Chaffinch)
Brambling	V	*Fringilla montifringilla*

Carduelini

Fire-fronted Serin	R	*Serinus pusillus* (Gold-fronted Serin, Red-fronted Serin)
Yellow-breasted Greenfinch	R	*Carduelis spinoides* (Himalayan Greenfinch, Himalayan Goldfinch)
Black-headed Greenfinch	r	*Carduelis ambigua*
Eurasian Siskin	V	*Carduelis spinus*
Tibetan Siskin	r	*Carduelis thibetana* (Tibetan Serin)
European Goldfinch	r	*Carduelis carduelis* (Goldfinch, Eurasian Goldfinch)
Twite	r	*Carduelis flavirostris* (Tibetan Twite)
Eurasian Linnet	w	*Carduelis cannabina* (Linnet, Common Linnet, Brown Linnet)
Plain Mountain Finch	r	*Leucosticte nemoricola* (Hodgson's Mountain Finch)
Brandt's Mountain Finch	r	*Leucosticte brandti* (Black-headed Mountain Finch)
Spectacled Finch	r	*Callacanthis burtoni* (Red-browed Finch)
Crimson-winged Finch	r	*Rhodopechys sanguinea* (Crimson-winged Desert Finch)
Trumpeter Finch	r	*Bucanetes githagineus* (Trumpeter Bullfinch)
Mongolian Finch	r	*Bucanetes mongolicus* (Mongolian Trumpeter Bullfinch, Mongolion Desert Finch)
Desert Finch	V	*Rhodospiza absoleta* (Lichtenstein's Desert Finch, Black-billed Finch)
Blanford's Rosefinch	r	*Carpodacus rubescens* (Crimson Rosefinch)
Dark-breasted Rosefinch	r	*Carpodacus nipalensis* (Nepal Rosefinch)
Common Rosefinch	rW	*Carpodacus erythrinus*
Beautiful Rosefinch	r	*Carpodacus pulcherrimus*
Pink-browed Rosefinch	r	*Carpodacus rodochrous*
Vinaceous Rosefinch	r	*Carpodacus vinaceus*
Dark-rumped Rosefinch	r	*Carpodacus edwardsii* (Large Rosefinch, Edward's Rosefinch)
Three-banded Rosefinch	V	*Carpodacus trifasciatus*
Spot-winged Rosefinch	r	*Carpodacus rodopeplus*
White-browed Rosefinch	r	*Carpodacus thura*
Red-mantled Rosefinch	r	*Carpodacus rhodochlamys* (Blyth's Rosefinch)

Streaked Rosefinch	r	*Carpodacus rubicilloides* (Eastern Great Rosefinch)
Great Rosefinch	r	*Carpodacus rubicilla*
Red-fronted Rosefinch	r	*Carpodacus puniceus* (Red-breasted Rosefinch)
Crimson-browed Finch	r	*Propyrrhula subhimachala* (Red-headed Rosefinch)
Scarlet Finch	r	*Haematospiza sipahi*
Red Crossbill	r	*Loxia curvirostra* (Crossbill)
Brown Bullfinch	r	*Pyrrhula nipalensis*
Orange Bullfinch	r	*Pyrrhula aurantiaca*
Red-headed Bullfinch	r	*Pyrrhula erythrocephala*
Grey-headed Bullfinch	r	*Pyrrhula erythaca* (Beavan's Bullfinch)
Hawfinch	w	*Coccothraustes coccothraustes*
Japanese Grosbeak	V	*Eophona personata*
Black-and-yellow Grosbeak	r	*Mycerobas icterioides*
Collared Grosbeak	r	*Mycerobas affinis* (Allied Grosbeak)
Spot-winged Grosbeak	r	*Mycerobas melanozanthos* (Spotted-winged Grosbeak)
White-winged Grosbeak	r	*Mycerobas carnipes*
Gold-naped Finch	r	*Pyrrhoplectes epauletta* (Gold-headed/Gold-crowned Black Finch)

Emberizinae

Crested Bunting	R	*Melophus lathami*
Yellowhammer	V	*Emberiza citrinella* (Yellow Bunting)
Pine Bunting	w	*Emberiza leucocephalos*
Rock Bunting	R	*Emberiza cia* (Eurasian Rock-Bunting)
Godlewski's Bunting	r	*Emberiza godlewskii*
Grey-necked Bunting	W	*Emberiza buchanani*
Ortolan Bunting	V	*Emberiza hortulana*
White-capped Bunting	rw	*Emberiza stewarti* (Chestnut-breasted Bunting)
House Bunting	r	*Emberiza striolata* (Striolated Bunting)
Chestnut-eared Bunting	rw	*Emberiza fucata* (Grey-headed Bunting)
Little Bunting	w	*Emberiza pusilla*
Rustic Bunting	V	*Emberiza rustica*
Yellow-breasted Bunting	w	*Emberiza aureola*
Chestnut Bunting	w	*Emberiza rutila*
Black-headed Bunting	w	*Emberiza melanocephala*
Red-headed Bunting	w	*Emberiza bruniceps*
Black-faced Bunting	w	*Emberiza spodocephala*
Pallas's Bunting	V	*Emberiza pallasi* (Pallas's Reed Bunting)
Reed Bunting	rw	*Emberiza schoeniclus* (Common Reed Bunting)
Corn Bunting	V	*Miliaria calandra*

INDEX OF COMMON NAMES

INDEX OF SCIENTIFIC NAMES